# International Conference on Recent Trends in Physics (ICRTP 2016)



Journal of Physics: Conference Series Volume 755

Indore, India
13 – 14 February 2016

**Editors:**

**Shashank N. Kane**
**Ashutosh Mishra**
**Anup K. Dutta**

ISBN: 978-1-5108-3224-4
ISSN: 1742-6588

**Printed from e-media with permission by:**

Curran Associates, Inc.
57 Morehouse Lane
Red Hook, NY 12571

**Some format issues inherent in the e-media version may also appear in this print version.**



Printed by Curran Associates, Inc. (2016)

For permission requests, please contact the Institute of Physics
at the address below.

Institute of Physics
Dirac House, Temple Back
Bristol BS1 6BE UK

Phone: 44 1 17 929 7481
Fax: 44 1 17 920 0979

techtracking@iop.org

**Additional copies of this publication are available from:**

Curran Associates, Inc.
57 Morehouse Lane
Red Hook, NY 12571 USA
Phone: 845-758-0400
Fax: 845-758-2633
Email: curran@proceedings.com
Web: www.proceedings.com

# Table of contents

## Volume 755

**International Conference on Recent Trends in Physics 2016 (ICRTP2016)**
**13–14 February 2016, Indore, India**

**Accepted papers received: 20 September 2016**
**Published online: 1 November 2016**

### Preface



### Papers

#### Invited talks

**Laser and Plasma Physics**

**Others**

Journal of Physics: Conference Series **755** (2016) 011001 doi:10.1088/1742-6596/755/1/011001

# Preface: International Conference on Recent Trends in Physics (ICRTP 2016)

The ICRTP conference series, a bi-annual affair, has been hosted by School of Physics, Devi Ahilya University, Indore – 452001 (India), since 2012. The third such conference in this series ICRTP2016 was held during 13 – 14 February 2016. The purpose of the conference was to provide a platform that would enable scientists, teachers, researchers and students to interact and share ideas, thoughts and scientific findings in all areas of Physics, including Condensed matter and Materials Physics, Laser and Plasma Physics, Nuclear Physics, Computational Physics, High Energy Physics.

Over 160 participants from India, Italy, Portugal and, Spain attended the conference that included an inaugural talk, a special talk, 16 invited talks, 3 oral presentations and, a poster session. Presented posters were judged by a panel of experts and, 3 best posters were awarded cash prize. We extend our heartfelt appreciation to all the conference committee members and, reviewers for their valuable time in making this conference a great success. We thank all the participants for making ICRTP 2016 an unforgettable event. Special thanks to Ms. Madhuri Shah for providing editorial assistance.

Financial support from University Grants Commission (UGC), New Delhi, India is gratefully acknowledged. Sincere thanks to university administration, faculty members, administrative staff volunteers and, students of the School of Physics for their support in organizing ICRTP 2016.

Thanks are due to Ms Sarah Toms, Conference Publishing Coordinator, *Journal of Physics Conference Series* (JPCS), IOP Publishing Ltd, UK, for all help in bringing out ICRTP 2016 proceedings in JPCS.

Indore, 10 September 2016

**Shashank N. Kane*, Ashutosh Mishra and, Anup K. Dutta**

School of Physics, Devi Ahilya University, Indore – 452001, India

*Guest Editors*

*Corresponding author, e mail : kane_sn@yahoo.com

Journal of Physics: Conference Series **755** (2016) 011001 doi:10.1088/1742-6596/755/1/011001

## Organized by:

School of Physics, Devi Ahilya University, Indore 452001, India
Tel: +91 – 731- 2467028

Web site: www.icrtp.com
www.physics.dauniv.ac.in

**Patron**
**Prof. (Dr.) A. MISHRA**
Hon`ble *Vice Chancellor*

**Conference Chairs**
Prof. (Dr.) A. Mishra
Prof. (Dr.) A. K. Dutta
Prof. (Dr.) Shashank N. Kane

**Advisory Committee**
F. Alves, LGEP, Orsay, France
J. P. Araujo, UP, Porto, Portugal
V. Basso, INRIM, Italy
K. S. Bhatt, DRDO, Bangalore, India
M. Coisson, INRIM, Torino, Italy
A. K. Dutta, DAVV, Indore, India
E. Fleury, UM, Metz, France
V. Ganesan, UGC-DAE CSR, Indore, India
A. K. Sinha, UGC-DAE CSR, Indore, India
S. N. Joshi, CEERI, Pilani, India
S. N. Kane, DAVV, Indore, India
S. N. Kaul, UOH, Hyderabad, India
T. Kulik, WUT, Warsaw, Poland
M. Lo-Bue, ENSC, Cachan, France
F. Mazaleyrat, ENSC, Cachan, France
A. Mishra, DAVV, Indore, India
C. Ragusa, POLITO,Torino, Italy
F. Renz, LUH, Hannover, Garmany,
A. K. Sinha, RRCAT, Indore, India
B. D. Shrivastava, VU, Ujjain, India
P. Tiberto, INRIM, Torino, Italy
L. K. Varga, RISSPO, Budapest, Hungary
M. Vazquez, ICMM-CSIC, Cantoblanco, Marrid, Spain
F. Vinai, INRIM, Torino, Italy

**Program Committee**
A. Mishra, DAVV, Indore (*Conference Chair*)
A. K. Dutta, DAVV, Indore (*Conference Chair*)
P. Sen, DAVV, Indore
D. Varshney, DAVV, Indore
G. Mishra, DAVV, Indore
M. Banerjee, DAVV, Indore
S. N. Kane, DAVV, Indore (*Conference Chair*)
Y. Choyal, DAVV, Indore
A. Sheikh, DAVV, Indore
R. P. Sharma, DAVV, Indore

**Local Organizing Committee**
A. Sheikh, S. Raghuvanshi, K. Gehlot, , S. Ninama,,
S. Bisen, K. Mahalaxmi, M. Shah

## CONFERENCE PHOTO

# Peer review statement

All papers published in this volume of *Journal of Physics: Conference Series* have been peer reviewed through processes administered by the proceedings Editors. Reviews were conducted by expert referees to the professional and scientific standards expected of a proceedings journal published by IOP Publishing.

# Synthesis and magnetism of modulated FeCo-based nanowires

**Ester M Palmero[‡], Cristina Bran, Rafael P del Real and Manuel Vázquez[*]**
Institute of Materials Science of Madrid, CSIC, 28049 Madrid, Spain

E-mail: [‡]epalmero@icmm.csic.es, [*]mvazquez@icmm.csic.es

**Abstract.** FeCo based nanowire arrays have been proposed for different applications as high-density magnetic storage or sprintronics. Depending on the application different magnetic properties are required and the electrochemical techniques are low-cost routes to tailor them, by tuning the nanowires composition and geometry. In this work two different types of nanowires are studied: multisegmented nanowires and nanowires with modulated diameter. The magnetic properties of multisegmented $[FeCoCu(y)/Cu(x)]_{10}$ nanowire arrays have been studied considering different Cu layer thicknesses ($x$) and FeCoCu segment lengths ($y$). The obtained results show that the hard magnetic properties of these nanowire arrays present a higher coercivity and remanence comparing to the continuous nanowires, for thicker Cu layers and shorter FeCoCu segments. On the other hand, for the nanowires with modulated diameter it has been stated that the magnetization reversal process takes place by a vortex domain wall and with the tailored modulations along the nanowire it is possible to induce the domain wall pinning. This opens the possibility of controlling the domain wall motion along magnetic nanowires.

## 1. Introduction

Novel nanoscaled magnetic systems are attracting much interest in basic research and for their potential in advanced technologies as high-density magnetic storage, spintronic media, magnetotransport, logic circuits, sensing devices and biomedicine [1-6]. The importance of the increased packing density of integrated circuits and functional nanodevices in future technologies has called for advanced research in nanowires [7-10]. Particularly, the template-assisted electrodeposition of metallic elements and alloys as highly ordered nanowire arrays has been proved to be an interesting and low-cost alternative to enhance the magnetic data storage capability in perpendicular recording media [11, 12].

Different alternatives have been proposed to tune, according to the application, the magnetic properties of the nanowires and, consequently, of the nanowire array. Essentially, we can gather them into two main groups: tailor (i) the composition (single element, alloy or multilayer/multisegmented nanowires) and (ii) the geometry (modulation in diameter) of the nanowires. Considering the first group, Co, Fe and CoFe based nanowire arrays are being investigated due to their relevant magnetic and magnetotransport properties [13-17]. They have been proposed as a novel family of rare-earth free permanent magnets as they present a high saturation magnetization as well as high Curie temperature, which implies a high working temperature [18, 19]. It is well-known that the addition of a small amount of elements (e.g. C,

Cr, Cu, B, P, Mn, or V) as alloy doping can modify the mechanical, electrical, thermal or magnetic properties of the alloy [20-25]. Previous reports show that for FeCo nanowires the highest values of coercivity and magnetization at remanence are obtained for around 65-70% of Co in the alloy [17]. In the particular case of these magnetic FeCo nanowire arrays, it has been reported that the doping with a small amount of Cu, followed by suitable thermal treatment, induces relevant magnetic hardening [14, 26, 27].

On the other hand, adding non-ferromagnetic elements to the nanowires not only as a component into the alloy but as interfacial layers between ferromagnetic segments has been also studied [28-36]. Such multilayer systems include usually Cu, Pd, Pt and Au as non-ferromagnetic elements, showing interesting results in the way of tailoring the magnetic properties of the nanowire arrays, like magnetic hardening or their functionalization. The magnetic behavior of the multilayer/multisegmented nanowire arrays depends on the non-ferromagnetic and ferromagnetic layers/segments thicknesses and the ratio between them [28, 29, 31, 36, 37]. The magnetic behavior will be determined by the net interactions in the nanowire array: (i) intrawire interactions generated between the magnetic segments coupled by the non-ferromagnetic layers along the nanowire; and (ii) interwire interactions among neighbor nanowires induced by the stray fields generated by the magnetic segments [28, 31].

An alternative way to tune the magnetic properties of the nanowires is to tailor their geometry by modulating their diameter along the nanowire length (i.e. segments with different diameter along the nanowire) [38-42]. The fabrication of anodic aluminum oxide (AAO) templates by hard anodization process has been proved to be a suitable alternative fabrication method which optimizes the nanopores growth rate and allows for a fast hexagonal highly ordering [43, 44]. For the nanopores with modulated diameter, the combination of mild and hard anodization is used, which allows us to obtain segments with different diameter along the nanopores [39, 41-43, 45]. Tailoring the diameter and length of the different segments is a promising new route which provides the possibility of control the domain wall displacement along the nanowire [42, 46-48].

In this article we review recent results obtained for FeCo-based nanowires with: (i) modulated composition (i.e. multisegmented nanowires) and (ii) modulated diameter along the nanowire length. We show how it is possible to tune the magnetic properties of FeCo-based nanowire arrays by tailoring their composition and geometry. We overview the nanowire geometrical and crystallographic structure as well as the magnetic properties of the nanowire arrays and the magnetization reversal process of individual FeCo-based nanowires with modulated diameter. We have paid particular attention to the influence of the ferromagnetic/non-ferromagnetic segments/layers thickness on the multilayer nanowire arrays behaviour and the influence of the tailored periodical diameter modulations.

## 2. Synthesis of FeCo-based nanowires into Anodic Aluminium Oxide (AAO) membranes

### *2.1. AAO membranes fabrication*

The AAO templates were produced by different anodization techniques as they present different geometries. The templates used to grow the multisegmented nanowires were fabricated by a two-step anodization process [49]. Those used to grow the modulated diameter nanowires were fabricated by hard [43, 44] and pulsed anodization [40, 41].

Previous to the anodization, the high purity (99.999%) Al foils were ultrasonically degreased in acetone for 10 min followed by 10 min in ethanol. The clean Al foils were electropolished in a mixed solution of perchloric acid and ethanol (1:3 in volume), under a constant voltage of 20 V for 5 min. Then, the polished foils were anodized.

The two-step anodization was carried out using a 0.3 M oxalic acid aqueous solution under a constant voltage of 40 V at 3 °C for 24 h and 20 h, first and second anodization step duration, respectively. The alumina barrier layer formed after the first anodization step was chemically etched by a mixed solution of 0.18 M $CrO_3$ and 0.72 M $H_3PO_4$ [17].

Templates with modulated diameter nanopores were fabricated by pulsed anodization. Additionally, hard anodization was used to fabricate the templates with non-modulated pores, to grow the non-modulated nanowires to be used as reference sample. The hard and pulsed anodization processes were both carried out using 0.3 M oxalic acid aqueous solution and 5% ethanol in volume at 0-1 °C. In the case of hard anodization, a pre-anodization was performed to create an alumina protective surface to avoid burning the Al foil. A constant potential of 80 V was applied for 5 min. Then, the voltage was increased using a voltage ramp (0.08 V/s) until reaching a voltage of 140 V which was kept constant for 1 h [50].

The pulsed anodization was performed in two different ways to obtain two different types of modulations. For both, firstly a pre-anodization was carried out (for the first type, constant voltage of 80 V for 15 min, followed by 5 min at 140 V; and 80 V for 10 min and after that 6 min at 100 V for the second one) and then pulses with different durations in time (pulses of 10 s at 80 V and 30 s at 140 V for the first type; and pulses at 130 V for 5 s and 150 s at 100 V for the second one). The modulations in the nanowire are obtained by controlling the duration and the amplitude of the hard anodization pulses [42, 50].

Once the porous membranes were obtained, they needed to be open at both sides to proceed with the nanowire electrodeposition. The residual Al was chemically etched by a mixed solution of 0.1 M $CuCl_2{\cdot}2H_2O$ and 1.6 M HCl. The alumina barrier layer was chemically removed and the modulated nanopores were enlarged by an aqueous $H_3PO_4$ solution (5 wt%).

### *2.2. Nanowires growth*

Previous to the nanowire growth, a Au layer was sputtered at the bottom of the pores followed by Au electroplating into the pores in order to serve as a working electrode.

All the nanowire arrays reviewed in this work were grown in a three-electrode cell using a single Watts-type bath which composition was 0.12 M $CoSO_4{\cdot}7H_2O$, 0.05 M $FeSO_4{\cdot}7H_2O$, 0.01 M $CuSO_4{\cdot}5H_2O$, 0.16 M $H_3BO_3$, and 0.06 M $C_6H_8O_6$, and the pH was 3.0.

FeCoCu/Cu multilayer nanowires were grown by pulsed electrodeposition. The potential used to grow the ferromagnetic segments (FeCoCu) was -1.5 V and -0.3 V for the non-ferromagnetic layers (Cu), both versus Ag/AgCl reference electrode. Two series of multilayer nanowire arrays were prepared, where each nanowire is composed by ten FeCoCu/Cu bilayers: (i) FeCoCu segment thickness was kept constant (25 s electrodeposition time) and the pulses of Cu electroplating were varied between 5 and 60 s; and (ii) constant Cu layer thickness (15 s) and the FeCoCu electroplating pulses were varied between 10 and 60 s. Additionally, a continuous FeCoCu nanowire array was prepared to be used a reference sample. It was electroplated using the same composition bath and applying a voltage of -1.5 V for 6 min [51].

The modulated nanowires were grown under an electroplating potential of -1.8 V vs. Ag/AgCl reference electrode at room temperature [50].

## 3. Geometrical, compositional and microstructure characterization

The nanowire geometrical characterization was performed by a Nova Nano 230 High Resolution Scanning Electron Microscope (SEM), the composition was determined by Energy

Dispersive X-ray Spectrometer (EDS) and the morphology and microstructure characterization was carried out by High- Resolution Transmission Electron Microscopy (HRTEM) and Selected Area Electron Diffraction in an image-corrected FEI Titan Cube 60–300.

### *3.1. Multisegmented nanowires*

All the multisegmented nanowire arrays presented a hexagonal highly ordered structure where the nanowire diameter ($d$) had a constant value of 35 nm and the distance between nanowire centres ($D_{int}$) was 105 nm (Figure 1(a)). The first set of multisegmented $[FeCoCu/Cu(x)]_{10}$ nanowire arrays were fabricated with constant FeCoCu segment length of 300 nm and with Cu layer thickness ($x$) ranging between 7 and 40 nm. The second series of multisegmented $[FeCoCu(y)/Cu]_{10}$ nanowire arrays where grown, keeping constant the Cu layer thickness with a value of 15 nm and varying the FeCoCu segment length ($y$) from 120 to 900 nm. An example is shown in Figure 1(b). The nanowire total length ($L$) for the $[FeCoCu(300\ nm)/Cu(x)]_{10}$ series varies from 3.07 to 3.40 μm, and for the $[FeCoCu(y)/Cu(15\ nm)]_{10}$ series ranges between 1.35 and 9.15 μm. The EDS analysis confirmed that the composition of the continuous FeCoCu nanowires and, consequently, that of the FeCoCu segments in the multilayer nanowires, was $Fe_{30}Co_{64}Cu_6$ [51].

The nanowires present a polycrystalline structure as is concluded from the granular-like contrast in the bright field image (see Figure 1(c)). Figure 1(d) shows a HRTEM image of a Cu layer. In this image multiple sets of fringes are found in different regions, related to the presence of grains with different crystallographic orientations. To determine the crystal structure of the FeCoCu segment and the Cu layer, it was performed a local Fourier analysis. Both present a polycrystalline structure. The Fast Fourier Transform (FFT) in the Cu area presents reflections at approximately 2.08 and 1.80 Å corresponding to *fcc* (111) and (200) planes (Figure 1(e)). The FFT for the FeCoCu area present reflections at around 2 Å, associated with a (110) *bcc* structure (Figure 1(f)). The detailed analysis of the chemical composition of the multilayer nanowires was performed by Scanning Transmission Electron Microscopy (STEM). The combination of High-Angle Annular Dark Field (HAADF) and Electron Energy Loss Spectroscopy (EELS) was used to acquire spectrum images (SI) of the different regions. The chemical map of a $[FeCoCu(300\ nm)/Cu(7\ nm)]_{10}$ multilayer nanowire is shown in Figure 1(g). We observe that the Cu layers (green colour) are perfectly segregated from the FeCoCu segments (red colour), and they present an irregular and rough shape, probably caused by their polycrystalline nature, the large grain size, and possible inhomogeneities during the growth. Additionally, an oxidation layer is observed all around the nanowire (see blue colour at the surface), evidenced by the high oxygen content at the nanowire surface. It was determined that the oxidation layer present a thickness of 9 nm at the FeCoCu surface, while at the Cu surface the oxidation layer thickness decreases to less than 4 nm [51].

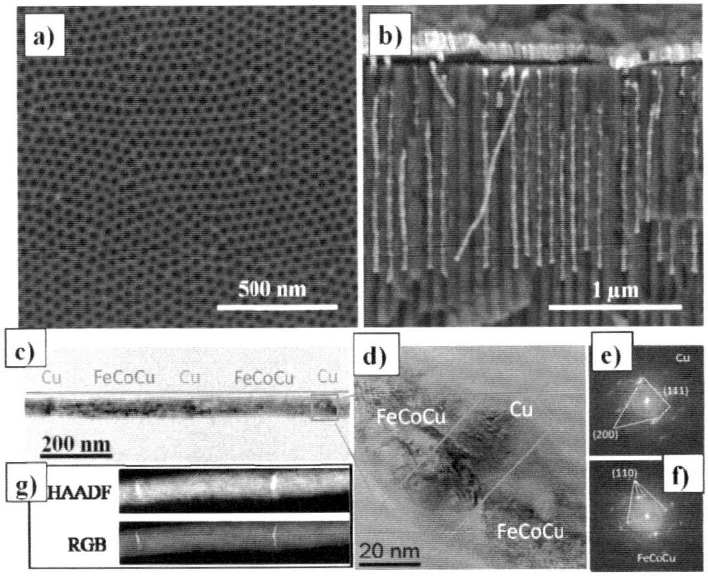


**Figure 1.** SEM images: (a) Top view of an AAO membrane and (b) cross-section of $[FeCoCu(120\ nm)/Cu(15\ nm)]_{10}$ multilayer nanowires inside the AAO membrane. TEM images of $[FeCoCu(300\ nm)/Cu(15\ nm)]_{10}$ multilayer nanowires: (c) Bright field image of an individual nanowire; (d) HRTEM image of a Cu spacer; (e) and (f) FFT at Cu and FeCoCu regions of the HRTEM image, respectively. (g) STEM-EELS chemical mapping of $[FeCoCu(300\ nm)/Cu(7\ nm)]_{10}$ nanowires: low magnification SI of one of the nanowires, including the HAADF and the RGB composition of the Co $L_{2,3}$ (red), Cu $L_{2,3}$ (green), and O K (blue) integrated intensities. TEM images adapted from Palmero et al. [51].

### *3.2. Nanowires with modulated diameter*

In order to characterize individually the modulated diameter nanowires, they were released from the templates by chemical dissolution of the membranes (see Figure 2). Firstly, the Au layer was chemically etched by a mixture of iodine and potassium iodide. Then, the alumina template was dissolved by a mixed solution of $CrO_3$ and $H_3PO_4$. After that, the modulated nanowires were cleaned, immersed in ethanol and dispersed on Si a substrate.

Nanowires with constant diameter (as reference sample) and two types of nanowires with two types of diameter modulations were fabricated, namely *Modulated A*: wide segments ($D_1$ in diameter and $L_1$ in length) shorter than the narrow ones ($D_2$ in diameter and $L_2$ in length); and *Modulated B*, with the narrow segments shorter than the wide ones (see Figure 2). The geometrical parameters (i.e. segment diameters and lengths, and nanowire total length, $L_T$) were measured by SEM and the values are collected in Table 1. The composition of the modulated nanowires was confirmed to be $Fe_{30}Co_{63}Cu_7$ by means of EDS technique [50].

Journal of Physics: Conference Series **755** (2016) 012001 doi:10.1088/1742-6596/755/1/012001

**Table 1.** Geometrical characterization of the FeCoCu nanowires shown in Figure 2. It also includes the fitted exchange constant, $A_{exc}$ (see Section 4.2 in the text).

| Sample | Type | $D_1$ (nm) | $L_1$ (nm) | $D_2$ (nm) | $L_2$ (nm) | $L_T$ (μm) | $A_{exc} \cdot 10^6$ (erg/cm) |
|---|---|---|---|---|---|---|---|
| S1 | Constant diameter | 150 | - | - | - | 10 | 3.45 |
| S2 | Modulated A (a) | 170 | 50 | 150 | 400 | 14 | 3.65 |
| S3 | Modulated A (b) | 170 | 50 | 150 | 800 | 13 | 3.09 |
| S4 | Modulated B | 130 | 1000 | 100 | 300 | 17 | 2.24 |

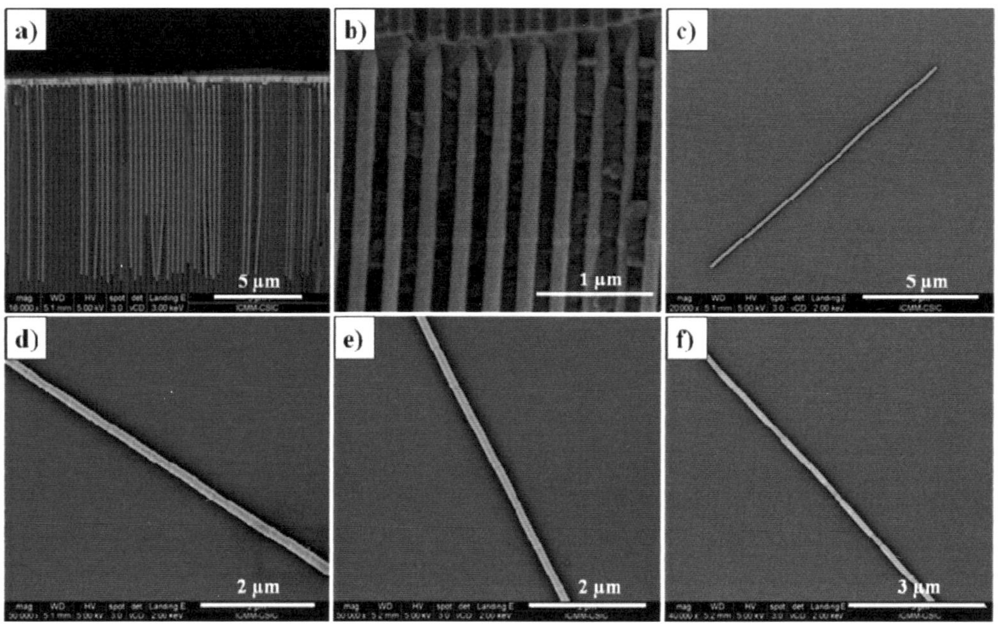


**Figure 2.** SEM images: cross-section of nanowires embedded into AAO membranes with (a) constant diameter and (b) modulated diameter, sample S2; single nanowires on silicon substrate: (c) constant diameter, sample S1; (d), (e) and (f) modulated diameter nanowires, samples S2, S3 and S4, respectively. Figures (d) to (f) adapted from Palmero et al. [50].

## 4. Magnetic characterization

The magnetic properties of the multisegmented nanowire arrays were determined by measuring the hysteresis loops using a Vibrating Sample Magnetometer (VSM), ADE system EV7 KLA-Tencor, applying a maximum magnetic field of ±18 kOe in parallel (|| to the nanowires) or perpendicular (⊥ to the nanowires) configuration.

On the other hand, the magnetic characterization of single nanowires with modulated diameter was performed by Magneto-Optical Kerr Effect (NanoMOKE™ 2), applying a maximum magnetic field of ±500 Oe in parallel configuration or making an angle (θ) with respect to the nanowire axis from θ = 0° (|| to the nanowire axis) to θ = 90° (⊥ to the nanowire axis), every 15°.

Each MOKE hysteresis loop was the result of 1000 averaged loops, in order to reduce the noise to signal ratio.

*4.1. Magnetic behaviour of multisegmented nanowires*

The normalized hysteresis loops (i.e. fractional magnetization ($M/M_s$) vs. applied magnetic field ($H$)) of [FeCoCu(300nm)/Cu($x$)]$_{10}$ and [FeCoCu($y$)/Cu(15nm)]$_{10}$ multilayer nanowire arrays and of the FeCoCu alloy nanowire array are plotted in Figure 3.

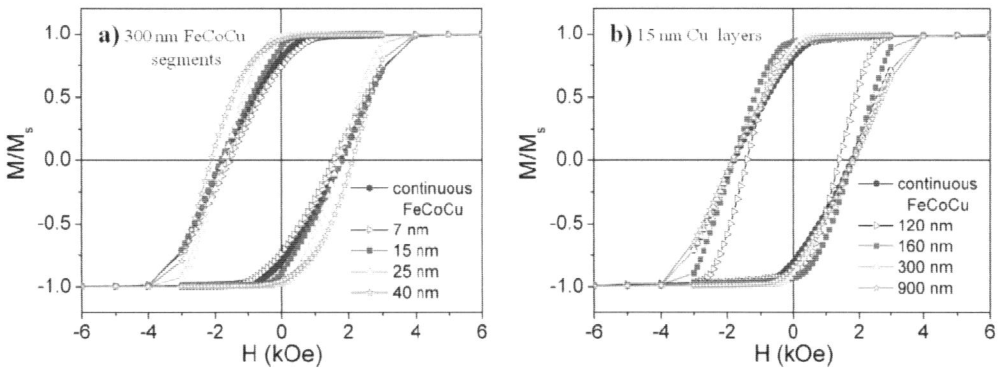


**Figure 3**. Hysteresis loops measured in parallel field configuration: (a) [FeCoCu(300nm)/Cu($x$)]$_{10}$ series for different Cu layer thicknesses and (b) [FeCoCu($y$)/Cu(15nm)]$_{10}$ series for different FeCoCu segment lengths. Adapted from Palmero et al. [51].

From the hysteresis loops, we determined the coercivity ($H_c$), fractional remanence ($M_r/M_s$) and differential magnetic susceptibility around the coercivity ($\chi=dM/dH$, for parallel field configuration). The values for both series of multilayer nanowire arrays as a function of the Cu layer thickness and FeCoCu segments length are plotted in Figure 4. The data for continuous FeCoCu nanowire array are also included for comparison. For the continuous FeCoCu nanowires, it was calculated a value for the saturation magnetization, $M_s$, of 1448 emu/cm$^3$, while for the multilayer nanowire arrays a maximum reduction of 11% in $M_s$ was determined from the comparison with the continuous nanowires.

A magnetization easy axis parallel to the nanowire axis is determined according to the high values of the fractional remanence under parallel configuration and the vanishing values for the perpendicular configuration (Figure 4). All the magnitudes (i.e. coercivity, $H_c$, fractional remanence, $M_r/M_s$, and susceptibility, $\chi$) show a general trend to increase with the Cu layer thickness for parallel field configuration, while for perpendicular field configuration, both $H_c$ and $M_r/M_s$ show small values (see Figure 4(a)). On the other hand, $H_c$ and $M_r/M_s$ when increasing the FeCoCu segments length for constant the Cu layer thickness ($x$=15nm) present a moderate and relatively complex variation (see Figure 4(b)). It is worth mentioning that the susceptibility shows a significant reduction for FeCoCu segment up to around 300 nm long while it remains almost constant for larger thicknesses [51].

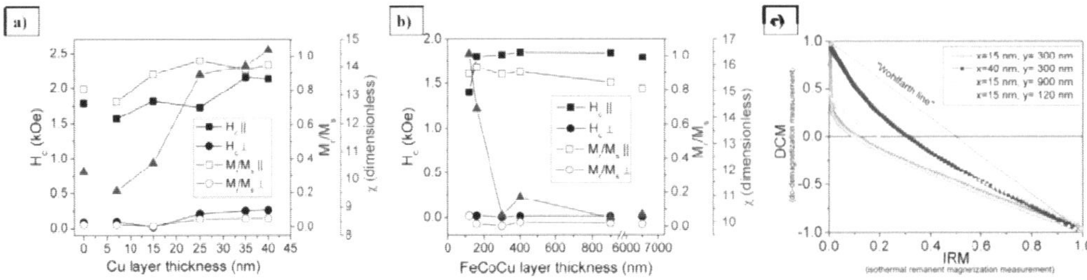


**Figure 4.** Coercive field, $H_c$, and fractional remanence, $M_r/M_s$, in parallel (||) and perpendicular (⊥) field configuration, and differential susceptibility, χ, for (a) [FeCoCu(300nm)/Cu($x$)]$_{10}$ and (b) [FeCoCu($y$)/Cu(15nm)]$_{10}$ series. The single scatters represent the data for continuous FeCoCu alloy nanowires. (c) Henkel plots for different [FeCoCu($y$)/Cu($x$)]$_{10}$ multilayer nanowires. The black line shows the behaviour of non-interacting nanocylinders. Adapted from Palmero et al. [51]

Henkel plots were obtained for FeCoCu/Cu multilayer nanowire arrays in order to get more information about the magnetic interactions between nanowires in the array. Henkel plots represent the values of dc-demagnetization (DCM) as a function of isothermal remanent magnetization (IRM), both normalized. A more detailed description of the measurement process can be found elsewhere [52]. In Figure 4(c) the Henkel plots are shown for several [FeCoCu($y$)/Cu($x$)]$_{10}$ nanowire arrays for parallel field configuration. The values for $x$ and $y$ for each nanowire array are indicated in the graph. A black line has been added to the graph, which represents a theoretical system of non-interaction nanoparticles (Wohlfarth model) [53-54].

In a multilayer nanowire array, we can consider two types of interactions: (i) intra-wire (positive) interactions between adjacent magnetic layers along each nanowire which favour the parallel alignment of the magnetization within the wire; and (ii) the inter-wire (negative) magnetostatic interactions between magnetic layers of adjacent nanowires which increase the effective demagnetization factor [28, 31]. According to the Henkel plots in Figure 4(c), the net magnetostatic interactions in all the multilayer nanowire arrays are demagnetizing since all the curves fit below the Wohlfarth model line. For nanowire arrays with the same FeCoCu segment length ($y$ = 300 nm), the one with thickest Cu layer ($x$ = 40 nm) is closer to the non-interacting line. On the other hand, for constant Cu layer thickness ($x$ = 15 nm), the nanowire array with the shortest FeCoCu segments ($y$ = 120 nm) is closer to the Wohlfarth line. According to these results, we can conclude that thicker Cu layers and shorter FeCoCu segments promote a decrease of the dipolar interaction field in the nanowire array [51].

### *4.2. Magnetic behaviour of individual modulated diameter nanowires*

Magneto-Optical Kerr Effect (MOKE) hysteresis loops measurements were used to characterize isolated nanowires with modulated diameter. Firstly, for each type of nanowire the hysteresis loop was obtained by focusing the laser spot on the centre of the nanowire (see Figure 5). For all the nanowires, the normalized longitudinal Kerr effect signal at remanence is close to 1, suggesting the existence of a nearly single domain structure with magnetization easy axis lying along the nanowire. In the case of the nanowires with constant diameter (Figure 5(a)), we observed a squared hysteresis loop with a sharp transition between two stable magnetic states at remanence, a single giant Barkhausen jump [44, 55]. In the same way, only one sharp transition is observed for the *Modulated A (a)* nanowire with shorter thin segments (Figure 5(b)). In contrast, for the *Modulated A (b)* nanowire with longer narrow segments, two abrupt and symmetric magnetization jumps are observed in each branch of the hysteresis loop (Figure 5(c)). In the case of the *Modulated B* nanowire, the hysteresis loop presents a main Barkhausen jump together with several additional ones of less amplitude (Figure 5(d)). The hysteresis loops measured for *Modulated A (b)* and *Modulated B* nanowires suggest the existence of metastable magnetic states during the magnetization reversal process, which can be correlated to their particular diameter modulations [38, 50].

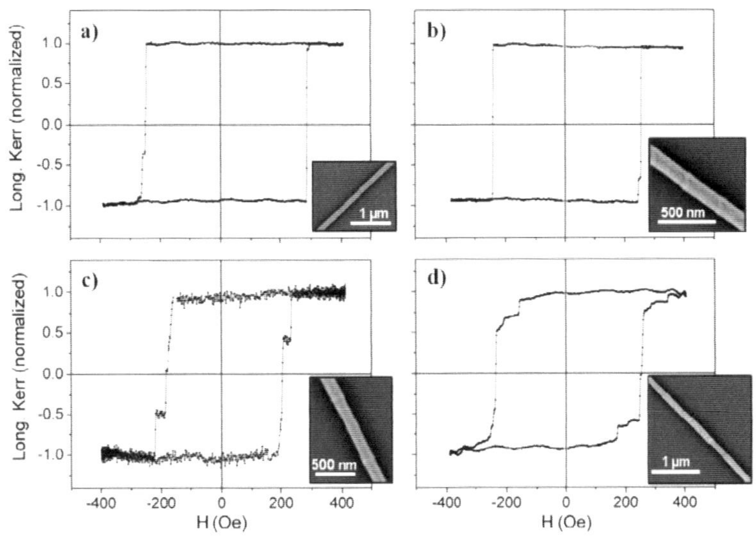


**Figure 5.** MOKE hysteresis loops for constant diameter S1 (a) and modulated diameter S2 (b), S3 (c) and S4 (d) nanowires. Inset in each figure correspond to SEM images of each type of nanowire. Adapted from Palmero et al. [50]

The variation of the magnetization process was also studied performing the MOKE measurements along the nanowire, obtaining the hysteresis loops at different positions along the nanowires length (see the obtained profile in Ref. [50]). The measurements were performed along the nanowire from one end to the other using steps of 1 μm. It was observed that the MOKE electrical signal obtained at the ends of the nanowire was several times smaller than when the measurement was performed focusing the laser spot at the nanowire centre. Along the nanowire, again several steps are observed that suggest metastable states in the magnetization reversal process that can be correlated to pinning centres for the domain wall propagation at local diameter modulations. For comparison, additional local MOKE measurements were carried out along a nanowire with constant diameter. In this case, the hysteresis loops present square shapes along the nanowire with single Barkhausen jumps. Similar coercivities were obtained at the different positions along the nanowire although, as similar to the modulated nanowire, at the ends the MOKE signal was reduced seemingly as a consequence of the formation of closure domain structures. The similar values of coercivity and hysteresis loop shape indicate that the magnetization reversal process takes place uniformly along the nanowire with constant diameter [56], allowing us to ascribe the small Barkhausen jumps observed in modulated nanowires to pinning sites correlated to diameter modulations [50].

A method to determine the coercivity mechanism of magnetic nanowire is the quantification of the angular dependence of the magnetization process [57]. Previous works about demagnetization process in nanowires by means of micromagnetic simulations proposed that this process in nanowires with constant diameter and effective longitudinal axial anisotropy takes place by the depinning of a domain wall from one nanowire end followed by its propagation along the wire [58]. The hysteresis loops obtained from the angular measurements performed on the *Modulated B* nanowires are shown in Figure 6(a). It was observed a similar behaviour for all the samples, obtaining an increasing switching field, $H_{sw}$, with the angle, $\theta$ (Figure 6(b)). Note that for $\theta > 60°$, the maximum magnetic field available in the setup was not enough to give rise to the reversal process.

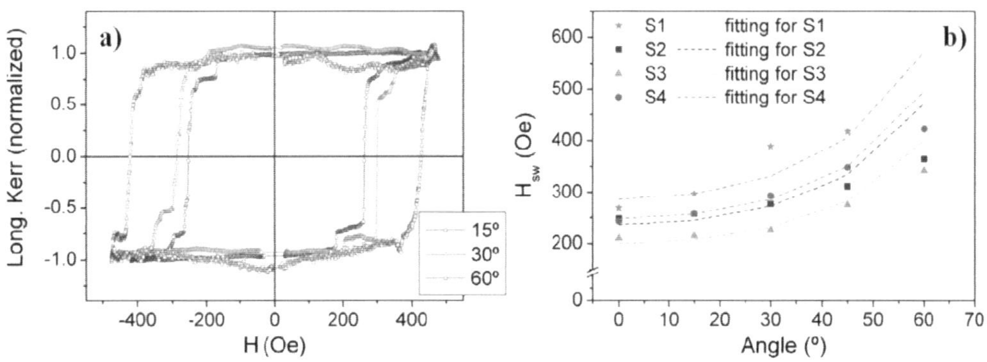


**Figure 6.** (a) Angular MOKE hysteresis loop for Modulated B nanowire (S4), and (b) obtained switching field for the nanowire with constant and modulated diameter together with the experimental data fitting curves. Adapted from Palmero et al. [50]

The increasing coercivity with the angle of the applied magnetic field with the nanowire axis as it is observed in Figure 6(b) can be ascribed to the propagation of a vortex domain wall [59]. In the case of a vortex magnetization reversal mode, the angular dependence of the coercivity for an isolated nanowire with high aspect ratio is given by the following expression [60]:

$$H_{sw} = \frac{\frac{8q^2A_{exc}}{M_sD^2}\left(2\pi - \frac{4q^2A_{exc}}{\pi M_s^2D^2}\right)}{\sqrt{\frac{16q^4A_{exc}^2}{\pi^2M_s^4D^4}\sin^2\theta + \left(2\pi - \frac{4q^2A_{exc}}{\pi M_s^2D^2}\right)^2\cos^2\theta}} \quad (1)$$

In this equation, the parameter $q^2$ takes a value of $1.08\pi$ for a cylindrical geometry [61], $D$ is the nanowire diameter, $A_{exc}$ is the exchange constant and $M_s$ is the saturation magnetization, which for this particular FeCoCu alloy is $M_s(Fe_{30}Co_{63}Cu_7)$ = 1448 emu/cm$^3$ [51]. The experimental data obtained for the angular dependence of the switching field for the different nanowires were fitted using Equation 1 (dashed lines in Figure 6(b)). For the modulated wires, the best fitting curves were obtained for the wide segments diameter ($D_l$). By fitting the experimental data, reasonable values of $A_{exc}$ were obtained (see Table 1) according to previous data for the nanowire alloy [62].

## 5. Conclusions

This work reports two different ways to tune the magnetic properties of nanowires by tailoring their composition or geometry. Both alternatives can be controlled by low-cost electrochemical processes modifying the fabrication parameters. For the first group, multilayer $[FeCoCu(y)/Cu(x)]_{10}$ nanowire arrays were fabricated with different FeCoCu segment lengths and Cu layer thicknesses, and constant diameter of 35 nm. The segments and the layers present both polycrystalline cubic structures, *fcc* and *bcc* structures, respectively. For all the samples, it was determined a magnetization easy axis parallel to the nanowires. For constant FeCoCu thickness, the coercivity, remanence and susceptibility increase with Cu layer thickness, while for the series with constant Cu layer thickness, a significant decrease of susceptibility with FeCoCu thickness was observed. A complementary Henkel curves study indicates that the net inter/intra nanowires magnetostatic interactions always contribute to demagnetization of nanowires. The results show a decrease in the effective magnetostatic interactions for multilayer nanowires with thicker Cu and thinner FeCoCu layers arising from stray fields at the surface of intra-wire and adjacent inter-wires magnetic layers. The hard magnetic characteristics of the multilayer nanowire arrays present a moderate improvement with regards to the continuous nanowire array in the case of thicker Cu layer and shorter FeCoCu segments.

For single modulated nanowires we can conclude from the MOKE measurements that the magnetization reversal process involves a giant Barkhausen jump preceded or followed by smaller jumps, which can be ascribed to the pinning of the domain wall at the diameter modulations as the smaller jumps are not present in the nanowires with constant diameter. Additionally, the experimental results from the angular dependence of the switching field and its analytical calculations, allow us to conclude that the coercivity mechanism corresponds to the propagation of a vortex domain wall. The domain wall is eventually pinned at local modulations along the nanowire resulting in metastable magnetic states during its propagation. This work opens new opportunities to further control the motion of domain walls through a further tailoring of nanowire diameter modulations.

## Acknowledgments

This work was supported by the Spanish Ministerio de Economia y Competitividad under Project MAT2013-48054-C2-1-R, by the European Community's 7th Framework Program under Project REFREEPERMAG, Grant Agreement 280670, and by the Spanish National Research Council under project 201460E014.

## References

[1] Mourachkine A, Yazyev O, Ducati C and Ansermet J-P 2008 *Nano Letters* **8** 3683
[2] Gao N, Wang H and Yang E-H 2010 *Nanotechnology* **21** 105107
[3] Kou X, Fan X, Dumas R K, Lu Q, Zhang Y, Zhu H, Zhang X, Liu K and Xiao J Q 2011 *Adv. Mat.* **23** 1393
[4] Chen X, Wong CK, Yuan CA and Zhang G 2013 *Sens. Act. B: Chem.* **177** 178
[5] Lavrijsen R, Lee J-H, Fernández-Pacheco A, Petit DC, Mansell R and Cowburn R P 2013 *Nature* **493** 647
[6] Lee D J, Kim E, Kim D, Park J and Hong S 2013 *ACS Nano* **7** 6906
[7] Allwood D A, Xiong G, Faulkner C, Atkinson D, Petit D and Cowburn R 2005 *Science* **309** 1688
[8] Terabe K, Hasegawa T, Liang C and Aono M 2007 *Sci. Tech. Adv. Mater.* **8** 536
[9] Parkin S S, Hayashi M and Thomas L 2008 *Science* **320** 190

[10] Chałupniak A, Morales-Narváez E and Merkoçi A 2015 *Adv. Drug Deliv. Rev.* **95** 104
[11] Losic D and Losic D Jr 2009 *Langmuir* **25** 5426
[12] Sousa C, Leitao D, Proenca M, Ventura J, Pereira A and Araujo J 2014 *App. Phys. Rev.* **1** 031102
[13] Gandha K, Elkins K, Poudyal N, Liu X and Liu J P 2014 *Sci. Rep.* **4** 5345
[14] Bran C, Palmero E M, del Real R P and Vazquez M 2014 *Phys. Status Solidi A* **211** 1076
[15] Kuz'min M, Skokov K, Jian H, Radulov I and Gutfleisch O 2014 *J. Phys. Condens. Matter* **26** 064205
[16] Gandha K, Tsai P, Chaubey G, Poudyal N, Elkins K, Cui J and Liu J P 2015 *Nanotechnology* **26** 075601
[17] Bran C, Palmero E M, Li Z-A, del Real R P, Spasova M, Farle M and Vazquez M 2015 *J. Phys. D: Appl. Phys.* **48** 145304
[18] Coey J 2011 *IEEE Trans. Magn.* **47** 4671
[19] Lewis L H and Jiménez-Villacorta F 2013 *Metall. Mater. Trans. A* **44** 2
[20] Chen Y, Tzeng H, Wei L, Wang L, Oung J and Shih H 2005 *Science* **47** 1001
[21] Fu J, Yan Z, Xu Y, Fan X and Xue D 2007 *J. Phys. Chem. Solids* **68** 2221
[22] Sharif R, Shamaila S, Ma M, Yao L, Yu R, Han X, Wang Y and Khaleeq-ur-Rahman M 2008 *J. Magn. Magn. Mater.* **320** 1512
[23] Baker T 2009 *Mater. Sci. Tech.* **25** 1083
[24] Liu Z, Li W, Jin P and Fei W 2013 *J. Magn. Magn. Mater.* **345** 96
[25] Hussein M A, Mohammed A S and Al-Aqeeli N 2015 *Materials* **8** 2749
[26] Bran C, Ivanov Y P, García J, del Real R P, Prida V, Chubykalo-Fesenko O and Vazquez M 2013 *J. Appl. Phys.* **114** 043908
[27] Rodríguez-González B, Bran C, Warnatz T, Rivas J and Vazquez M 2014 *J. Appl. Phys.* **115** 133904
[28] Carignan L-P, Lacroix C, Ouimet A, Ciureanu M, Yelon A and Ménard D 2007 *J. Appl. Phys.* **102** 023905
[29] Béron F, Carignan L-P, Menard D and Yelon A 2008 *IEEE Trans. Magn.* **44** 2745
[30] Jeon I T, Yoon S J, Kim B G, Lee J S, An B H, Ju J-S Wu J H and Kim Y K 2012 *J. Appl. Phys.* **111** 07B513
[31] Krimpalis S, Dragos O-G, Moga A-E, Lupu N and Chiriac H 2011 *J. Mater. Res.* **26** 1081
[32] Elbaile L, Cubero I, Crespo R, Vega V and Garcia J 2012 *J. Alloy Compd.* **536** S359
[33] Qi K, Li X, Zhang H, Wang L, Xue D, Zhang H, Zhou B, Mellors N J and Peng Y 2012 *Nanotechnology* **23** 505707
[34] Jaleh B, Koosha F and Dezfuli A O 2014 *J. Supercond. Nov. Magn.* **27** 1065
[35] Yang W, Cui C, Liu Q, Cao B, Liu L and Zhang Y 2014 *J. Cryst. Growth* **399** 1
[36] Hnida KE, Żywczak A, Gajewska M, Marciszko M, Sulka G D and Przybylski M 2016 *Electrochim. Acta* **205** 29
[37] Clime L, Zhao S, Chen P, Normandin F, Roberge H and Veres T 2007 *Nanotechnology* **18** 435709
[38] Pitzschel K, Bachmann J, Martens S, Montero-Moreno J M, Kimling J, Meier G, Escrig J, Nielsch K and Görlitz D 2011 *J. Appl. Phys.* **109** 033907
[39] Esmaeily A, Venkatesan M, Razavian A S and Coey J 2013 *J. Appl. Phys* **113** 17A327
[40] Salem M S, Sergelius P, Corona R M, Escrig J, Görlitz D and Nielsch K 2013 *Nanoscale* **5** 3941
[41] Minguez-Bacho I, Rodriguez-López S, Vázquez M, Hernández-Vélez M and Nielsch K 2014 *Nanotechnology* **25** 145301
[42] Bran C, Berganza E, Palmero E M, Fernandez-Roldan J, del Real R P, Aballe L, Foerster M, Asenjo A, Fraile-Rodríguez and Vazquez M 2016 *J. Mater. Chem. C* **4** 978
[43] Lee W, Ji R, Gösele U and Nielsch K 2006 *Nat. Mater.* **5** 741
[44] Vega V, Böhnert T, Martens S, Waleczek M, Montero-Moreno J M, Görlitz D, Prida V M and Nielsch K 2012 *Nanotechnology* **23** 465709

[45] Pitzschel K, Montero-Moreno J M, Escrig J, Albrecht O, Nielsch K and Bachmann J 2009 *ACS Nano* **3** 3463
[46] Yuan H and Wang X 2014 *Phys. Rev. B* **89** 054423
[47] Chen T-C, Kuo C-Y, Mishra AK, Das B and Wu J-C 2015 *Physica B* **476** 161
[48] Lage E, Urs N O, Siddiqui S, Baldo M, McCord J and Ross C A 2016 *IEEE Magn. Lett.* **7** 1
[49] Masuda H and Fukuda K 1995 *Science* **268** 1466
[50] Palmero E M, Bran C, del Real R P and Vázquez M 2015 *Nanotechnology* **26** 461001
[51] Palmero E M, Bran C, del Real R P, Magen C and Vazquez M 2014 *IEEE Magn. Lett.* **5** 6700304
[52] García-Otero J, Porto M and Rivas J 2000 *J. Appl. Phys.* **87** 7376
[53] Wohlfarth E P 1958 *J. Appl. Phys.* **29** 595
[54] Samwel E O, Bissel P R and Lodder J C 1993 *J. Appl. Phys.* **73** 1353
[55] Vilanova-Vidal E, Ivanov Y P, Mohammed H and Kosel J 2015 *Appl. Phys. Lett.* **106**, 032403
[56] Burn D M, Arac E and Atkinson D 2013 *Phys. Rev. B* **88**, 104422
[57] Kronmüller H and Fähnle M 2003 Micromagnetism and the Microstructure of Ferromagnetic Solids, Cambridge Univ. Press, Cambridge
[58] Vázquez M (ed) 2015 Magnetic Nano- and Microwires (Cambridge: Woodhead) ch 14, 22, 25
[59] Tejo F, Vidal-Silva N, Espejo A P and Escrig J 2014 *J. Appl. Phys.* **115**, 17D136
[60] Aharoni A 1997 *J. Appl. Phys.* **82**, 1281
[61] Shtrikman S and Treves D 1963 Magnetism, Vol. 3 (Eds: G.T. Rado, and H. Suhl), Academic, New York
[62] Liu X, Sooryakumar R, Gutiérrez C J and Priz G A 1994 *J. Appl. Phys.* **75**, 7021

# Development and calibration of a MFM-based system for local hysteresis loops measurements

**M. Coïsson, G. Barrera, F. Celegato, P. Tiberto**
INRIM, strada delle Cacce 91, 10135 Torino (TO), Italy
E-mail: m.coisson@inrim.it

**Abstract.** A measurement technique derived from a field-dependent magnetic force microscope (MFM) is presented for the measurement of local hysteresis loops on patterned micrometric and sub-micrometric magnetic structures. The technique exploits the synchronisation of the applied field variations with the end-of-line signal of the microscope, while keeping the slow scan axis disabled. In this way, a single MFM image contains the whole field evolution of the magnetisation processes in the sample along a user-defined profile. An analysis procedure is presented for the subsequent determination of local hysteresis loops on magnetic dots. The system has been calibrated for what concerns the applied field values. No significant artifacts induced in the measurements by the applied field have been observed up to applied fields of $\approx$ 1000 Oe.

## 1. Introduction

In recent years, the need has emerged to reliably measure the physical properties of materials at the micrometer or sub-micrometer scales exploited by the new lithographic techniques available in nanotechnology [1, 2]. Among the important properties that need to be investigated, magnetisiation processes as a function of the applied magnetic field in patterned magnetic structures are of utmost importance in a variety of applications, including magnetic recording, spintronics, sensors [3]. However, available magnetometry techniques, including high-sensitivity SQUIDs, or optical Kerr-effect microscopes, lack the required sensitivity to investigate individual patterned structures or space resolution respectively [4, 5]. Therefore, the solution so far has been to investigate large arrays of nominally identical structures in order to overcome these limitations. Experimental techniques such as SEMPA are actually able to provide access to the sub-micrometer scale on magnetic samples, but are not widely available [6].

Magnetic force microscopy (MFM) is a technique that offers excellent space resolution but limited possibility of investigating the magnetic-field dependence of the magnetisation of the sample [7, 8, 9, 10, 11]. In fact, each MFM image is quite time-consuming, and it is not possible to acquire hundreds of images of the same sample at different applied magnetic fields with a single tip. Recently, an MFM-derived setup has been presented [12] that overcomes this limitation by allowing the acquisition of a single magnetic image, made of several hundreds of lines, each acquired at a different applied magnetic field. In order to achieve this result, a single profile of the measured patterned structure is repeatedly acquired as a function of the applied field; from this information, the magnetisation evolution of the whole patterned structure is inferred, therefore providing a convenient way to measure "local hysteresis loops" on patterned magnetic systems with a widely available setup.

In this paper, on overview of this technique is presented, with particular emphasis on the development of the experimental setup and on its calibration. Examples of local hysteresis loops on $Ni_{80}Fe_{20}$ dots with a lateral size of $2\mu$m are reported.

## 2. MFM-based local field loops measurement setup

A system aimed at the measurements of local hysteresis loops based on a magnetic force microscope (MFM) has been developed. The selected microscope (a Bruker Nanoscope V Multimode 8) is equipped with a fully non magnetic scanner, head and tip holder, and is coupled with an electromagnet surrounding the head. The sample is located in the geometrical centre of the magnet, whereas the Hall probe used for the measurement of the applied magnetic field is placed close to one of the poles. The difference between the measured field and its value at the position of the sample will be taken into account with a suitable calibration procedure discussed in the following section. The field is applied in the sample plane along the direction of the fast scan axis.

In order to measure local hysteresis loops, the procedure summarised in [12] is discussed here in more details. A patterned magnetic sample, i.e. a $Ni_{80}Fe_{20}$ dot with a lateral size of $2\mu$m and a thickness of 30 nm, is imaged with the microscope in intermittent contact mode, while the magnetic signal is acquired in lift mode using the phase channel. An AFM representation of a selected dot is shown in Fig. 1. A profile of the dot is arbitrarily chosen, to be repeatedly scanned by disabling the slow scan axis of the microscope. In Fig. 1 an example of a possible choice of scanned profile is shown by the dashed line.

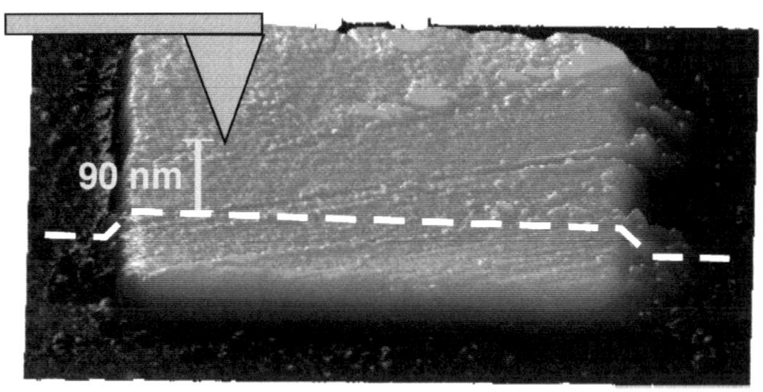


**Figure 1.** 3-dimension atomic force microscopy image of a $Ni_{80}Fe_{20}$ dot with a lateral size of $2\mu$m and a thickness of 30 nm. The dashed line represents a profile that is repeatedly scanned during local hysteresis loops measurements. The tip is also schematically drawn indicating the lift scan height at which MFM data is acquired.

The field-dependence of the magnetisation of the dot can be obtained by submitting the sample to a different magnetic field intensity each time the same profile is acquired. In order to achieve this result, the end-of-line (EOL) signal of the microscope controller is processed by custom electronics, whose diagram is shown in Fig. 2. This electronic circuit counts the transistor-to-transistor logic (TTL) impulses that the MFM controller sends each time the tip has finished scanning a line. Since the magnetic signal is acquired in pass 2, the microscope sends 2 EOL signals for each repetition of the profile scansion, one for the height channel (AFM data) and one for the phase channel in pass 2 (MFM data). The electronic circuit counts the impulses and sends the result to a control PC that communicates through a RS232 interface. In this way, the control PC can synchronise the variations of the applied magnetic field with the

EOL signals, ensuring that each time the dot profile is acquired, a new value of the magnetic field is applied and is stable throughout the whole acquisition process.

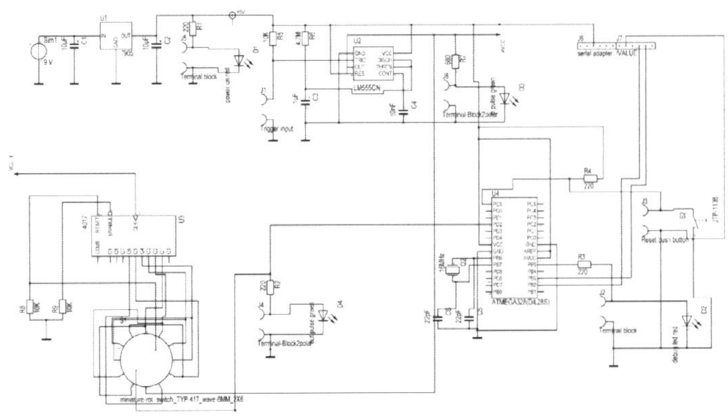

**Figure 2.** Circuit scheme of the custom electronics used count the EOL signals of the MFM controller and communicate with the control PC.

Figure 3 shows an example of local hysteresis loop acquisition on a profile similar to the one marked in Fig. 1 with the dashed line. In panel (a), the morphology (height channel, pass 1) is reported; as the slow scan axis is disabled, the same profile is repeatedly acquired until the whole frame is filled. The dashed lines clearly mark the dot edges, indicating a negligible drift during the scanning. As each line is acquired, the magnetic field changes, resulting in the MFM signal (phase channel, pass 2) shown in Fig. 3(b). In this case, the resulting image is no longer independent on the line, as in panel (a), because the magnetic signal is affected by magnetisation structure in the dot, that reflects the variations of the applied magnetic field. In panel (c), in false colours, the applied magnetic field is reported: it can be seen that after the acquisition of a few lines at the top of the image, the field is suddenly switched on at its maximum value, then gradually decreased through zero down to the minimum value, and then back again to its initial value. The last third of the figure in panel (c) shows that the field has been switched off. The same information is reported in panel (d), but with an indication of the actual field values. The field dependence in Fig. 3(d) is actually just a vertical profile along Fig. 3(c). The switching off of the field at the end of the hysteresis loop is clearly visible in the MFM signal, and very weakly visible in the height channel.

Figure 3 has been obtained by choosing a profile close to an edge of the dot; however, this choice is not critical, as in principle any profile could be picked up to measure local hysteresis loops. Indeed, the choice should be driven by the amount of information that can be extracted from the MFM image along the acquired profile. In the case of a magnetic dot displaying a vortex magnetisation at the remanence, a profile close to the symmetry axis of the dot (i.e. passing through its centre) is not particularly indicated, as relatively weak phase variations occur as a function of the applied field; therefore, a profile closed to a dot edge has been chosen, but it is not necessary to precisely know its position.

## 3. Calibration and validation

Before analysing Fig. 3, it is necessary to perform a few validation steps. As discussed in Section 2, the Hall probe used for measuring the field is placed close to a magnet pole, and not to the sample that is located in the centre of the gap. It is therefore required to calibrate the magnetic field in order to apply the calcuated correction factor to the signal displayed in Fig. 3(c). To this purpose, two Hall probes have been used during the calibration process, the second one

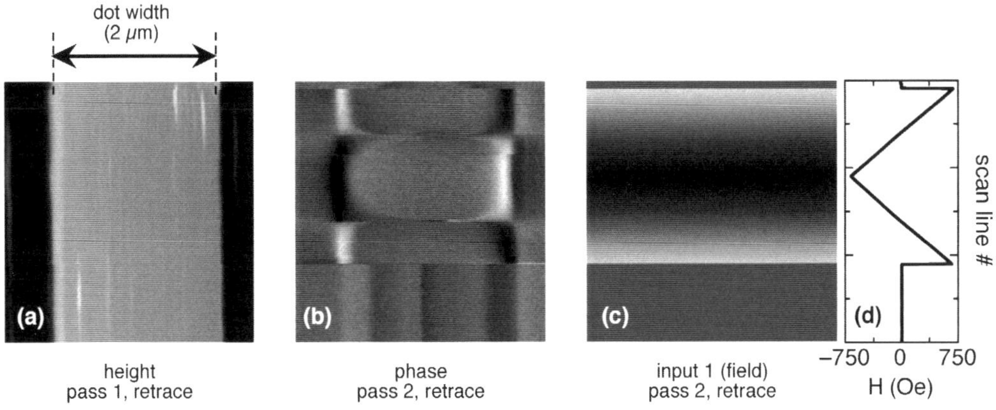


**Figure 3.** (a) Dot profile acquired with the height channel with the slow scan axis disabled. The dashed lines mark the dot edges. (b) Corresponding MFM image (phase, pass 2) at a lift scan height of 90 nm. (c) Applied magnetic field (in false colours) as a function of the acquired scan line. (d) Same as (c), in Oe.

being positioned at variable distance from a magnet pole, without the microscope head. As the head is totally non magnetic, it has no shielding effect and can be safely removed to perform the field calibration. The results are displayed in Fig. 4, where the difference between the two probes readings is reported as a function of $H_{nominal}$ (the field measured by the Hall probe close to the magnet pole and used during the experiments) and of the distance of the second probe with respect to the same magnet pole.

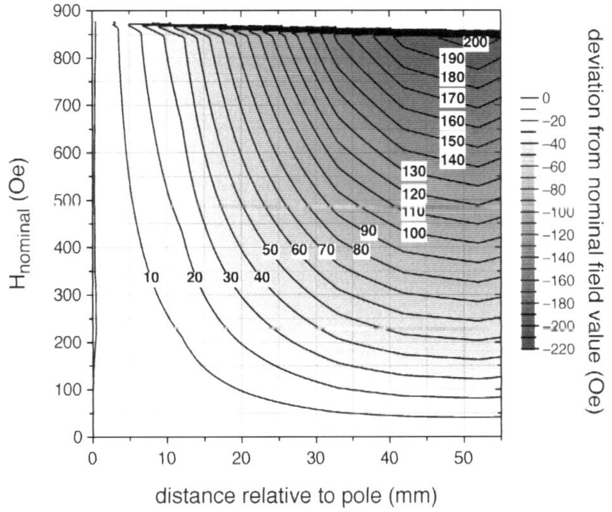


**Figure 4.** Difference between the magnetic field measured by the Hall probe close to the magnet pole ($H_{nominal}$) and the one measured by the second Hall probe (contour and colour plots and labels) as a function of $H_{nominal}$ and of the distance of the second Hall probe with respect to a magnet pole.

The contour plots shown in Fig. 4 indicate that at $\approx 52$ mm from the pole (where the sample is located) the deviation between the two field measurements has a significant dependence on the second probe position especially at high applied field values. As a consequence, it is important to place the magnet around the head as much as possible always in the same position. Conversely, as the sample only moves by a few tens of micrometers during scanning in the microscope, its movement is not critical for the determination of the actual magnetic field. The final calibration curve applied at the $\approx 52$ mm position is reported in Fig. 5, that can be used to calculate the actual field value in the magnet centre (where the sample is located), for any given nominal magnetic field measured by the probe close to the magnet pole.

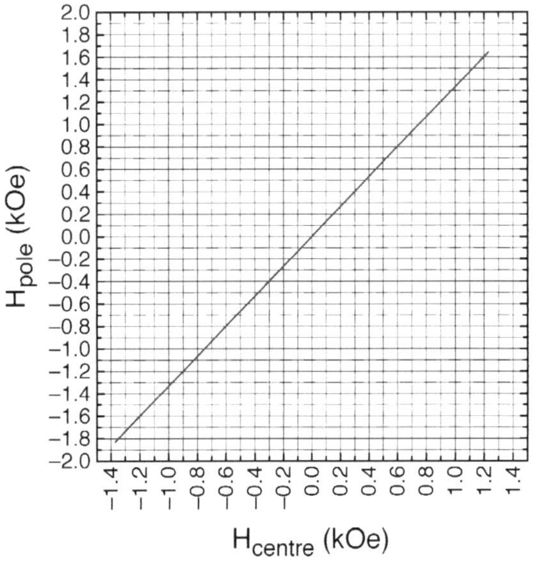


**Figure 5.** Relationship between the actual field value at the centre of the magnet and the value measured by the Hall probe close to one of its poles.

The field values reported in Fig. 3(d) have already been rescaled from the raw data of Fig. 3(c) according to the calibration curve of Fig. 5.

Even though the microscope head, scanner and tip holder are not magnetic, the tip used for MFM imaging is of course coated with a magnetic material, and its response to an applied magnetic field has to be investigated, as spurious effects may lead to misinterpretation of the magnetic data. MFM tips usually report nominal coercivity values that refer to the field that is necessary to reverse their magnetisation along their axis. In our setup, the magnetic field is applied along the sample plane, i.e. perpendicular to the tip axis. Even in this case, it is safer to choose a tip whose nominal coercive field is larger than the maximum applied field. In our case, we opted for Co-Cr coated HR-MESP tips whose coercivity is $\approx 900$ Oe. The absence of artifacts in MFM images induced by the application of an in-plane magnetic field has been verified with a Cu sample displaying a sharp step. The results, obtained with the same procedure exploited for the data reported in Fig. 3, are shown in Fig. 6.

As it can be seen in Fig. 6(b), the magnetic signal does not show any dependence on the applied field; it has to be remarked that the image is more noisy than Fig. 3(b) because, being the imaged sample non magnetic, the phase variations are much more limited and the colour scale is much more magnified; this further supports the absence of artifacts induced in the MFM image by the applied field. As a consequence, the phase data obtained in Fig. 3(b) can be considered reliable and further analysed to investigate the magnetisation reversal processes occurring in the studied dot. A further proof that the application of the in-plane magnetic field has not significantly perturbed the tip is given by the fact that when the field is switched off the

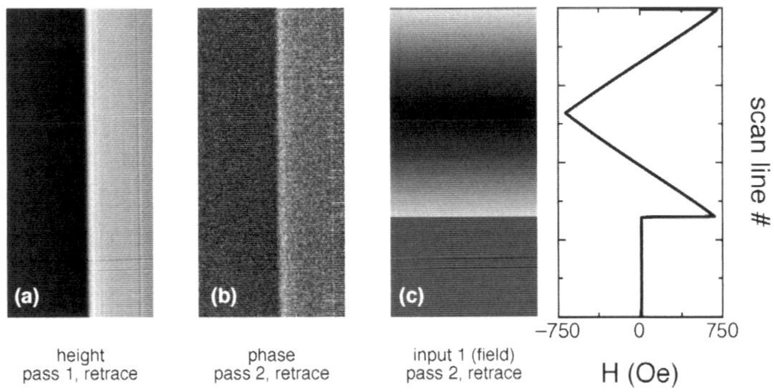


**Figure 6.** Height, MFM and field profiles of a Cu sharp step as a function of the applied magnetic field.

magnetic contrast is not lost (see bottom of Fig. 3(b)): if the tip magnetisation had flipped out of its axis during the application of the field, the magnetic contrast would have been lost since the microscope is operated in the normal vertical oscillation of the cantilever, that is sensitive only to the second derivative of the vertical component of the magnetic field generated by the underneath magnetisation of the sample along the vertical direction, provided that the tip is magnetised along the same axis.

In order to proceed with the investigation of the reversal processes occurring in the studied dot, the MFM signal reported in Fig. 3(b) must be analysed as a function of the applied field (Fig. 3(c) and (d)). To this purpose, the procedure described in Fig. 7 is applied.

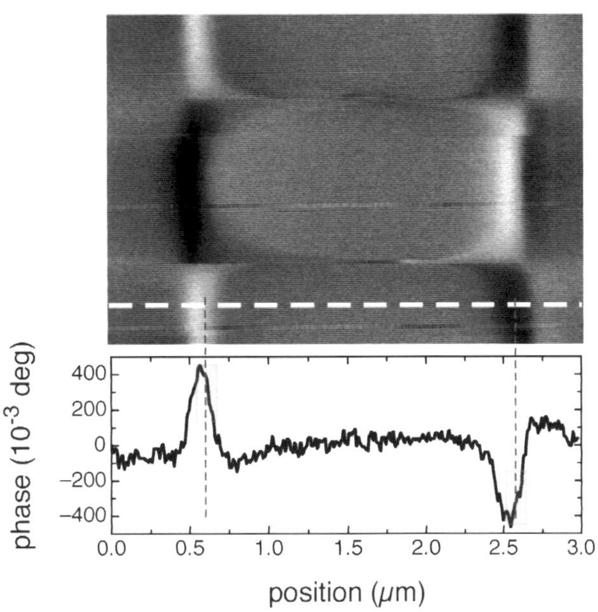


**Figure 7.** Top: same as Fig. 3(b). Bottom: phase profile along the yellow dashed line. The blue dashed lines indicate the positions of the left and right dot edges. The shadowed areas mark the data that have been used for the analysis described in the text.

Each line of the MFM image is extracted and the two dot edges are identified: this information is available through the height channel that is simultaneously acquired. Then, a subset of the phase data is extracted around the position of the two dot edges, with a lateral size of a few tens of nanometers. The phase signal is then averaged independently for each of the two selected portions of the line (shaded areas in Fig. 7), and the two results subtracted. In the end, for each line of the MFM image a "phase contrast" value is obtained, being e.g. positive for the line reported in Fig. 7 and negative when the dominant colour close to the dot edges is reversed. The "phase contrast" can then be plotted as a function of the applied field, resulting in the so-called "local hysteresis loop", shown in Fig. 8.

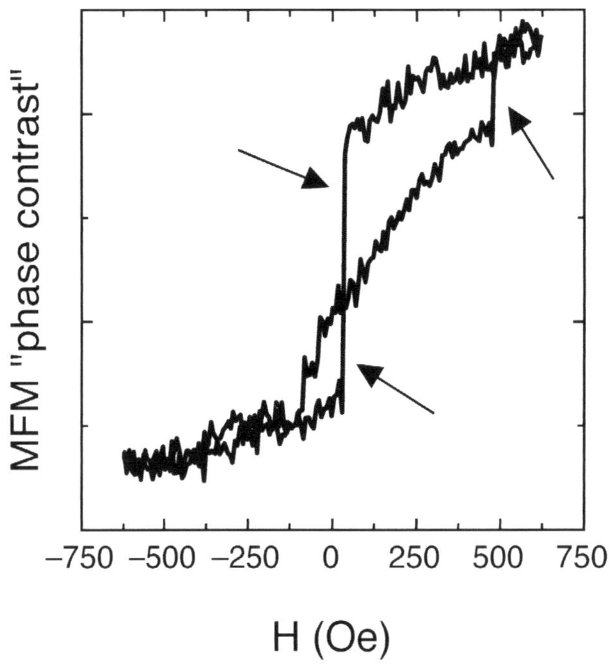


**Figure 8.** Local hysteresis loop of the $Ni_{80}Fe_{20}$ dot having a side of 2$\mu$m and a thickness of 30 nm. The arrows indicate irreversible features in the magnetisation reversal processes.

The curve displayed in Fig. 8 reports, in the vertical axis, the "phase contrast", that is expressed in arbitrary units since the response of the MFM tip has not been calibrated to a traceable standard giving a reference value of the field gradient to which the tip is sensitive. However, should a quantitative measurement be required, such calibration can be performed as described in [13, 14, 15]. In our case, it is not necessary to know the magnetisation value of the dot, as our interest lies in the mechanisms induced by the varying magnetic field. These are clearly visible in Fig. 8 in the form of abrupt jumps marked by the arrows. A suitable comparison with micromagnetic simulations allows to unequivocally assign these jumps to irreversible magnetisation processes occurring in the dot, in the form of magnetic vortex nucleation and expulsion [12]. These field values can be obtained with other experimental techniques, but magnetometers (e.g. SQUID) require rather large arrays of identical and non interacting dots, whereas magneto-optics setups do not have the space resolution to distinguish the details of the magnetisation patterns in structures having sizes smaller than $\approx 1 - 2\mu$m. Conversely, the present MFM-derived setup provides both space and field resolution required for the investigation of magnetisation reversal processes in micrometric and sub-micrometric individual patterned structures.

## 4. Conclusions

An experimental setup for the measurement of local hysteresis loops on micro-and nanometric patterned structures, based on a field-dependent magnetic force microscope, has been presented and calibrated. The setup allows to image the magnetic field evolution of the magnetisation in a patterned system (e.g. a magnetic dot) by acquiring a single MFM profile as a function of the applied field. A suitable image analysis, exploiting the "phase contrast" at the patterned structure edges, provides access to all the relevant irreversible magnetisation reversal processes that occur in the sample. The system has been properly calibrated for what concerns the accurate field values applied to the sample. Conversely, the proposed setup does not require calibration of the response of the MFM tip, unless a quantitative measurement of the sample magnetisation is required. The presented data support a field-dependent response of the technique that is sensitive to the variation of the magnetic properties of the sample without significant artifacts up to applied field of $\approx$ 1000 Oe.

## Acknowledgments

This work has been partially performed at NanoFacility Piemonte, an INRIM laboratory supported by Compagnia di San Paolo.

## References

[1] *Nanoelectronics and information technolgy: Advanced electronic materials and novel devices*, R. Waser ed., Wiley-VCH, Weinheim (2003).
[2] *Magnetic nanostructures*, H.S. Nalwa ed., American Scientific Publishers, USA (2002).
[3] *Nanomagnetism and Spintronics*, T. Shinjo ed., Elsevier (2009).
[4] H.J. Williams, F.G. Foster, E.A. Wood, Phys.Rev. **82** 119 (1951).
[5] D.A. Allwood, G. Xiong, M.D. Cooke, R.P. Cowburn, J. Phys. D: Appl. Phys. **36** 2175 (2003).
[6] M.R. Scheinfein, J. Unguris, M.H. Kelley, D.T. Pierce, R.J. Celotta, Rev. Sci. Instrum. **61** 2501 (1990).
[7] U. Hartmann, J. Magn. Magn. Mater. **157** 545 (1996).
[8] P. Grütter, H.J. Mamin, D. Rugar, *Magnetic force microscopy (MFM): Scanning Tunneling Microscopy vol 2*, H.-J. Güntherodt and R Wiesendanger ed, Springer, Berlin, pp 151-207 (1992).
[9] T.G. Sorop, C. Untiedt, F. Luis, M. Kroll, M. Rasa, L.J. de Jongh, Phys. Rev. B **67** 014402 (2003).
[10] M.V. Rastei, R. Meckenstock, J.P. Bucher, Appl. Phys. Lett. **87** 222505 (2005).
[11] M. Jaafar, L. Serrano-Ramón, O. Iglesias-Freire, A. Fernández-Pacheco, M.R. Ibarra, J.M. DeTeresa, A. Asenjo, Nanoscale Res. Lett. **6** 407 (2001).
[12] M. Coïsson, G. Barrera, F. Celegato, E. Enrico, A. Manzin, E.S. Olivetti, P. Tiberto, F. Vinai, J. Phys. D: Appl. Phys. **47** 325003 (2014).
[13] L. Kong, S.Y. Chou, Appl. Phys. Lett. **70** 2043 (1997).
[14] P.J.A. van Schendel, H.J. Hug, B. Stiefel, S. Martin, and H.-J. Güntherodt, J. Appl. Phys. **88** 435 (2000).
[15] T. Göddenhenrich, H. Lemke, M. Mück, U. Hartmann, C. Heiden, Appl. Phys. Lett. **57** 2612 (1990).

Journal of Physics: Conference Series **755** (2016) 012003 doi:10.1088/1742-6596/755/1/012003

# Opportunities of research in multiferroic materials using Angle Dispersive X-ray Diffraction (ADXRD) beamline on Indus-2 synchrotron source

**A K Sinha[1,2], M N Singh[1], A Upadhyay[1] and A Sagdeo[1,2]**
[1]Indus Synchrotrons Utilisation Division
Raja Ramanna Center for Advanced Technology, Indore - 452013, India
[2]Homi Bhabha Naational Institute
Raja Ramanna Center for Advanced Technology, Indore - 452013, India

E-mail: anil@rrcat.gov.in

**Abstract.** Synchrotron beamlines have advantages of higher flux and wide tunability of photon beam compared to laboratory based equipment for performing x-ray diffraction and x-ray absorption spectroscopy (XAS). In this paper we report capabilities of angle dispersive x-ray diffraction beamline (BL-12) on Indus-2 synchrotron source for structural and spectroscopic characterisation of multiferroic materials. Brief description of the beamline along with the photon beam specifications at the experimental station is given. Results on low temperature XRD measurements on mixed spinel system ($Fe_{1.5}Co_{1.5}O_4$) between 30K and 300K and subsequent Reitveld refinement reveal that there are no phase changes but the lattice parameter show anomalous changes between 100 and 150K. It has been explained how XANES spectra on a type II multiferroic, $Co_3TeO_6$, and $Fe_{1.5}Co_{1.5}O_4$ can be used for the determination of charge and spin states of transition metal ions.

## 1. Introduction

Single phase compounds and multiphase composites, which posses simultaneously, two or more ferroic orders among ferroelectric, ferromagnetic (FM), ferroelasticity or ferrotoroidicity are called multiferroic (MF) materials. An important class of MF materials are magnetoelectric (ME) MF materials, which consists of ferromagnetic and ferroelectric orders. In modern day definition of ME MF materials, other magnetic orders like antiferromagnetic (AFM) and ferromagnetic are also included, in addition to FM order. From application point of view of ME MF materials, the coupling between the magnetic and the ferroelectric orders (the ability to control the ferroelectricity using the magnetic field and vice versa) is important. ME MF having coupling between FE and FM orders are called type II or proper MF materials. More often than not, type II MF materials have weaker magnitude of individual orders and also the MF properties are observed at very low temperatures. Current research efforts in MF research are directed towards enhancing the magnitudes of magnetic and ferroelectric orders, increasing the onset temperature of coexistence of the two ferroic orders and physics understanding of the magnetic transition as well as appearance of the two orders simultaneously in these materials. Bulk (both single crystals and powder) as well as thin film multilayers are being explored by various groups worldwide to achieve the above mentioned goals [1-5].

In order to understand the ME properties and magnetic transitions at low temperatures, the structural as well as the spectroscopic properties are important inputs. Structural techniques like low and high temperature x-ray diffraction (XRD), anomalous XRD and extended x-ray absorption fine structure (EXAFS) give changes in lattice parameters as a function of temperature, exact positions of the atoms in unit cell [6], distortion in lattice and neighborhood of a chosen atom [7]. These techniques along with neutron diffraction, magnetic and ferroelectric measurements have been used to solve a number of very interesting phenomena like charge ordering [8], Verway transition [9] and Jahn- Teller transition [10], in different systems. The behavior of lattice parameters as a function of temperature reveals the origin of magnetic transition. The refinement of large Q (momentum transfer vector) XRD data using Reitveld refinement technique is useful in revealing small distortions in the unit cell [11]. In addition, the low temperature lattice parameters have been used for first principle theoretical calculations for density of states [12]. Spectroscopic techniques like x-ray absorption near edge spectroscopy (XANES), soft x-ray absorption spectroscopy (Soft-XAS) and x-ray photoemission spectroscopy (XPS) are used to study the charge/spin states of transition metal ions [13], electronic structure, hybridization of various levels etc. XANES, in particular, has been applied to estimate average charge state of Co ions in $Co_3TeO_6$ [13]. It has been reported that Co is in mixed oxidation state (both $Co^{2+}$ and $Co^{3+}$) in $Co_3TeO_6$. This explains the possibility of small ferromagnetic nano regions in $Co_3TeO_6$ in majority AFM matrix, giving rise to the so called Griffiths phase [14]. Also the pre-edge of the XANES spectra have been used to comment on the spin states of Co ions [13]. Therefore, it is important to perform temperature dependent high Q XRD, anomalous XRD and XANES measurements on MF materials. These measurements in wide temperature range (3 - 450K) can be performed in ADXRD beamline, in order to correlate the structural and spectroscopic properties with magnetoelectric properties.

In this paper, we present brief description of ADXRD beamline on Indus-2 synchrotron source [15]. Its capabilities for performing high resolution XRD measurements both at low and high temperatures will be discussed with the help of some examples. The measurements can be done for polycrystalline and single crystal samples in the form of powder, bulk as well as thin films. High resolution XANES as a function of temperature can also be performed in the same beamline. The capability of XANES measurements to estimate charge and spin states of transition metal ions in MF and magnetic systems will also be discussed with the help of some examples.

## 2. Description of ADXRD beamline

The beamline is the instrumentation which processes the photon beam emitted by synchrotron source according to the requirement of the experiment. ADXRD experiments require monochromrtic beam with high energy resolution ($E/\Delta E = 10^4$), wide tunabilty (5 – 25 keV), high collimation and high photon flux. Ideally the polycrystalline bulk/single crystal and powder samples need different experimental stations. We have designed the beamline in such a way to accommodate two experimental stations, for the two different experiments. ADXRD beamline (BL-12) consists of Si (111) crystal pair based double crystal monochromator (DCM). A pair of plane Pt coated Si mirrors, bendable in the form of cylinders and coated with 400 – 500 Å of Pt are used as collimating/focusing elements. These mirrors have dimension of 1200mm (L) x 60mm (W) x 40mm (H) and are used for collimating/focusing the photon beam in the meridional direction. The bender placed with the second crystal of the DCM is used for focusing the photon beam in the sagittal direction. DCM, thus, is used for the dual purpose of monochromatisation as well as for focusing in sagittal direction. Decoupled meridional and sagittal focusing reduces optical aberration and therefore improves both focusing and energy resolution. In figures 1, we show the optical layout of the beamline. Important parts of the beamline are marked in the figures. In table I, we summarize important photon beam parameters delivered at the experimental station of the beamline.

The experimental hutch of the beamline consists of two experimental stations in tandem. First, a six circle diffrcatometer (Huber 5020) along with a scintillation point detector is a high angular resolution set up and is used mostly for single crystal diffraction. The diffractometer is equipped with Eulerian cradle (χ- axis) and also φ axis, to orient the single crystal samples or the epitaxial films. The diffractometer is also equipped with Mythen strip detector for fast and high resolution measurements. Another experimental station in tandem with the difractometer is the Image plate area detector (MAR 345 dtb). The setup is quick but has moderate angular resolution and is used for powder diffraction. For low temperature XRD (LTXRD), we have liquid He based cryostat which operate between 3K and 450K. The set up contains beryllium (Be) heat shield and Be vacuum enclosure. Innovative arrangements have been made to stop Be diffraction peaks in the set up. The beamline is also equipped with a Cry free closed cycle He refrigerator (CCR) with Kapton window. The CCR works between 30K and 450K. The photographs of two LTXRD set ups are shown in figures 2 and 3. The software for interfacing the DCM for energy scan and data acquisition has been developed in-house. More details of the beamline can be found in our earlier publication [16].

Table 1: Photon beam parameters at the experimental station

| Parameter | Value |
|---|---|
| Spectral Range | 5-25 keV |
| Energy Resolution | $10^4$ |
| Beam Size | 0.5 x 0.5 mm$^2$ |
| Flux | $10^9$ to $10^{10}$ /sec |
| Angular Resolution | 15" (single crystal) |

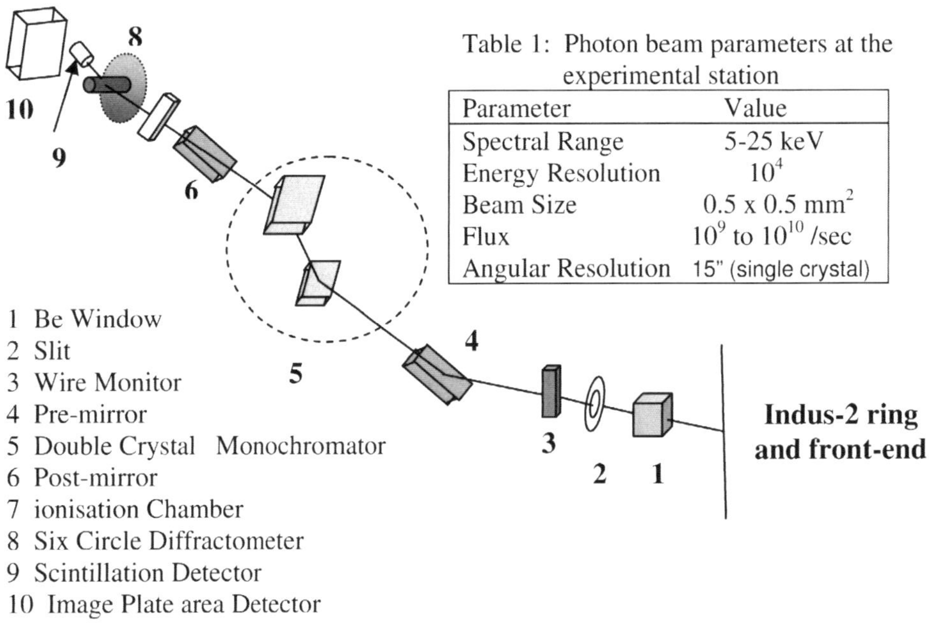


Fig. 1: Schematic representation of the optical lay-out of ADXRD beamline on Indus-2

## 3. Typical Results

### 3.1. *Low temperature XRD studies in Cobalt Ferrite $Co_{1.5}Fe_{1.5}O_4$*

$Co_3O_4$ and $Fe_3O_4$ are direct and inverse spinels, respectively. The formula unit for the two may be written as $(Co^{2+})[Co^{3+}]_2O_4$ and $(Fe^{3+})[Fe^{3+}Fe^{2+}]O_4$, where ( ) and [ ] represent the tetrahedral and octahedral sites, respectively. Magnetic properties of the two spinel structures are also quite different. $Co_3O_4$ shows antiferromagnetic behavior, whereas, $Fe_3O_4$ shows ferrimagnetic interaction. $Co_{1.5}Fe_{1.5}O_4$ is partially inverted spinel and in a recent publication [6], based on anomalous x-ray diffraction technique we propose its formula unit to be $(Co^{2+}_{0.45}\ Fe^{2+}_{0.15}\ Fe^{3+}_{0.40})[\ Co^{2+}_{0.30}\ Co^{3+}_{0.75}\ Fe^{3+}_{0.95}]O_4$. These compounds show very interesting properties like charge ordering [8], Verway transition etc. [9] at low temperatures. Here we present the low temperature XRD measurements and subsequent Reitveld analysis of LTXRD data on $Co_{1.5}Fe_{1.5}O_4$. Figure 4 shows the measured and calculated XRD profiles at various temperatures between 30 K and 300K. Also shown in the figure is the difference (between the measured and the calculated) curves. The theoretical XRD pattern was generated using Reiveld analysis software FulProf [17]. For calculation, Pseudo-Voit (P-V) profile shape was assumed and the lattice parameter, occupancy and position of the atoms, P-V profile parameters were refined. The background was manually picked and the background between the manually chosen points is extrapolated using a high order polynomial. We find that all the XRD patterns can be fitted with Fd3m space group with no extra peaks. Therefore, there are no phase changes in the sample in the measured temperature range. However, the values of lattice parameters plotted as a function of temperature (shown in figure 5), show anomalous behavior between 100K and 150K. Actually the sample shows negative thermal expansion in this temperature range. This clearly shows that the lattice is undergoing some rearrangement like charge ordering, J-T distortion or some transition. More experiments are required to be done to ascertain the exact reason for negative thermal expansion.

Fig. 2: Liquid He based Cryostat For LTXRD

Fig. 3: Close Cycle He refrigerator based Cryostat for LTXRD

*3.2 X-ray absorption near edge spectroscopy (XANES)*

Figure 6 shows Co K-edge normalized XANES spectra for $Co_{1.5}Fe_{1.5}O_4$ along with the spectra of two standards, CoO (oxidation state +2) and $Co_3O_4$ (oxidation state +8/3). The spectra consists of four features marked "A", "B", "C" and "D" in the figure. The feature "A" represents Co 1s-3d quadrupole allowed transitions. Since this is a dipole forbidden transition, these are normally weak but they get enhanced because of hybridization between TM 3d and Oxygen s and p states and also because of the distortion in oxygen polyhedra. The feature marked "B" is monopole transitions (1s-4s) and are used to estimate average charge state of the absorbing atom. Although we attribute these transitions to 1s-4s, there is some controversy regarding the assignment of these transitions. Some authors attribute this feature to 1s-4p transitions [18]. The edge shifts towards higher energies with the increase in the oxidation states. The feature "C" is called the white line and has important information on the disorder present in the system. The feature marked "D" in the figure represents the absorption at unoccuopied density of states beyond 4s and is modified by the multiple scattering. We have used the main edge part of the spectra (feature "B") to estimate average oxidation state of Co and also Fe using Fe K-edge XANES spectra. It is found that in sample $Co_{1.5}Fe_{1.5}O_4$ the average Co and Fe oxidation states are 2.5 and 2.9, respectively.

In our earlier publication [19], we use the pre-edge part (feature "A") of Co K-edge XANES spectra of type II multiferroic system $Co_3TeO_6$ to obtain information on the spin states of Co ions. We observe from the analysis of main edge (feature "B" in figure 6) that Co is in mixed oxidation state of +2 and +3. The pre edge shows two absorption bands attributed to splitting of 3d band of Co into t2g and eg bands due to crystal field splitting. The observed splitting is 1.26 eV between these two bands (see Fig. 4 of Ref 19) prefers high spin state of $Co^{2+}$ as well as $Co^{3+}$. It is known from literature that the splitting in the case of Low spin $Co^{3+}$ is 2.2 eV, much larger than 1.23 eV observed by us. Also the value of crystal field splitting in the case of high spin $Co^{2+}$ in tetrahedral site is 1.14 eV.

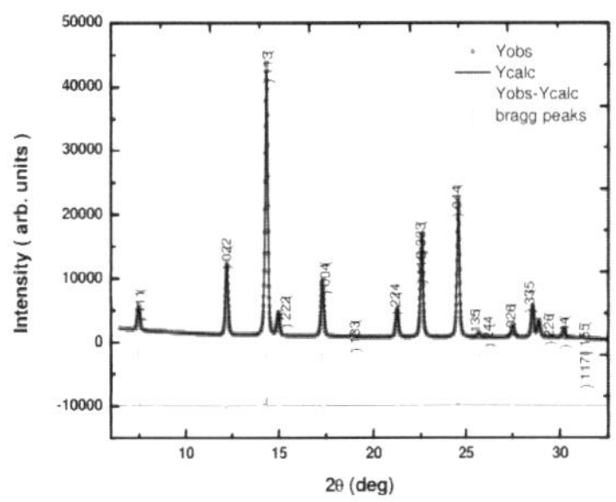


Fig. 4: Calculated, measured and difference XRD pattern at 30K for $Fe_{1.5}Co_{1.5}O_4$

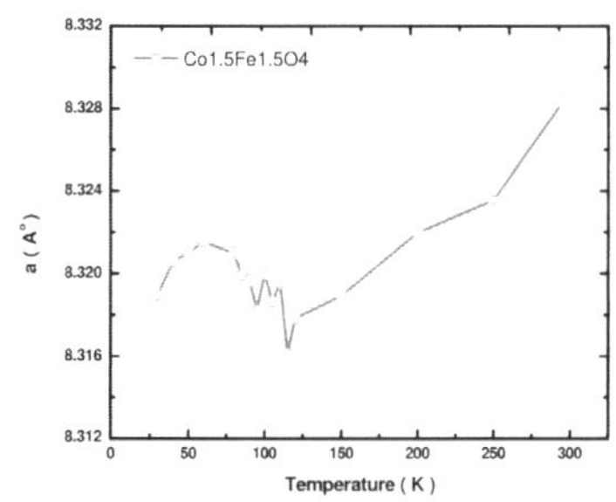


Fig. 5: Lattice parameter as a function of temperature

Finally the white line intensity has been used by us to to study Pt incorporation in $LaMn_{0.96}Pt_{0.04}O_{3-\delta}$ and $LaFe_{0.94}Pt_{0.06}O_{3-\delta}$ perovskite [20]. Synchrotron XRD data pattern for the system $LaMn_{0.96}Pt_{0.04}O_{3-\delta}$ showed that Pt does get incorporated in the matrix but remains in the metallic form, as the XRD peaks corresponding to metallic Pt is observed in the XRD pattern. No Pt peaks are observed in $LaFe_{0.94}Pt_{0.06}O_{3-\delta}$. This gives indication that Pt is incorporated in the matrix. However, to confirm this we have performed Pt L–edge XANES measurements (shown in Fig. 2 of Ref 20). The white line of $LaMn_{0.96}Pt_{0.04}O_{3-\delta}$ sample is very small indicating metallic form of Pt, in agreement with XRD data. On the other hand the white line for the sample $LaFe_{0.94}Pt_{0.06}O_{3-\delta}$ is very prominent suggesting the fact that Pt has been incorporated in the matrix and therefore, we don't observe XRD peaks of Pt metal. White line intensity also increases when defects are induced by some perturbation like Ar implantation or $\gamma$ ray irradiation, which results in donor density of states.

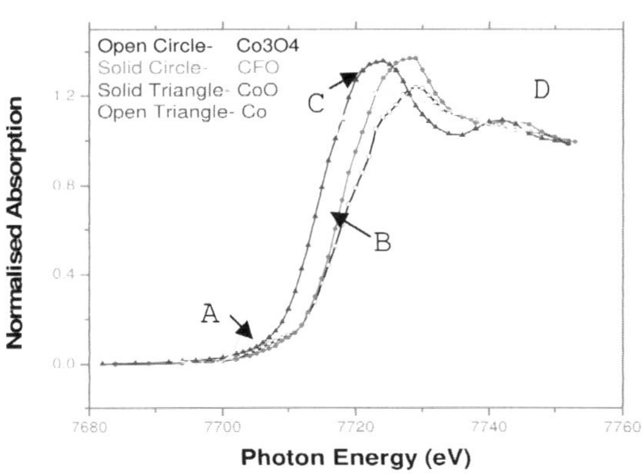


Fig. 6: XANES spectra at Co K-edge for $Co_{1.5}Fe_{1.5}O_4$ (CFO) along with those of standards $Co_3O_4$, CoO and Co metal foil. The metal foil XANES is used for photon energy calibration

## 4. **Conclusions**

Structural and spectroscopic characterization like Low and high temperature XRD, XANES are extremely useful in the field of research in multiferroic materials. ADXRD beamline on Indus-2 synchrotron source is well equipped with these measurements. In this paper, we provide a brief description of the beamline and experimental station. The photon beam specifications at the experimental station are given for the benefit of potential users. Anomalous changes in mixed spinel cobalt ferrite with composition Fe1.5Co1.5O4 between temperatures 100 K and 150 K have been discussed. The pre-edge and main edge features of XANES spectra have been used to estimate the average charge state and spin state of transition metal ions. The white line has been used to comment of the incorporation of Pt in $LaMn_{0.96}Pt_{0.04}O_{3-\delta}$ and $LaFe_{0.94}Pt_{0.06}O_{3-\delta}$ perovskite systems.

**Acknowledgement**: Authors would like to thank Dr. P. D. Gupta and Dr. P. A. Naik for their support and encouragement.

References

[1] Ramesh R and Spaldin N A 2007 *Nature materials* **6** 21
[2] Spaldin N A, Cheong S W and Ramesh R 2010 *Phys. Today* **63** 38
[3] Singh H, Sinha A K, Ghosh H, Singh M N, Rajput P, Prajapat C L, Singh M R and Ravikumar G 2014 *J. Appl. Phys.* **116** 074904
[4] Singh H, Ghosh H, Chandrasekhar Rao T V, Sarma G, Saha J and Patnaik S 2016 *J. Appl. Phys.* **119** 04410
[5] Martin L W, Chu Y H and R Ramesh 2010 *Materials Science and Engineering: Reports* **68** 89
[6] Singh M N, Sinha A K and Ghosh H 2015 *AIP Advances* **5** 087115
[7] Als-Nielsen J and McMorrow D 2001 Elements of Modern X-ray Physics, *John Wiley and Sons*
[8] Zhang T, Wang X P, Fang Q F and Li X G 2014 *Appl. Phys. Rev.* **1**, 031302
[9] Wright J P, Bell A M T and Attfield J P 2000 *Solid State Sciences* **2** 747
[10] Ruiz-Fuertes J, Bernert T, He M, Winkler B, Vinogard V L and Milman V 2014 *Appl. Phys. Lett.* **105** 071911
[11] Singh A, Pandey V, Kotnala R K, and Pandey D 2008 Phys. Rev. Lett. 101, 247***602***
[12] Sen S, Ghosh H, Sinha A K and Bharathi A 2014 *Supercond. Sci. Technol.* **27** 122003
[13] Singh H, Ghosh H, Chandrasekhar Rao T V, Sinha A K and Rajput P 2014 *J. Appl. Phys.* **116** 214106
[14] Deisenhofer Braak J D, Krug von Nidda H A, Hemberger J, Eremina R M, Ivanshin V A, Balbashov A M, Jug G, Loidl A, Kimura T, and Tokura Y 2005 *Phys. Rev. Lett.* **95** 257202
[15] Deb S K, Singh G and Gupta P D 2013 J. Phys.: Conf. Series **425** 072009
[16] Sinha A K, Sagdeo A, Gupta P, Upadhyay A, Kumar A, Singh M N, Kane S R, Verma A and Deb S K 2013 J. Phys.: Conf. Series **425** 072017
[17] Rodriguez-Carvajal J 1993 Physica B **192** 55
[18] Berry A J and O'Neill H St. C 2004 *American Mineralogist* **89** 790
[19] Singh H, Sinha A K, Singh M N, Tiwari P, Phase D M and Deb S K 2014 *J. Phys. Chem. Solids* **75** 397
[20] Rajesh T, Upadhyay A, Sinha A K, Deb S K and Nandini Devi R 2014 *J. Mol. Catal. A: Chemical* **395** 506

Journal of Physics: Conference Series **755** (2016) 012004 doi:10.1088/1742-6596/755/1/012004

# Practicing DSAM in aberrant domain: use of multi-disciplinary techniques

**S Das[1], S Samanta[1], R Bhattacharjee[1], R Raut[1], S S Ghugre[1], A K Sinha[2], U Garg[3], R Chakrabarti[4], S Mukhopadhyay[4], N Madhavan[5], S Muralithar[5], R P Singh[5], J Sethi[6], S Saha[6], R Palit[6]**

[1] UGC-DAE Consortium for Scientific Research, Kolkata Centre, Kolkata 700098
[2] UGC-DAE Consortium for Scientific Research, Indore 452001
[3] Department of Physics, University of Notre Dame, Notre Dame, Indiana 46556, USA
[4] Nuclear Physics Division, Bhabha Atomic Research Centre, Mumbai 400085, India
[5] Inter University Accelerator Centre, Aruna Asaf Ali Marg, New Delhi 110067, India
[6] Tata Institute of Fundamental Research, Mumbai 400005, India

E-mail: `ssg@alpha.iuc.res.in`

**Abstract.**
Measurement of level lifetime of nuclear states is of relevance in nuclear structure research as it provides us with an unique probe into the underlying microscopic structure of these states . Of the several experimental techniques for lifetime measurements, the Doppler Shift Attenuation Method (DSAM) is the one adopted for measuring lifetimes typically in the range of few tens of fs to few ps. The technique is based on the analysis of the observed Doppler affected gamma rays emitted by the recoils in flight. The crucial component in the related analysis is the simulation of the stopping process, of the residues of interest, in the target and the backing media. This requires calculation of the corresponding stopping powers and the same has been identified as one of the principal uncertainties in the extracted lifetime in DSAM. Traditionally the method is pursued with a thin target, for production of nuclei of interest, on a thick elemental backing wherein stopping process is perceived to occur. The present work in light of it's objectives uses a setup which is in sharp variance with the conventional scenario, such as the use of a thick molecular target, which contributes both to the production of the residues as well as their subsequent slowing down. This demanded extensive developments in the analysis procedure particularly in the domain of simulating the stopping process with due incorporation of the nuances of nuclear reaction kinematics besides subjecting the molecular medium to a detailed structural characterization, routinely carried out in the domain of material science. These developments have been used to extract the level lifetimes of nuclei at the interface of the sd & pf shells such as $^{26}$Mg, $^{29}$Si, and $^{32}$P.

## 1. Introduction

The nucleus provides us with a unique laboratory to investigate the two extreme degrees of freedom *viz.* single particle and collective. One of the primary objectives of any contemporary experimental nuclear structure endeavour is to populate high angular momentum states in the nucleus of interest, using an appropriate nuclear reaction, and study its de-excitation through emission of $\gamma$-rays that are detected using large arrays of high resolution detectors supported by state-of-the-art pulse processing electronics and data acquisition systems. Such efforts lead to detailed information on the level structure (excitation pattern) of the nucleus of interest that includes the energies, spin-parities, and lifetimes of the states along with the $\gamma$-ray transitions connecting them. These are eventually compared with the model calculations for interpretation of the underlying degrees of freedom based on the predicted microscopic configurations.

The lifetime of a level encodes information on its microscopic structure (wavefunction) and its overlap with other available states, thus providing a stringent to constraint on the model calculations. Measurement of level lifetimes is, thus, of much significance in nuclear structure research and is zealously pursued in many of the related endeavours in the domain. The experimentally measured lifetimes and the transition probabilities extracted therefrom are often compared with the theoretical predictions to validate the model interpretation of the level structure. Several procedures [1] of lifetime measurements have been practiced for decades and choice of the same depends on the range of lifetimes being addressed in a particular investigation.

The Doppler Shift Attenuation Method (DSAM) is adopted for measurement of level lifetimes in the range of $\sim$ few tens of fs ($10^{-15}$ s) to $\sim$ few ps ($10^{-12}$ s). Nuclei of interest produced in a reaction recoil through the target and the backing media, slowing down and eventually stopping in the process. In conjunction, these recoiling nuclei also de-excite from the state they are initially populated (entry state) through the characteristic level sequences, by $\gamma$-ray emissions, finally into the ground state. If the lifetime of a level is less than or of the order of the aforementioned stopping time, the de-exciting $\gamma$-ray from the level may be emitted from a recoiling nucleus in-flight and exhibit Doppler effect, observed in the resulting spectrum, as per the equation,

$$E_\gamma(\theta) = E_{\gamma 0}(1 + \beta cos\theta) \qquad (1)$$

where $E_{\gamma 0}$ is the actual energy, $E_\gamma(\theta)$ is the Doppler affected energy of the $\gamma$-ray, $\beta$ is the velocity of the recoil during emission, and $\theta$ is the angle of observation. In case of very short lived levels (lifetime $\ll$ stopping time) and detector at angles $\neq 90^o$, the $\gamma$-ray transition peak in the spectrum is shifted, compared to the actual energy, to higher (lower) energies when observed at forward (backward) angles with respect to the direction of the recoiling nuclei (close to the beam direction). For levels with lifetime $\lesssim$ stopping time, the peak has a stopped component at the actual energy and an accompanying shape, extending to higher (lower) energies for detectors at forward (backward) angles. Typically observed Doppler shapes and shifts in the $\gamma$-ray spectrum, obtained in the present work, are illustrated in Fig. 1. The DSAM pertains to analyzing these shapes and shifts to extract lifetimes of the respective levels.

The cardinal components of analyzing data from a DSAM experiment are as follows.

(i) The slowing down and stopping of the recoils in the target and the backing media is simulated using the Monte Carlo techniques. This requires, apart from other inputs, stopping powers of the target and the backing for the recoils of interest. Results of the simulation are typically the velocity ($\beta$) and the trajectory information (direction cosines) of a chosen number of recoils, in uniform time steps, starting from the production of the recoils through their slowing down and eventual stopping in the medium.

(ii) The velocity profiles of the recoils, as viewed by the detectors at different angles, are calculated using the results of the above simulations and an input of the detector geometry, such as the detector radius and the target-to-detector distance.

(iii) The expected Doppler shape / shift of a given $\gamma$-ray transition at different angles is calculated using inputs on the level structure, such as the energy of the transition being analyzed, initial estimate of the level lifetime, and side feeding description (intensity, feeding time) of the level. This is combined with the velocity profiles as viewed by the detectors

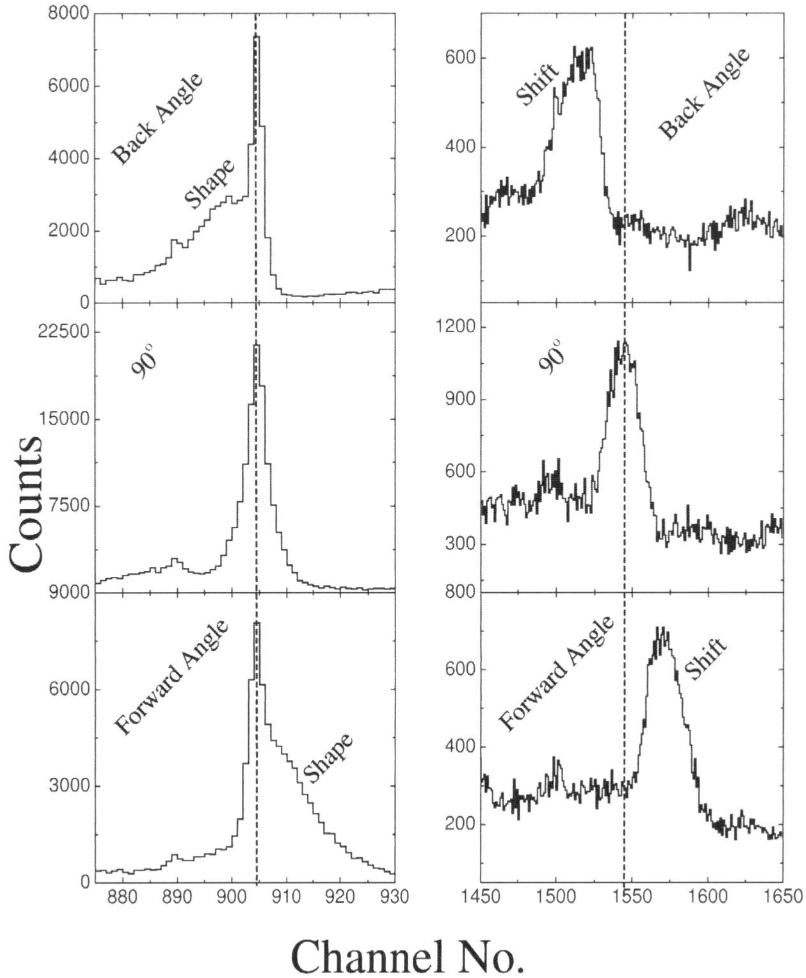


**Figure 1.** Typical Doppler shape and shift observed in $\gamma$-ray spectrum following a DSAM measurement reported in the present paper.

at different angles, from the preceding step, along with the information on the intrinsic resolution of the detector, to generate the calculated Doppler affected shaped/shifted $\gamma$-ray transition peak.

(iv) The calculated shape is fitted to the experimental spectrum through least square minimization, in order to extract the level lifetime of interest. The crucial parameters of fitting are essentially the level lifetime and the side feeding time.

The principal uncertainties in the DSAM analysis have been identified to stem primarily from (i) the estimation of the stopping powers that govern the slowing down process of the recoils and (ii) the side feeding models adopted in the analysis. The uncertainties owing to the side feeding can, in principle, be eliminated through use of gates set on transitions above the level of interest [5]. Thus, the major source of uncertainty in lifetime determination using the DSAM may be attributed to the knowledge of the stopping powers. The problem is further compounded in case the measurement requires deploying molecular target and backing wherein the estimation of stopping powers, using theoretical models, is even more difficult.

Of the several codes used for analyzing data from DSAM measurements, the LINESHAPE package [2] is one that is extensively used in many of the contemporary endeavours. Stopping powers are calculated using one of the earlier models by Ziegler [3] or Northcliffe and Schilling [4] that are based on the proton and $\alpha$ stopping data and uses different scaling algorithm for the heavy-ions. The stopping simulation for the recoils in the target and the backing is carried out by the DECHIST program of the LINESHAPE package, using Monte Carlo techniques. The approach holds good for a conventional DSAM measurement that customarily uses a thin target on a thick elemental backing so that the energy loss of the recoils in the target can be neglected and the stopping process can be perceived to be occurring entirely in the backing medium wherein the stopping powers, calculated from the above models, are reasonably accurate. Certain lifetime measurements, however, require setups that are incongruent with the traditional scenario and uses target that are molecular in nature. Moreover, depending on the method of target fabrication, the molecular target may be substantially thick so as to operate both as target as well as backing which, owing to the molecular nature, is beyond the description allowed by the conventional LINESHAPE-based analysis.

The neutron rich $^{18}O$ target, extensively used in the spectroscopy of *sdpf* nuclei [5, 6], come in the form of metallic oxides, such as $Ta_2O_5$. The present work reports developments and modifications pertaining to the analysis codes and methodology for application of DSAM in such unconventional setup. The nuclei of interest in this work were populated in a series of experiments using the $^{18}O(^{18}O,xpynz\alpha)$, $^{18}O(^{13}C,xpynz\alpha)$ and $^{18}O(^{16}O,xpynz\alpha)$ reactions at $E_{lab}$ = 30-34 MeV. The beam was provided by the Pelletron-LINAC Facility at the Tata Institute of Fundamental Research (TIFR), Mumbai and the Pelletron Facility at the Inter University Accelerator Center (IUAC), New Delhi. The neutron-rich $^{18}O$ target was prepared by heating a Ta foil of thickness $\sim$ 50 mg/cm$^2$ , in an enriched $^{18}O$ atmosphere, resulting in the formation of $Ta_2O_5$ layer on either side of the foil. The thickness of the $Ta_2O_5$ on either side of the foil was estimated to be 9.25 mg/cm$^2$ which was sufficient to stop the recoils. For the experiment with the $^{18}O$ beam, an array of seven Compton-suppressed Clover detectors was used as the detection system while that in the experiments with the $^{13}C$ and the $^{16}O$ beams, it was the Indian National Gamma Array (INGA), then consisting of 15-16 Compton-suppressed Clovers. Further details on the experimental setups can be found in Refs. [5, 6].

The acquired data were sorted into angle dependent $\gamma$-$\gamma$ matrices for lifetime analysis, using the LINESHAPE [2] package. However, as indicated earlier, the programs / sub-programs of the package employed for simulating the slowing down process of the recoils, were replaced as detailed in the next section.

## 2. Modifications of the LINESHAPE Code for Thick Molecular Targets

The conventional LINESHAPE program does not allow for molecular media in simulation of the stopping process of the recoils in the target and the backing. Moreover, as indicated above, it is suited to a thin-target-on-thick-backing setup wherein the energy of the beam can be assumed to remain largely constant throughout the target thickness and consequently, the yield of the residual (recoiling) nuclei is uniform therein. The DECHIST program within the LINESHAPE code is used for these stopping simulations and it was modified to incorporate the characteristics of the thick molecular target setup.

Updated stopping powers, for heavy ions in elemental as well as molecular media are available in contemporary software like SRIM [7], which have been exhaustively benchmarked with

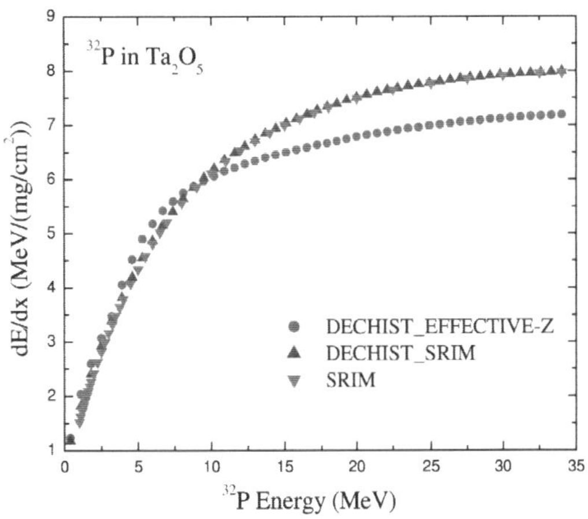


**Figure 2.** (Color online) Stopping powers of $^{32}$P ions in $Ta_2O_5$ using SRIM (labelled as SRIM in the legend), DECHIST with SRIM input and interpolation (DECHIST_SRIM) and DECHIST with effective $Z$ and effective $A$ for the $Ta_2O_5$ target ( DECHIST_EFFECTIVE-Z).

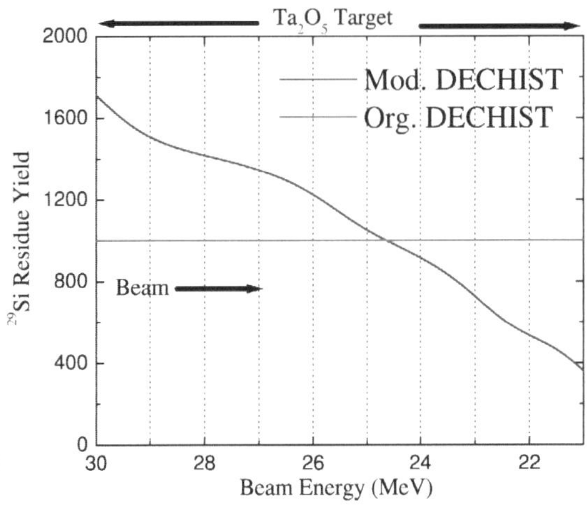


**Figure 3.** (Color online) Cross-section of the residue as a function of beam energy. The original DECHIST program assigns uniform cross section (red line) of the residue while the modified DECHIST program incorporates beam energy dependent cross section (blue line) along the target thickness.

respect to the experimental data and are considered to be more reliable. The stopping powers of $^{32}$P ions, one of the nuclei studied in these investigations, in $Ta_2O_5$ from SRIM are included in Fig. 2. In the light of the reliability of the SRIM stopping powers, these have been incorporated in the modified DECHIST program. The DECHIST program calculates the stopping powers at 61 hard-coded energies, ranging from 1 keV to 1000 MeV, for the residue in the target and the backing, as well as for the beam in the target. The stopping power required for all the intermediate energies are calculated through an in-built interpolation subroutine. In the modified version of the program the user is required to provide these stopping powers, calculated from the SRIM code, at these 61 energies through an input file. The interpolated stopping powers for $^{32}$P + $Ta_2O_5$ case, as calculated by the modified DECHIST program, are also presented in the Fig. 2 (DECHIST_SRIM). Nearly complete overlap between the stopping powers calculated by SRIM and DECHIST, with SRIM inputs and appropriate interpolation, validates the modification. Fig. 2 also illustrates the stopping powers calculated using an effective $Z$ and an effective $A$ to represent the molecular stopping medium in the conventional DECHIST program. The deviations in the stopping powers, calculated using the latter, at intermediate and higher energies indicate unacceptability of the approach.

Use of a thick target results in a substantial change (decrease) in the energy of the incident projectile as it traverses the target thickness and thus affect the production cross section of the residues along the depth of the target. Further, in a thick target, one needs to determine the thickness of the target that actually contributes to the residue production while the rest of the target thickness acts only as the stopping medium. The knowledge of target thickness contributing to the residue production follows from the information on the cross section depen-

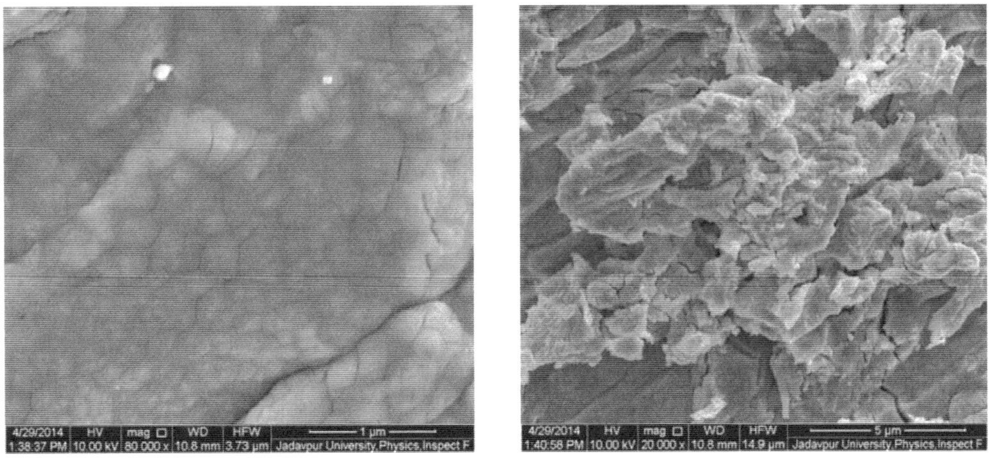


**Figure 4.** Scanning Electron Micrograph images for $Ta_2O_5$ target. The left panel illustrates the crystalline component while the right panel indicates the presence of amorphous component [5].

dence on the beam energy and the evolution of the latter along the target thickness. The original version of the DECHIST program, which simulates the stopping of a given number of recoiling residues, divides the target thickness into bins equal to the (user) given number of recoil nuclei and treats each bin as the origin of one recoil nucleus. This is a valid approximation for a thin target through which the beam energy can be assumed to remain unaltered and thus produce the residues with uniform probability (cross-section). In the modified version of the DECHIST program used in the present work, changes were implemented to incorporate the cross-section dependence of the residues on the changing beam energy along the thickness of the target. In this version, upto 10 bins can be defined along the target thickness and relative cross-sections for the residue, in arbitrary units, can be specified therein, as illustrated in Fig. 3. It is preferable to specify the cross sections with reference to available experimental data or, in its absence, from statistical model calculations. It follows that the total thickness of the aforesaid bins is the thickness of the target that contributes in the production of the residues while the remaining thickness contributes only to the stopping process.

The use of molecular target, introduces one more degree of uncertainty in the analysis. Most of the molecular targets are prepared as oxides and it is possible that these oxides appear in several structural phases. Some of these phases may also occur as a consequence of the temperature dependent preparation technique. Therefore, it is imperative to carry out a detailed structural evaluation of the target using well established structural evaluation techniques such as X-Ray Diffraction (XRD) and Field Emission Scanning Electron Microscopy (FESEM) as detailed in Ref. [5]. These techniques were applied to the $Ta_2O_5$ target used in the present work and they revealed the presence of multiple crystalline phases, coupled to an amorphous phase in the sample. The SEM images of the $Ta_2O_5$ target are presented in Fig. 4. The presence of multiple phases would manifest as a modification in the density of the target medium, that will have a significant impact on its stopping characteristics. The consistent analysis of several known level lifetimes in a host of nuclei populated in the experiments indicated that a density of 5.76 g/cm$^3$ produces the best agreement with the reported lifetimes and hence this value has been used in the subsequent analysis.

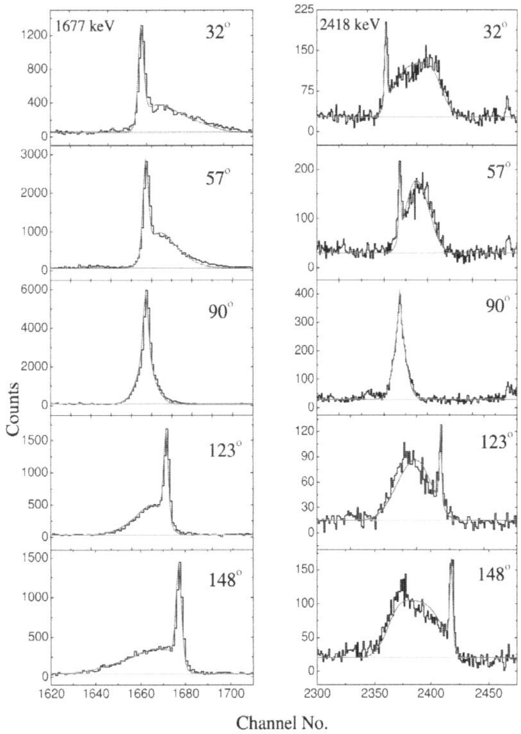


**Figure 5.** (Color online) Representative fits of the Doppler shapes of the 1677 keV and the 2418 keV transitions from the $^{32}$P nucleus.

Representative fits to the observed Doppler shapes of the 1677 and 2418 keV transitions in the $^{32}$P nucleus, studied in the present work, using the modified LINESHAPE code, are illustrated in Fig. 5. The lifetime obtained for the 1755 keV level, de-exciting by the 1677 keV transition, is $616^{+65}_{-46}$ fs that is in excellent overlap with the literature value of $620^{+40}_{-40}$ [9]. An upper limit of $629^{+82}_{-61}$ fs was obtained on the lifetime of the 5862-keV level, de-populated by the 2418-keV transition and was reported for the first time [5]. The compliance [5, 6] of the lifetimes following the present analysis with those reported previously essentially validates the present procedure.

## 3. Methodology for inclusion of TRIM simulations in lifetime analysis

It may be noted that even though the stopping powers in the preceeding analysis were being calculated through the SRIM code, the Monte Carlo simulations of the stopping process was still dependent on the DECHIST program of the LINESHAPE package. This might be replaced with contemporary stopping simulation codes such as TRIM that uses SRIM calculated stopping powers and updated statistical algorithms to simulate the transport of ions through material media. Use of such state-of-the-art simulation tools is also expected to render immense flexibility in the definition of the target and / or backing media that may be elemental or molecular, thin or thick, single or multi-layered, as allowed by the modern simulation packages such as TRIM.

The TRIM program, acronymed for TRansport of Ion in Matter, finds extensive use in the domain of material science research wherein it is applied to investigate the effects of energetic ions passing through a medium that might be elemental or molecular and might consist of multiple layers of varying depths. It is a common understanding that the code was never intended for use in lifetime analysis. It is an efficient simulation tool through use of statistical algorithms and thus is an alluring option for simulating the slowing down process of the recoiling nuclei in the target and the backing media as is required in analysis for lifetime measurements using DSAM.

To begin with, we consider a thick elemental or molecular target operating as the target, for production of the nuclei of interest, as well as the stopping medium. Our objective is to simulate the stopping process of the recoiling nuclei using the TRIM code and use the results of the simulation for subsequent analysis. The output of the TRIM simulations, however, cannot be naively used for DSAM studies. A set of programs have been developed for elaborate processing of the TRIM generated output to conform to the requirements of lifetime analysis. These pertain to correcting the trajectories with the reaction kinematics associated with the production of the residue, extracting the velocity profile (variation of $\beta$ with time) of the residues moving in the target-backing media, and incorporating the possible changes in the cross section for residue formation along the target thickness. Various aspects of the exercise are detailed in this section.

The TRIM code characteristically assigns a common starting point for all the simulated trajectories. The trajectories are all assumed to start from the origin (0,0,0) on the first surface of the stopping medium. The actual experimental scenario of lifetime measurements with thick target is, however, different. In case of a thick target, the incident beam traverses the thickness of the target and looses energy in the process. Consequently, the cross section for production of the residue of interest evolves along the thickness of the target and a varying number of residues are produced at different target depths, each with different starting points for their respective trajectories. In the current analysis, the cross section dependence of the residue of interest on the beam energy was determined either based on available experimental data or from statistical model calculations. This was combined with the calculations for change in the beam energy, along the target depth, using the SRIM code. It is understood that for a thick target, only a part of the same would contribute to the production of the residue of interest while the remaining thickness would operate only as the stopping medium. The effective thickness of the target, contributing to population of the residue, can be divided into bins representing the changing beam energy and the evolving cross section therein. Upto ten such bins can be considered in the present scheme. Each bin is characterized by a thickness (width), a beam energy and a yield (arbitrary units) of the nucleus being investigated.

The residues produced in each bin would have different initial energy-angle distribution and different starting points for their trajectories. The distribution in the initial energies and angle of the residues follows from the evaporation of the particles $(p, n, \alpha)$, de-exciting the compound nucleus (Fig. 6) and has to be incorporated at the starting point of the trajectories in the stopping medium. The trajectories obtained from TRIM simulations, however, initiate aligned to the X-direction, with direction cosines (1,0,0), and does not include this distribution. The two-dimensional energy-angular distribution of the residues was calculated using the statistical model code PACE4 [8] for each beam energy corresponding to the individual bins, along the depth of the target. The number distibution of the residue in the energy-angle system, obtained from the statistical model code, is converted into a probability distribution for subsequent use.

The (stopping) simulation using the TRIM code is carried out for a residue energy greater than the highest possible energy, indicated by the statistical model calculations, typically for 10000 ions. The results of the TRIM simulations appear as trajectory information, in the form of the energy of the recoiling nuclei and the corresponding co-ordinates, recorded at certain

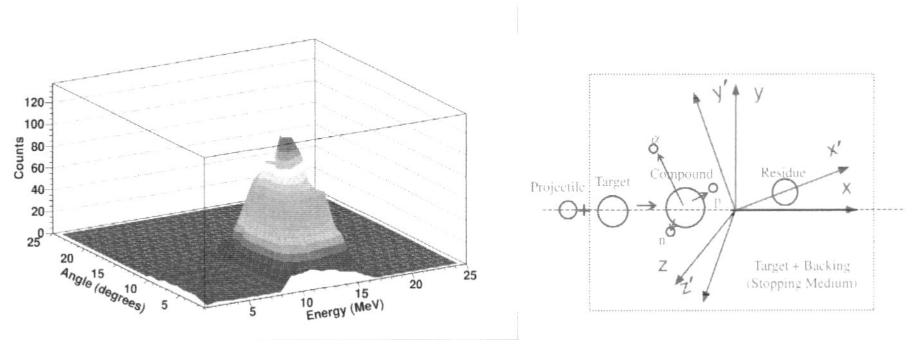


**Figure 6.** (Color online) Typical energy-angle distribution of residues generated in the statistical model calculation [8]. Schematic representation of a nuclear fusion-evaporation reaction producing residue (of interest) following particle evaporation from the compound nucleus. The TRIM trajectories, initiating aligned to the X-direction are rotated to match the angle of the residue chosen under the constraint of the energy-angle distribution.

energy interval. This output file containing the simulated trajectories serves as one of the input to our program. The program also reads-in, for each of the aforementioned bins, the respective energy-angle probability distribution file, relative yield of the residue (normalized to 10000) and the distance (in Angstrom) of the geometrical center of the bin from the surface of the target.

During execution of the program, an energy and an angle is chosen for each residue under the constraint of the probability distribution of the respective bin. The corresponding TRIM trajectory, chosen sequentially from the set of trajectories generated with identical incident energy of all the ions, is inspected for the point where the energy of the residue is nearly equal (or just lower than) to the chosen energy from the distribution. This point in the trajectory is then treated as the origin of the trajectory of this particular residue in the stopping medium. The origin is appropriately offset to initiate at the center of the respective cross section bin. All the subsequent points in the trajectory are subjected to the same offset. In conjunction, the angle chosen from the distribution is used to modify the trajectory through application of a rotation matrix, so as to incorporate the angular entry of the residue, deviated from the beam direction (following particle evaporation), into the stopping medium. The rotation aligns the initial direction of the simulated trajectory to the direction of the residue. The same rotation is subsequently applied to all the direction vectors in the simulated trajectory.

The stopping simulations for lifetime analysis are typically in the form of trajectory information recorded in uniform time steps. A given trajectory (history) is logged in terms of the $\beta$ and the direction cosines at each time step, the size of which is specified by the user and is chosen to be much less than the minimum lifetime being expected. The history is tracked from the initial $\beta$ to the point where the $\beta$ drops by 1/1000 of the initial value. The trajectory information obtained from TRIM and modified with inclusion of reaction aspects thus needs to be converted into a description wherein a trajectory is represented by the time profile of $\beta$ and the direction cosines, logged in uniform time steps, in compliance with the requirements of lifetime analysis. This conversion is implemented by considering the consecutive points in a trajectory, to calculate the change in the energy and the accompanying intermediate distance, from which the associated time is extracted. This is followed by re-binning the change of the $\beta$ and the direction cosines, between the consecutive points, in uniform time steps, through linear interpolation. The latter is a reasonable approximation owing to the small change in energy (typically $\sim 1\%$) between the consecutive points in the record of the simulated trajectory. The size of the time step is user specified and is usually $\sim$ 0.001-0.002 ps.

In case of the more conventional scenario of a thin target on a thick backing, the working with the TRIM simulations is slightly different from that described above. This is particularly relevant if the residues loose an amount of energy in the target medium, before reaching the backing, that is non-negligible. In this case, the TRIM simulation is separately executed for the target and for the backing. Modified versions of the above program are used to individually process the simulation outputs for the target and the backing media. The program for the target uses the energy-angle probability distributions for the residues, generated using the PACE4 code, and produces, apart from trajectory info in terms of the $\beta$ and direction cosines, a file containing energy and direction information for the trajectories at the target boundary. This file is read-in by the program to process the TRIM simulations for the backing medium. The energy and the direction of each residue at the target boundary is used by this program to modify the TRIM simulated trajectories through the backing medium and generate uniform time-binned $\beta$ and direction cosines for these trajectories. The processed trajectories in the target medium and the backing medium are then combined in a common file. It may be noted that there may be trajectories in the target medium, particularly for not-very-thin target, that do not make it to the backing. Such trajectories are separately recorded and eventually combined with the common file with all the trajectories.

The output of this exercise are simulated trajectories, using experimentally benchmarked stopping powers from the SRIM database and state-of-the art Monte Carlo algorithms of the TRIM code with due incorporation of the details of the nuclear reaction and associated kinematics. This output is used for the subsequent analysis for extracting the level lifetimes. In the present work, the output of the stopping simulations is read-in by the HISTAVER program of the LINESHAPE package, to calculate the velocity profiles as viewed by the detectors at different angles. The subsequent steps for determining the level lifetimes are similar to the standard practice detailed in Ref. [5]. It should also be noted that the LINESHAPE package is being referred here in a very generic sense. The result of the stopping simulations using the TRIM code can well be made compatible with any code for DSAM analysis that may use these trajectory information for determination of level lifetimes.

Fig. 7 illustrates the representative fits to the Doppler shape / shift obtained through use of TRIM simulated stopping trajectories in the $^{34}$P and $^{26}$Mg nuclei. The corresponding lifetime results for the 3352-keV level, de-excited by the 1047-keV transition, in $^{34}$P is an upper limit of 475 fs that is consistent with the previously reported lifetime of $520^{+173}_{-115}$ fs [9]. For the 1274-keV transition, de-populating the 3083-keV level, in the $^{26}$Mg nucleus, the lifetime results from the present analysis is an upper limit of 1339 fs, and determined for the first time in this investigation.

## 4. Conclusions

The present work attempts to address the uncertainties in the stopping powers used in the lifetime measurements with DSAM and extend its application beyond the conventional thin-target-on-thick-backing setups to the atypical thick molecular targets. The latter, as it is known, is indispensable in the spectroscopic pursuits of certain nuclei such as those in the $A \sim 30$ region, around *sdpf* shells, that require $^{18}$O target in the form of metallic $Ta_2O_5$ for investigations of their level structure. Stopping powers calculated from the updated and experimentally validated SRIM code has been used to analyze the Doppler shape / shifts observed in the spectra of these nuclei populated using heavy-ion induced fusion-evaporation reaction. While in the first step

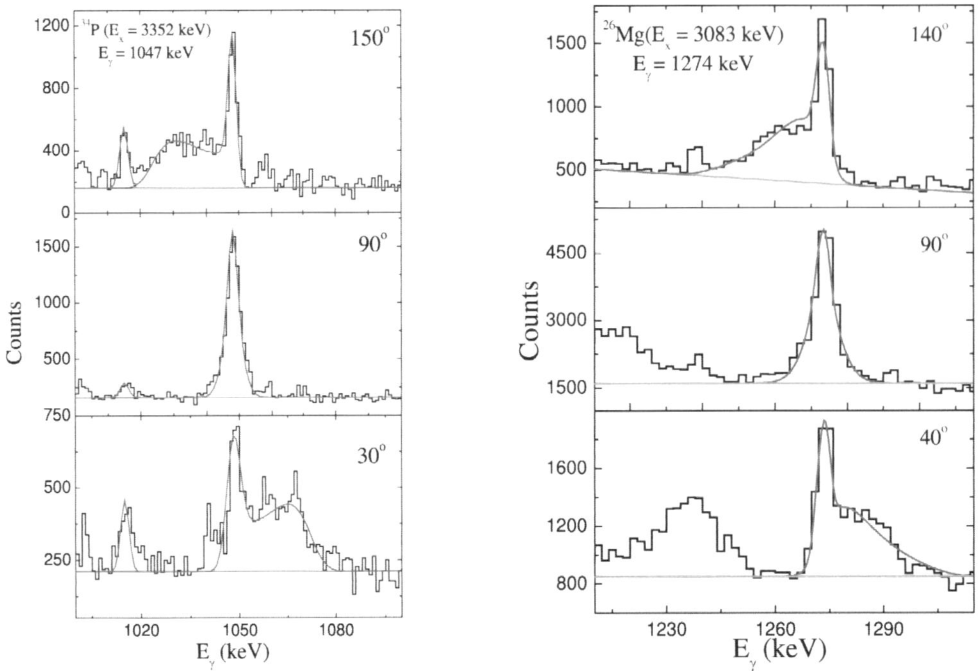


**Figure 7.** (Color online) (Left Panel) Fitted Doppler shape of the 1047-keV transition, de-exciting the 3352 keV level, in the $^{34}$P nucleus. (Right Panel) Fitted Doppler shape of the 1274-keV transition, de-exciting the 3083 keV level, in the $^{26}$Mg nucleus.

these stopping powers were applied within the existing code for DSAM analysis, LINESHAPE, in a further improvement the entire stopping simulation was carried out with the TRIM code using the stopping powers calculated by SRIM, and using the LINESHAPE code only for the subsequent analysis. The latter development is expected to extend the applicability of the DSAM technique to a wider range of target and/or backing media, as may be required for specific life-time measurements.

## 5. ACKNOWLEDGMENTS

The authors thank all the participants of the INGA collaboration for their help in setting up the facility at IUAC, New Delhi. We are thankful to J. P. Greene (ANL, USA) for the $^{18}$O target. The authors appreciate the help and support from Mr. Kausik Basu (UGC-DAE CSR, KC) during various phases of the project. Help received from S.S. Bhattacharjee during the data analysis is also acknowledged. The authors also thank the Pelletron staff at IUAC, New Delhi for their excellent support during the experiment. This work has been supported in part by the U.S. National Science Foundation (Grants No. PHY-1068192 and PHY-1419765).

## References

[1] Nolan P J, Sharpey-Schafer J F 1979 *Rep. Prog. Phys.* **42** 1
[2] Wells J C, Johnson N R 1991 *ORNL Report* **6689** 44
[3] Ziegler J F, 1980 *The Stopping and Ranges of Ions in Matter, Pergamon Press*
[4] Northcliffe L C, Schilling R F 1970 *At. Data Nucl. Data Tables* **7**

[5] Bhattacharjee R, Bhattacharjee S S, Basu K, Rajesh P V, Raut R, Ghugre S S, Das D, Sinha A K *et al.* 2014 *Phys. Rev.* C **90** 044319
[6] Bhattacharjee S S, Bhattacharjee R, Raut R, Ghugre S S, Sinha A K *et al.* 2015 *Phys. Rev.* C **91** 044306
[7] Ziegler J F www.srim.org
[8] Gavron A, 1980 *Phys. Rev.* C **21** 230
[9] www.nndc.bnl.gov

# Accelerator Mass Spectrometry – an ultrasensitive dating and tracing technique

**Pankaj Kumar**[1]

[1]Inter-University Accelerator Centre, New Delhi

E-mail: pkb@iuac.res.in

## 1. Introduction:

Nature is composed of various elements. The isotopes of elements are mainly of two types – stable and unstable. Isotopes which decay with time on a specific rate are known as unstable isotopes, they have their own decay constant or half life. Depending on the half life unstable isotopes or radioisotopes can be subdivided into four categories: ultra short lived (μs-seconds), short lived (days - few years), long lived (103 years to 107 years) and primordial radionuclide (>108 years). Radioisotopes are very useful in various dating and tracing applications and it is done by counting their numbers present in the sample at given time [1-5]. Counting of radioisotope is possible by simply counting the radiation (α, β or γ) emitted in the decay of radioisotope. This method is called decay counting. However, number of radiation emitted in decay is proportional to the number of radioisotope atoms present at that time and half life of radioisotope. The natural abundance of long lived radioisotopes is usually in the range of 10-10 to 10-16. Therefore, decay counting will take a very long counting time and need large amount of sample material for counting long lived radioisotopes to achieve considerable statistics. Decay counting is best suited for short lived radio isotopes [2, 4].
Other ways of estimating long lived radioisotopes could be counting atoms directly without waiting for their decay. Low energy mass spectrometers (LEMS), works on the principle of M/q selection, can serve the purpose. However, sensitivity of LEMS is limited (10-8-10-10) and isobaric interferences cannot be removed. High energy mass spectrometry or accelerator mass spectrometry (AMS) is the best suitable method of quantifying long lived radioisotopes [4-6].
A typical tandem accelerator based AMS system is composed of an ion source, an injector system with dipole magnets, a tandem accelerator, post acceleration analysis system and a detector. The ion source produces pre-accelerated negative ions, which under goes selection procedure by injector system through ME/q2 filtering. These selected negative ions are accelerated and converted into positive ions with the help of accelerator. Tandem accelerators have stripper system at the middle of their tank, which converts negative ions to positive and also helps in reducing molecular background by breaking the molecule. Due to high energy achieved by accelerator (order of MeV) atom counting is possible using particle detectors. High energy also helps in removing isobaric and isotopic interferences. Furthermore various other filters like velocity filter (E/M selector), electrostatic analyzer (E/q selector), gas filled magnet etc. are used in AMS to reduce background. Use of high energy accelerator in combination with various filters increases the sensitivity of AMS to 10-14-10-16 and makes it versatile and powerful method to accurately measure ultra trace elements [1-7].
Usually, AMS determines the ratio of radioisotope to stable isotope, as absolute counting of particular nuclide is difficult. However, two isotopes may not be ionized or transmitted through the system in

exactly similar manner and therefore, measured ratio of radioisotope to stable isotope may not reflect the true ratio in unknown sample. To solve this problem standard sample having known ratio of radioisotope to stable isotope is measured along with the unknown sample to normalize the results [5-7].
Inter-University Accelerator Centre (IUAC), New Delhi has established India's first facility for AMS measurement of 10Be (T1/2= 1.38 million years), 14C (T1/2= 5730 years) and 26Al (T1/2= 0.75 million years) radioisotopes [6, 7].

**Source of long lived radioisotopes:**
Earth's atmosphere is constantly bombarded by high energy particle (called cosmic rays) Galactic cosmic rays (originating from outside solar system) consists of 87% of protons, 12% alpha particles and remaining fraction of heavier nuclides. The energy of GCR ranges from 0.1GeV to 10GeV with an average of 2GeV. When these cosmic rays interact with the nuclei of atoms present in the atmosphere or on earth's surface, radioisotopes are produced [4]. For example when neutrons present in cosmic rays spectrum interact with nitrogen present in the atmosphere 14C (T1/2= 5730years) is produced following the nuclear spallation reaction 14N(n,p) 14C. Thus formed 14C is quickly oxidized into 14CO2 and equilibrates with volatile carbon species. The 14C/C ratio in the atmosphere is homogeneous with natural value of about $1.2 \times 10-12$. The distribution of atmospheric 14C in hydrosphere and biosphere is done by gaseous exchange and by photosynthesis respectively which is a part of carbon cycle [3, 4]. Once 14C is mixed in nature, it present everywhere like in plants, animals, hydrosphere and biosphere. As long as plants or animals are alive they stay in equilibrium with nature and natural 14C value remains constant. But when they are dead natural value starts decreasing and older is the sample fewer is the number of 14C atoms in the sample. Therefore, 14C can acts as natural clock present in natural archive.
Due to high sensitivity of AMS technique, it has been used in multi-disciplinary research including archaeology, geology, environmental science, biology, pharmacology, and bio-medicine [3, 4, 10, 11]. This short review focuses few applications of AMS in archaeology and biomedicine.

## A few applications of AMS:

### Radiocarbon dating in archaeology:

Archaeology is one the most explored area, where radiocarbon dating using AMS has revealed truths and has solved disputed problems. Few interesting examples are listed here:

1. **Shroud of Turin:**
   A linen cloth of about 13.5 Feet long and 4.5Feet wide is kept in a church of Italian city Turin. This cloth bears a faint outline of man [Fig. a] and since many years it is purported that this cloth was used to wrap the crucified body of Jesus Christ. Three different AMS laboratories request to Vatican to provide only few threads of Shroud to perform radiocarbon dating using AMS. Radiocarbon dating from three different laboratories dated few threads and concluded that this cloth material is originated between 1262 and 1312 AD [3, 8]. This indicates that this cloth is much younger than it is believed and seems unauthentic.

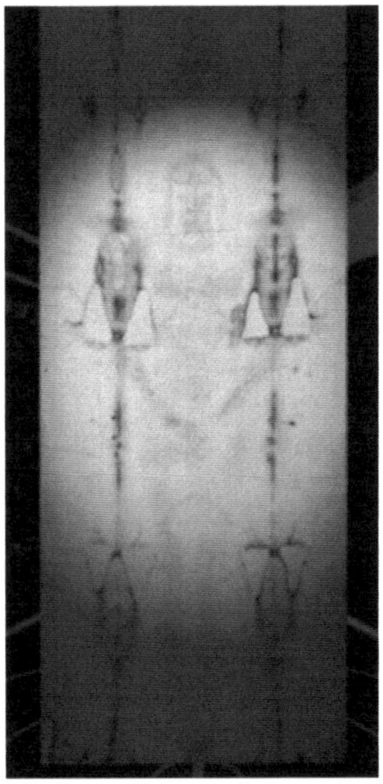

*Fig1. a: Picture of shroud of Turin. Lightened part is showing image of man*

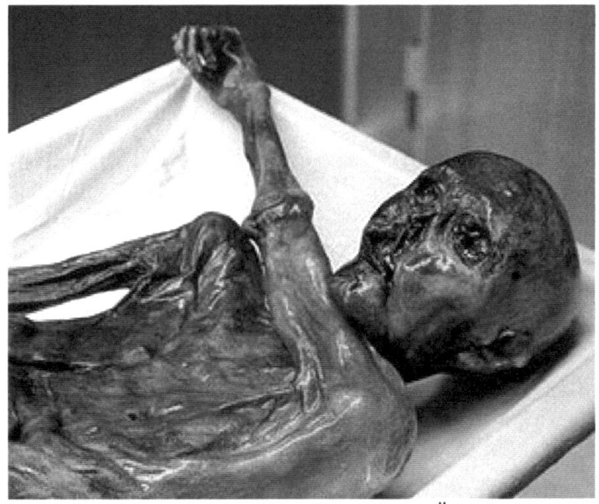

*Fig.b: Picture of Iceman (Named as Ötzi in German language)*

## 2. The Iceman "Otzi":

In September, 1991, two hikers discovered dead body of a man in Alps near Austria- Italy border. This body was very well preserved because it might have been buried in ice shortly after man died [Fig. b]. Radiocarbon dating using AMS at three different laboratories confirmed that he was died about 5200 years ago [3, 9]. Due to accurate determination of age of Iceman, scientists were able to learn about the life 5000 years ago. Body of Iceman and his belongings found from Alps are displayed in the South Tyrol Museum of Archaeology in Bolzano, South Tyrol, Italy.

## AMS in biomedical sciences:

Since, last two decades AMS has been used as a biomedical tool due to its unmatchable sensitivity, selectivity, and precision to measure the radioisotopes. In the beginning it was used in life science to study the effect of low levels of carcinogen and pollutants to cause the damage at cellular level. For example, AMS can be used to analysis and quantitate metabolite at the picomole level or lower than that. It was also used to identify macromolecular targets for drugs and allergic compounds [12-15].

Sensitivity of AMS allowed researchers to address important issues in relation to nutrition, pharmacology, cell biology and comparative medicine. It also helps to study the exfoliated tissues, isolated cell subpopulations, and precious tissues of human or animal origin.

Few years back an interesting method to determine retrospectively the birth of cells in humans was developed using AMS at the Karolinska Institute in Stockholm [16]. The principle of this method was based on the constant flux of the most of the molecule in the cell except genomic DNA, which is not exchanged after its last division. Hence, the level of $^{14}C$ integrated into genomic DNA should be same as that of atmosphere at that time, and determination of $^{14}C$ levels in genomic DNA can be used to retrospectively know the birth time of cells. Not only $^{14}C$/ $^{12}C$ ratio is playing important role in biomedicine but other radionuclide like $^{26}Al/^{27}Al$ and $^{41}Ca/^{40}Ca$ isotope ratios are being helpful to have control on the disease. As, $^{26}Al/^{27}Al$ and $^{41}Ca/^{40}Ca$ ratios is very low ranging to $10^{-14}$, with longer half-lives than $^{14}C$, strong radionuclide signals can be administered to humans. The $^{41}Ca$ can be used to study the reason behind the occurring of osteoporosis or bone related disease [13], while $^{26}Al$ can be implemented to correlate the kidney related problems [13].

The sensitivity of AMS also enables to study the ligand—macromolecule interactions at very low concentrations or for studying effects of hormones and for studying early events in host and pathogen interaction [13]. This Resource seeks to both develop the methods and instrumentation to make AMS a general use tool for biomedical researchers and to make it available for investigators needing access to techniques for the ultra-trace analysis of radioisotopes in biological studies.

Drug discovery is another area where AMS provides unmatchable advantages. The amount of radiotracer tagged drug dose to study the behaviour of drug on human body is drastically reduced to micrograms level by the advent of AMS technology [13-15]. Micro-dosing reduces the risk of radiotracers on human body and faster measurements by AMS reduce the drug trial time.

In nutshell, the sensitivity and accuracy of AMS allows, it to be used in any field where ultra-sensitive analysis of long-lived radionuclides can be of interest to resolve the asked questions. The further development in AMS to distinguish among isobars and use of samples in micrograms can undoubtedly expand the use of AMS in almost each and every field of research. The applications of AMS can also be widen by making more radionuclide available for AMS to be exploited in interdisciplinary areas.

**References:**

1. Chen J, Guo Z, Liu K and Zhou L 2011 *Reviews of Accelerator Science and Technology* **4** 117.
2. Fifield L K 1999 *Rep. Prog. Phys.* **62** 1223.
3. Kutschera W 2005 *Int. J. Mass Spectrometry* **242** 145.
4. Tuniz C, Bird J R, Fink D, and Herzog G F, Accelerator mass spectrometry: Ultrasensitive analysis for global science: Boca Raton, Florida, CRC Press, 371 p. (1998)
5. Litherland A E 1980 *Ann. Rev. Nucl. Part. Sci.* **30** 437.
6. Kumar P, Pattanaik J K, Ojha S, Gargari S, Joshi R, Roonwal G S, Balakrishnan S, Chopra S and Kanjilal D 2011 *J. Radioanal. Nucl. Chem.* **290** 179
7. Kumar P, Chopra S, Pattanaik J K, Ojha S, Gargari S, Joshi R and Kanjilal D 2015 NIM B **361** 115.
8. P E Damon, Donahue D J, Gore B H, Hatheway A L, Jull A J T, Linick T W, Sercel J, Toolin L J, Bronk C R, Hall E T, Hedges R E M, Housley R, Law I A, Perry C, Bonani G, Trumbore S, Woelfli W, Ambers J C, Bowman S G E, Leese M N and Tite M S 1998 *Nature* **337** 611.
9. Fornvagner R P, Niklaus Th.R 1994 NIM B 92 282
10. Khare N, Govil P, Kumar P, Mazumdar A, Chopra S, Pattanaik, J K Balakrishnan S and Roonwal G S 2011 *J. Radioanal. Nucl. Chem.* **290** 197.
11. Kumar P, Pattanaik J K, Khare N, Chopra S, Yadav S, Balakrishnan S and Kanjilal D 2014 J. Radioanal. Nucl. Chem. **302** 903
12. Freeman S P H T, Vogal J S 1995 *International Journal of Mass Spectrometry and Ion Processes* 143 247.
13. Jackson G S, Weaver C and Elmore D 2001 *Nutrition Research Reviews* 14 317.
14. K. Brown, Tompkins E M, White I N H 2005 *Mass Spectrometry Reviews* **25** 127.
15. Vogal J S, McAninch J, Freeman S P H T 1997 NIMB **123** 241.
16. Spalding K L, Arner E, Westermark P O, Bernard S, Buchholz B A, Bergmann O, Blomqvist L, Hoffstedt J, Näslund E, Britton T, Concha H, Hassan M, Rydén M, Frisén and Arner P 2008 *Nature* **453** 783.

# Efficiency and stability aspects of CdS photoanode for solar hydrogen generation technology

**Alka Pareek, Arthi Gopalakrishnan and Pramod H. Borse** [1]

International Advanced Research Centre for Powder Metallurgy and New Materials, Balapur P.O., Hyderabad-500005, Telangana, India

E-mail: pborse@physics.org; phborse@arci.res.in

**Abstract.** Photoelectrochemical (PEC) cell based technology is expected to be one of the easiest green technologies to harness and to convert available solar energy into hydrogen [1]. Among the known systems of GaAs, Si, GaP and CdS- Cadmium sulphide is one of the best suited PEC materials that display a balance between its efficiency and stability. It has capability to absorb the visible light photons (E~1.5eV – 3eV), and displays the band-energetics that suits for water-splitting reaction ($H_2O \rightarrow H_2+O_2$), that ultimately is based on the electronic and optical structure of the sulphides. However, the photo-induced dissolution of CdS in an electrolyte during its *photo-illumination* in PEC cell is its major drawback [2]. Though arsenides and phosphides show higher efficiency however CdS exhibits significant stability. In contrast though $TiO_2$/ ZnO show good stability but CdS displays good optical response towards visible light photons as compared null response of titanate like systems. This necessitates one to identify the practical way to inhibit the photocorrosion in case of CdS photoanodes which is mainly facilitated due to interaction of photogenerated holes with CdS lattice. In past, Pt/ $RuO_2$/ Ru modified CdS surface were found to control the unwanted photocorrosion [3]. Commercially, usage of such materials is un-economic option for any technological usage. The present work discusses that with advent of present day new synthetic routes how the dynamics of photo generated holes and electrons can be controlled to improve the stability and efficiency of the sulphide photoanodes, which in turn shows an an improvement in the performance and stability of the PEC cell for desirable technological applications.

## 1. Introduction

A renewable hydrogen generation by photoelectrochemical water-splitting reaction is highly desirable for ever-growing energy demands. Photoelectrochemical (PEC) hydrogen production through splitting of water under sunlight is one of the most promising technologies for a future hydrogen economy [1, 4]. The cadmium chalcogenides comprise one of the most important class of materials in the photoelectrochemical research. CdS exhibits a direct band gap of ~2.42eV, and is considered to be an excellent visible light responsive material [5]. Interestingly, it possesses suitable band-edge positions desirable for cleavage of water molecule into hydrogen and oxygen. Despite these advantages,

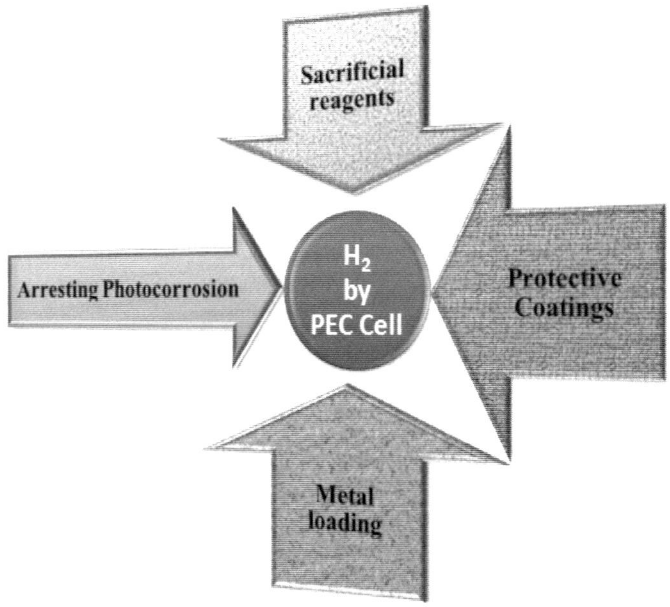


**Figure 1**. Schematic displaying different strategies to inhibit the photocorrosion process in photocatalysts

chalcogenides are associated with a problem of photocorrosion that hampers their long term stability in the PEC application. In past, there have been different strategies, adopted to curb the photocorrosion *viz*. (1) Use of the sacrificial agent in an electrolyte which acts as a scavenging agent and consumes the holes participating in the photocorrosion reaction [3,6]; (2) Loading some inert metal on the semiconductor surface so that it behaves as a catalyst to increase the rate of oxygen or hydrogen evolution [7]; (3) Modification of the semiconductor surface with metal oxide/ polymers to protect the semiconductor from electrolyte [8]. This is schematically illustrated in figure 1. Among all the strategies the most attractive is the protection of the semiconductor surface by metal-oxide thin layer that resist the electrolyte, and at the same time allows the sunlight to be absorbed by the electrode.

With the advent of nanostructuring technologies, one is able to tailor the electrode surface using nano-dimensional adsorbates. This is expected to render a favourable and constructive physical mechanism such as quantum size effects, quantum tunnelling, *etc*, for the modified configuration of the electrode. Ultimately, we expect these modified structures lead to a dramatically improved optoelectronic performance of the photoelectrodes. In order to demonstrate the expected "nanostructure induced performance" of CdS electrode, we selected two model metal oxides (MO; M = Ti, Nb) *viz*., $TiO_2$ and $Nb_2O_5$. $TiO_2$ and $Nb_2O_5$ are considered as the most stable metal oxides under wide *p*H range. $TiO_2$ has been regarded as the most promising system for passivation of chemically unstable semiconductors (GaAs, Si, GaP and CdS) due to its high photochemical stability, non-toxicity and low cost. Similarly, $Nb_2O_5$ (+V oxidation state) shows higher stability in the *p*H range 0.1-12.2 rather than $NbO_2$ (+II oxidation state) [9]. In present work, we have used these metal oxides as "nanoparticle model" systems to modify the CdS photoanode to improve its stability. We have attached metal oxide nanoparticles using Thioglycerol (TG) as an organic linker by making use of a simple adsorption technique, over the photoanode surface. These electrodes were investigated in depth to understand the underlying physics and the electrochemical aspects those are responsible to attain stable photoanode for PEC cell.

Journal of Physics: Conference Series **755** (2016) 012006 doi:10.1088/1742-6596/755/1/012006

## 2. Experimental Section

*2.1. Materials*

$CdCl_2$ (Sigma-Aldrich, 99.999%), $(NH)_2CS$ (Sigma-Aldrich, 99.9%), $NH_4[NbO(C_2O_4)_2(H_2O)_2].nH_2O$ (CBMM, Brazil), $TiCl_4$ (Sigma-Aldrich, ≥ 99.0%), $H_2O_2$ (Sigma-Aldrich, 30% w/w), Thioglycerol (Sigma-Aldrich, ≥ 97%), Fluorine doped Tin oxide, FTO (Pilkington, TCO-15) glass substrate (12-15Ω/sq.) and Deionized water (18.2 MΩ.cm) were used.

*2.2. Deposition of CdS thin film and fabrication of photoelectrode*

The CdS thin films were deposited onto preheated FTO substrate by computer automated spray pyrolysis deposition technique. Its schematic is shown in figure 2. Nanostructured CdS thin films were deposited by optimization of various parameters like flow rate, number of deposition cycles, duration of the deposition, temperature and area of deposition as reported in our earlier work [10]. The precursor solution containing mixture of $CdCl_2$ (0.1M) and $(NH)_2$ CS (0.1M) were used for the deposition onto FTO substrate at 350°C with flow rate of 5ml/min. The nozzle to substrate distance of ~23cm was maintained throughout the experiment.

The CdS electrodes were fabricated from the films by exposing active area of unit dimension over the film. The electrical contact was made over the undeposited area of FTO substrate, for applying desired biasing voltage. The undesired film area was electrically insulated using an epoxy. Thus obtained photoelectrode was used for the surface modification and further for the PEC hydrogen generation.

*2.3. Preparation of $TiO_2$ and $Nb_2O_5$ oxide nanoparticles*

In order to synthesize nanoparticles of metal oxides, hydrolysis and hydrothermal techniques were used for $TiO_2$ and $Nb_2O_5$ respectively. $TiO_2$ nanoparticles were synthesized *via.* known procedure in the literature [11]. A precise concentration of titanium tetrachloride ($TiCl_4$) is dissolved in the deionized water at ice-cold temperature and is mixed with an ammonia solution. The resulting mixture was kept under stirring condition for an hour and refluxed for ~18h. The precipitate from the refluxed mixture was collected by centrifugation and washed with deionized water for the removal of un-

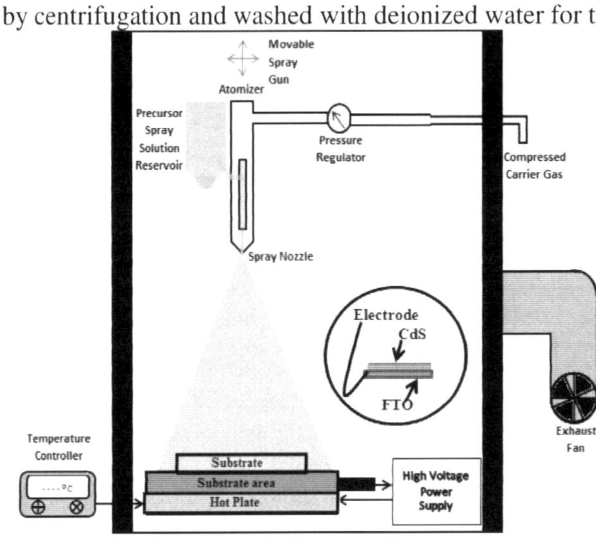


**Figure 2.** Schematic of spray pyrolysis deposition technique used for deposition of nanostructured CdS films; *inset* illustrates fabricated photoelectrode.

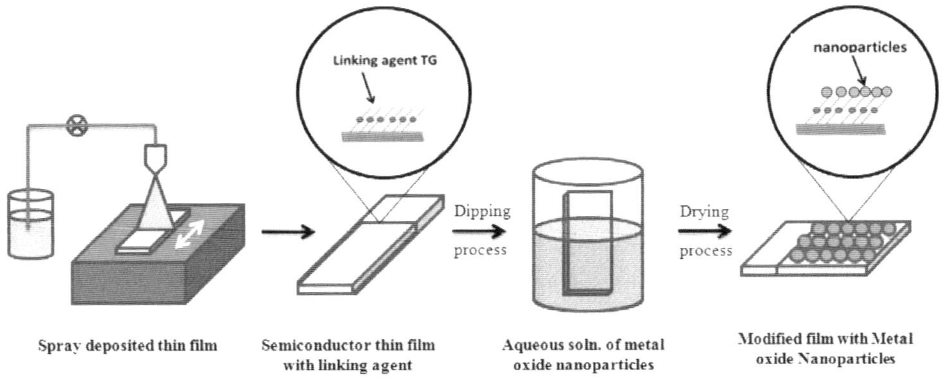


**Figure 3.** Schematic of surface modification process of semiconductor photoanode

reacted impurities. Thus obtained precipitate was air-dried to obtain the free standing powder of nanoparticles. This yielded 2.5nm sized $TiO_2$ nanoparticles.
Further, $Nb_2O_5$ nanoparticles were synthesized by hydrothermal method [12]. In brief, 8g of ammonium niobium oxalate powder was dissolved in deionized water (100 mL) to obtain a transparent solution. To this, $H_2O_2$ (30% by vol.) was added that led to the formation of niobium peroxo complex, a transparent yellow solution. The obtained solution mixture was then transferred to a teflon lined autoclave (200 mL) and subjected to 120°C for 9hrs. After the reaction, finally obtained white precipitate was washed thoroughly with water-ethanol mixture and separated by centrifugation. This yielded $Nb_2O_5$ nanoparticles of 5nm size. These MO nanoparticles were used for the photoanode surface modification.

### *2.4. Surface modification of CdS photoanode by metal oxide nanoparticles*
The surface of CdS photoanode was modified by MO nanoparticles by adsorption mechanism *via.* linking agent. This was reported to yield a stable and efficient photoanode during the PEC measurements [13]. Figure 3 schematically illustrates the procedure used for surface modification. TG was used as linking agent for the surface modification with 0.1 v/v % concentration, after optimization that yielded best nanoparticles adsorption and thus an improved PEC performance. An organic linker helps in yielding uniform surface adsorption of nanoparticles over the CdS surface. The CdS surface linked by TG was further treated by nanoparticles by immersing the photoanode in nanoparticle dispersion. Thus nanoparticle dispersion in aqueous medium was ultrasonicated for 20 min prior to electrode immersion. An immersion time of 7 min was used for this purpose. Thus obtained electrode was dried and used for the further measurements.

### *2.5. Physico-Chemical characterization of the photoanode*
The crystal structure of the nanostructured films were investigated using X-ray diffraction analysis. For this, Bruker AX D8 XRD diffractometer equipped with Cu $K_\alpha$ ($\lambda$ = 1.514Å) X-ray gun was used. The scan was performed in the range of $2\theta$ = 10-70° with an increment of 0.01°/s at the grazing angle of incidence 0.2°. A detailed elemental and compositional analysis of the bare and surface modified CdS films was done by X-ray photoelectron spectroscopy (Omicron model). The UV-Vis absorption analysis was done by using Perkin Elmer Lambda 650 spectrometer. The PEC analysis was carried out using an electrochemical workstation (PARSTAT model 2273) to obtain photocurrent density of the photoanode of unit area. A PEC reactor cell with three electrode configuration consisting of photoanode as working electrode (WE), standard calomel electrode (SCE) as reference electrode (RE) and graphite as counter electrode (CE) was used. An aqueous mixture of Na2S (0.1M) and Na2SO3 (0.1M) was used as an electrolyte to maintain inherent electrode stability against photocorrosion of the CdS film. In order to study the PEC performance and solar hydrogen production, solar simulator

(Oriel Model 91160) equipped with A.M.0, A.M.1.0 and an AM1.5 Global (Newport) filter was utilized. The irradiance of 80mW/$cm_2$ was used in the present case. The incident photon current conversion efficiency (IPCE) measurements were carried out using an Oriel monochromator capable of generating desired wavelengths (accurate up to ±2nm) in the range of 200-900 nm. Quantitative estimation of evolved hydrogen was done by using gas chromatograph-GC (Schimadzu 2010 AFT) equipped with thermal conductivity detector was used for hydrogen detection and analysis.

## 3. Results

The bare and modified CdS photoanode has been used throughout, to the study structural, elemental, optical and charge transfer properties.

### *3.1. Structural and elemental characterization of bare and modified CdS photoanodes*

Figure 4 shows X-ray diffraction spectra of bare and the modified CdS photoelectrodes. The CdS films exhibited hexagonal phase of CdS that belongs to P63MC (JCPDS file No. 00-006-0314) space group. The background showed weak low intensity peaks of $SnO_2$ (FTO) due to the substrate. The existence of $TiO_2$ and $Nb_2O_5$ nanoparticles over the surface is not detected as it is beyond the detection limit of XRD. This confirms that the modification does not alter the XRD pattern of CdS. It is important to mention that the particle size of metal oxide nanoparticles was estimated separately from XRD and TEM (Transmission Electron Microscopy) of nanoparticle powders. Nanoparticle characterization is not included here, but the details can be found elsewhere [12-13]. Briefly, the nanoparticle sizes were calculated using Debye Scherer's formula [14] and were found to be 2.5 and 5 nm for $TiO_2$ and $Nb_2O_5$ respectively.

$$D = \frac{0.9\lambda}{\beta \cos\theta} \quad (1)$$

where, D-particle size (nm), λ-wavelength (nm), θ-diffraction angle, β-FWHM.

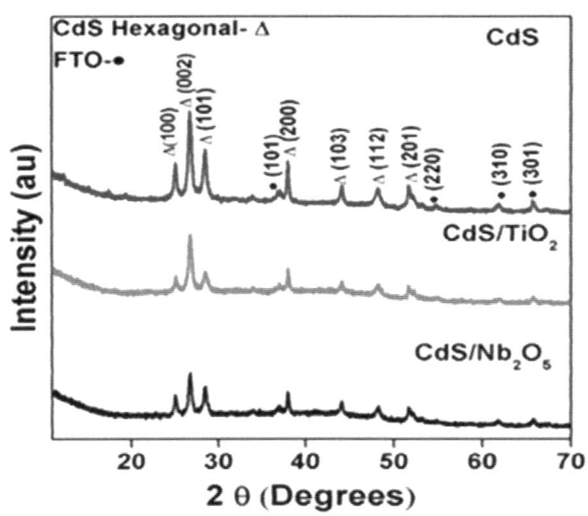


**Figure 4.** XRD patterns of bare CdS and MO nanoparticles modified CdS films.

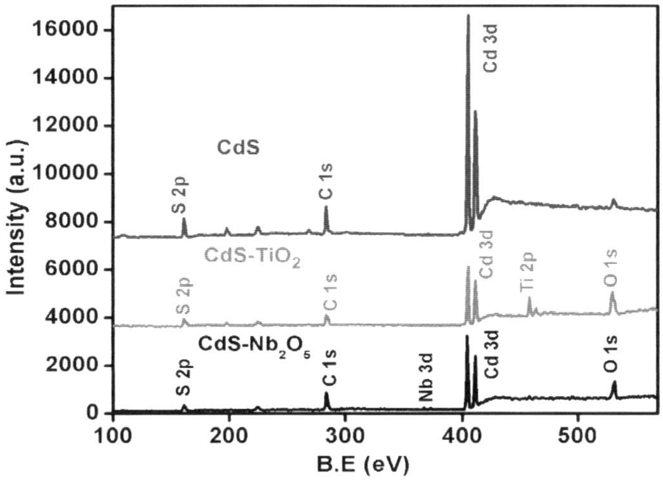


**Figure 5.** XPS survey spectra of bare CdS and MO nanoparticles modified CdS films.

Figure 5 shows XPS analysis of bare and modified CdS photoelectrode before PEC measurements in the form of survey scan. It indicates the presence of Cd 3d and S 2p lines at B.E. ~160.9eV and 404.69eV respectively along with carbon peak. This clearly shows that the film consists of a stoichiometric composition of CdS. For modified CdS film with $TiO_2$ nanoparticles, the presence of Ti 2p at 457.17eV and O 1s at 529.56eV is confirmed. Further, the spectrum of $Nb_2O_5$ modified CdS displays the peaks due to Nb 3d and O 1s. An in depth analysis of region wise scan of Ti/Nb/Cd/S/O by deconvolution [15], confirms that MO nanoparticles are linked to the CdS via.TG linking agent.

### *3.2. Optical characterization of bare and modified CdS photoanodes*

The UV-Vis absorption spectra of bare and modified CdS photoanode are shown in the figure 6. It can be clearly seen from the UV spectrum that as compared to bare CdS spectrum, the modified one shows blue shift in the absorption due to the existence of nano-sized particles of larger band gap (~3.2eV). The study reveals that the optical property of the CdS film remains unaffected after the adsorption of TG and the presence of these particles screens the overall surface which leads to a decrease in the absorption of the CdS film in visible range.

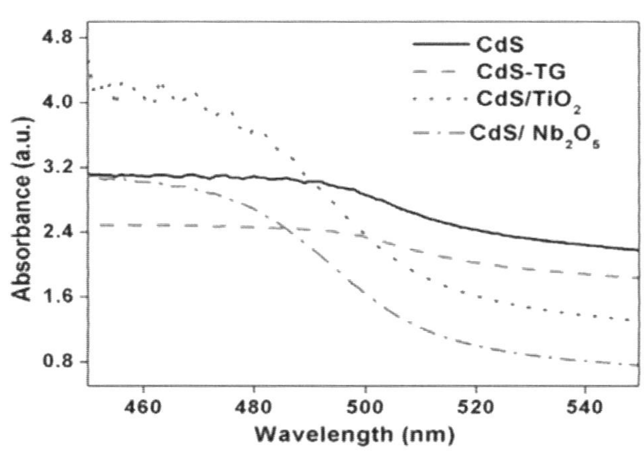


**Figure 6**. UV-Vis absorbance spectra of bare CdS film, TG, $TiO_2$, $Nb_2O_5$ nanoparticles modified CdS films.

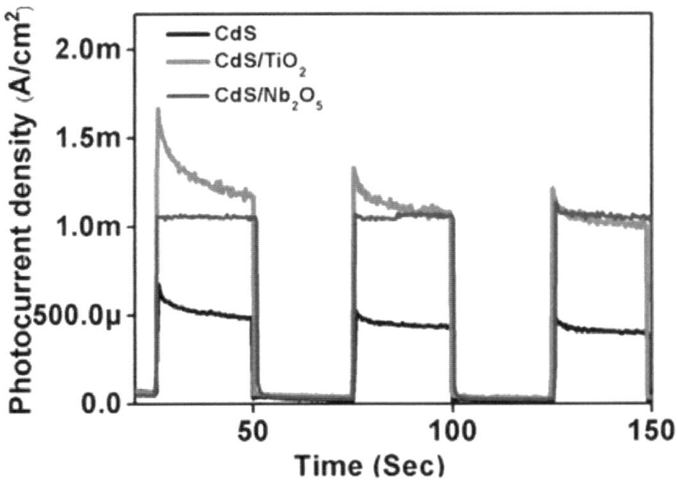


**Figure 7. Comparison of chronoamperometric curves of bare CdS photoanode with MO modified CdS photoanode.**

*3.3. Photoelectrochemical measurements of modified CdS photoanodes*

The surface modification of CdS with nanoparticles is done to improve the efficiency of photoanode. The photocurrent density, charge transfer mechanism, efficiency and hydrogen rate were analysed with PEC measurement. This yields the performance of respective photoanodes. Figure 7 shows the comparative chronoamperometric curves for the modified CdS *w.r.t.* bare CdS photoanode at 0.2V. The photocurrent density of bare CdS electrode was found to be 480μA/cm$^2$ while that of $TiO_2$ and $Nb_2O_5$ modified CdS was 1.2mA/cm$^2$ and 1mA/cm$^2$. The reduced photocurrent of CdS is due to the fast recombination of photogenerated electron-hole pairs that largely depends on the density of surface states in the semiconductor that acts as an electron trapping center. But, the MO nanoparticle modified CdS has enhanced photocurrent. It is attributed to the formation of passivating layer on the surface which in turn reduces the recombination rate of the photogenerated charge carriers.

*3.3.1. Incident photon to current conversion efficiency*

It is one of the most important characterizations which provides information about how efficiently a cell converts photons of a particular frequency into electrons and is known as wavelength dependent efficiency. An incident light from monochromator falling on the photoanode generates an output current that can be used to calculate the number of electrons generated. IPCE is given by the relation [16]

$$IPCE = \frac{1240 * I}{\lambda * P} \times 100\% \quad (2)$$

Where I – photocurrent density (A/cm$^2$), P - source power (W/cm$^2$) and λ- wavelength (nm). Figure 8 shows that the calculated IPCE for MO nanoparticles modified CdS photoanode was found to be 22% at 470nm and 18% at 480nm respectively. These values are found to be enhanced as compared to bare CdS photoanode (8%). This high efficiency improvement in turn generates hydrogen from water during the hydrogen evolution reaction.

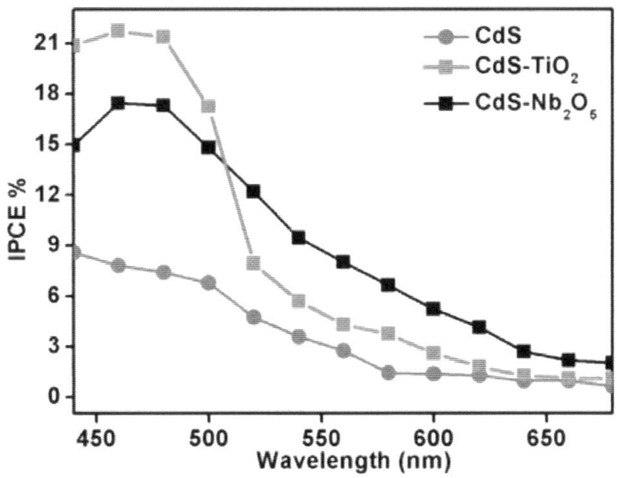


**Figure 8.** IPCE curve of bare and MO modified CdS photoanode.

*3.4. Photoelectrochemical stability studies of modified CdS photoanodes*

Apart from the PEC activity and stability, long-term PEC performance potrays main concerns for the development of efficient photocatalystic electrode. Specifically, this needs the study of related parameters, such as degradation of electrode-corrosion that changes the electrode surface composition as well as releases mineral impurities in the electrolyte solution. Hence to detect the deterioration of photocatalyst due to photocorrosion, a long hour chronoamperormetric measurement is desirable. Figure 9 shows the stability measurements for bare and modified CdS carried out for 9hrs at 0.2 biasing voltage in photocurrent Vs time plot. It can be observed that the photocurrent of bare CdS decreases abruptly with the time, whereas the modified CdS photoanode shows a constant photocurrent during 9hrs of measurement indicating a drastic improvement and a sustained photocurrent over the time of measurement.

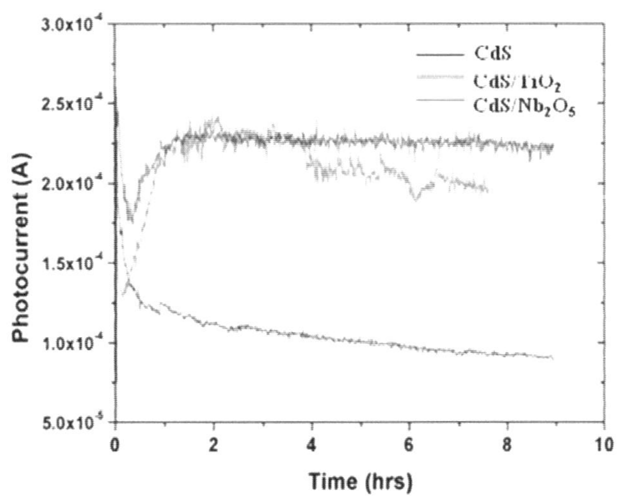


**Figure 9.** Chronoamperometric curve recorded for 9hrs of stability for bare and modified CdS photoanodes.

## 4. Discussions

Earlier section has clearly demonstrated that the nanoparticles induced photocurrent enhancement and stability is due to the modification of CdS photoanode. It is important to understand the underlying mechanism that has led to such drastic improvement in the efficiency and stability. Accordingly, we have analyzed and explained this effect in terms of, (1) adsorption of nanoparticles on CdS surface; (2) charge transfer mechanism occurring at electrode/substrate and electrode/electrolyte interfaces.

### *4.1. Mechanism of surface adsorption process*

The surface modification of the CdS photoanode was carried out using chemisorption process, which involves the use TG as organic linkers. As shown in figure 10, this linking agent chemically links two compounds with preferred functional groups. Hence the formation of hybrid structure is due to affinity of CdS and $TiO_2$ and $Nb_2O_5$ nanoparticles towards thiol(R-SH) and hydroxyl (-OH) groups.

### *4.2. Charge transfer mechanism in the modified CdS photoanode*

Nanoparticle adsorption over the substrate reduces the charge trapping states over the surfaces of CdS which in turn leads to the changes in the electrode/electrolyte interface. Thus, there is a difference in band energies of the modified CdS *w.r.t.* bare CdS photoanode. Basically, the charge transfer mechanism in the CdS photoanode involves the following process [17].

- The incident light is absorbed by the CdS photoanode
- When incident light energy is greater than band gap (Eg) of photocatalyst, it absorbs and further excites the electrons from valence band (VB) to conduction band (CB) leaving holes in VB.
- Thus, the photogenerated electrons and holes migrate over the surface of CdS, while some charge carriers recombine to release photons.
- Since the valence band edge of metal (Ti and Nb) oxide is more positive than CdS, hole transfer from CdS to electrolyte can take place only through the "quantum tunnelling" process.
- The electrons at counter electrode produce hydrogen gas whereas holes at working electrode combines with water molecule to produce oxygen gas.
- Due to adsorbed nanoparticles, the surface states on the CdS gets passivated and hence with enhanced photocurrent could be achieved due to the reduced recombination of the photogenerated charge carriers.

### *4.3. Stability factor*

Stability is the most important factor that dictates the life of a photoanode, which in turn yields to an efficient PEC cell. We have defined the stability factor to quantify the photoelectrode performance. It has been defined in depth in our past report [12]. Briefly, it depends on several, physical, chemical and

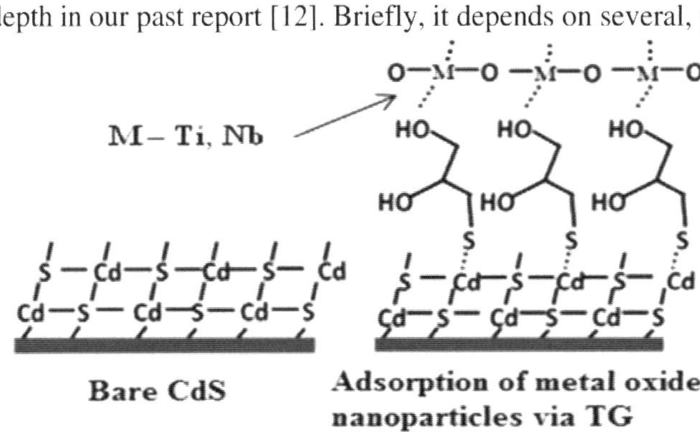


**Figure 10.** Schematic of adsorption mechanism of MO nanoparticles on CdS surface using TG as linking agent.

other factors. We have defined this new parameter ($\Sigma_{photoelectrode}$) which is function of several dependent parameters (physical/chemical/others) as described below:

$$\Phi_{photoelectrode} = \Phi\text{ (physical)} \bullet \Phi\text{ (chemical)} \bullet \Phi\text{ (others)} \quad (3)$$

Where, Φ (physical) – physical properties of electrodes correlated to stability viz. crystal structure, orientation, thickness, area, geometry.
Φ (chemical) – pH of electrolyte, electrolyte composition, concentration of electrolyte.
Φ (others) – temperature, time, stirring and photocurrent density.

It is imperative that the stability ($\Sigma_{photoelectrode}$) is multiple-parameter dependent function involving various processes, and thus it is difficult to describe it quantitatively. However, in order to simplify its quantitative estimation in present study, we have considered the variation of photocurrent with time as a means to estimate the time-dependent stability of a photoanode; accordingly, we propose "photoanode-stability" to be a simple function of photocurrent density and illumination time, assuming other parameters constant. This has led to a stability factor as:

$$\sum_{photoelectrode} \% = \frac{\Delta_{Cds} - \Delta_{modified}}{\Delta_{Cds}} * 100\%$$

$$\Delta_{modified} \rightarrow \Delta_{CdS}, \Sigma_{photoelectrode} \rightarrow 0$$

$$\Delta_{modified} \rightarrow 0, \Sigma_{photoelectrode} \rightarrow 1 \quad (4)$$

Where $\Delta_{photoelectrode}$ is slope of "I *vs.* t" curve of the photoelectrode, given by:

$$\Delta = \frac{\partial I}{\partial t} = \frac{I(t_2) - I(t_1)}{t_2 - t_1} \quad (5)$$

Here, I ($t_1$) is the photocurrent at time '$t_1$' and I ($t_2$) is the photocurrent at time '$t_2$'.

Accordingly, the slope of I - t curve for CdS ($\Delta_{CdS}$) and modified photoanode ($\Delta_{modified}$) were calculated from figure 9. Subsequently, stability (Σ) is estimated for each modified photoanode *w.r.t.* the bare CdS photoanode. Figure 9 clearly reveal that, if the decay in the photocurrent of photoanode is comparatively faster as in case of CdS, it indicates a poor stability. However, if there is no decay in the photocurrent as seen in modified photoanode, then it indicates the photoanode has 100% stability. Photocurrent performance of photoanode dictates the photoefficiency and stability of hydrogen evolution reaction. Accordingly, we studied quantitatively the long-term PEC hydrogen evolution over

Table 1: List of various parameters values for bare and modified CdS photoanodes

| | Band gap, Eg (*eV*) | Photocurrent density (*mA/cm²*) | $H_2$ Rate (*μmol/h*) | Slope Δ | Stability Factor (*Σ in %*), at illumination time [T] |
|---|---|---|---|---|---|
| **CdS** | 2.42 | 0.48 | 1.7 | 12.7 | 0 [1 day] |
| **CdS/Ti** | 2.50 | 1.20 | 22 | 0.0 | 100 [1 day] |
| **CdS/Nb** | 2.52 | 1.00 | 10 | 5.5 | 99.8 [0.4 day] |

bare and modified electrodes. This study clearly indicated modified CdS performs excellently *w.r.t.* efficiency and stability in contrast to the poor performance of less stable bare CdS electrode. Table 1 summarizes the variation in the stability i.e. photocurrent density, hydrogen rate, slope of "I vs. t" and stability factor for all electrodes.

## 5. Conclusion

This addresses the main issue of stability improvement of Cd chalcogenides i.e. mainly CdS by using nanoparticle induced surface modification of the electrode. Chemically synthesized MO nanoparticles ($TiO_2$ and $Nb_2O_5$) when linked *via.* TG rendered a stability improvement to the modified CdS electrode. This also led to increased hydrogen evolution of one order as compared to the bare CdS photoanode. This also demonstrated that $TiO_2$ modified CdS films are more stable than $Nb_2O_5$ modified CdS films. This study clearly indicates that MO nanoparticle induced surface modification is best & economic way of stabilizing the CdS photoanode for hydrogen production technology.

## Acknowledgments

Permission granted by Director, ARCI to publish the work is gratefully acknowledged.

## References

[1] Fujishima A and Honda K 1972 *Nature* **238** 37
[2] Meissner D, Memming R and Kastening B 1986 *Chem Phys. Lett.* **127** 5
[3] Kalyanasundaraman K, Boragarello E, Duonghong D and Gratzel M 1981 *Angew Chem. Int. Ed. Engl.* **20** 987-988
[4] Graetzel M 2001 *Nature* **414** 51-59
[5] Li Y, Hu Y, Peng S, Lu G and Li S 2009 *J. PhysChem.* **113** 9352-9358
[6] Inoue T, Watanabe A, Fujishima K Honda and Koshyakawa 1977 *Japan. J. Electrochem. Soc.* **124** 719-722
[7] Darwent J R and Porter G 1981 *J. Chem. Soc.ChemCommun.* 145-146
[8] Navarro R M, Valle F D, Fierro J L G 2008 *Int. J. Hydrogen Energy* **32** 4265-4273
[9] Scaife D E 1980. *Sol. Energy* **25** 41-54
[10] Pareek A, Dom R, Borse P H 2013 *Int. J. Hydrogen Energy* **38** 36-44
[11] Borse P H, Kankate L S, Dassenoy F, Wogel W, Urban J, Kulkarni S K. Urban, Kulkarni S K 2002 *J. Mater. Sci. Mater. Elect.* **13** 553-59
[12] Pareek A, Paik P, Borse P H 2014 *Langmuir* **30** 15550-15549
[13] Pareek A, Purbia R, Paik P, Hebalkar N Y, Kim H G, Borse P H 2014 *Int. J. Hydrogen Energy* **39** 4170-4180
[14] Culity B D 1978 *Elements of X-ray diffraction* **2$^{nd}$** edition Reading (MA: Addison-Wesley publishing company, Inc)
[15] Kulkarni S K, Winkler U, Deshmukh N, Borse P H, Fink R, Umbach E 2001 *Appl. surf. Sci.* **169,** 438-446
[16] Chen Z, Jaramillo T F, Deutsch T G, Shwarsctein A K, Forman A J, Gaillard N et al. 2006 *J Mater Res* **110,** 18088-18090
[17] Chen H and Wang L 2014 *J. Nanotechnol.* **6** 696-710

# Fusion-fission study at IUAC: Recent results

**Sugathan Pullanhiotan**

Inter University Accelerator Centre(IUAC), Aruna Asaf Ali Marg, New Delhi 110067, INDIA

E-mail: `sugathan@iuac.res.in`

**Abstract.** Several properties observed in heavy ion induced fission led to the conclusion that fission is not always originated from fully equilibrated compound nucleus. Soon after the collision of two nuclei, it forms a di-nuclear system than can fission before a compound nucleus is formed. This process termed quasi-fission is a major hurdle to the formation of heavier elements by fusion. Fission originated before complete equilibration showed anomalously large angular anisotropy and mass distribution wider than what is expected from compound nucleus fission. The standard statistical model fails to predict the outcome of quasi-fission and currently no dynamical model is fully developed to predict all the features of quasi-fission. Though much progress has been made in recent times, a full understanding of the fission dynamics is still missing. Experiments identifying the influence of entrance channel parameters on dynamics of fusion-fission showed contrasting results. At IUAC accelerator facility many experiments have been performed to make a systematic study of fission dynamics using mass distribution, angular distribution and neutron multiplicity measurements in mass region around $A \sim 200$. Recent measurement on mass distribution of fission fragment from reaction $^{19}F +^{206,208} Pb$ around fusion barrier energy showed the influence of multi-mode fission in enhancing the mass variance at low excitation energy. In this talk I will present some of these results.

## 1. Introduction

Fission of heavy nuclei is a complex process involving rapid re-arrangement of nucleons and large shape changes. Soon after the experimental discovery of neutron induced fission of $^{235}$U in 1939[1], the detailed understanding and the physics concepts behind the fission phenomenon was explained by Bohr and Wheeler[2]. Since then several decades of fission research have led to many applications and wealth of new knowledge in nuclear physics. Despite significant progress in fission studies, the dynamics of the fission process is still not yet fully understood in terms of quantum microscopic physics. The standard statistical model of fission was successful in explaining many of the features observed in light ion induced fission. However, it fails to explain many features of heavy ion induced fission. In recent years, there have been many experimental and theoretical efforts made towards better understanding of the dynamics of fission. These studies are important for understanding the formation of super heavy elements in laboratory. It is well known that a major hurdle in the formation of super heavy element is the presence of non-compound nuclear process such as fast fission or quasi fission(QF) that are dominant in heavy nuclear systems[3]. Near the fusion barrier energies the non compound process exhibit their own reaction characteristics which are distinguished by various measurements such as mass and angular distributions of fission fragments, mass-angle correlation, and pre-scission particle multiplicities etc. For example, presence of excess of pre-scission particle multiplicity, large anisotropy in angular distribution, strong mass-angle correlation and broadened mass distribution for fission fragments(FF) in some of the systems indicated strong evidence of non compound nuclear process and hindrance to fusion-fission process[4]. A large number of such experimental data using heavy ion fusion-fission reaction have been collected in the past and such measurements are still continued in many laboratories. At IUAC accelerator facility, we have been performing research on fusion-fission studies using heavy ion reaction and dedicated facilities. In the following, some of our existing experimental facilities for fission research and highlights of recent experiment will be discussed.

## 2. Facilities for fission study at IUAC

The accelerator facility at inter university accelerator centre(IUAC) consist of a 15UD Pelletron accelerator followed by super conducting LINAC[5]. Heavy ions with energy ranging from below barrier to above barrier are used to study fusion-fission dynamics in reactions involving stable target-projectile combination. The distinct features of fusion-fission reactions are studied in many measurements including fission excitation functions, angular distributions, mass-energy distributions, mass-angle distribution, mass-angle correlation and fusion evaporation residue (ER) measurements. Detection of light particles in coincidence with fission fragments also provide crucial information on the time scale and dynamics of the process. As each measurement demand efficient detector system specific for the type of particles to be detected, these measurements are carried out employing different kind of experimental facilities. For example, to measure angular distribution of fission fragments, array of charged particle detector telescopes covering angular range from $90°$ to $180°$ are typically used. Mass distribution of fragments is deduced from the velocity of the fragment measured from time of flight(TOF) and energy of the fragments. For velocity measurement, TOF setup using indigenously developed large area 2D position sensitive multi-wire proportional counters(MWPC) of dimension $20cm \times 10cm$ are being used[6]. By placing the detectors in folding angle and detecting fragments in kinematic coincidence, full momentum transfer fusion-fusion events can be separated from incomplete fusion-fission events. At IUAC, these detectors are placed on two rotatable arms inside the 1.5 m diameter general purpose scattering chamber(GPSC), that has been extensively used for fission fragment angular and mass distribution measurements. Recently a new facility consisting 100 neutron detectors coupled with fission fragment detectors has been commissioned further enhancing our experimental research programs at IUAC[7]. Over the years we have performed several experiments devoted to the systematic study of the fission dynamics in nuclei around mass 200 region. Figure 1 shows photograph of typical set up used for FF mass distribution and angular distribution studies in GPSC[8].

**Figure 1.** Picture of time of flight setup consisting two large area MWPC detectors mounted inside scattering chamber facility and array of charged particle telescope detectors ($\Delta E$ (gas) - E(Silicon)) detectors used for angular distribution measurement.

## 3. Experimental signatures of quasi-fission

### *3.1. fission Fragment Angular distribution*

The fission fragment angular distribution measurements have been used to investigate the dynamics of fission in heavy ion reactions for along time. According to the standard transition state model(TSM) of fission theory, the fragments are emitted along the nuclear symmetry axis and their properties are determined at the transitional saddle point[9]. In this formalism, the the angular anisotropy defined as the ratio of the yield at 180° or 0° to that at 90°($A = W(180°)/W(90°)$ is given by approximate expression

$$A \sim 1 + \left( \frac{\langle l^2 \rangle}{4K_0^2} \right). \tag{1}$$

with $K_0^2 = T \times \frac{I_{eff}}{\hbar^2}$ where T is the temperature, $I_{eff}$ is the effective moment of inertia at the saddle point and $\langle l^2 \rangle$ the mean square angular momentum of the fissioning system. Measurements of the fission anisotropy therefore gives the information about the effective moment of inertia of the fissioning nucleus at the saddle point of deformation.

For light charged particle induced reactions, observed anisotropy have been found to be in agreement with the TSM predictions. However it has been observed that in large number of reactions using actinide targets, angular anisotropy values larger than those predicted by TSM. These anomalously high anisotropies were attributed to the presence of pre-equilibrium fission(PEF) or orientation dependent quasi fission. In PEF fission[10], if fission barrier height is comparable to its temperature, the fission takes place before the K degree of freedom is fully equilibrated (K is the projection of angular momentum of compound nucleus on the symmetry axis). It is suggested that entrance channel mass asymmetry, $\alpha$ plays major role in the occurrence of PEF; reactions having $\alpha$ smaller than the Businaro-Gallone mass asymmetry $\alpha_{BG}$ showed larger anisotropy where as for reactions with $\alpha$ more than $\alpha_B G$, anisotropies were normal[10, 11]. Experimental evidence of entrance channel dependence was observed in several systems such as $^{11}$B + $^{237}$Np, $^{12}$C + $^{236}$U, $^{13}$C + $^{235}$U and $^{16}$O + $^{232}$Th, $^{19}$F + $^{232}$Th etc having different entrance channel mass asymmetry values [12, 13, 14, 15]. However it was found that for actinide targets, due to channel couplings and consequent shift in $\alpha_{BG}$ towards higher mass asymmetry the PEF occur irrespective of the $\alpha$ on either side of $\alpha_{BG}$[16]. It was shown that apart from mass asymmetry, the target deformation and the intrinsic spin of target or projectile can also enhance the angular anisotropy at sub barrier energies[17, 18, 19]. Hinde *et al* observed that fragment anisotropy increases below barrier energy in reaction $^{16}$O+$^{238}$U and this was interpreted due to presence of nuclear orientation dependent quasi-fission[17]. Here it was shown that below barrier energy, the projectile interaction with tips of the deformed target nucleus lead to quasi fission events whereas collisions with the sides lead to fusion-fission.

It is generally believed that quasi fission is expected in heavy systems with high fissility and charge product $Z_p Z_t \geq 1600$. However Berriman *et al.* found that even low fissility system such as $^{19}$F+$^{197}$Au showed signatures of quasi-fission[20]. They have observed broader mass distribution and suppression of evaporation residues in reactions $^{19}$F+$^{197}$Au and $^{30}$Si+$^{186}$W as compared to $^{12}$C+$^{204}$Pb, all three reactions forming same compound nucleus $^{216}Ra$. This was attributed to presence of quasi fission in former two reactions. However the measurement of angular distribution of $^{19}$F+$^{197}$Au system by Tripathi *et al* showed normal anisotropy consistent with TSM prediction implying absence of quasi-fission[21]. At IUAC also we have performed angular distribution measurements for $^{11}$B+$^{204}$Pb and $^{18}$O+$^{197}$Au forming the nearby compound nucleus $^{215}$Fr and found that both systems exhibit normal values of anisotropies consistent with the TSM predictions, implying the absence of any PEF or quasi-fission in these reactions[22]. Similarly for the reaction $^{16}$O+$^{238}$U, Nishio *et al* measured the evaporation residues cross sections for the energy range from above- to extreme sub barrier and found that the cross section data could be well reproduced by statistical model calculation implying the absence of any quasi-fission in this reaction[23]. Thus there are contrasting results and a very clear distinction of quasi fission from complete fusion-fission process becomes more challenging.

### *3.2. Fission fragment mass distribution*

Mass distribution of fission fragments is another observable by which one can study the property of fission process. In recent times, mass variance and mass-angle correlation of fragments have been used as a sensitive tool for identifying the presence of quasi fission events. Generally in heavy ion induced fission where fission is occurring from a fully equilibrated system, the mass distribution is expected to be a symmetric Gaussian with variance of the distribution given by

$$\sigma_m^2 = \lambda T + \kappa \langle l^2 \rangle. \qquad (2)$$

where T is the temperature of compound nucleus(CN) at saddle, $\langle l^2 \rangle$ mean square angular momentum and $\lambda$ and $\kappa$ are constants. It shows that the mass width increases linearly with temperature the compound nucleus. In the case of quasi fission process, the mass distribution becomes more asymmetric and exhibit strong mass-angle correlation[24, 25]. It was observed that for a CN formed through different entrance channel reactions, even though no mass angle correlation is observed, the reaction dominated by quasi fission events exhibited large mass width compared to the pure fusion-fission reaction[24, 26, 27]. For example, the fission fragment mass distribution from two reactions $^{16}$O+$^{194}$Pt and $^{24}$Mg+$^{186}$W forming the same CN $^{210}$Rn was measured at IUAC facility[28]. Though neither of the two reactions showed any noticeable mass-angle correlations, the reaction $^{24}$Mg+$^{186}$W showed significantly different mass distribution with larger mass width as compared to reaction $^{16}$O+$^{194}$Pt where pure compound nucleus fission is expected. The broad mass distribution in reaction $^{24}$Mg+$^{186}$W was attributed to the presence of quasi-fission in the reaction. For this system, though the $Z_pZ_t$ and fissility values are less than the classical limit, it shows a clear indication of onset of quasi fission in low fissile systems.

Even though large anisotropy and wider mass distribution in low fissile system may suggest the presence of quasi-fission, the unambiguous identification of quasi-fission still remains a challenging problem. For example, the mass distribution of fragments in reaction $^{16}$O + $^{238}$U showed large deviation at below barrier energies[29]. Here the target is deformed and the orientation dependent quasi -fission has been attributed to be the reason for enhanced mass variance at low excitation energies. It is to be noted that, for same reaction, large angular anisotropies were also observed around barrier energies[17]. However, for the same reaction $^{16}$O + $^{238}$U, measurement of ER cross section at barrier and deep sub-barrier energies did not show any evidence of quasi-fission[23]. Similar anomalous increase in width of mass distribution at below barrier energy was also observed in systems with the deformed target $^{232}$Th[30]. It should be noted that, in the case of actinide nuclei, the increase in width of mass distribution $\sigma_M^2$ at low energy could also be possible due to the presence of asymmetric fission manifested through the shell effects. For example, fragment mass distribution in reaction $^{12}$C + $^{235}$U showed increasing yield of mass-asymmetric components at low excitation energies[31]. Here the presence of asymmetric mass division was attributed to the role of shell effects. Though the reaction involved deformed target, no significant amount of quasi-fission was observed in this reaction. Hence, it is important to identify the contribution of asymmetric fission events in the mass distribution measurement using deformed targets.

## 4. FF mass distribution from reactions $^{19}$F + $^{206,208}$Pb

Recently we have performed the measurements of mass distribution of FF from reactions $^{19}$F + $^{206,208}$Pb forming $^{225,227}$Pa compound nuclei over a range of excitation energies[32]. Since both $^{206,208}$Pb targets are spherical, the orientation dependent quasi fission effects are not expected to play any role here. These reactions have low $Z_pZ_T$, low fissility ( see Table 1) and it is expected to follow normal fusion-fission path. The fission fragment angular distribution data for $^{19}$F + $^{208}$Pb system did not show any anomalously large anisotropy implying absence of any non-compound process in the reaction[33]. In the absence of any non-compound process, the mass distribution measurement from these systems are expected to follow the results of standard normal fusion-fission dynamics only.

The experiment was carried out using pulsed $^{19}$F beams from the 15UD Pelletron accelerator at different beam energies. Pulsed beam had 250 $ns$ repetition rate and $\sim$ 1.5 $ns$ time width. Beam was bombarded on isotopically enriched $^{206,208}$Pb targets of $\sim$ 110 $\mu$g/cm$^2$ thickness deposited on $\sim$ 20 $\mu$g/cm$^2$ thick $^{12}$C backings forming CN $^{225,227}$Pa. Beam energy steps were chosen to form compound nucleus(CN) with similar excitation energies. For the detection of complimentary fission fragments, two large area($20cm \times 10cm$) position sensitive multi-wire proportional counters (MWPCs) were used [6]. They were mounted on two arms of the GPSC and placed at folding angles. The detectors were operated with iso-butane gas typically at about two Torr pressure.

**Table 1.** Entrance channel reaction parameters for reactions populating $^{225,227}$Pa nuclei. Columns shows the reaction, compound nucleus (CN), $Z_PZ_T$, $\alpha_{bg}$, $\alpha(=\frac{A_T-A_P}{A_T+A_P})$, fissility ($\chi$), and the neutron number N of CN.

| Reaction | CN | $Z_PZ_T$ | $\alpha_{bg}$ | $\alpha$ | $\chi$ | N |
|---|---|---|---|---|---|---|
| $^{19}$F + $^{206}$Pb | $^{225}$Pa | 738 | 0.87 | 0.831 | 0.773 | 134 |
| $^{19}$F + $^{208}$Pb | $^{227}$Pa | 738 | 0.88 | 0.832 | 0.786 | 136 |

### *4.1. Experimental results*

The TOF and energy loss signals from MWPC detectors were used to discriminate fission events from elastic recoil events. Fig. 2a displays the kinematic coincidence plot showing correlation between the time signals in two MWPC detectors. Clean separation of fission events can be seen in the plot. The data analysis was performed by velocity reconstruction of events based on method given by [34]. From calibrated position and timing information, the fragment velocities and center-of-mass angles of were reconstructed for each event. The velocity vector components (parallel ($V_{\|}$) and perpendicular ($V_{\perp}$) to beam direction) of the fissioning nucleus were also determined for each event. Complete fusion events were selected by applying the condition for full momentum transfer (FMT) using the correlation of velocity components. This eliminates any possible contamination from transfer induced fission events. From the ratio of the velocities in the center-of-mass frame, the mass ratios defined as $M_R = \frac{m_1}{m_1+m_2}$ ($m_1$ and $m_2$ are two fragment masses) were determined. Fig. 2b displays the spectra showing the correlation between mass ratio and $V_{\|}$ -$V_{c.m.}$ (where $V_{c.m.}$ is the center-of-mass velocity) for fission events from the reaction $^{19}$F + $^{208}$Pb at below barrier energy. The intense region shown as black rectangle in the plot, corresponds to the events originated from FMT fission and a software gate around these events was used in the analysis of mass angle correlation and mass ratio distribution.

*4.1.1. Mass angle correlation* The mass-angle correlation is the distribution of mass ratio as a function of emission angle in center of mass frame ($\theta_{c.m.}$). This correlation is a direct evidence for presence of quasi fission and provides information on its time scale[24]. For fusion-fission reaction, the mass ratio yield is independent of the emission angle of fragments whereas for fission coming from non-compound processes the yield depends on fragment angle $\theta_{c.m.}$. The

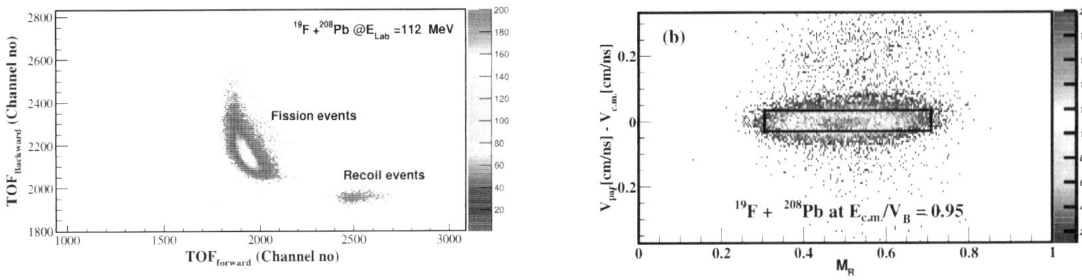


**Figure 2.** (a) Time correlation between two MWPCs showing fission events separated from recoils and (b) mass ratio distribution plotted against velocity components at below barrier energy for the reaction $^{19}F$+$^{208}Pb$. The events within rectangular box corresponds to FMT fission.

mass-angle distribution were generated for both reactions and figure 3 shows selected part of the mass-angle-correlation observed in $^{19}$F + $^{208}$Pb at below and above barrier energies. As one can see from Fig.3, the mass ratio yield is independent of emission angle $\theta_{c.m.}$ within the range from $\approx$ 110 ° to 150 °. Thus the absence of any mass-angle correlation in this case demonstrate that there is no significant presence of any non-compound process in this reaction around the barrier energies.

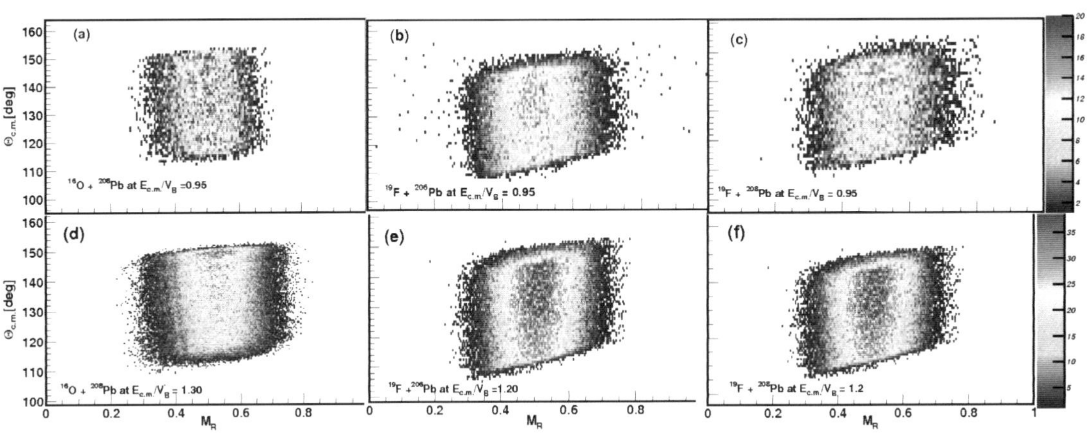


**Figure 3.** Mass angle correlation of fission fragments from reaction $^{19}F$+$^{208}$Pb at various excitation energies

*4.2. Variation of mass variance with $E_{c.m.}/V_B$*

In the present work, the mass distribution measurements have been performed at energy range of $0.93 \leq \frac{E_{c.m}}{V_B} \leq 1.29$. The mass distributions of FMT events measured in the reaction $^{19}$F+$^{206}$Pb$\longrightarrow$$^{225}$Pa$^*$ for different excitation energies are given in Figure 4. In all cases, the measured mass distribution appears to be symmetric centred around mass $M \sim \frac{A_{CN}}{2}$. The shape of the mass distribution could be described by fitting single Gaussian except at lowest excitation energy where the distribution appears to be non-symmetric. To understand how the distribution varies with energy, the standard deviation $\sigma_M$ of the mass distribution was extracted at each excitation energy after making best Gaussian fit to the data. The variance of the mass distribution $\sigma_M^2$ as a function of energy(reduced barrier energy) is shown in the right panel of figure 4. Above the barrier energy, the widths of the FF mass distributions show a gradual increase over the energy. This is expected distribution from equilibrated system due to the influence of the increasing nuclear temperature. However at beam energies below the fusion barrier, the width increases again strongly with decreasing energy.

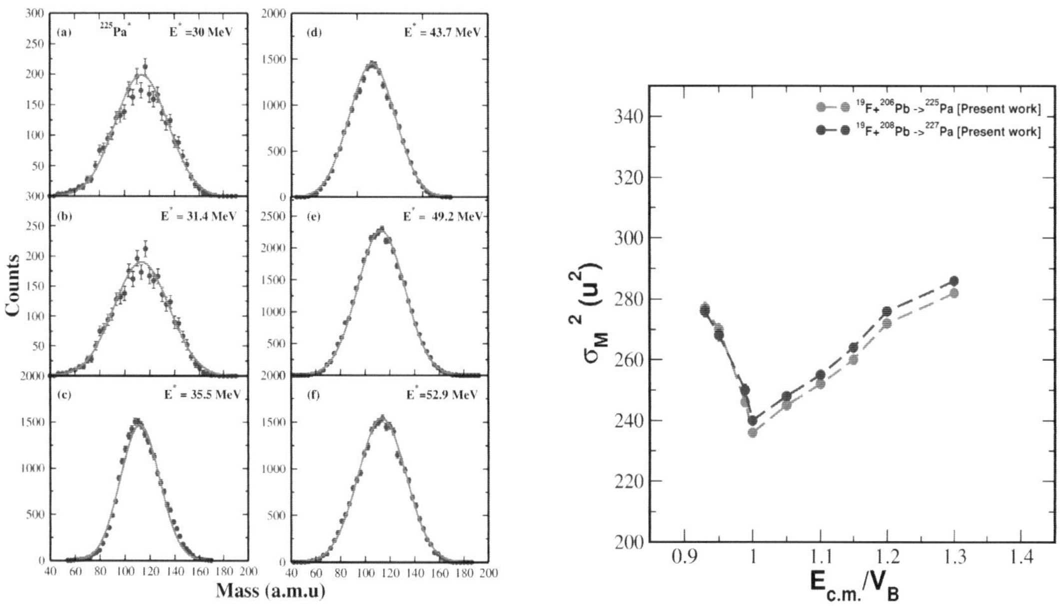


**Figure 4.** Mass distribution from reaction $^{19}$F+ $^{206}$Pb at various excitation energies(left panel) and the mass variance $\sigma_M^2$ as a function of reduced Coulomb barrier $E_{c.m.}/V_B$ (right panel)

## 5. Discussion

The sudden increase in mass variance below barrier energy was earlier observed in fission induced by $^{16}$O and $^{19}$F projectile on deformed targets such as $^{232}$Th and $^{238}$U[30, 29]. This was attributed to orientation dependent quasi fission process dominating at below barrier energies. The presence of quasi fission increases the width of mass distribution. However, fission induced by similar projectiles on spherical targets such as $^{208}$Pb or $^{209}$B did not show any such evidence of quasi fission at below barrier energies. For the reaction $^{19}$F + $^{208}$Pb, experimental angular anisotropy as well as the theoretical calculation suggested that the reaction follows mainly fusion-fission path at near and sub-barrier energies[33, 35]. Comparatively low $Z_P Z_T$ value, low fissility and absence of any mass-angle correlation also support that reaction is mainly fusion-fission only.

Considering quasi fission events are not dominant in these reactions, the observed enhancement in mass variance at low excitation energies signify some different characteristics of low energy fission. It is interesting to note that, an increase in mass variance at lower excitation energies have already been reported in earlier works on heavy ion induced fission in this region of the chart of nuclides where multi-mode fission was observed[36, 37, 38]. For example, mass distribution measurement from fission of $^{228}$U in the reaction $^{19}$F+$^{209}$Bi showed rapid rise of $\sigma_M^2$ near below the barrier energy and it was attributed to the possible contribution from asymmetric fission mode at low excitation energy[36]. Similarly, multi-mode fission were observed in induced fission of $^{224,226}$Th nuclei produced in the reaction $^{16,18}$O + $^{208}$Pb[37, 38]. In fact four fission

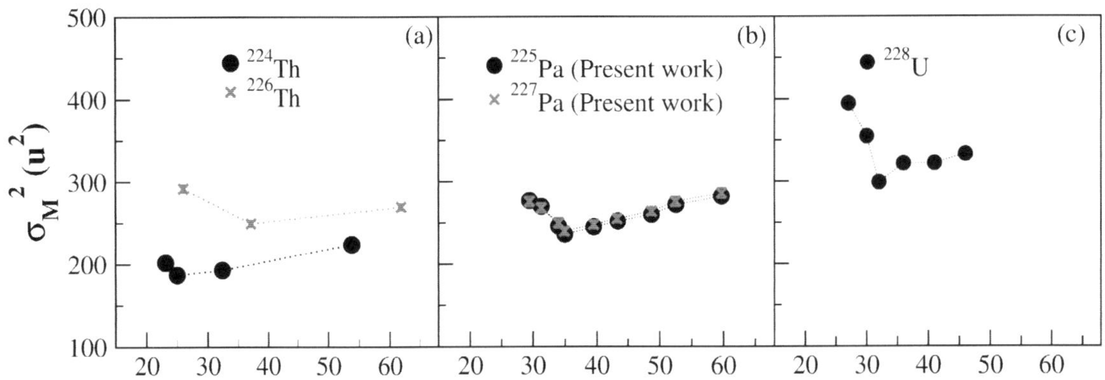


**Figure 5.** Mass variance as a function of CN excitation energy for present system[32] as well as nearby systems from Ref[36, 37, 38]

modes were observed in $^{226}$Th at 26 MeV of excitation energy[38]. In figure 5 the variation of $\sigma_M^2$ as a function of excitation energy for these nuclei are shown along with present data. It is to be noted that, all these reactions were studied at near barrier energies and they show similar characteristics of fragment mass distributions with respect to excitation energy. The $\sigma_M^2$ values decreases monotonically as energy is lowered, but rises again at lower excitation energies. The systematic analysis of present data shows that the increase in $\sigma_M^2$ values at low excitation energies indicates the presence of multi mode fission in present system also[32]. At low excitation energy, the shell effects come into play and superposition of two fission modes, one following the normal symmetric fission and the other following the asymmetric fission broadens the mass distributions. With excitation energy increasing the asymmetric mode disappears and the symmetric mode become dominant. But at low excitation energy the contribution of asymmetric mode will come into play leading too enlarged widths of mass distributions.

Multi-mode fission has been experimentally verified in low energy fission of light actinide nuclei[39, 40, 41, 42, 43]. Measurement by Schmidt *et al.* suggested that in this region of nuclei, a transition from the symmetric to the asymmetric fission mode takes place around mass A≈ 226[44]. Theoretically, the origin of multi-mode fission has been attributed to fragment shell structures affecting the potential energy surface (PES) of the deformed fissioning nucleus[45, 46, 47]. Multi dimensional potential energy calculations applying standard Strutinsky shell correction method [48] have been successfully used in explaining existence of different modes that corresponds to distinct valleys in the potential energy landscape of the deformed system [49, 50, 51, 52]. For the present system also, a theoretical investigation has been made by calculating the adiabatic potential energy in four dimensional deformation space using NRV code[53]. The code uses liquid-drop model for calculating macroscopic energy and two-center shell model(TCSM) for applying microscopic shell correction[53, 54]. The results of the calculation is shown in Fig. 6.

Figure 6 (left panel) shows the adiabatic potential energy ($V_{Adb}$) of deformed $^{225}$Pa nucleus plotted as a function of mass asymmetry parameter $\eta$ ($\eta = \frac{m_2-m_1}{m_2+m_1}$) and elongation R. Clearly one can see multiple valleys in the potential energy surface of the system. The global minimum valley corresponds to mass asymmetry $\eta \sim 0.17$ favouring mass asymmetric fission with heavy fragment mass $\sim$ 132 and light fragment mass $\sim$ 93. The calculation indicates the potential energy landscape being influenced by the shell closure in the heavy fragments due to its proximity to doubly magic $^{132}$Sn (Z=50, N=82). Similar results have also been observed in multi dimensional PES calculation by other authors [55, 38]. Experimentally measured mass and charge distribution of fission fragment from $^{225}$Pa in reaction $^{16}O$ + $^{209}$ Bi also

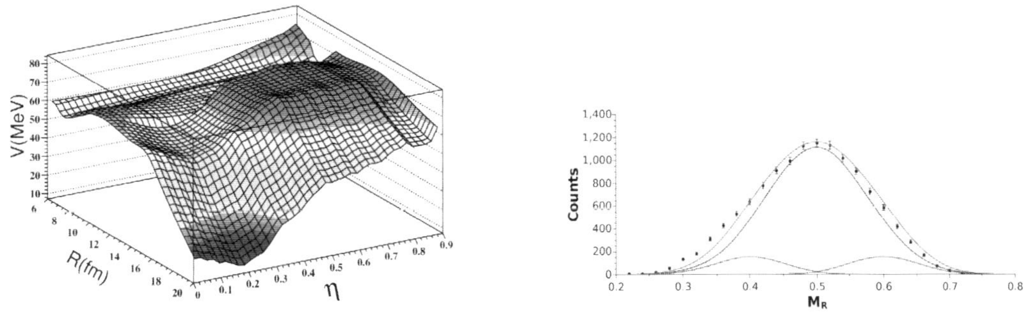


**Figure 6.** Calculated potential energy surface of fissioning nucleus $^{225}$Pa as a function of mass asymmetry $\eta$ and elongation R. On right side is the measured mass ratio distribution from reaction $^{19}F$ + $^{206}$Pb at 31MeV excitation energy best described by superposition of three peaks[57].

showed presence of asymmetric fission components with relative yield about 10% of total fission yield[56]. Figure 6(right panel) shows the mass ratio distribution of $^{225}$Pa measured at the lowest excitation energy fitted by three Gaussian distributions. The experimental data closely matches to the description of superposition of three distributions, one centred around symmetric mass ($M_R \sim 0.5$) and others centred around light fragment ($M_R \sim 0.41$) and heavy fragments ($M_R \sim 0.59$)respectively[57].

## 6. Summary & Conclusion

Using the TOF setup at IUAC, fission fragment mass distribution have been measured for fission induced by $^{19}$F projectile on spherical target $^{206,208}$Pb at different excitation energies. No mass-angle correlation have been observed in the reaction. Experimental data on angular anisotropy as well as theoretical prediction suggest that the reaction is dominated by complete fusion-fission events. However, increase in widths of mass distribution have been observed in low excitation energy data. Contribution from QF is found to be negligibly small for present reaction. Systematic analysis of mass variance data in nearby actinide nuclei suggest that multi-mode fission manifested at low energy could broaden the mass distribution due to presence of asymmetric fission modes. At lower excitation energies nuclear shell effects comes into effect and asymmetric fission mode becomes dominant. As the excitation energy is increased, shell effects wash out and a transition from multi mode fission to symmetric fission is observed. Superposition of symmetric and asymmetric modes can increase the width of fragment mass distribution at low excitation energy. In conclusion, it appears that fission induced by projectiles with mass $A_p < 20$ on spherical targets and $Z_P Z_T$ ($< 800$) are dominated by fusion-fission events only, but the mass distribution at low energy could possibly contain contribution from multi-mode fission events also. Theoretical calculation of multi-dimensional potential energy surface also shows shell structure influencing the multi-modal nature of fission in actinide nuclei.

## Acknowledgments

Author thank group members and all collaborators for their help and inputs. The author acknowledge R. Dubey for his help in preparing this manuscript.

## References

[1] O Hahn and F Strassmann, Naturwiss. 27,11(1939).

[2] N. Bohr and J. A. Wheeler, Phys. Rev. 56,426 (1939).
[3] Yu. Ts. Oganessian and Yuri A Lazarev, Treatise on heavy ion scie nce, Volume 4, page 3 (1985).
[4] Birger B. Back, Journal of Physics: Conference Series 282, 012003 (2011).
[5] G.K. Mehta and A.P. Patro, Nucl. Instr. and Meth. A 268, 334 (1988).
[6] A. Jhingan *et al*, Rev. Sci. Instrum. 80, 123502(2009).
[7] P. Sugathan *et al*, Pramana Journal of Physics 83, 807 (2014).
[8] Akhil Jhingan, Pramana Journal of Physics 85, 483 (2015).
[9] R.Vandenbosch and J.R. Huizenga, Nuclear Fission( academic New York,1973).
[10] V.S. Ramamurthy and S.S. Kapoor, Phys. Rev. Lett. 54 , 178 (1985).
[11] Z. Liu *et al*, Phy. Lett. B 353,173(1995).
[12] V.S. Ramamurthy *et al*, Phys. Rev. let 65, 25(1990).
[13] J. P. Lestone *et al*, J. Phys. G 23, 1349 (1997).
[14] J. C Mein *et al*, Phys. Rev. C 55, R995 (1997).
[15] B.P. Ajithkumar et al, Phys. Rev. C 77, 021601(R)(2008).
[16] R.G. Thomas, R.K. Choudhary, A.K.Mohanty, A. Saxena and S.S. Kapoor, Phys. Rev.C 67, 041601(R)(2003).
[17] D. J. Hinde *et al*, Phys. Rev. Lett.74, 1295 (1995).
[18] J. P. Lestone, A. A. Sonzogni,M. P. Kelly, and R. Vandenbosch, Phys. Rev. C 56, R2907 (1997).
[19] B. K. Nayak *et al*, Phys. Rev. C 62, 031601(R) (2000).
[20] A. C. Berriman *et al*, Nature (London) 413, 144 (2001).
[21] R. Tripathi *et al*, Phys. Rev. C 71, 044616 (2005).
[22] S. Appannababu *et al*, Phy. Rev. C 80, 024603(2009).
[23] K. Nishio *et al*, Phys. Rev. let 93, 162701(2004).
[24] J. Toke *et al*, Nucl.Phys. A 440,327(1985).
[25] B.B. Back *et al*, Phys.Rev. C 41,1495(1990).
[26] W.Q. Shen *et al*, Phys.Rev. C 36,115(1987).
[27] R.Rafiei *et al*, Phys.Rev. C 77,024606 (2008).
[28] E. Prasad *et al*, Phys.Rev. C 81,054608(1990).
[29] K. Banerjee *et al.*, Phys. Rev. C 83, 024605 (2011).
[30] T.K.Ghosh *et al.*, Physics Letters B 627, 26 (2005).
[31] J. Khuyagbaatar *et al.*, Phys. Rev. C 91, 054608 (2015).
[32] R. Dubey *et al.*, Physics Letters B 752, 338(2016).
[33] D. J. Hinde *et al.*, Nuclear Physics A 685, 72 (2001).
[34] D. J. Hinde *et al.*, Phys. Rev. C 53, 1290 (1996).
[35] A. Nasirov *et al.*, The European Physical Journal A-Hadrons and Nuclei 34, 325 (2007).
[36] L. M. Pant, R. K. Choudhury, A. Saxena, and D. C. Biswas, Eur. Phys. J. A 11, 47 (2001).
[37] M.G. Itkis *et al.*, Proceedings of the Low Energy Nuclear Dynamics Conference, St. Petersburg,Russia , 177 (1995).
[38] A. Rusanov *et al.*, Physics of Atomic Nuclei 71, 956 (2008).
[39] J. P. Balagna *et al.*, Phys. Rev. Lett. 26, 145 (1971).
[40] D. C. Hoffman and M. R. Lane, Radiochim. Acta 70, 135 (1995).
[41] E. K. Hulet *et al.*, Phys. Rev. Lett. 56, 313 (1986).
[42] T. Ohtsuki *et al.*, Phys. Rev. Lett. 66, 17 (1991).
[43] Y. Nagame *et al.*, Physics Letters B 387, 26 (1996).
[44] K. H. Schmidt *et al.*, Nucl. Phys. A 665, 221 (2000).
[45] A. Turkevich and J. B. Niday, Phys. Rev. 84, 52 (1951).
[46] V.V. Pashkevich, Nucl. Phys. A 169, 275 (1971).
[47] U. Brosa, S. Grossmann, A. Müller, Phys. Rep. 197, 167u (1990).
[48] V. M. Strutinsky, Nucl. Phys. A 95, 420 (1967).
[49] P. Möller and J. R. Nix, Nucl. Phys. A229, 269 (1974).
[50] B.D. Wilkins, E.P. Steinberg, R.R. Chasman, Phys. Rev. C 14,1832 (1976).
[51] M.G. Mustafa, U. Mosel, H.W. Schmitt, Phys. Rev. C 7,1519 (1973).
[52] V. Pokrovsky, M. G. Itkis, J. M. Itkis, *et al.*, Phys.Rev. C 62, 014615 (2000).
[53] V.I. Zagrebaev *et al.*, Physics of Particles and Nuclei 38,469 (2007); NRV codes for driving potential,http://nrv.jinr.ru/nrv/.
[54] J. Maruhn, W. Greiner, Z. Phys. A 251,431 (1972).
[55] P. Möller, D. G. Madland, A. J. Sierk, and A. Iwamoto, Nature(London) 409,785 (2001).
[56] I. Nishinaka *et al.*, Phys. Rev. C 56,891 (1997).
[57] R. Dubey *et al.*, Acta Physica Polonica B,47,953(2016)

# Electron-phonon coupling in perovskites studied by Raman Scattering

**V G Sathe**[1], **S Tyagi and G Sharma**

UGC-DAE Consortium for Scientific Research
University Campus, Khandwa Road, Indore, M.P. INDIA

E-mail: vasant@csr.res.in

**Abstract.** Raman scattering is an unique technique for characterization and quantification of electron-phonon, spin-phonon and spin-lattice coupling in many of the currently prominent compounds like multiferroics and manganites. In manganites, it is understood now that a phase separated landscape with coexisting metallic and insulating regions exist in most of the compounds and application of small external perturbation causes an alteration in this landscape. In such scenario, local metallic regions grow suddenly at the expense of insulating regions below the magnetic ordering temperature. Such regions can be characterized effectively using Raman scattering measurements where delocalized electrons couple with the adjacent phonon peaks giving a Fano resonance in the form of asymmetric line shape.

## 1. Introduction

Perovskites, currently are the hot topic of research as they show very fascinating properties like, superconductivity, colossal magneto-resistance, colossal dielectric properties, multiferroic behaviour and thus results in a wide variety of applications. Raman spectroscopy is an ideal tool to detect electron-phonon coupling, spin-phonon coupling and spin-lattice coupling and therefore has been extensively used in detecting above phenomena in perovskites. In this paper, examples of electron-phonon coupling detected by Raman spectroscopy in metallic $LaMnO_{3+\delta}$ films is described.

Manganites show very strong coupling between the charge carriers, the localized spin moments of the manganese ions and the lattice structure. Such a coupling is manifested in the structural modification on application of magnetic field [1]. The main cause of such a structural modification is attributed to the changes in tolerance factor of the perovskite structure [2]. The tolerance factor gives a measure of Mn-O-Mn bond angle deviation from 180°. For example, the deviation is larger in orthorhombic structure leading to charge localization than compared to rhombohedral lattice where charges are de-localized. In such a way, the lattice degree of freedom is coupled with the kinetic energy of the conduction $e_g$ electrons and holes. To understand the mechanisms of the colossal magnetoresistance, many attempts were made taking into account the strong interplay of magnetism, electron-lattice coupling, and orbital and charge ordering in these materials. It was shown earlier that double exchange alone cannot explain the large magneto-resistance in these materials [3]. Present understanding about the phenomena revolves around a phase separation scenario, [4, 5, 6, 7] the collapse of a phase coexisting and competing with other phases, into an energetically more favourable state that is triggered by either temperature or an external stimuli like magnetic field, electric field, etc. [8]. The insulator-metal transition induced by small perturbations like magnetic field [9], x-rays [10], visible photons [11, 12], current [13] and even acoustic pulses [14] signifies that the system is electronically phase separated. A large number of mechanisms of matching energies are simultaneously active in the system, involving strong correlations among lattice and electronic degrees of freedom.

$LaMnO_{3\pm\delta}$ compounds show range of oxygen nonstoichiometry [15]. The oxygen excess '$\delta$', produces La and Mn vacancies, because of the trouble in accommodating excess oxygen atoms interstitially in the closed packed perovskite structure [16]. The excess oxygen produces $Mn^{4+}$ ions along with $Mn^{3+}$ ions in the system. It is known that the presence of $Mn^{4+}$ is accompanied with the decrease in degree of distortion of oxygen octahedra leading to structural alteration [16]. Also $Mn^{4+}$ transforms the system to ferromagnetic-metallic state. Therefore, the Mn-vacancies are likely to be more efficient for the electronic delocalization than the substitution on the La site. The Mn-vacancies leads to a randomization of the magnetic interaction and this inhomogeneous magnetic state is related to the local structural distortion. It is shown earlier that the local lattice distortion is highly coupled with the magnetic order and a less distorted pseudo cubic structure supports ferromagnetic ordering and metal-like conductivity [17]. A direct correlation between Raman mode intensities induced due to Jahn-Teller distortion and metallic-like conductivity is observed in $LaMnO_{3+\delta}$ films [18]. Therefore in these compounds, there should be a possibility of co-existing continuum scattering due to metallic regions along with phononic scattering. In many cases, these electronic and phononic scattering processes interacts strongly leading to Fano line shape. "Fano resonance" effect is reported previously in many compounds due to onset of metallicity in the system [19, 20, 21].

In this report we show strong electron-phonon coupling in conducting $LaMnO_{3+\delta}$ films reflected in the form of Fano asymmetric line shape in one of the Raman modes.

## 2. Experimental:

The parent $LaMnO_3$ (LMO) bulk compound was prepared by using standard solid-state reaction method by taking the stoichiometric amount of precursors having 99.99 % from Aldrich. $LaMnO_{3+\delta}$ films were deposited epitaxially on $LaAlO_3$ (LAO) (001) substrate using pulsed laser deposition technique equipped with an excimer laser $\lambda$ = 248 nm. During the preparation, the substrate temperature was kept at ~650°C. All the films were prepared in 250 mTorr oxygen pressure. After deposition the films were annealed for 30 minutes in-situ at ~650°C in different oxygen environments 150 mTorr, 450 mTorr, 1000 mTorr and 1 atmosphere (atm) respectively and then the heater was slowly cooled down to room temperature. The thickness of all the films was kept at ~200 nm. After deposition the films were characterized by X-Ray diffraction (XRD) technique using a Rigaku make powder diffractometer fitted on a rotating anode (18 KW) X-Ray generator with Cu target. Resistivity of the films was measured using standard four-probe method. Raman spectra were collected in backscattering geometry using a $Ar^+$ excitation source ($\lambda$=488 nm) coupled with a Labram-HR800 micro Raman spectrometer equipped with a ×50 objective, appropriate notch filter and peltier cooled charge coupled device detector. For the low temperature Raman measurements, the sample was mounted on a THMS600 stage from Linkam UK, with temperature stability of ±0.1 K.

## 3. Results and discussions:

The x-ray diffraction of the films are shown in figure 1. The pattern around (002) pseudo cubic unit cell is only shown for better comparison. The peaks showed a regular trend, the 2$\theta$ value shows an increase with increasing oxygen content ($\delta$) in the film. Larger 2$\theta$ value means shorter out of plane lattice parameter which in turn signifies larger in plane lattice parameter. This indicates that the film relaxes with increasing oxygen content. Generally, relaxation process leads to *strain disorder* [4,5] which supports phase separation.

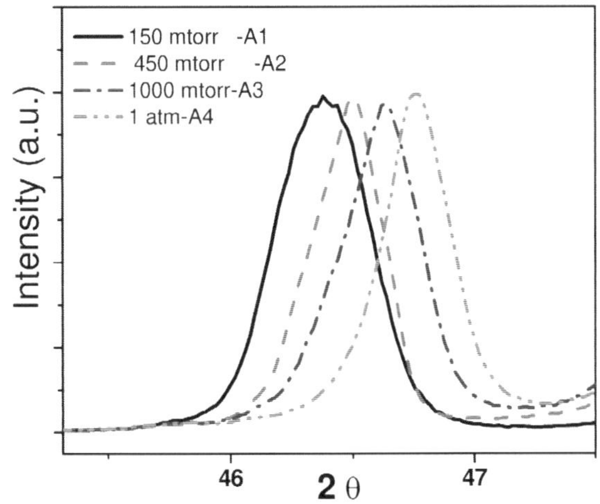


Figure 1: X-ray diffraction of films collected with Cu Kα radiation. The diffraction pattern is normalized and only region around (002) (pseudo-cubic) peak of LaMnO3 is shown.

The resistivity of all the film as a function of temperature is shown in figure 2. It can be seen that film A1 and A2 are insulating with small kink showing a hint to the presence of insulator to metal transition at 186 K. On the other hand film A3 and A4 show a clear insulator to metal transition around 206 K and 270 K respectively. From the resistivity data it can be concluded that all the films are phase separated with predominant insulating phase fraction in the films A1 and A2, dominant metallic fraction in the film A3 and mostly metallic fraction in the film A4.
Figure 3 shows the doping dependence of Raman spectra of $LaMnO_{3+\delta}$ films with increasing $\delta$ (A1 – A4) in parallel (XX) polarization at 280 K and 240 K respectively.

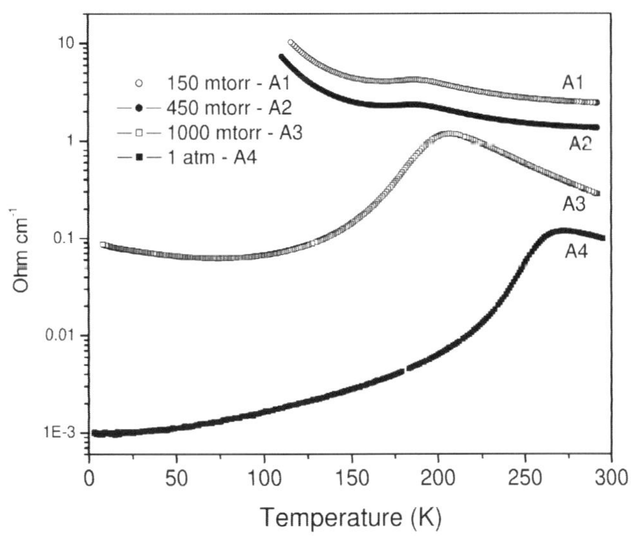


Figure 2: Resistivity as a function of temperature for the $LaMnO_{3+\delta}$ films with different oxygen doping.

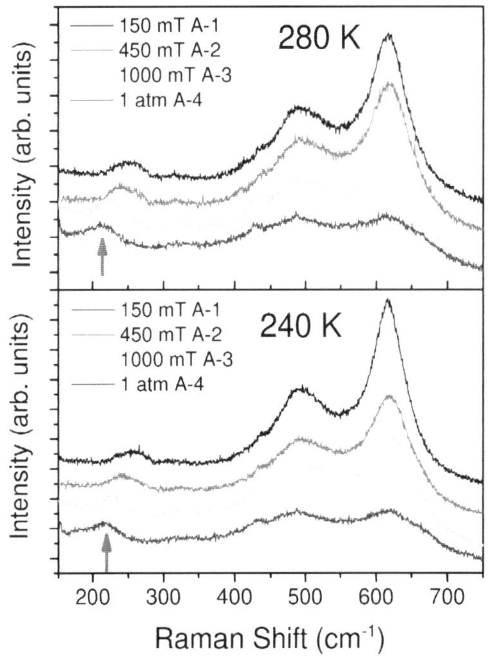


Figure 3: Comparison of Raman spectra for the films A-1, A-2, A-3, and A-4 collected at 280 K and at 240 k. The spectra are shifted vertically for clarity. The arrow marks the peak showing asymmetric line shape.

The Raman peaks around 490 $cm^{-1}$ and 620 $cm^{-1}$ shows behaviour typical of a metallic and insulating phases and are discussed in ref. [18] and will not be a part of discussion in this paper. The low frequency peaks between 200 and 300 $cm^{-1}$ shows an interesting trend and that will be the focus of this report. The Raman spectra in this region for insulating samples i.e. the samples A-1 and A-2 are significantly different than for the metallic samples A-3 and A-4. The peak in insulating sample is broad, probably resulting from a combination of two peaks. It falls at 250 $cm^{-1}$ and 270 $cm^{-1}$ that is typical of an orthorhombic structure. In A-3, the peaks shifts to lower frequency around 225 $cm^{-1}$ and is sharper when compared with the samples A-1 and A-2. This peak further shifts to 215 $cm^{-1}$ for the mostly metallic sample A-4. Such a peak shift clearly suggests that all the samples are phase separated with mixture of orthorhombic and rhombohedral structures. The red shift and sharpening of the peak in Raman spectra of mostly metallic (A-4) film also suggests that the oxygen doping is reducing the tilt angle of the $MnO_6$ octahedra along y-axis, making it rhombhohedral [22]. According to ref. [22] this mode in the metallic sample is assigned to in-phase rotational vibration of the $MnO_6$ octahedra about the y-axis with $A_g$ character while the same mode in the insulating sample is assigned to out-of-plane rotational of the $MnO_6$ octahedra around the x-axis. This shows that the tilt angle of the $MnO_6$ octahedra decreases significantly with increasing oxygen doping.

The most important result which is the focus of this report is the asymmetric peak shape of the $A_g$ mode (~215 $cm^{-1}$) observed only for dominant metallic (A-3 and A-4) samples. The Raman spectra collected at 160 K, 120 K, and 80 K for the four films is compared in figure 4. It can be noticed that the asymmetry in $A_g$ mode is absent for the samples A-1 and A-2 while it is more for A-4 compared to A-3. The Raman intensity falls significantly after this Raman mode and the electronic background remains suppressed after this peak. This clearly indicates the presence of strong interaction between electronic background and the phonon modes. Observation of this phenomenon only in the samples with metallic character further strengthens this argument.

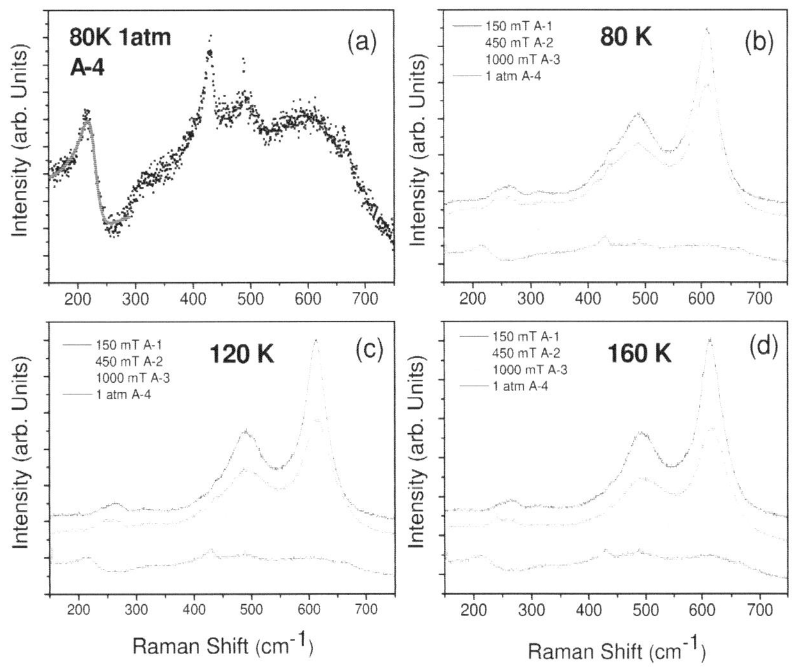


Figure 4: (a) The Fano line shape fit to the assymetric peak ~215 cm-1 of sample A-4 collected at 80K. The dots represent experimental points while the line represents fit to the data using Fano line shape. Comparison of Raman spectra for the films A-1, A-2, A-3, and A-4 collected at 80 K (b), at 120 k (c), and at 160 k (d). The spectra are shifted vertically for clarity.

The peak shape has been fitted with Fano function:

$$I(\omega) = I_C \frac{|q+\varepsilon|^2}{1+\varepsilon^2} + I_b(\omega),$$

and is shown in figure 4 (a). Here q is the asymmetry parameter, $\varepsilon = (\omega - \omega_p)/\Gamma$, $\omega_p$ is renormalized phonon frequency, and $\Gamma$ is the line width. As the peak becomes more asymmetric with decreasing temperature, 'q' decreases. The inverse of the asymmetry parameter (1/q) is a measure of metallicity in the system [23]. It is observed that the Fano parameter increases as the temperature is decreased for both A-3 and A-4. When compared, the asymmetry parameter is higher for A-4 from that for A-3 at all the temperatures. The asymmetry parameter (q) deduced from the Fano profile fit of $A_g(1)$ mode as a function of temperature is plotted for both the films in Figure 5 (a,b), here the solid line is guide to the eye.

A decrease in the parameter $q$ with increasing temperature is observed for both the compounds signifying enhancement of metallicity in this phase separated system with decreasing temperature. Such an increment of asymmetry in Fano line shape of Raman mode was reported in Mott-Hubbard insulating titanates where asymmetry evolves from insulating $GdTiO_3$ and become maximum for nearly metallic $LaTiO_3$ [24]. In the present system, the asymmetry parameter $q$ shows a sudden change below ferromagnetic transition temperature (see Figure 5) establishing the connection between electron-phonon coupling and the spin degree of freedom.

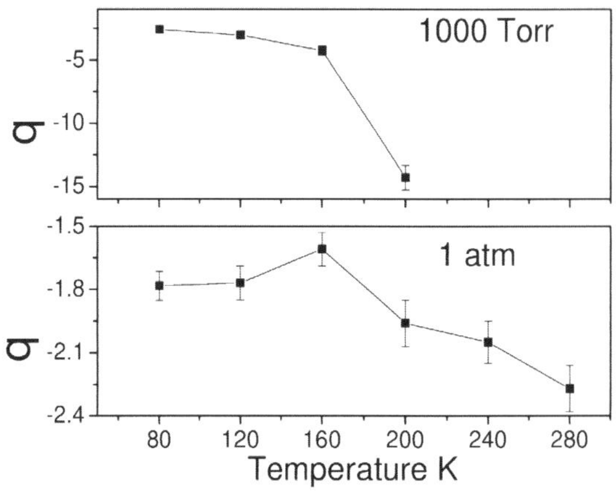


Figure 5: The Fano resonance parameter "*q*" obtained from fitting the Ag mode in films A-3 and A-4 at various temperatures.

## 4. Summary:

In summary; Raman spectroscopic studies as a function of temperature have been carried out on $LaMnO_{3+\delta}$ films. Evidence of Fano resonance certifying coupling between discrete phonon modes and electronically originated continuum has been observed signifying the presence of strong electron-phonon coupling. The existence of Fano line shape in only metallic films establishes phase separated metallic and insulating regions as predicted due to charge and orbital order/disorder transition in this system. Temperature dependence of asymmetry parameter derived from Fano profile reaffirms the development of the metallic region with ferromagnetic order in the metallic films.

## References:


[1] Asamitau A, Moritomo Y, Tomioka Y, Arima T and Tokura Y 1995 *Nature* **373**, 407

[2] Rini M, Tobey R, Dean N, Itatani J, Tomioka Y, Tokura Y, Schoenlein R W & Cavalleri A 2007 *Nature* **449** 72

[3] Millis A J, Littlewood P B and Shraiman B I 1995 *Phys. Rev. Lett.* **74**, 5144

[4] Mishra D K, Sathe V G, Rawat R, Ganesan V, Kumar R, Sharma T K 2015 *Appl. Phys. Lett.* **106**, 072401

[5] Mishra D K, Sathe V G, Rawat R and Ganesan V 2013 *J. Phys.: Condens. Matter* **25** 175003

[6] Rawat R, Kushwaha P, Mishra D K, and Sathe V G 2013 *Phys. Rev. B* **87** 064412

[7] Mathur N and Littlewood P 2003 *Phys. Today* **56** 25

[8] Dagotto E Science 2005 **309** 257

[9] Campbell A J, Balakrishnan G, Lees M R and McKPaul D and McIntyre G J 1997 *Phys. Rev. B* **55** R8622

[10] Kiryukhin V, Casa D, Hill J H, Keimer B, Vigliante A, Tomioka Y & Tokura Y, 1997 *Nature* **386** 813

[11] Fiebig M, Miyano K, Tomioka Y and Tokura Y 1998 *Science* **280** 1925

[12] Miyano K, Tanaka T, Tomioka Y and Tokura Y 1997 *Phys. Rev. Lett.* **78** 4257

[13] Asamitsu A, Tomioka Y, Kuwahara H and Tokura Y 1997 *Nature* **388** 50

[14] Prabhakaran D, Coldea A I, Boothroyd AT and Blundell S J 2002 *J. Crystal Growth* **237-239** 806

[15] Tofield B C and Scott W R 1974 *J. Solid state Chem.* **10** 183

[16] Subias G, Garcia J, Blasco J, and Proietti M G 1998 *Phys. Rev. B.* **58**, 9287

[17] Dubey A and Sathe V G 2007 *J. Phys.: Condens. Matter* **19** 346232
[18] Dubey A, Sathe V G and Rawat R 2008 *J. Appl. Phys.* **104** 113530
[19] Baldini M, Struzhkin V V, Goncharov A F, Postorino P, and Mao W L 2011 *Phys. Rev. Lett.* **106**, 066402
[20] Marrocchelli D, Postorino P, Di Castro D, Arcangeletti E, Dore P, Cestelli Guidi M, Ray S and Sarma D D 2005 *Phys. Rev. B* **76**, 102405
[21] Mishra D K and Sathe V G 2012 *J. Phys.: Condens. Matter* **24** 252202
[22] Iliev M N and Abrashev M V 1998 *Phys. Rev.* B **57** 2872
[23] Fano U 1961 *Phys. Rev.* **124** 1866; Hartinger Ch, Mayr F, Loidl A and Kopp T 2004 *Phys. Rev. B* **70** 134415
[24] Reedyk M, Crandles D A, and Cardona M 1997 *Phys. Rev. B* **55** 1442

# Nonlinear dynamics of bimodality in vehicular traffic

**Arjun Mullick**[1] **and Arnab K Ray**[2]

[1] Audible.com, Floor 26, 10 Exchange Pl, Jersey City, NJ 7302, USA

[2] Department of Physics, Jaypee University of Engineering and Technology, Raghogarh, Guna 473226, Madhya Pradesh, India

E-mail: `arjun.mullick@gmail.com`, `arnab.kumar@juet.ac.in`

**Abstract.** We provide a global model for the bimodal distribution of a one-dimensional vehicular traffic flow. Our model captures the essential features of bimodality, namely, asymptotically decaying tails, asymmetry of the bimodal peaks, and their oscillatory exhange in a twenty-four cyle of traffic flows. We analyse our model from the perspective of nonlinear dynamics, and show that in a phase portrait, bimodality is implied by fixed points, closed loops, local periodicity and homoclinic paths. We also find the conditions for the asymmetry in a bimodal function, and for a bimodal-to-unimodal transition.

## 1. Introduction

Bimodal distributions have two distinct modes in a single frequency distribution [1]. In standard statistical treatments, such distributions are seen as a mixture of two normal distributions, but here we develop a mathematical model that gives a single global description of bimodal phenomena. Although our modelling work is founded on the bimodal distribution of data pertaining to the case of one-dimensional vehicular traffic flows, its implications go much beyond any specific context, so long as the basic principle of bimodality, observed in diverse phenomena (see [2, 3] and all references therein), is maintained. Studying the dynamics of a bimodal distribution, by applying standard methods of nonlinear dynamics to our model [4, 5], we see how the symmetry of a bimodal system can be broken and how the controlling parameters can be adjusted for a bimodal-to-unimodal transition (like a bifurcation)

## 2. Bimodality in traffic flows and its mathematical model

Studies of vehicular traffic flows aim to understand interactions among streams of vehicles, with the practical motive of eliminating the problem of traffic congestion and devising effective methods of traffic control. In our model, we take a macroscopic view of the one-dimensional vehicular traffic of a city at a given hour in the day — in effect, the bulk of traffic as a function of time. The data we need are taken from the repository of the Alabama Department of Transportation (ALDOT) [1], pertaining to a city named Jackson in Alabama State, USA. This city is comparatively small in size, and so the data about its traffic, free of the complications associated with the traffic of very large cities, have been systematically recorded. The data we have chosen are the average values of all the Wednesday traffic data of Jackson city in the year, 2007. Wednesday is in the middle of the week, and it is in this period of the week that all city traffic rids itself of any weekend effect and reaches a local mid-week equilibrium. Taken together, these factors give us the advantage of simplicity in building our mathematical model.

[1] `http://aldotgis.dot.state.al.us/atd/default.aspx`

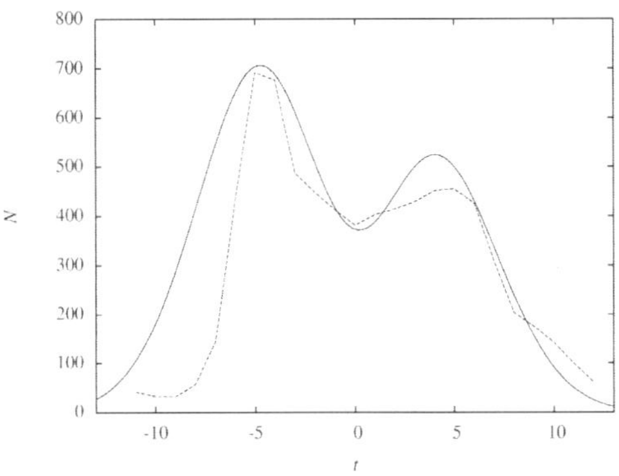


**Figure 1.** Bimodal distribution of traffic flow due west, plotting traffic volume, $N$, at time, $t$ (hours). The time is scaled to set $t = 0$ at midday, 12:00 hours. The dotted broken curve joins the real data points, and the continuous curve traces the model given by Eq. (1), with $A = 44.0$, $\mu = 8.53$, $\lambda = 0.19$ and $\beta = -0.09$.

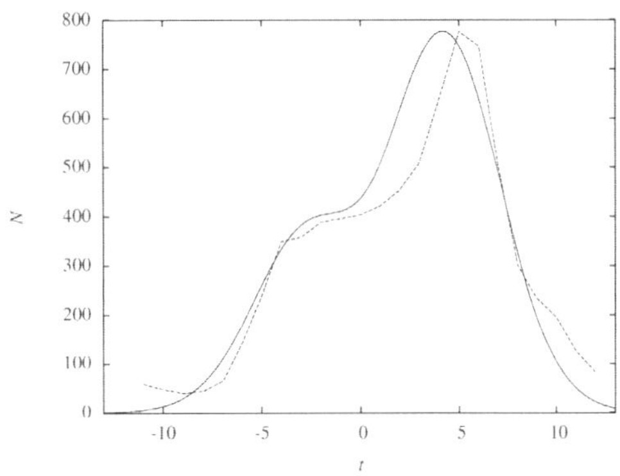


**Figure 2.** Bimodal distribution of traffic flow due east. The plotting follows the method used for the westward traffic, with $A = 44.1$, $\mu = 10.5$, $\lambda = 0.22$ and $\beta = 0.24$. The sign and the absolute magnitude of $\beta$ make the most significant difference between the two bimodal distributions.

Vehicular traffic in Jackson city is bi-directional, i.e. its flows are only in the eastward and westward directions. The data for the westward traffic is plotted in Fig. 1, with the broken line. Similarly, the plot of the eastward traffic is shown in Fig. 2. In both these plots, the time, $t$, at which the traffic volume, $N(t)$, is given, is scaled to coincide $t = 0$ with midday, 12:00 hours. This allows us look for any possible symmetry in the distribution of the data, especially if there is a symmetry about $t = 0$. In both Figs. 1 & 2, however, only asymmetric bimodal distributions are seen, with two peaks of unequal height in either plot. Fig. 1 suggests that the morning traffic volume is much higher in the western direction than in the reverse direction, while the behaviour of the bimodal curve in Fig. 2 is the opposite, with the eastward traffic being disproportionately high in the evening. Evidently, there is a clear commuter preference in the traffic flow as a function of time, and the traffic volume oscillates back and forth between the eastern and western limits of the city, once in a cycle of twenty four hours.

We plot the continuous curves in Figs. 1 & 2 with the model function,

$$N(t) = A\left(\mu + t^2\right)\exp\left[-\left(\lambda t - \beta\right)^2\right]. \tag{1}$$

The foregoing function captures the essential features of bimodality, the first of which is that there should be three turning points, with a local minimum flanked by two peaks. This is a fundamental property of any bimodal function, and in Eq. (1), this is controlled by the parameter, $\mu$. Secondly, for $t \longrightarrow \pm\infty$, there is an asymptotic decay of $N$, which is given by the Boltzmann tail of Eq. (1). Finally, there is a broken symmetry in the function about $t = 0$, in consequence of which, the two peaks of the bimodal distribution have uneven heights. In addition, we see that the uneven peaks also exchange their respective positions in Figs. 1 & 2. In Eq. (1) both the breaking of symmetry and the exchange of the peaks in the bimodal curve are controlled by the parameter, $\beta$. The former condition is given by the absolute magnitude of $\beta$, while the latter is given by its sign.

## 3. The dynamics of the bimodal model

The form of Eq. (1) is $N \equiv N(t)$, in which $\mu$ and $\beta$ collectively play a crucial part to cause an asymmetric bimodality. This part can be known more clearly from the perspective of a dynamical system, and accordingly we cast Eq. (1) in a first-order non-autonomous form as $\dot{N} = f_1(t)N$, where $f_1(t) = 2[t(\mu + t^2)^{-1} - \lambda(\lambda t - \beta)]$ and the overdot on $N$ implies a time derivative. Further, the second

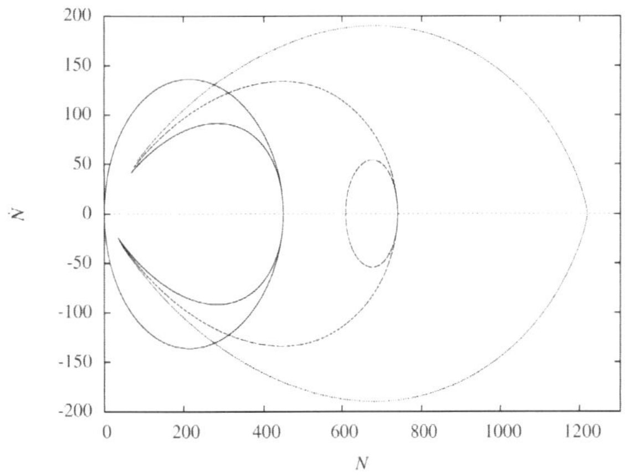


**Figure 3.** A symmetric phase portrait with $\beta = 0$. When $\mu = 0$, the phase trajectories are shown by the continuous curve at the extreme left. The closed loop is the largest in this case. When $\mu = 0.5\lambda^{-2}$, the loop size shrinks, as the dotted curve shows in the centre of the plot. When $\mu = \lambda^{-2}$, the loop vanishes as the curve on the extreme right shows. The closed loops are a signature of bimodality. When $\mu = \lambda^{-2}$, a bimodal-to-unimodal transition occurs, and the closed loop disappears, i.e. bimodality is lost. There is a fixed point at $N = \dot{N} = 0$, joined to itself by a homoclinic trajectory.

derivative of $N$ gives a second-order non-autonomous equation as $\ddot{N} = f_2(t)N$, where $f_2(t)$ is another polynomial function of $t$. The turning points of $N$ are obtained when $\dot{N} = 0$, and the fixed points in the phase portrait of $N$–$\dot{N}$ are obtained when both $\dot{N} = \ddot{N} = 0$. The simplest way to understand these features is to consider the symmetric limit of $\beta = 0$ and to solve for $\dot{N} = 0$. This delivers three roots of $t$ at $t = 0$ and $t = \pm(\lambda^{-2} - \mu)^{1/2}$. It is easy to show that none of these roots satisfies the condition for $\ddot{N} = 0$, and so we conclude that there is only one fixed point in the bimodal system at $N = N^\star = 0$. Any other solution of $\dot{N} = 0$ gives only turning points. At $t = 0$ a turning point of $N$ is given by $N_0 = A\mu$, as Eq. (1) shows, while the two non-zero roots of $t$ give two coinciding turning points, $N_{\rm T} = A\lambda^{-2}\exp(\mu\lambda^2 - 1)$. All three roots of $t$ and their corresponding turning points are real under the condition, $\mu < \lambda^{-2}$, by means of which, we can also argue that $N_{\rm T} > N_0$. As $\mu \longrightarrow \lambda^{-2}$, the positions of the turning points approach each other, and when $\mu = \lambda^{-2}$, there is a bifurcation-like behaviour, due to which, the turning point, $N_0$, and one of the turning points, $N_{\rm T}$, annihilate each other. What survives this mutual annihilation is the second "concealed" turning point, given also by $N_{\rm T}$. Subsequently, for $\mu > \lambda^{-2}$, only this turning point continues its existence, and the distribution becomes unimodal.

This behaviour of the turning points over the range, $0 < \mu < \lambda^{-2}$, is demonstrated in Fig. 3. We see clearly from this plot that bimodality is closely related to the formation of a closed loop by a phase trajectory in the $N$–$\dot{N}$ portrait. When $\mu \longrightarrow 0$, bimodality is very pronounced, and when $\mu \longrightarrow \lambda^{-2}$, bimodality becomes progressively enfeebled, disappearing altogether when $\mu = \lambda^{-2}$. In the former case, the sole fixed point at $N = N^\star = 0$ coincides with the turning point, $N_0 = 0$, while in the latter case, bimodality is eradicated because of a mutual annihilation between the turning points, $N_0(\neq 0)$ and $N_{\rm T}$. Once bimodality disappears and the distribution becomes unimodal, the phase-portrait also shows the existence of a homoclinic trajectory, that joins the fixed point, $N^\star = 0$, to itself. Further, the closed loop in the phase portrait indicates a locally periodic behaviour (implying that the two peaks in the bimodal distribution give a local scale), and when $\mu < \lambda^{-2}$, the time scale of this local periodicity is $\lambda^{-1}$.

We now look into the breaking of symmetry, when $\beta \neq 0$. What happens in this case is that the two coinciding turning points at $N = N_{\rm T}$ are separated from each other. If $|\beta| \ll 1$, then by perturbing about the turning point $N_{\rm T}$ (determined for $t^2 = \lambda^{-2} - \mu$), we linearise all terms involving $\beta$ in Eq. (1), and obtain the two separated roots as $N_{\rm T\pm} \simeq N_{\rm T}[1 \pm 2\beta(1 - \mu\lambda^2)^{1/2}]$. As $\beta$ increases in magnitude, the position of $N_{\rm T+}$ shifts to higher values of $N$ on $\dot{N} = 0$, while the position of $N_{\rm T-}$ slides in the opposite direction. This pair of roots corresponds to the two asymmetric peaks in the bimodal curve, with $N_{\rm T+}$ tracing the locus of the higher peak, and $N_{\rm T-}$ the lower. Between them, these peaks flank a local minimum in the bimodal distribution, which itself turns out to be the third turning point on $\dot{N} = 0$. When $\beta = 0$, this point is at $N = N_0$, with $N_0 < N_{\rm T}$, but when $\beta \neq 0$, the position of this third unpaired turning point shifts to larger values of $N$, and for sufficiently high values of $\beta$, it suffers a separate pair

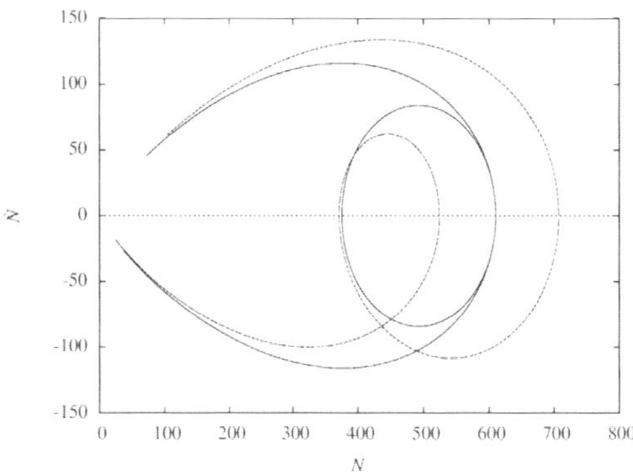

**Figure 4.** Phase plot of the bimodal distribution of the traffic flow due west. When $\beta = 0$, the resulting plot is shown by the continuous symmetric curve. When $\beta = -0.09$, asymmetry sets in, as shown by the dotted curve, in which a closed loop implies that bimodality prevails nonetheless.

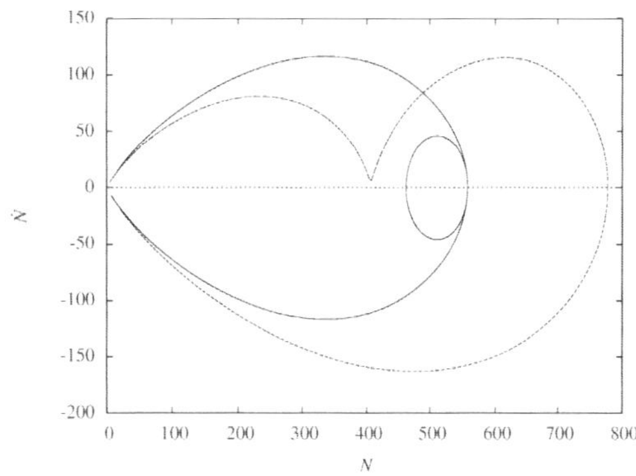

**Figure 5.** Phase plot of the bimodal distribution of the traffic flow due east. The continuous curve is for $\beta = 0$, while the asymmetric dotted curve is for $\beta = 0.24 (> 0.21)$. A bimodal-to-unimodal transition has occurred, and instead of a closed loop on the dotted curve, there is a cusp just above $\dot{N} = 0$.

annihilation with the shifting turning point at $N_{T-}$. We support this reasoning by solving for $f_1(t) = 0$, and getting a third-order polynomial in $t$, which is $t^3 - (\beta/\lambda)t^2 - (\lambda^{-2} - \mu)t - (\mu\beta/\lambda) = 0$, whose analytical roots are extracted by the application of the Cardano-Tartaglia-del Ferro method. Transforming $t = \tau + h$ with $h = \beta(3\lambda)^{-1}$, gives the canonical form of a cubic equation, $\tau^3 + P\tau + Q = 0$, in which $P$ and $Q$ depend on $\beta$. The discriminant of this cubic equation also depends on $\beta$, and is given as $D = (Q^2/4) + (P^3/27)$. If $D > 0$, $\tau$ has a single root, giving a unimodal condition. If, however, $D < 0$, then $\tau$ has three real roots, providing the bimodal condition. Knowing $\tau$, leads to $t$, which, by Eq. (1), gives the value of $N$, as a turning point on $\dot{N} = 0$.

The critical condition for a bimodal-to-unimodal transition is $D = 0$, for $\beta = \beta_b$. With the values of $\mu$ and $\lambda$ used to calibrate the data of both the east and west-moving traffic, we find that $|\beta_b| \simeq 0.21$ for the eastward traffic, and $|\beta_b| \simeq 0.46$ for the westward traffic. In Fig. 1, the value of $|\beta| = 0.09$ is much less than $|\beta_b| \simeq 0.46$, while in Fig. 2, the reverse is true ($|\beta_b| \simeq 0.21 < |\beta| = 0.24$). Therefore, as opposed to Fig. 1, there is no conspicuous secondary peak in the bimodal distribution of Fig. 2. The implications of both these conditions have been depicted in Figs. 4 & 5. These plots show the dynamics of bimodality for the west-moving and the east-moving traffic, respectively. In both cases, when $\beta = 0$, there is a point on $\dot{N} = 0$, where the curve loops over itself. In either case, when $\beta \neq 0$, this point shifts away from the line of $\dot{N} = 0$, suggesting asymmetric bimodality. Stated concisely, $\mu$ causes bimodality, and $\beta$ causes a breaking of symmetry. Taken together, the result is an asymmetric bimodality. Changing the values of both these parameters can cause a bimodal-to-unimodal transition, like a bifurcation.

## Acknowledgments

The authors thank R. Atre, A. Basu, J. K. Bhattacharjee and T. Naskar for discussions.

## References

[1] Ross S M 2004 *Introduction to Probability and Statistics for Engineers and Scientists* (San Diego: Academic Press)
[2] Mullick A and Ray A K 2012 Dynamics of Bimodality in Vehicular Traffic Flows *Preprint* arXiv:1205.2314
[3] Mullick A and Ray A K 2014 *J. Appl. Nonlinear Dynamics* **3(1)** 17-26
[4] Strogatz S H 1994 *Nonlinear Dynamics and Chaos* (Reading, MA: Addison-Wesley Publishing Company)
[5] Jordan D W and Smith P 1999 *Nonlinear Ordinary Differential Equations* (Oxford: Oxford University Press)

# Tungsten Deposition on Graphite using Plasma Enhanced Chemical Vapour Deposition.

**Uttam Sharma[1], Sachin S Chauhan[1], Jayshree Sharma[2], A K Sanyasi[3], J Ghosh[3], K K Choudhary[4] and S K Ghosh[5]**

[1]Department of Physics, Shri Vaishnav Institute of Technology and Science, Indore,
[2]Department of Physics, M.B. Khalsa College, Indore 452009, India
[3]Institute for Plasma Research, Bhat, Gandhinagar 382 428, India
[4]Indian Military Academy Uttrakhand, India
[5]School of Studies in Physics, Vikram University, Ujjain

E-mail: druttamsharma1971@gmail.com, amulya@ipr.res.in

**Abstract:** The tokamak concept is the frontrunner for achieving controlled thermonuclear reaction on earth, an environment friendly way to solve future energy crisis. Although much progress has been made in controlling the heated fusion plasmas (temperature ~ 150 million degrees) in tokamaks, technological issues related to plasma wall interaction topic still need focused attention. In future, reactor grade tokamak operational scenarios, the reactor wall and target plates are expected to experience a heat load of 10 $MW/m^2$ and even more during the unfortunate events of ELM's and disruptions. Tungsten remains a suitable choice for the wall and target plates. It can withstand high temperatures, its ductile to brittle temperature is fairly low and it has low sputtering yield and low fuel retention capabilities. However, it is difficult to machine tungsten and hence usages of tungsten coated surfaces are mostly desirable. To produce tungsten coated graphite tiles for the above-mentioned purpose, a coating reactor has been designed, developed and made operational at the SVITS, Indore. Tungsten coating on graphite has been attempted and successfully carried out by using radio frequency induced plasma enhanced chemical vapour deposition (rf -PECVD) for the first time in India. Tungsten hexa-fluoride has been used as a pre-cursor gas. Energy Dispersive X-ray spectroscopy (EDS) clearly showed the presence of tungsten coating on the graphite samples. This paper presents the details of successful operation and achievement of tungsten coating in the reactor at SVITS.

## 1. Introduction

Recent trends on consumption and demands of energy among developing as well as in developed countries have alarming effects on global climate. Till now, world population is mostly dependent on thermal power, hydropower, natural gases, natural hydrocarbons, nuclear power (fission), wind energy, and very recently solar energy for their energy needs. Global climatic changes has adversely affected the pattern of rain fall, wind flow and the natural resources like coal, natural gas, fuel hydrocarbon has started depleted at an alarming rate and expected to be finished by the end of this century. This trend is going to create a chaotic situation in future in order to maintain the energy needs. Scientists from interdisciplinary fields all over the world are working tirelessly to device new sources of energy for meeting the energy demands suitably met and several international collaborations are initiated by extending the scientific as well as technical know-how to each other. Nuclear power is known as the most efficient way to produce huge amount of energy. Nuclear fusion has definite advantage over fission and promises clean and environment friendly solution to present energy demands. Fusion of two light hydrogen isotopes (D-T) in a magnetically confined reactor can generate energy equal to mass difference which can be efficiently extracted from the fusion reactor and convert it into electrical energy. To fuse the two light hydrogen isotopes, they need to be heated to very high temperature ~ $10^8$ K. But, maintaining such high temperature in fusion reactors is not an easy task. In present day scenario, virtually no material exist that can directly withstand this temperature and hence a magnetic cage is required to hold this superheated gas known as plasma. In the past few decades several machines such as tokamaks, stellarators, spheromaks, reverse field pinch etc. have come into existence [1-9]. Out of them, tokamak is the frontrunner to be converted into a fusion reactor and has yielded a fusion power with D –D phase reactions. In a larger step, USA, Russia, China, South Korea, Japan, India and European Union has decided to collaboratively fund and develop world's largest tokamak called International Thermonuclear Experimental Reactor (ITER) which is in its construction phase in Cadarache, France [10]. The machine is expected to be operational in mid 2020s and expected to yield a fusion power ten times to its input power. Although, the superheated plasmas are trapped in the magnetic cage, the trap is not full proof and the heated particles leave the trap and fall on the plasma facing components damaging them seriously. Hence, the first wall and the target material choice is very crucial for any fusion reactor and ITER is designed with Beryllium as its first wall and tungsten as target material i.e. divertor. The plasma-material interaction effects influence the operation, safety and performance of the fusion reactors. Till now, most of the tokamaks are operated with stainless steel and carbon as first wall and target materials as Molybdenum, carbon, tungsten, titanium, Beryllium, carbon etc. and researchers have gained enough experience on the behaviour of these materials in plasma environment [11-12]. However, suitable choice of first wall materials still remains a subject of research with material as well as fusion scientists.

As fusion power remains a candidate for supplying future energy needs to the human community, reactor design and its interruption free working is being aimed with high priority. A fusion machine depends upon all branches of physics to engineering knowledge to be put together for its success. Starting from thick stainless steel reactor body, support structures, superconducting magnetic coils, cryogenic fluids, major electrical components (DC & Rf), beam sources for heating, vacuum pumps, plasma facing surface etc. are required to be put into order for successful operation of the reactor. The most crucial engineering and physics challenge to the fusion community has been identified as the interaction between plasma and wall material (PWI) [12]. The core plasma is required to be maintained at high temperature and with minimal impurity penetration from the walls. This requires edge plasma to be acting as a good thermal as well as physical barrier so that impurities from the walls cannot travel to the core plasma. The wall material also experiences high heat flux, particle flux, transients like edge localised modes (ELM), disruptions. This leads to substantial damage of the plasma facing components in forms of physical erosion i.e. particle erosion, layer melting, and material evaporation and chemical erosion i.e. formation of macro hydrocarbons amongst many other processes.

Plasma facing components (PFCs) are primarily divided into two categories depending upon their functionalities. They are classified as 1) target materials, i.e. limiter and divertor, 2) first wall materials. In the limiter configuration, a solid surface is used to define the plasma radius and to avoid direct contact of hot plasma on the vacuum vessel wall. In the divertor configuration, the plasma boundary is defined by specially designed magnetic fields and the particle flux coming out of the main plasma is diverted to a remote location of the tokamak. The limiter and divertor plates are called the target plates which see the majority heat flux and particle that are diffused cross field in the Scrape Off Layer (SOL) from the core plasma and direct neutron bombardment generated as a fusion by product. Although the target plates take the maximum plasma heat and particle load, the plasma particles and heat also fall on vacuum vessel wall, generally known as the first wall. The choice of the target and the first wall materials are therefore very important in order to run the fusion reactor successfully. In section 2, we will discuss on the different choice of suitable material for plasma facing component (PFC) and review choice of tungsten, section 3 will describe the methodologies that can be adopted to carry out an in-situ or ex-situ repair of the tungsten and will explain plasma assisted chemical vapour deposition technique suitability, section 4 will describe the radio frequency assisted coating reactor at SVITS, in section 5, we shall be presenting preliminary results on tungsten coating characterization.

## 2. Material choice for fusion reactors

Cleanliness and the purity of the core plasma are largely governed by the choice of the target and wall materials. The criterion for choice of target materials is different than the wall materials as the nature of particle and heat flux seen by them is different. The target materials experience a high level of heat flux than the wall materials and hence they need special design and material characteristics. The plasma interaction with the target and wall material impact the life time of the plasma facing components and fusion fuel retention in these materials remain a safety concern. In this section, we shall briefly discuss the life time of the plasma facing components under different plasma conditions. The lifetime of the plasma facing components is determined by the erosion processes and the erosion is initiated by physical sputtering, chemical erosion, material loss due to uncontrolled plasma impact (ELM), plasma disruptions leading to the bulk melting and evaporation. It is well known that in order to sustain the fusion reaction, the concentration of species with high mass number in the core plasma should be very low. So obvious choice for wall materials is the materials having low mass number and carbon is a prospectus candidate. However, there are several limitations with carbon i.e. carbon reacts chemically with oxygen, hydrogen, deuterium and tritium thus fuel retention is high and reactor depletes the fuel. Being low atomic mass its sputtering yield is high compared to the metals. For example, in ITER plasma discharges (Q = 10), typical particle flux reaching the divertor region will be $10^{24} m^{-2} s^{-1}$ leading to a large particle incidence ( $\sim 10^{26} m^{-2}$) per plasma shot with divertor plasma densities $\sim 10^{21} m^{-3}$. This could lead to the local temperature rise on the divertor to $\sim 1500°K$. With the given statistics, various codes have predicted the life time of the wall and target materials. Table 1 shows the estimated numbers predicted by DIVIMP code[1] for wall material and ERO code[2] for divertor. During steady state operation the walls and target will experience intensive particle and heat loads ~ 10 MW/m$^2$ and transient loads ~20 MW/m$^2$.

Table 2 shows the properties of the target materials presently under consideration although ITER has finalized its first wall and target materials. Hence, proper choice is still remains a burning subject of research. Numerical calculations have predicted that 280 tons of iron will be eroded from ITER under it's normal operation conditions and this rate would be of much higher order if carbon, boron and beryllium is used.

Table 1. Estimated life times for plasma facing components.

| Material | Erosion rate (nm/s) | Erosion Source (atoms/s) | Eroded material (g/shot) | Life time (Shot) |
|---|---|---|---|---|
| Be[1] | 0.1 | $8 \times 10^{21}$ | 48 | 20000 |
| W[1] | 0.01 | $2 \times 10^{20}$ | 26 | 200000 |
| CFC[2] | 100 | $4 \times 10^{22}$ | 330 | 200 |
| W[2] | 2 | $4 \times 10^{20}$ | 48 | 10000 |

Table 2. Probable candidates for target materials in a fusion reactor [13]. (1) Hot pressed Beryllium, (2) EK98 fine grain graphite, (3) hot pressed boron carbide, (4) CVD pyrolitic β – SIC, (5) oxygen free high conductivity copper (OHFC), (6) austenitic steel 316L, (7) Molybdenum based alloy (Mo – 0.5%, Ti – 0.1%, Zr), (8) pure tungsten.

| Material | Density ($10^3$ kg/m$^3$) | Melting temp (°C) | Thermal Conduc. (W.m$^{-1}$.K$^{-1}$) | Therm. Expa. Coeff ($10^{-6}$. K$^{-1}$) | Tensile Strength (MPa) |
|---|---|---|---|---|---|
| Be(1) | 1.85 | 1283 | 187 | 11.3 | 350 |
| Graphite(2) | 1.85 | --- | 80 | 3.4 | 50 |
| $B_4C$(3) | 2.51 | 2450 | 29 – 67 | 5.6 | 350 |
| SIC(4) | 3.2 | ---- | 67 | 2.9 | 360 |
| Cu(5) | 8.25 | 1083 | 391 | 16.7 | 206 |
| SS316L(6) | 7.96 | 1400 | 14.6 | 16.2 | 525 |
| TZM(7) | 10.2 | 2620 | 125 | 5.3 | 600 |
| W(8) | 19.3 | 3410 | 144 | 4.4 | 600 |

## 3. Choice of tungsten (W)

For many reasons, tungsten (W) is considered as superior material amongst the various materials suitable for the target plates in tokamaks. Tungsten has a high melting point and could act as the best plasma facing component if plasma temperature near the target plates remains less than 50 eV. Plasma contamination is suppressed with tungsten especially with low edge temperatures and having a very low chemical sputtering affinity with hydrogen as fuel. Thereafter, sputtered tungsten has a lower ionization potential and lower velocity than sputtered carbon thus ionized tungsten tends to re-deposit near to their point of origin. Tritium inventory in tungsten is expected to be very small. Tungsten re-crystallization occurs at fairly low temperatures (1100 °C – 1300 °C) than its melting point and its melting is only anticipated by thermal quench loads with type I Edge Localized Modes. Current plasma wall research on tungsten focuses on the melt layer behavior under ITER like disruption scenarios.

Despite its superior properties tungsten will suffer erosion under unfavorable plasma conditions. Repairing of the affected area ex-situ i.e. inside the tokamak without breaking its vacuum, will consume a lot of machine operation time. Moreover, in small scale machines pure tungsten component addition generates complications in PWI studies. For example, ASDEX – upgrade uses graphite tiles with few micron coating of tungsten for PWI studies [14]. Similar experiments are being conducted in JET – ITER like wall experiments [15]. First wall of the test blanket modules (TBM) for ITER may be considered to be coated with tungsten for better stability and performance. For such several reasons tungsten coating on various substrate materials are being tried for its future requirements [16].

## 4. Tungsten coating methodology

Tungsten coating can be suitably employed using physical vapor deposition (PVD), chemical vapor deposition (CVD), magnetron sputtering, plasma spraying both in vacuum and atmosphere, and plasma assisted chemical vapor deposition techniques (PACVD). PVD is a tedious and most difficult process to perform on tungsten as it has a high melting temperature. Chemical vapor deposition requires a precursor material preferably a halide (F) based, having low dissociation temperature but the substrate temperature requires being elevated. CVD in presence of a reactant gas to F i.e. Hydrogen tend to cool the system. Magnetron sputtering is an effective technique but in a laboratory scale a very small area only could be coated but it is efficient in producing a higher coating thickness. JET divertor tiles have been coated with magnetron sputtering technique. Plasma spraying is one of the effective methods to deposit tungsten melted grains on a substrate but it is associated with several issues. It is a violent process involving arc plasmas in atmosphere or in a vacuum chamber and required very high process temperature. The coating process is rapid thus micro as well as macro cracks develop through out the coating volume and coating becomes non-uniform. Difference of thermal conductivity of the substrate and the tungsten leads to weak interlayer bonding as the deposition process itself leads to sputter the deposits. In atmospheric plasma spraying, impurity impregnation cannot be avoided and maintaining surface uniformity will be challenging. Plasma assisted chemical vapor deposition is considered one of the effective method as it can be done at a fairly low temperature plasmas produced mainly by application of radio frequency waves. Plasma it self acts as a catalyst to the chemical reaction and ionized particles can be preferentially guided by electric fields and deposited on the substrates. It's a fairly slow process but degree of uniformity and sticking coefficient of the surface is high.

*4.1 Plasma enhanced chemical vapor deposition (PECVD)*

Fractionally ionized plasmas have great interests in thin film coating and material processing. In such plasmas, the energy exchange between the electron and neutral gas is very inefficient. So, the bulk plasma is more positive than any object (substrate) present and in contact with plasma. The voltage between the plasma and substrate drops at the sheath region and the ions in the sheath feel an electrostatic force and are accelerated towards the substrate. Thus, the substrate exposed to plasma receives energetic ion bombardment. CVD performed in a plasma environment leads to increase in the nucleation, growth kinetics, and hence density of the deposited film and also plays an important role in the process of sputtering of the impurities called Plasma Enhanced Chemical Vapor Deposition (PECVD).

PECVD has been developed to combine the good adhesion of CVD and low temperatures of physical vapor deposition (PVD). This process avoids typical drawbacks like requirement of high temperature, temperature related deformations, poor adhesion etc. The plasma environment in PECVD performs mainly two basic functions; reactive chemical species are formed by cracking of relatively stable molecules via electron impact collisions and supplies energetic radiation such as positive ions, meta stable species, electrons and photons. Increased ion bombardment tends to make the film denser and cause film stress to become more compressive. PECVD performed with RF plasma is called rf-PECVD. Furthermore, PECVD can be performed at low pressure and temperature than CVD. In PECVD, ions suffer more collisions in a gas phase and lose most of their energy before they reach the cathode. Hence, effects due to sputtering on the substrate are minimized. Substrate heating due to the PECVD process is very small and the substrate temperature is mainly controlled by external heating. In PECVD, higher flow rate and RF power can increase the deposition rate and also the uniformity but at the same time there will be waste of reactant gas because of non ionization. Thus process parameters play a critical role in film formation, growth and characteristics.

In plasma assisted chemical vapor deposition, low temperature plasma is produced using different plasma production methods. For tungsten coating a precursor gas/solid containing tungsten is used. The precursor gases/solids are broken down in the plasma, which acts as a catalyst to the reaction. Tungsten ions form in the plasma get attracted towards the cathode and get deposited. The rate of deposition can be controlled or enhanced by the input RF power and applied negative bias on the substrate plate. Tungsten ions get deposited on the substrate with good adhesion and get redistributed over the surface with good step coverage in PACVD process. In this process a fairly large area can be coated by choosing a pair of larger diameter capacitive plate in the reactor chamber.

### 4.2 Coating reactor at SVITS

Under the National Fusion Program (NFP) of Board of Research Fusion Science and Technology (BRFST) [17], tungsten coating program using PECVD has been taken up at SVITS [18] for developing tungsten coated graphite tiles to be used for Indian tokamaks, ADITYA upgrade tokamak and SST-1 tokamak. Glow discharge plasma is produced by applying a radio frequency between two parallel capacitive plates in this reactor. For deposition of tungsten, tungsten hexafluoride ($WF_6$) gas is used as precursor gas, which is mixed with hydrogen ($H_2$) neutral gas and fed to a reaction chamber.

$$3H_2(g) + WF_6(g) \xrightarrow{Plasma} W + 6HF\ (g) \uparrow$$

The coating system is designed, fabricated and installed in Shri Vaishnav Institute of Technology & science (SVITS), Indore. The vacuum chamber is a stainless steel (SS304) cylindrical chamber with diameter 360 mm and height 300 mm. The vacuum vessel consists of various CF ports (35CF & 63CF) used for plasma viewing, fixing of vacuum gauges, diagnostics etc. The system is supported on a support structure made of Aluminum hollow square bars. The system is being pumped by a turbo and rotary pump combination. The turbo pump is having a pumping speed of 60 litres/sec and is connected to the vacuum vessel through a gate valve. The turbo pump is roughed through a diaphragm pump inbuilt in the pumping unit itself. The main vacuum system is pumped to the roughing vacuum with a rotary pump of capacity 12 $m^3$/hr. The vacuum on the system is being monitored by combination of pirani and cold cathode ionization gauges. The system is provided with an electro - magnetically actuated vent valve for venting. The vacuum chamber consists of two circular metallic discs (diameter = 15 cm, thickness = 1 cm) placed at the top and bottom flanges and are used as plasma electrodes. The disks are connected to a DC power supply (1kV, 1Amp) as well as to a RF source (13.56 MHz, 600W) with auto matching network used for plasma generation. The electrical connections are taken out with appropriate high voltage RF feed-through. A high precision quadrupole mass analyzer (QMA 300 AMU, mass resolution of 0.5 AMU) is attached to the system for monitoring of impurities as well as the reactant gas presence. Mass flow controller for the $WF_6$ gas is installed for calibrated gas flow to the reaction chamber and hydrogen gas is introduced via a precision leak valve. The system is pumped down to 2 x $10^{-6}$ mbar and coating pressure regime is maintained at ~ $10^{-3}$ to $10^{-4}$ mbar. Suitable arrangement with KOH solution is used to neutralise the HF gas. Figure 1 shows the schematic of the experimental system with the connected subsystems.

Figure 1. Experimental coating reactor installation at SVITS showing vacuum chamber, table top turbo pump, quadrupole mass analyser, vacuum gauge, rf power supply etc.

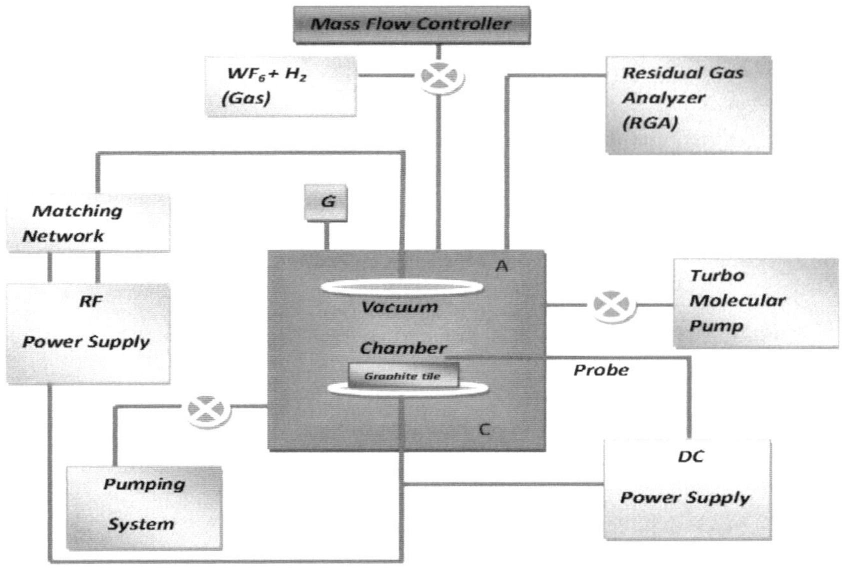


Figure 2. Schematic of the experimental arrangement used for coating experiment.

## 5. Experiment

Table 3 shows the plasma and system parameters in the experimental system.

| |
|---|
| Test gas: Argon |
| RF power: 0 – 200W |
| Frequency: 13.56MHz |
| Bias: 200 VDC |
| Substrate temperature: 250 °C |
| Ultimate pressure: 2 x $10^{-6}$ mbar |
| Working pressure: 5 x $10^{-4}$ mbar |
| Plasma density: ~$10^{10}$ – $10^{11}$ $cm^{-3}$, |
| Coating Gas mixture: $H_2$ + (5%) $WF_6$ |

Before initiation of the coating process, a uniform plasma region is identified between the parallel electrodes by producing Argon plasma. Potential and density measurements were carried out using a single Langmuir probe compensated for 13.56 MHz. It is observed from the figure that the plasma has a uniform region between -8cm < R < 8cm. Figure 3 shows the sample arrangement on the cathode i.e. bottom plate electrode. The substrate is pure polished graphite tiles to be used as limiter tiles in Aditya tokamak [19]. The sample tiles are cleaned with detergent solution and rinsed thoroughly with water before final cleaning with iso-propyl alcohol. The samples were initially baked to 250 °C under high vacuum in the coating reactor so that the trapped moisture is evaporated. The evaporation process is monitored on the QMA and it is allowed to bake until the $H_2O$ traces are bare minimum to the initially recorded values. In the second step, hydrogen plasma is created at RF power of 60W and the samples are exposed to the plasma for 60 minutes for plasma surface cleaning and micro level surface modification. The plasma is simultaneously monitored with a spectrometer working at visible spectral range. In the third stage, tungsten hexafluoride ($WF_6$) gas is injected to the reactor chamber in a controlled manner through a calibrated mass flow controller. Controlled gas injection is required because $WF_6$ is a heavy gas and it may lead to a substantial change in the plasma resistance between the electrodes and hence plasma may get disrupted. With the injection of the gas, $WF_x$ (x = 1 to 5) isotope ion traces are observed on the QMA and a typical spectrum is shown in figure 4. Figure 4 shows the traces of tungsten found in natural abundance. Figure 5 shows the picture of mixture of tungsten hexafluoride and hydrogen plasma. During the coating process hydrogen reacts with fluorine gas and produces a huge amount of acidic hydrogen fluoride (HF) gas which is required to be efficiently removed form the reactor volume before opening the chamber for removal of the samples. HF signal is continuously monitored through out the experiments. The exhaust line of the pumping system is put into a potassium hydroxide (chemical base) solution and the HF is efficiently decomposed to solid potassium fluoride which is not so reactive.

$$KOH + HF = KF + H_2O$$

Figure 3. Figure showing the arrangement of the graphite tiles on the heater surface mounted on the bottom electrode. The heater assembly and the bottom plate are electrically conducting.

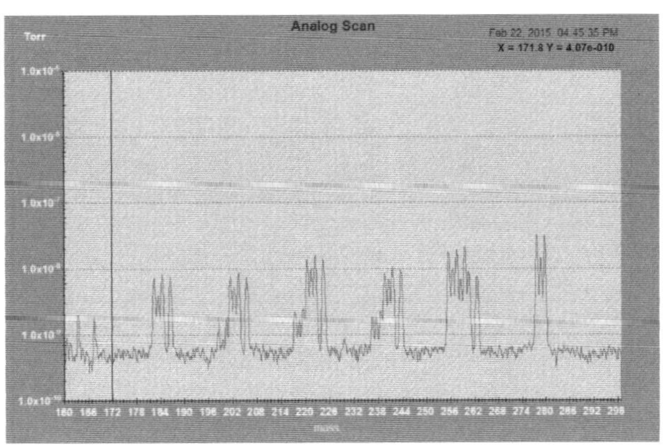


Figure 4. Figure shows the analog scan for the traces of tungsten isotopes detected by the high precision residual gas analyser.

Figure 5. Picture shows the first plasma of $WF_6 + H_2$ gas mixture and the sample pieces are exposed to the plasma.

## 6. Results & Discussion

The coated samples are tested for quality of tungsten deposits and stability. Figure 6 shows the EDX trace for elemental analysis and tungsten rich signal is observed. Apart from the tungsten traces, oxygen, carbon and fluorine traces are also present in the coating. The origin of the carbon and oxygen signals is due to the beam interaction with graphite and from the plasma system respectively. The fluorine is embedded from the precursor gas. The tungsten coating thickness achieved is the order of 300 nm. It is considered that with increase in the coating thickness (few μm), the beam interaction with substrate will decrease and the carbon trace will be reduced. The fluorine trace suppression would require more concentration of hydrogen into the coating reactor and reduced inflow of the $WF_6$. This is more a process optimization engineering.

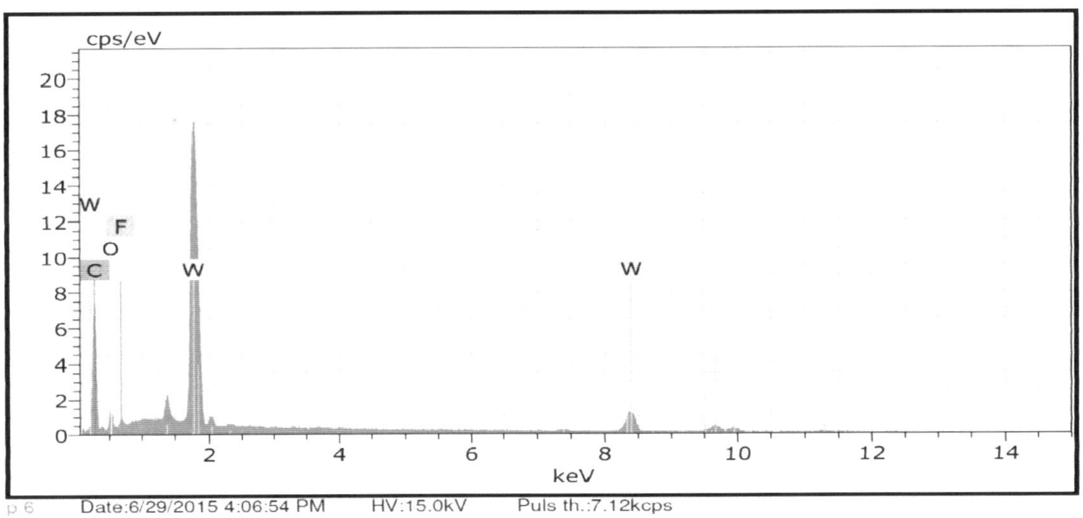


Figure 6. EDX scan of the tungsten coating with beam power maintained at 15 kV.

The above tests on the coated samples clearly establish the successful tungsten coating with the help of plasma enhanced chemical vapour deposition at SVIT, Indore, which is achieved first time in India. The state of art technology has been established and the process requires minor optimization for improvisation of the coating properties and thickness. Initial characterization results are very satisfactory and promising to utilising this technique further for our own in-house requirements. Cross-sectional analysis of the coatings suggests achievement of a thickness of ~ 300 nm and it is possible to produce micro – meter thick coatings by enhancing the reaction rate by increasing the RF power and deposition time. The coatings will be further tested for its stability in IPR LASER and electron beam facility for its stability.

**Acknowledgement**

The authors would like to thank National Fusion Program (NFP), Board of Research for Fusion Science & Technology (BRFST) for funding the project and their support at the times of need.

**References**

[1] Petersen P I et. al. 2003 *Nucl. Fusion* **43** 812
[2] Gruber O et. al. 2009 *Nucl. Fusion* **49** 115014.
[3] Winter J et. al.1984 *J. Vac. Sci. Technol. A* **2** 679
[4] Khilhacker M et. al.1990 *Phys. Fluids B* **2** 1291
[5] Xiao B J et. al. 2007 *34th EPS Conference on Plasma Phys. Warsaw* ECA **31F** O-4.014
[6] Hong S – H et. al.2015 *Fus. Sci. Technol.* **68** 36
[7] Beidler Craig D et. al. 2002 *J. Plasma Fusion Res. Series* **5** 149
[8] Skinner C H et. al.2002 *Nucl. Fusion* **42** 329
[9] Brunsell P R et. al. 2001 *Plasma Phys. Control. Fusion* **43** 1457
[10] https://iter.org
[11] Federici G 2006 *Phys. Scr. T* **124** 1
[12] Federici G et. al.2001 *Nucl. Fusion* **41** 12R
[13] Bolt H, Linke J et. Al. 1986 *KFA Report, Juelich,* 2086
[14] Tabasso A, et. al. *Nucl. Fusion* **40** 8
[15] Matthews G F, et. al.2011 *Phys. Scr. T***145** 014001.
[16] Philipps V et. al.2011 *Phys. Scr. T* **145** 014030.
[17] www.nfp.psi.in
[18] www.svits.ac.in
[19] Tanna R L et. al.2015 *Nucl. Fusion* **55** 063010.

Journal of Physics: Conference Series **755** (2016) 012011 doi:10.1088/1742-6596/755/1/012011

# Investigation of magnetoresistance and localization behavior in nanobulk assembly of $Bi_2Te_3$

**R.Venkatesh, S. Shanmukharao Samatham, and V.Ganesan**

Low Temperature Laboratory, UGC-DAE Consortium for Scientific Research, University Campus, Khandwa Road, Indore 452 001, M.P, India

E-mail: rvenkatesh@csr.res.in

**Abstract.** Magnetoresistance (MR) measurements are used to investigate the topological nature of nanobulk assembly of two dimensional (2D) $Bi_2Te_3$ nanoflakes prepared by microwave assisted solvothermal reactions. Average edge to edge length of individual hexagonal nanoflakes is found to be of ~3-4 microns as demonstrated by SEM micrographs. Nanobulk assemblies prepared out of this 2D nanoflakes exhibit rhombohedral structure and have a high surface to volume ratio that is advantageous for probing the signatures of surface states through bulk measurements. At low temperatures, magnetic field dependent MR measurements demonstrate a clear positive MR with a prominent dip below ±1T. Significant changes happening in magnetic field dependent MR measurements are analyzed and discussed in the light of Weak anti localization (WAL) effect arising likely from surface states in nanobulk $Bi_2Te_3$ materials which on proper optimization can open up a new gateway for large scale application of these topological insulating materials in NEMS and MEMS technology.

## 1. Introduction

The topologically protected gapless edge/surface states in Topological insulators (TI) along with their bulk states is being investigated rigorously for the past few years through various measurement probes for nanoelectronics and thermoelectric applications[1]. Many of the reported TI materials are either single crystals or nanowires [2, 3]. Few reports are available regarding the investigation of surface states in TI thinfilms grown by molecular beam epitaxy [4] and significantly low amount of literature is available for polycrystalline thinfilms [5]. Dissipation-less electronic transport arising in surface states of these TI materials are detected through Shubnikov - de Haas oscillations (SDH) at high magnetic field - low temperatures and by the WAL/WL effects at low magnetic field - low temperatures [6]. In this work, we are reporting the signature of surface states with detailed MR measurements in nanobulk assembly of two dimensional $Bi_2Te_3$ nanoflakes. These materials on optimization assures the possibility of large scale application in NEMS/MEMS technology. Interesting behavior observed in MR measurements at low temperatures are discussed and presented in the light of 3D WAL arising likely from the hybridization of surface states in intrinsic nanopores formed on the surface of flakes.

## 2. Experimental details

Two dimensional $Bi_2Te_3$ nanoflakes were prepared by microwave assisted wet chemical synthesis [7]. In a typical synthesis, analytical grade $BiCl_3$, Telluric acid ($Te(OH)_6$), Thioglycolic acid, 1,5 Pentanediol, were used without any further purification. The microwave reactions were carried out in

Journal of Physics: Conference Series **755** (2016) 012011 doi:10.1088/1742-6596/755/1/012011

microwave oven (1250 W). The black precipitate of nano flakes obtained finally is centrifuged and washed several times in ethanol, acetone and finally stored in acetone. These nanoflakes were dried in room temperature and were cold pelletized followed by vacuum annealing at 200$^{O}$C. SEM images and chemical composition analysis using EDS measurements were performed with JEOL SEM EDS instrument. Atomic force microscope (AFM) images of nanoflakes were obtained using nanoscopeE AFM in contact mode. MR measurements were performed in Quantum design made PPMS in 12T/1.8K range.

## 3. Results and Analysis

Powder Xray diffraction (PXRD) pattern of $Bi_2Te_3$ (figure1) shows that it crystallizes in rhombohedral structure with $\bar{R}3m$ space group with no observable evidence for secondary phase. This is in accordance with the reported literature [8]. SEM image of $Bi_2Te_3$ nanoflakes as shown in figure 2 demonstrates the surface morphology of cluster of flakes. The white rod like structure seen above the flake is the C axis growth which on giving more microwave doses can develop into a three dimensional spherical nanoparticle. The flake size and morphology can be systematically varied by increasing the dosage of microwave given to the sample. Chemical composition as analyzed from EDS analysis confirmed that the Bismuth and tellurium exists in the ratio of 2:3. Temperature dependence of resistance for $Bi_2Te_3$ in various applied magnetic field is shown in figure 3. Resistivity measurements were carried out in ~0.5mm thin rectangular pellet using linear four probe method. R vs T at zero applied magnetic field and upto 10T shows a metallic behavior from 2 to 300K. On application of magnetic field a negative MR is observed upto a magnetic field of 0.5T which becomes positive MR from 0.5 to 10T as shown in the inset of figure 3. Magnetic field dependence of resistance for $Bi_2Te_3$ is as shown in figure 4. It indicates a positive MR in all the measured temperatures. A parabolic behavior is observed in R vs. H down to a temperature of 8K. Below that temperature a pronounced dip appears around 0T as shown for 5K in figure 4 which broadens significantly at further lower temperatures. At 1.8K the dip broadens upto a magnetic field of ±1T as shown in figure 4. The interesting features observed in MR measurements of $Bi_2Te_3$ are analyzed and discussed in the forthcoming section.

### 4. Discussion

Two dimensional single crystalline [9] nanoflakes of $Bi_2Te_3$ with very high surface to volume ratio is advantageous for probing the signatures of surface states through various measurements. In the present work, these 2D nanoflakes with negligible thickness when compared to their surface area are put together by cold pressing and annealed in vacuum at 200$^{O}$C. This annealed sample shows a metallic behavior (fig 3) in all the applied magnetic field which is in contrast to the semiconducting nature observed in transport measurements of single flake [10].

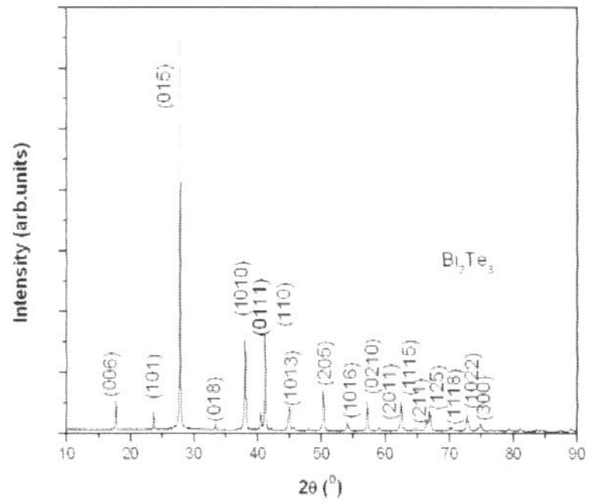


**Figure 1.** Powder XRD of $Bi_2Te_3$

**Figure 2.** SEM image of cluster of 2D nanoflakes and white rods like structures are c-axis growth

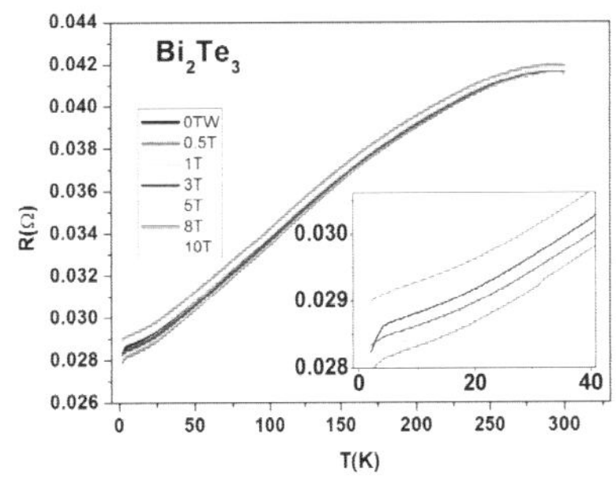


**Figure 3.** Temperature dependent resistance in different applied magnetic field (inset) R vs T at low temperature

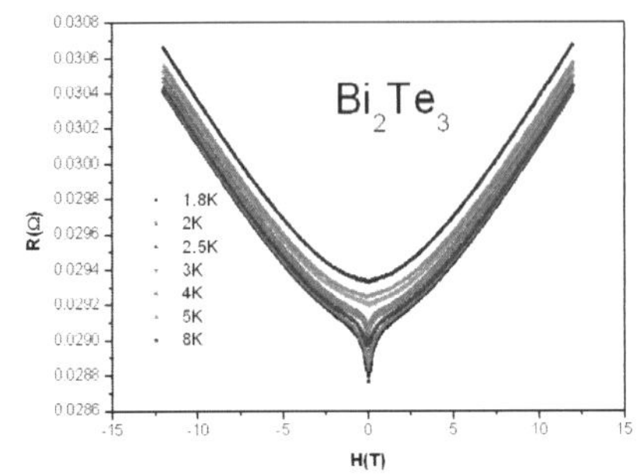


**Figure 4.**Magnetic field dependence of resistance for nanobulk assembly of $Bi_2Te_3$ in different temperatures

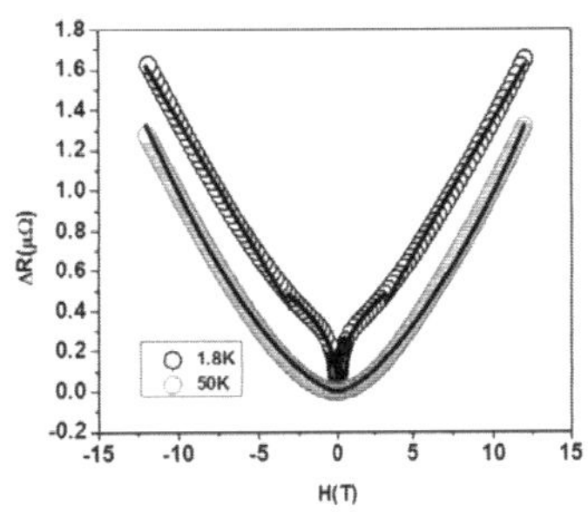


**Figure 5.** Magnetic field dependence of ΔR at 50K and 1.8K represented by red and black circles respectively while Black solid line is the power law fit

This contradiction demonstrates that the annealing treatment in 200$^{O}$C leads to the formation of more number of conducting channels due to the fusion of nanoflakes in a direction perpendicular to their surface area. Magnetic field dependence of MR at 50K is shown with red circles in figure 5. The solid line is the power law fit [11] for the equation

$$\Delta R \propto |B|^{\gamma} \quad (1)$$

Where ΔR is R(B)-R(0), B=$\mu_0$H and γ is the power factor which determines the nature of MR as a function of applied magnetic field. The value of γ obtained from fitted curve is found to be 2.41 that can be satisfactorily approximated as ΔR proportional to $B^2$ which denotes the parabolic fit for R vs H at 50K. This parabolic fit indicates the classical MR exhibited by most of the metals due to Lorentz force. On reducing the temperature below 8K, a prominent dip started appearing nearer to the zero Tesla. A prominent cusp in magneto conductivity or a dip in MR is considered to be the significant signature of WAL especially in TI materials due to the presence of π Berry's phase acquired by the surface states. Suppression of backscattering due to destructive quantum interference in Berry's phase leads to this WAL. This feature is one of the valuable indications for the transport that is happening along the surface state. The value of γ is found to be 0.45 as inferred from the power law fit in the narrow region below ±3T at 1.8K. Interestingly, the value of γ is nearer to 0.5. The value of γ points out the presence of 3dimentional (3D) WAL [12] instead of expected 2D WAL in TI materials. In the literature, quantum percolation is a phenomenon proposed theoretically and realized experimentally [12] using 3DWAL in $Bi_2Te_3$ sample. According to it, dot like structures are created artificially on the surface of TI materials which led to hybridization of 2D surface states along the walls of the dots. A detailed Transmission electron microscope (TEM) study [8] suggests that the $Bi_2Te_3$ flakes prepared by microwave assisted solvothermal reactions have intrinsic natural nanopores of 5nm along the surface of the flakes. These nanopores might be mimicking the artificially made percolation centers leading to the observed 3D WAL. The dip is not found in temperatures above 8K and MR vs H is broadened considerably as shown at 50K by the red symbol in figure 5. It can be attributed to the decrease in phase coherence at higher temperature as observed in many of the TI materials.[5]. Interestingly, $\gamma$ obtained above $\pm$3T using power law fit shows a value of 0.99 which shows that the obtained MR as a function of magnetic field is almost linear and it is very much beneficial for the MR device applications. The linearity in MR is consistent with the reported values in TI materials [11]. It can be attributed to the domination of dissipation-less 2D surface transport over the 3D bulk conductivity in low temperatures which might have modified the overall nature of MR at 1.8K when compared to MR at 50K. More measurements using various microprobes such as Hall- Effect as well as an analysis of high magnetic field MR in various magnetic field orientation is in progress which would be published else were.

## 5. Conclusions

In conclusion, we have probed the surface electrical transport properties of nanobulk assembly of $Bi_2Te_3$ nanoflakes using electrical resistivity and MR measurements which unambiguously support the signature of WAL at low temperature. Signature such as the prominent dip at low temperatures is attributed to be arising from the 3D WAL due to the intrinsic nanopores formed on the surface of the flakes which interrupt the 2D surface states in the form of percolation centers. This nanobulk assembly of TI materials is a promising candidate for spintronic /MR devices which on further optimization can open- up a new gateway for large scale application in NEMS and MEMS technology.

## Acknowledgments

Authors thank Dr. A. K. Sinha for his encouragement. The staff of low temperature and cryogenic laboratories especially Mr. P.Saravanan is thanked for their technical support. Authors would like to thank Dr. D.M.Phase and Mr. Vinay K Ahire for SEM/EDAX measurements and Mr.Mohan kumar Gangrade for AFM measurements. DST-India is thanked for their initial support for the LTHM facilities at UGC-DAE CSR Indore.

## References

[1] Hasan, Zahid M and Charles L Kane 2010 *Rev.mod.phys.* **82** 3045.
[2] Qu D - X, Hor Y S, Xiong J, Cava, R J and Ong N P 2010 *Science* **329** 821.
[3] Ning W, Haifeng D, Fengyu K, Jiyong Y, Yuyan H, Mingliang T, and Yuheng.Z 2013 *Scientific reports* **3.**
[4] He, Hong-T, Gan W, Tao Z, Iam-K, George KL W, Jian-Nong W, Hai-Z, Shun-Q, and Fu 2011 *Phys. Rev. Lett.* **106**, 166805.
[5] Zhang, Yu H, Bao D H, SLi, Wang C X, and Yang G W 2012 *Phys. Rev.B* **86** 075102.
[6] Pavlosiuk, Orest, Dariusz Kaczorowski, and Piotr Wiśniewski 2015 *Scientific reports* **5**
[7] Mehta, Rutvik J., Zhang Y, Karthik C, Singh B, Siegel R W, Borca-Tasciuc T, and Ramanath G 2012 *Nat. Mater.* **11** 233.
[8] Zhang, Yanliang, Rutvik M, Matthew B, Liang H, Ganpati R, and Theodorian B 2012 *Appl. Phys. Lett.* **100** 193113.
[9] Mehta, Rutvik J., Zhang Y, Karthik C, Singh B, Siegel R W, Borca-Tasciuc T, and Ramanath G 2010 *ACS nano* **4** 5055.
[10] Chiu S P and Lin J J 2013 *Phys.Rev. B* **87** 0351221.
[11] Roy A, Samaresh G, Sushant S, Rik D, Tanmoy P, Amritesh.R, Hema C P, Luigi C, and Sanjay K. 2013 *Appl. Phys. Lett.* **102** 163118.
[12] Song M., Chu J H, Zhou J, Tongay S, Liu K, Suh J, Chen H, Kang J S, Zou X. and You L. 2015 *Nanotechnology* **26** 265301.

# Determination of bond lengths from extended X-ray absorption fine structure in cobalt(III)-oxo cubane-like clusters

**N Nitin Nair[a], B D Shrivastava[a], Birinchi Kumar Das[b]**

[a] School of Studies in Physics, Vikram University, Ujjain-456010, India.
[b] Department of Chemistry, Gauhati University, Guwahati 781014, India.

E-mail: nnitinnair@gmail.com

**Abstract.** The extended X-ray absorption fine structure (EXAFS) at the K-edge of cobalt has been studied in two cobalt complexes having Co(III)-oxo cubane-like clusters of the type $Co_4O_4(O_2CR)_4L_4$ where R is $CH_3$ and L is pyridine (py) in one of the complex and ammonia ($NH_3$) in the other complex. The spectra have been recorded at BL-9 scanning EXAFS beamline at the 2.5-GeV INDUS-2 Synchrotron, RRCAT, Indore, India. The positions of EXAFS maxima and minima have been reported. Using this data, the bond length has been determined by simple bond determination methods, viz., Levy's, Lytle's and Lytle, Sayers and Stern's (LSS) methods. The normalized EXAFS spectra have been Fourier transformed and the value of the bond length has also been determined from the position of the first peak in the Fourier transform. This distance is the phase-uncorrected bond length. LSS method also gives such bond length. The results obtained from Fourier transform and LSS methods have been found to be comparable to each other. For the pyridine complex, the value obtained from Levy's method has been found to be in agreement with the available crystallographic value.

## 1. Introduction

X-ray absorption spectra at the K-edge of cobalt have been studied in two cobalt complexes having Co(III)-oxo cubane-like clusters of the type $Co_4O_4(O_2CR)_4L_4$ where R is $CH_3$ and L is pyridine (py) in one complex and ammonia ($NH_3$) in other complex. Das and Chakrabarty [1], have discussed the synthesis and physicochemical characteristics of these complexes and also the role of these complexes in catalytic oxidation under homogeneous conditions. The present paper deals with the determination of bond lengths from extended X-ray absorption fine structure (EXAFS) data of these complexes using four different simple bond determination methods, without doing rigorous theoretical calculations.

## 2. Experimental

The complexes $Co_4O_4(CH_3COO)_4(Py)_4$ and $Co_4O_4(CH_3COO)_4(NH_3)_4$ were prepared according to the standard methods reported in literature [1] and their purity was checked by standard methods. X-ray absorption spectra at the K-edge of cobalt have been recorded at the BL-9 scanning type EXAFS beamline at the 2.5-GeV INDUS-2 synchrotron, RRCAT, Indore, India. The key component of the beamline is double crystal monochromator (DCM) which is used for energy selection from the white synchrotron beam. The optics of this beamline consists of two Rh and Pt coated meridional cylindrical

**Table 1.** EXAFS data for Co complexes

| EXAFS Structure | n (for LSS method) | Q ( for Lytle's method) | $Co_4O_4(CH_3COO)_4(Py)_4$ | | $Co_4O_4(CH_3COO)_4(NH_3)_4$ | |
|---|---|---|---|---|---|---|
| | | | k ( Å$^{-1}$) | E (eV) | k ( Å$^{-1}$) | E (eV) |
| A | 0 | 2.04 | 2.35 | 21.00 | 1.85 | 13.01 |
| α | 1 | | 3.20 | 38.94 | 3.10 | 36.54 |
| B | 2 | 6.04 | 4.40 | 73.61 | 4.10 | 63.92 |
| β | 3 | | 5.63 | 120.31 | 5.23 | 103.80 |
| C | 4 | 12 | 6.95 | 183.66 | 6.70 | 170.68 |
| γ | 5 | | 7.80 | 231.33 | 7.30 | 202.62 |

For explanation of n and Q, see text.

mirrors and two Si(111) crystals. First one is plane crystal and second one is mechanically bendable (Sagittal) crystal which helps to focus beam horizontally at the fixed sample position.

## 3. Data Analysis

The X-ray absorption spectra, recorded at the beamline, have been processed and analyzed using the available computer software package, *Athena* [2]. The spectrum was first normalized by regressing a linear function to the pre-edge region (-200 to -30 eV before the edge energy) and by regressing a linear or quadratic function to the post-edge region (50–1000 eV above the edge energy). Normalized μ(E) spectrum was produced by subtracting the pre-edge line from the entire spectrum data and then dividing the spectrum by the step height [3]. Normalized μ(E) data were converted to the EXAFS signal χ(k) data, which were then Fourier transformed. The χ(k) spectra obtained from normalized μ(E) spectra and the Fourier transforms are given in figure 1 and figure 4 respectively. The inflection point on the K-edge was taken as the reference point for the measurements of the extended fine structure. The maxima in the EXAFS have been designated by capital Latin letters A, B , C, etc., whereas the minima by small Greek letters α,β,γ, etc.. The positions of these maxima and minima are given in table 1 in energy E (eV) as well as in wave vector k (Å$^{-1}$).

## 4. **Results and Discussion**

The aim of the present work is to determined bond lengths in the complexes from the EXAFS data by four different methods and to compare the results with the available crystallographic data in case of the pyridine complex as well as to compare and discuss the results obtained from the different methods for both the complexes. The methods of determining bond length from the different methods are given below.

### *4.1 Levy's method:*

Levy [4] has given a method for obtaining the bond lengths from the analysis of EXAFS. According to Levy, the distance $\Delta E$, from the maximum at B to the minimum at β should be a measure of the radius $R_1$ of the first coordination sphere through Bragg relation $R_1 = (151 / \Delta E)^{1/2}$ Å . The values of $R_1$ determined by this method are given in table 2. $R_1$ calculated in this way represents the average bond length. The value of bond length for the pyridine complex is available from crystallographic measurements and hence we have compared the crystallographic value [1], given in table 1, with that determined from Levy's method. It is clear from table 2 that the two values are in agreement with each other in the pyridine complex.

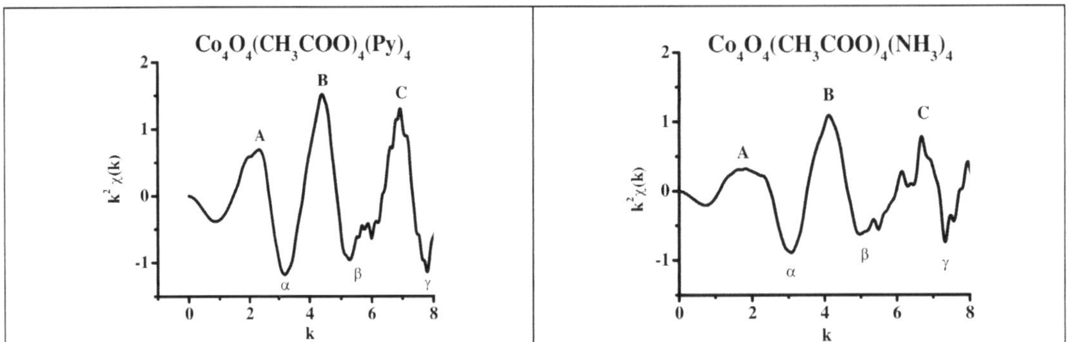


**Figure 1.** χ(k) vs. k spectra of cobalt complexes.

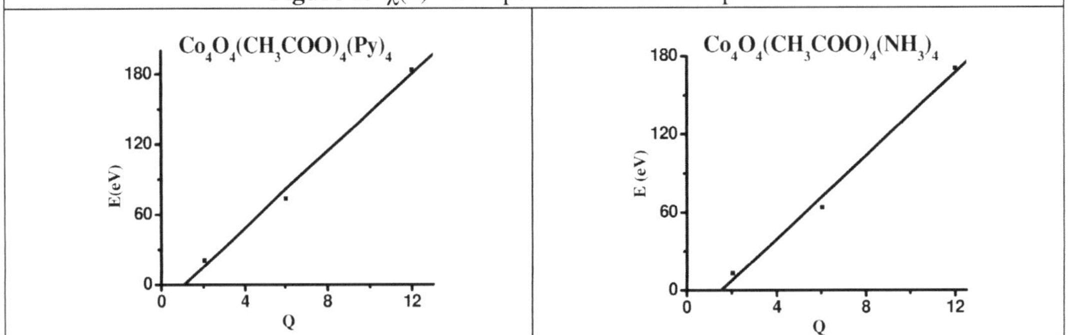


**Figure 2.** E vs Q plot (Lytle plot) for the cobalt complexes.

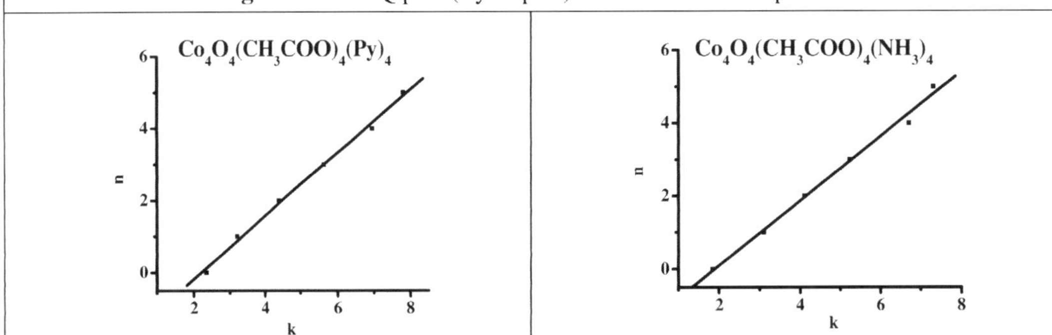


**Figure 3.** n vs k (in $Å^{-1}$) plot (LSS plot) for the cobalt complexes.

*4.2 Lytle's method:*

Lytle [5] has proposed a theory in which a nearly spherical polyhedron is constructed in the lattice and approximated by a sphere of equivalent volume with radius $R_s$ .In this method photoelectron ejected as a result of absorption of X-rays are thought of as if they are in a particle in a box. Lytle has found that an absorption maximum will occur whenever E satisfies the equation= $(h^2/ mR^2_s)Q$, where h is Planck's constant, m is the electron mass, and Q represents the zero roots of the half-order Bessel function, given in table 1. The plots of E vs. Q for the complexes studied are found to be linear and are shown in Fig 2. If M is the slope of this straight line, $R_s$ can be determined from the plot using the relation $R_s= (37.60/M)^{1/2}$Å . The values of $R_s$ are given in table 2. It may be remarked here that the interatomic distance obtained by this method has to be multiplied by a factor greater than unity and appropriate to the geometry of the system to obtain the correct bond length.

Journal of Physics: Conference Series **755** (2016) 012012 doi:10.1088/1742-6596/755/1/012012

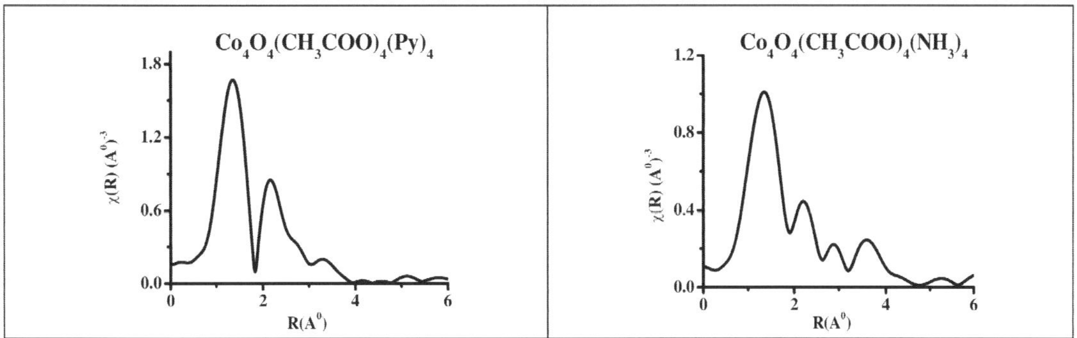


**Figure 4.** Fourier transformed EXAFS spectra for cobalt complexes

**Table 2.** Values of bond lengths (in Å )

| S. No. | Complex | Levy's method ($R_1$) | Lytle's method ($R_s$) | LSS method ($R_1$- $\alpha_1$) | Fourier Transform method | Crystallo-graphic value ($R_1$) |
|---|---|---|---|---|---|---|
| 1 | $Co_4O_4(CH_3COO)_4(Py)_4$ | 1.80 | 1.51 | 1.38 | 1.35 | 1.86 |
| 2 | $Co_4O_4(CH_3COO)_4(NH_3)_4$ | 1.95 | 1.53 | 1.39 | 1.35 | - |

*4.3 Lytle, Sayers and Stern's (L.S.S.) method:*
According to this method [6], plot of n vs k for the different maxima (n= 0, 2, 4, ….) and minima (n= 1, 3, 5, …) should results in a straight line. n vs k graphs for the complexes studied are found to be linear and are given in figure 3. The slope (M') of this line is given by $M' = 2(R_1 - \alpha_1)/\pi$. The value of ($R_1$ - $\alpha_1$) can be obtained from the slope of the plot. Once $\alpha_1$ is evaluated for a standard, $R_1$ for unknown materials can be determined. It is found that for chemically similar systems, the value of $\alpha_1$ remains more or less the same. The values of $R_1$ thus obtained by this method are given in table 2. The distance $R_1$ obtained from this method is phase uncorrected bond length.

*4.4 Fourier transform method:*
The position of the first peak in the Fourier transform (figure 4) corresponds to the distance between the absorbing atom and its nearest neighboring atoms, i.e., the radius of first coordination sphere. However, the bond length thus obtained is less than the actual bond length, because it is phase uncorrected. The values for the phase uncorrected bond lengths, as determined from the Fourier transforms, for the two complexes are given in table 2. The phase corrected bond length can be determined by fitting appropriate theoretical model to the experimental Fourier transform curve. This has not been done in the present work.

## 5. Conclusions
The values of the bond length as determined from Lytle's method are found to be slightly lesser than the corresponding bond lengths obtained by Levy's method. This may be ascribed to the fact that Levy's method gives the radius of coordination sphere directly where as Lytle's method does not. In the later method, the inter-atomic spacing is obtained by multiplying the obtained bond length by an appropriate factor. The value of bond length for the pyridine complex is available from crystallographic measurements and hence we have compared the crystallographic value for this complex, with that determined from Levy's method. It has been found that the two values are in agreement with each other in the pyridine complex, showing that the Levy's method is suitable method for determining bond length, when it is not possible to determine phase corrected bond length

from the Fourier transform by fitting an appropriate theoretical model to the experimental Fourier transform curve. Regarding the values of the phase-uncorrected bond length obtained from LSS method, it has been found that they are comparable with the values of phase-uncorrected bond length obtained from Fourier transformation method.

**References:**

[1] Das B K and Chakrabarty R 2011 *J. Chem. Sci.* **123** 163.
[2] Ravel B and Newville M 2005 *J. Synchrotron Rad.* **12** 537
[3] Kelly S D, Hesterberg D and Ravel B 2008 *Methods of Soil Analysis, Part 5 Mineralogical Methods*, Soil Science Society of America, Madison, WI p. 387.
[4] Levy R M 1965 *J. Chem. Phys.* **43** 1846
[5] Lytle F W 1966 *Adv. X-ray Annl.* **9** 398
[6] Lytle F W, Sayers D E and Stern E A 1975 *Phys. Rev. B* **11** 4825

# EXAFS study of Fe (II) complexes using Dispersive EXAFS beamline

**Samrath Ninama[1], A.Mishra[2]**

[1]Kasturbagram Rural Institute, Kasturbagram, Indore
[2]School of Physics, VigyanBhawan, Devi Ahilya University, Indore - 452001, India

Email: samrathn@gmail.com

**Abstract**: Extended X-ray absorption fine structure (EXAFS) spectra have been recorded at the K-edge of Fe (II) complexes using the energy dispersive EXAFS beam line at 2.5GeV Indus -2 synchrotron source at RRCAT, Indore, India. A theoretical EXAFS data analysis is also carried out by Fourier analysis of experimental EXAFS data of the two iron (II) complexes. This analysis includes details of the Fourier transform of the data and the extraction of metal-ligand bond length. Bond lengths determined from data analysis methods are compared with the bond lengths obtained from several other known existing techniques. These data have also been calibrated by derivative method and bond lengths have also been obtained from Fourier transformation method and the results have been compared with the each other. The EXAFS data have been analysed using the computer software *Athena.*

## 1. Introduction

In the present work, the X-ray absorption spectra have been recorded at the K-edge of Iron in two macro-cyclic complexes. The coordination chemistry of macro-cyclic ligand is a fascinating area of research. The extended X-ray absorption fine structure (EXAFS) spectroscopy is an effectivetechnique for selectively investigating the local coordination environment around the metal active site
of complexes [1]. The EXAFS K-edge of Fe (II) complexes were recorded at 7112eV and alsodetermined the bond length between metal-ligand of (a) Fe 1-Fe(II)2-methyl-3[(bis-orthonitophenyl)]-3H 1, 5 benzodiazepine chloride , (b) Fe 2-Fe(II)2-methyl-3[(bis-meta nitrophenyl)]-3H
1, 5 benzodiazepine chloride.
A method for determining the metal-ligand bond length by an appropriate Fourier analysis of theK-edge data has been presented. This is illustrated by a detailed analysis of the data for a series ofmetal organic based on pyridine ligand systems, which model active centres of natural coppercontaining metal proteins [1]. The Fourier transform of the experimental EXAFS data is done usingthe IFEFFIT program [2] and metal-ligand bond length is determined. Applications of EXAFS can befound in literature on a variety of fields, namely, chemical reactions at the solid state [3], surfacetreatments [4] and structural studies of thin films [5]. A structural analysis of X-ray absorption finestructure (XAFS) spectroscopy has been used for a local structural refinement on an atom of interest[6]. In general, the small pre-edge peaks of the K-edge absorption spectra for transition metalcompounds have been assigned to the transition from 1s to *n*d orbital, even though it is a dipoleforbidden transition. Since the transition is very sensitive to chemical environment, the pre-edgefeature in XAFS studies has been used to infer the local structure around the central atom. Somemolecular orbital calculation studies have been reported that the pre-edge peak intensity is related to3d and 4p orbital mixing by

perturbation of site symmetry and increases gradually with the departurefrom a centro-symmetric environment [7]. In this context, the extended X-ray absorption fine structure(EXAFS) spectra have provided many quantitative structural characterizations such as inter-atomicdistance, coordination number and Debye-Waller factor. Since the EXAFS refinement does notdepend on the long-range order of atomic arrangement and is very sensitive to atomic local sites, it is apowerful technique for local structural analysis of partially substituted inorganic compounds.

## 2. Material Synthesis

***Preparation of ligand (L)***

Iron Complexes have been prepared by chemical root method. The chemicals used are of puregrade. O-phenylenediemine (0.05 mol, 5.0g), substituted aniline(R) (0.05 mol, 6.9g) dissolved inethanol (25 mL), and ethyl acetoacetate (0.05 mol, 6.7g) dissolved in ethanol (25 mL). The mixtures ofboth reactants were refluxed for 6 hour. The volume concentrated on the steam bath until whitecolored crystals began to form. The product was purified by crystallization from hot alcohol. Thuswhite crystalline pure macro-cyclic ligand was collected.

***Preparation of [Fe (II)-L-Cl2]***

Hot ethanol solution of ligand (L) (0.01 m. mol, 0.2 g), hot ethanol solution of Iron chlorides [FeCl3.2H2O] with few drops of hydrochloric acid was added. The solution mixture was then refluxed for 4h. On cooling, colored precipitates were then filtered and washed with cold water and dried in vacuum [8].

## 3. Experimental

The complexes were finely powdered and then absorption screens were prepared by carefullyspreading calculating amount of the powdered on 1 cm area ofkapton tape. The Iron K-edge EXAFSspectra of these absorption screens have been recorded at BL-8 dispersive EXAFS beamline at 2GeVIndus -2 synchrotron radiation source at Raja RamannaCenter for Advanced Technology (RRCAT),Indore, India [9-12]. The energy calibration of beam line for a particular setting of the polychromatorcan be done by recording the absorption spectra of two standards, whose absorption edge energies arewell established. For example for recording the absorption spectra of Fe sample, the two standards canbe Fe metal foil and Gd oxide powder. The absorption spectra of Fe K edge and Gd2O3 L3-edge havebeen recorded under the same setting of crystal bender and goniometer. In the present work we aretaking the values of the energies of Fe K-edge in metal and Gd L3 –edge in Gd2O3 as 7709 eV and7243eV respectively [13].

## 4. Results and discussion

The X-ray absorption spectra, i.e., normalized $\mu$ (E) vs. E curve of Iron complexes is shown infigure 1. The bond lengths have been determined using the following four methods, i.e., Levy's, Lytle,Lytle, Sayers and Stern's (LSS) and Fourier transformation methods and the results were obtained.

### 4.1 Levy's method

The bond lengths According to Levy's method [14], are given by Bragg relation $R1 = [151/\Delta E]^{1/2}$ Å, where $\Delta E$ is the difference in eV of the energies of the EXAFS maximum B and minimum. R1 isthe radius of the first coordination sphere. The R1 calculated bond length value of Fe foil, Fe1complex and Fe2 complex are 2.25, 2.22 and 2.20 are respectively.

### 4.2 Lytle's method

According to Lytle [15] method the bond length (Rs) is evaluate by the radius Rs of equivalentpolyhedron through the relation Rs = [37.60 / M]1/2.where M is slope of E vs Q plot ,energy values (E)is EXAFS maxima and Q = 2.04, 6.04, 12.0, and 20.0 are constant . The Rs calculated bond length value of Fe foil, Fe1complex and Fe2 complex are 1.85, 1.82 and 1.80 are respectively.

### 4.3 Lytle, Sayers and Stern's (LSS) method

In the LSS method [16], for determination of the nearest neighbour distances, gives the value of $(1/2+n)\pi = 2k(R1 - \alpha 1)+2\beta 1$ ,where R1 is the bond length. $(R1 - \alpha 1)$ is the phase uncorrected bond length value of Fe foil, Fe1complex and Fe2 complex are 1.40, 1.32 and 1.30 are respectively.

Although, Fourier transform of EXAFS data is a general and powerful technique todetermine many

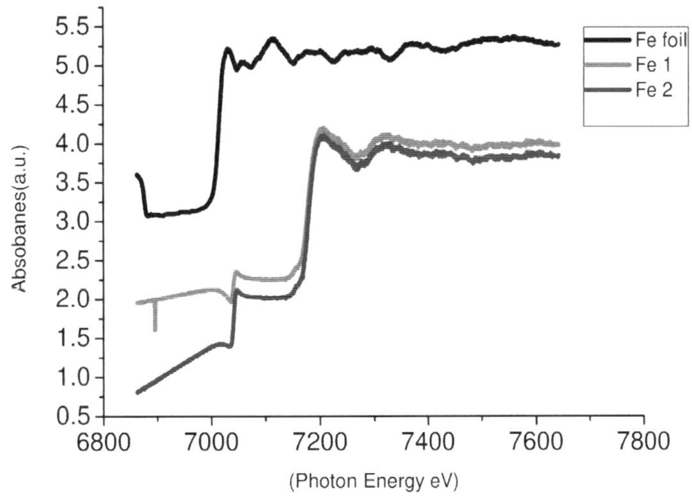


**Figure 1-** EXAFS spectra of Fe (II) complexes.

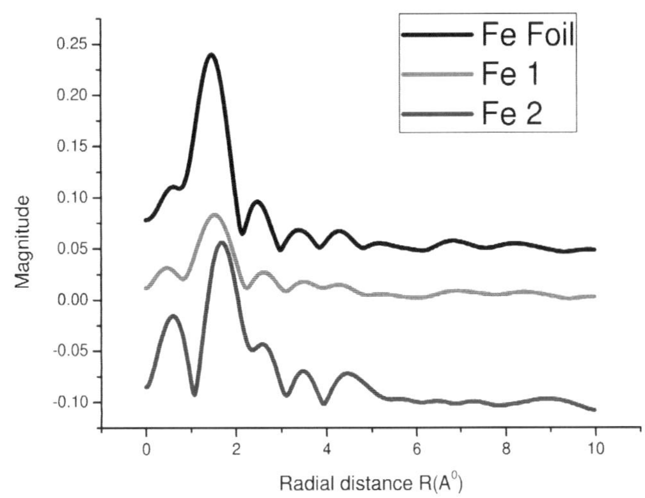


**Figure 2-** Magnitude of Fourier transformation in first shell maxima for Fe (II) complexes.

physical parameters; however, it is possible to determine nearest-neighbour distances and phase shifts from a simple graphical technique. This is possible because the first shell scattering usually dominates the EXAFS curves; then, if only the major EXAFS peaks are used, the period will be that of the first coordination shell. In this method, the energy positions of the maxima and mini main EXAFS are measured with respect to the inflection point on the main absorption edge, which signifies the Fermi level in metals and the first unoccupied level of suitable symmetry available for theabsorption of the ejected electron in the case of compounds.

**4.4 Fourier transformation method**

The magnitude of Fourier transform spectra are shown in figure 2. The Fourier transforms peaks at the radial distances of the neighbouring atoms from the absorbing atom. The bond length value of Fe foil, Fe1complex and Fe2 complex are 1.95, 1.92 and 1.90 are respectively. The distances found in Fouriertransform are, however, shorter by 0.2-0.5 Å than the actual distances due to energy dependence of thephase factors in sine function of the theoretical expression for EXAFS, which is known as the EXAFSequation [17]. The peaks in the Fourier transform are shifted towards the origin by an amount αj andhence the peaks are at distances Rj-αj. For the first peak j=1 and hence the position of the first peakdetermines the distance R1-α1.There are some subtleties associatedwith doing the FT of EXAFS data but techniques for dealing with these are well developed. Any signalthat is a sum of sine waves can be Fourier transformed to separate each frequency component indistance R, i.e., a Fourier transform separates a signal into its Fourier components. Each Fouriercomponent is defined by an amplitude and phase. The Fourier transform is a complex function containing both real and imaginary parts. The magnitude of the Fourier transform is an envelope of thereal and imaginary parts of the Fourier transform.

## 5. Conclusion

Calibration of particular setting of the polychromator has to done and it has been shown by derivative method, and after this EXAFS spectra has been recorded and comparison of the values of bond length Rs is seen to have slightly lesser values than the corresponding bond lengths obtained by Levy'smethod .This may be described to the fact that Levy's method gives the radius of coordination sphere
directly whereas Lytle method does not. In later method, the inter atomic spacing is obtained bymultiplying Rs with a factor Appropriate to the geometry of the system .the value of the bond length
obtained from LSS method are closer to obtained by Levy's method and Fourier transformationmethod.

## Acknowledgement

The authors are thankful to Dr. ShambhuNathJha, RRCAT, Indore for their help in recording the spectra at RRCAT, Indore, India.

## References

[1] Vlasenko V G, Uraev A I, ZubavichusYa V, Chernyshov A A, Garnovskii A D, &Mamin R K 2008 *Bull. Russian Acad. Sci.: Phys.* **72** 468.
[2] Newville M 2001 *J. Synch. Rad.* **8** 322.
[3] d'Acapito F, Ghigna P, Alessandri I, Cardelli A, &Davoli I 2003 *NuclInstrum Methods: Phys ResB.* **200** 421.
[4] Lutzenkirchen and Frahm R 2005 *Physica B.* **357** 213.
[5] d'Acapito F , Milita S, Satta A, and Colombo L 2007 *J ApplPhys.* **102** 043524.
[6] Rehr J J, Leon J M, Zabinsky S I, Albers R C 1991 *J Am ChemSoc.* **113** 5135.
[7] Westre T E, Kennepohl P, DeWitt J G, Hedman B, Hodgson K O, Solomon E I 1997 *J Am ChemSoc.* **119** 6297.
[8] Mayura Panchbhai A et al. 2008 *E-Journal of Chemistry.* **S1**5.
[9] Bhattacharyya D, Poswal A K, Jha S N, Sangeeta ,Sabharwal S C 2009 *Bull. Mater.Sci.* **32** 103.
[10] Bhattacharyya D, Poswal A K, Jha S N, Sangeeta , Sabharwal S C 2009 *Nucl. Instrum.Meth. A*; **609** 286.
[11] Das N C, Jha S N, Bhattacharyya D, Poswal A K, Sinha A K and Mishra V K 2004 *Sadhana* **29** 545.
[12] Lee P L, Beno M A, Jennings G, Ramanathan M, Knapp G S, Huang K, Bai J, Montano P A 1994 *Rev. Sci. Instrum.* **65**.
[13] Matthew A. Marcus, Andrew J. Westphal and Sirine C. Fakra 2008 *J. Synchrotron Rad.* **15**, 463.
[14] Levy R M 1965 *J. Chem. Phys.***43** 1846.
[15] Lytle F W 1966 *Advances in X-ray Analysis.* **9** 398.
[16] Lytle F W, Sayers D E and Stern E A 1975 *Phys. Rev. B.* **11** 4825.
[17] Stern E A , Sayers D E and Lytle F W 1975 *Phys. Rev. B.* **11** 4836.

# Calculation of Coster-Kronig energies and transition probabilities by linear interpolation method

**R K Trivedi[1], Uma Shrivastava[1], V K Hinge[2], B D Shrivastava[1]**

[1]School of Studies in Physics, Vikram University, Ujjain – 456010 M.P.
[2]Physics Department, Ujjain Engineering College, Ujjain 456010 M.P.

Email: rajeevtrivedi80@gmail.com

**Abstract.** The X-ray emission spectrum consists of two types of spectral lines heaving different origins. The diagram lines originate because of transitions in singly ionized atom, while the non-diagram lines or satellites originate due to transitions in doubly or multiply ionized atom. The X-ray satellite energy is the difference between the energies of initial and final states which are both doubly or multiply ionized. Thus, the satellite has a different energy than the energy of the X-ray diagram line. Once the singly ionized state has been created, it is the probability of a particular subsequent process that will lead to the formation of two-hole state. The single hole may get converted through a Coster-Kronig transition to a double hole state. The probability of formation of double hole state via this process is written as $\sigma.\sigma'$, where $\sigma$ is the probability of creation of single hole state and $\sigma'$ is the probability of the Coster-Kronig transition. The value of $\sigma'$ can be taken from the tables of Chen *et al.* [1], who have presented the calculated values of $\sigma'$ for almost all possible Coster-Kronig transitions in some elements. The energies of the satellites can be calculated by using the tables of Parente *et al.* [2]. Both of these tables do not give values for all the elements. The aim of the present work is to show that the values for other elements, for which values are not listed by Chen *et al.* and Parente *et al.*, can be calculated by linear interpolation method.

## 1. Introduction

The spectrum of the radiation emitted from an x-ray tube consists of characteristic spectrum of the target superimposed over the continuous spectrum. The spectral lines of the characteristic spectra resulting from transitions between atomic states, involving single vacancy are called diagram lines, because energy of such lines can be expressed as the difference of two terms in the single vacancy energy level diagram. The diagram lines are generally found to be accompanied by groups of lines of slightly different energies and usually smaller intensities. The energies of these lines do not correspond to the energy difference between any two states of the normal single vacancy energy level diagram of the element concerned. These lines are known as X-ray satellite lines or non-diagram lines.

Since the discovery of the X-ray satellites, the problem of finding the origin of these satellites has attracted the attention of many workers and various theories have been put forward from time to time. The most widely accepted theory of X-ray satellites is the multiple ionization theory. The parent diagram line is emitted by a transition in the singly ionized atom. The X-ray satellite line is emitted due to transition in the multiply ionized atom. The X-ray satellite energy is the difference between the energies of initial and final states which are both doubly or multiply ionized. Thus, the satellite has a different energy than the energy of the X-ray diagram line.

After the creation of a single-hole state, it is the probability of a particular subsequent process that leads to the formation of initial two-hole state, required for the emission of a satellite. The single-hole state can get converted into the two-hole states by Auger transition. The process of radiationless reorganization of an atom ionized in an inner shell is known as the Auger effect. After an Auger transition an atom may be left doubly ionized in inner shells.

Earlier attempts of explaining the origin of X-ray satellites and assigning transitions to them were mainly confined to energy considerations alone. In the recent years, however, intensities of the various transitions have also been considered and attempts have been made to compute theoretical X-ray satellite spectra. This can be done by calculating the theoretical X-ray satellite spectrum. To obtain such a spectrum, the requirements are:

(i) The energies of the different transitions which are allowed between the initial and final double vacancy states,

(ii) The relative probabilities of such transitions.

Ab initio calculations of energies and intensities of various double hole state transitions, considering multiplicity of energy levels, is a difficult and laborious task. It is also not necessary because, alternatively, calculations can be done with the help of several tabulations, in which necessary data are available. Such tabulations for calculations of energies of doubly ionised atoms are the following tables of Auger-electron energies:

(i) Tables of Larkins [3] The energies have been calculated by means of a general semi empirical method developed within an intermediate coupling framework. He has used the data derived from atomic Hartree-Fock calculations and has combined it with experimental electron subshell binding energies data.

(ii) Tables of Parente et al. [2] which contain the energies of L-series X-ray satellites that arise from the presence of one 'spectator hole' in the M-shell or N- shell of the emitting atom, for 11 elements, namely $_{65}$Tb, $_{67}$Ho, $_{70}$Yb, $_{74}$W, $_{78}$Pt, $_{80}$Hg, $_{85}$At, $_{88}$Ra, $_{90}$Th, $_{92}$U and $_{95}$Am. The energies of initial and final atomic states have been computed separately to allow for complete relaxation. Dirac-Hartree-Slater (D.H.S.) wave functions have been used with a local approximation to the atomic potential.

Since, the tables of Parente et al. are relativistic calculations and those of Larkins are non relativistic calculations, hence it is preferable that the tables of Parente et al. are used for calculations of the energies of the L emission satellites.

To calculate the probability of creation of the initial double vacancy state by the process of Coster-Kronig transition, the cross section for creation of single vacancy state $\sigma$ is first calculated by the formulae given by Moores et al. [5] and then this cross section is multiplied by the transition rates for the Coster-Kronig transition $\sigma'$ as given in the following tables Coster-Kronig transition probabilities. After multiplication, one gets the Coster-Kronig transition probability for the creation from the single vacancy state to the double vacancy state, i.e., $\sigma.\ \sigma'$.

(i) Tables of McGuire [4] which contain non-relativistic calculations for Coster-Kronig transition probabilities for creation of double vacancy states.

(ii) Tables of Chen et al. [1] which contain relativistic calculations of the rates of all energetically possible radiationless transitions with intensities above 10-5 milli atomic units with initial K- and $L_{1,2,3}$ - shell vacancies in elements with $18 \leq Z \leq 96$.

Since, the tables of Chen et al. are relativistic calculations and those of McGuire are non relativistic calculations, hence it is preferable that the tables of Chen et al. are used for calculations of the probabilities of creation of two hole-states which are required for the L emission satellites.

We have been interested in the study of origin of L emission satellites in the L$\gamma$ and L$\alpha$. For this, both the energies and intensities of the various transitions which can give rise to a particular satellite in doubly ionised atom have to be calculated and then the theoretical satellite spectrum has to be synthesised. By comparing the theoretical spectrum with the experimental spectral energies it is possible to assign the transition to a particular satellite accurately. Since we want to find the origin of the satellites in all the elements, we require energies and transition probabilities of all the elements which are not available in the tables of Parente et al. and Chen et al. respectively. In the present paper we have shown that these values can be obtained from the tables mentioned above by linear interpolation method.

## 2. Method of calculation

### *2.1 Calculation of energies*

Chen et al.[1] have given in their tables probabilities of Coster - Kronig transitions for elements $_{70}$Yb, $_{74}$W, $_{80}$Hg, $_{83}$Bi, and $_{88}$Ra. In the present work, the values for other elements have been obtained by linear interpolation method. The justification for linear interpolation can be seen from a sample graphs presented in figs. 1(a) and 1(b) for K - $L_1M_1$, K - $L_1M_2$, K - $L_1M_3$, K - $L_1M_4$ and K- $L_1M_5$ Auger transitions. In this figure the values of transition probabilities of elements given by Chen et al. have been plotted against Z values. The graphs are found to be linear in small regions, justifying the use of linear interpolation method for finding transition probabilities for intermediate elements.

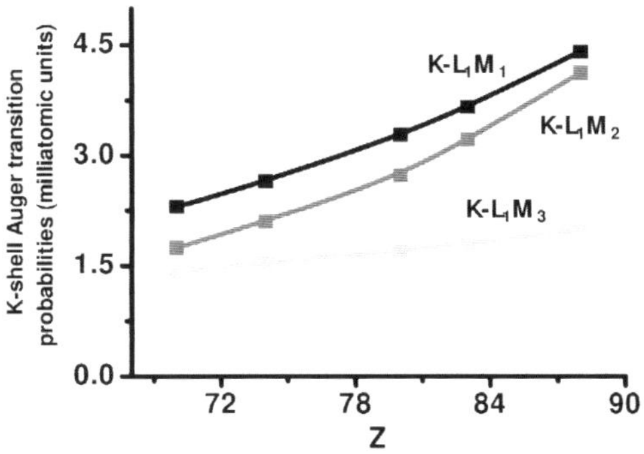


**Fig. 1(a) Variation of K-shell Auger transition probabilities (from tables of Chen et al.) with Z. This graph shows that linear interpolation can be used to find values for intermediate elements.**

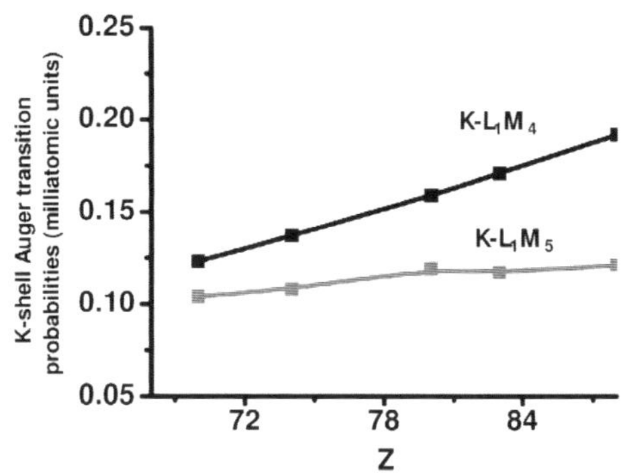


**Fig. 1(b) Variation of K-shell Auger transition probabilities (from tables of Chen et al.) with Z. This graph shows that linear interpolation can be used to find values for intermediate elements.**

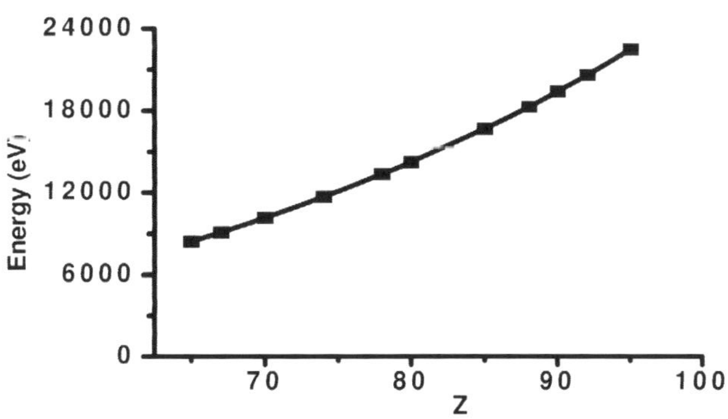


**Fig. 2 Variation of energy of $L_1M_1\,^1S_0$-$M_1N_2\,^1P_1$ transition (from tables of Parente et al.) with Z. This graph shows that linear interpolation can be used to find values for intermediate elements.**

### 2.2 *Calculation of relative intensities of transitions*

The relative intensities of the transitions can be calculated by first calculating the probability of initial single hole creation in the K shell by following the method of Moores et al. [5]. For calculating the probability of convertion of single hole K state to double hole state $L_1M_x$ through Coster - Kronig transition (K - $L_1M_x$) the tables of Chen et al. have to be used as mentioned above. The second method of calculation used in the present work is by employing the tables of Parente et al. As also pointed out above, Parente et al. have tabulated energies of all the possible transitions which can be assigned to L-emission satellites. However, they have tabulated the energies only for 11 elements namely, $_{65}$Tb, $_{67}$Ho, $_{70}$Yb, $_{74}$W, $_{78}$Pt, $_{80}$Hg, $_{85}$At, $_{88}$Ra, $_{90}$Th, $_{92}$U and $_{95}$Am. The transitions included in this table are for all the electric dipole and electric quadrapole L X-ray transitions for which the final state comprises the vacancies in the M and N shells. The probabilities of the transitions are mentioned in the table for all the elements in general and not for individual elements. Hence, we have proposed that by linear interpolation the energies of all the possible transitions of the transition schemes $L_1M_x$ - $M_xL_{2,3}$ (x = 1-5) can be obtained for all the elements from the tables of Parente et al. The justification for using linear interpolation method for finding energies for different elements not mentioned in these tables can be seen from a sample graph reproduced in fig. 2 for a sample transition $L_1M_1\,^1S_0$ - $M_1N_2\,^1P_1$. In this figure the values of energies of this particular transition is plotted against Z values for the 11 elements for which Parente et al. have reported the energy values. The graph is linear, specially in small regions, justifying the use of linear interpolation method.

## 3. Results and discussions

For calculating the theoretical X-ray satellite spectrum, the energies and the relative probabilities of the different transitions which are allowed between the initial and final double vacancy states are required. The transition probabilities can be taken from the tables of Chen et al. [1], who have presented the calculated values for almost all possible Coster-Kronig transitions in some elements. The energies of the satellites can be calculated by employing the tables of Parente et al. [2]. Both of these tables do not give values for all the elements. In the present work we have shown that the values for other elements, for which values are not listed by Chen et al. and Parente et al., can be calculated by linear interpolation method.

## References

[1] Chen M H, Crasmann B and Mark H 1979 *At. Data. Nucl. Data Tables* **24** 13
[2] Parante F, Chen M H and Craseman B 1981 *At. Data. Nucl. Data Tables* **26** 383
[3] Larkins F P 1977 *Atom. Data Nucl. Data Tables* **20** 311
[4] McGuire E J 1971 *L-Shell Auger, Coster-Kronig and radiative matrix elements, and Auger and Coster-Kronig transition rates in j-j coupling* (Report No. Sc-RR-710075) (New Mexico: Sandia Laboratories).
[5] Moores D L, Golden L B and Sampson D H 1980 *J. Phys. B* **13** 385

# Comparative XAFS studies of some Cobalt complexes of (3-N-phenyl -thiourea-pentanone-2)

**Namrata soni[1], Neetu Parsai[1], Ashutosh Mishra[2]**
[1]Shri Vaishanav Institute of Management, Indore
[2]School of Physics DAVV, Indore

E-mail: namrata.123soni@gmail.com

**Abstract.** XAFS spectroscopy is a useful method for determining the local structure around a specific atom in disordered systems. XAFS study of some cobalt complexes of (3-N-phenyle-thiourea-pentanon-2) is carried out using the latest XAFS analysis software *Demeter with Strawberry Perl*. The same study is also carried out theoretically using *Mathcad* software. It is found that the thiourea has significant influence in the spectra and the results obtained experimentally and theoretically are in agreement. Fourier transform of the experimental and theoretically generated XAFS have been taken to obtain first shell radial distance. The values so obtained are in agreement with each other.

## 1. Introduction

EXAFS spectroscopy provides structural information about a sample by way of the analysis of its X-ray absorption spectrum. It allows determining the chemical environment of a single element in terms of the number and type of its neighbors, inter-atomic distances and structural disorders. XAFS spectroscopy is a useful method for determining the local structure around a specific atom in disordered systems. XAFS study of some cobalt complexes of (3-N-phenyle-thiourea-pentanon-2) is carried out using the latest XAFS analysis software *Demeter with Strawberry Perl*. The same study is also carried out theoretically using *Mathcad* software using method given by N Parsai et al [1, 2]. It is found that the thiourea has significant influence in the spectra and the results obtained experimentally and theoretically are in agreement. Fourier transform of the experimental and theoretically generated XAFS have been taken to obtain first shell radial distance. The values so obtained are in agreement with each other.

The five cobalt complexes are:

| *S. No.* | *Complexes* | *Abbreviations* |
|---|---|---|
| 1 | Co [3-N-(3nitro)-thiourea-pentanone-2] | Co [3N3NTU2P] |
| 2 | Co [3-N-(4nitro)-thiourea-pentanone-2] | Co [3N4NTU2P] |
| 3 | Co [3-N-(2-chloro)-thiourea-pentanone-2] | Co [3N2CTU2P] |
| 4 | Co [3-N-(3-chloro)-thiourea-pentanone-2] | Co [3N3CTU2P] |
| 5 | Co [3-N-(4-chloro)-thiourea-pentanone-2] | Co [3N4CTU2P] |

## 2. Methodology

The phenomenon of extended X-ray absorption fine structure (EXAFS) refers to the oscillation of the X-ray absorption coefficient as a function of X-ray energy above threshold. It has been attributed to an interference of the ejected photoelectrons at the site of the absorbing atom. For K-Shell excitation the modulating part of the absorption coefficient called the extended X-ray absorption fine structure (EXAFS) is described by

$$\chi(k) = \frac{m}{4\pi\hbar^2 k}\sum_j \frac{N_j}{R_j^2} t_j(2k)\exp\left|-2R_j/\lambda\right| \times \sin 2[kR_j + \delta_j(k)]\exp\left|-2k^2\sigma_j^2\right| \quad (1)$$

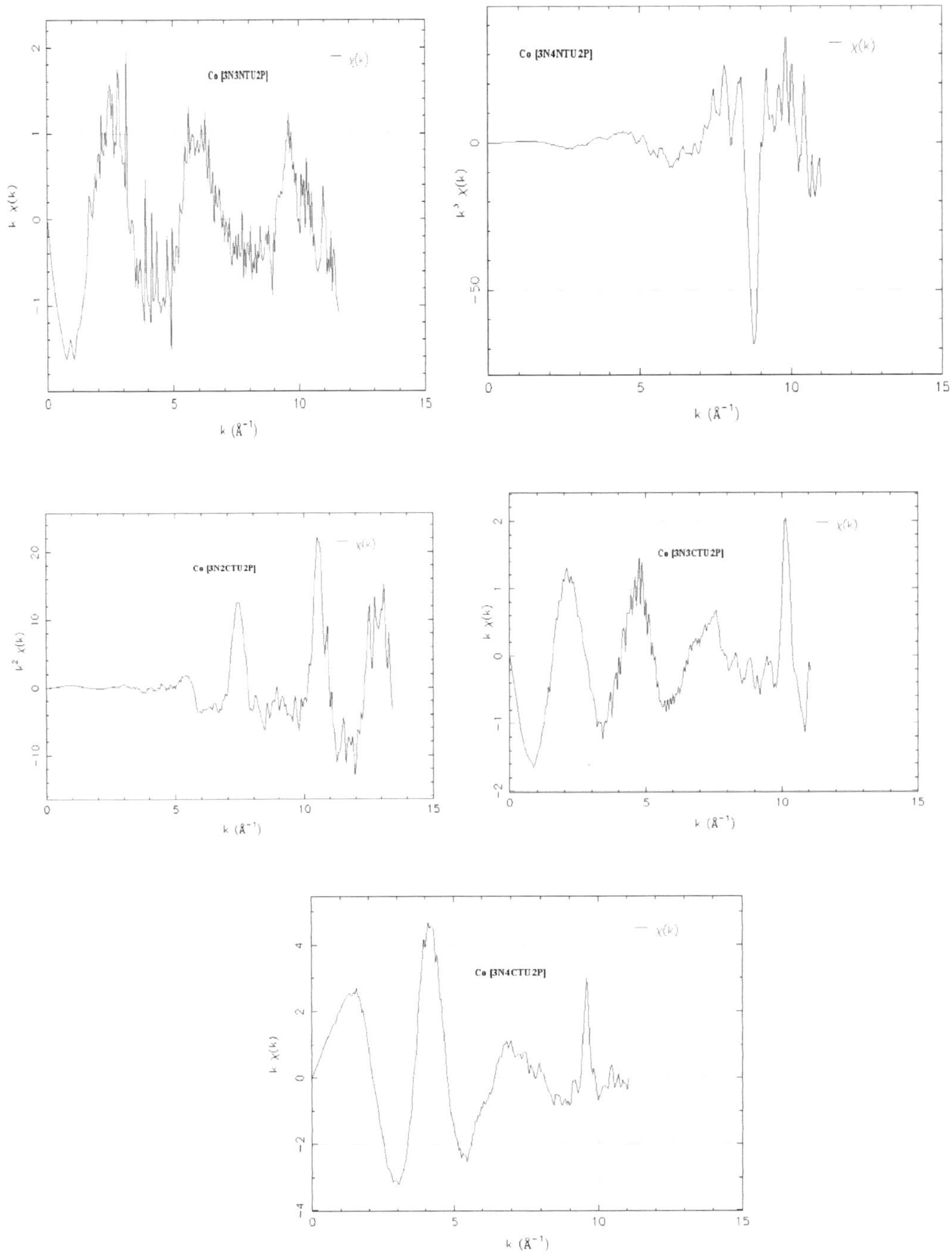


**Figure 1. χ(k) versus k curves for cobalt complexes.**

Journal of Physics: Conference Series **755** (2016) 012015 doi:10.1088/1742-6596/755/1/012015

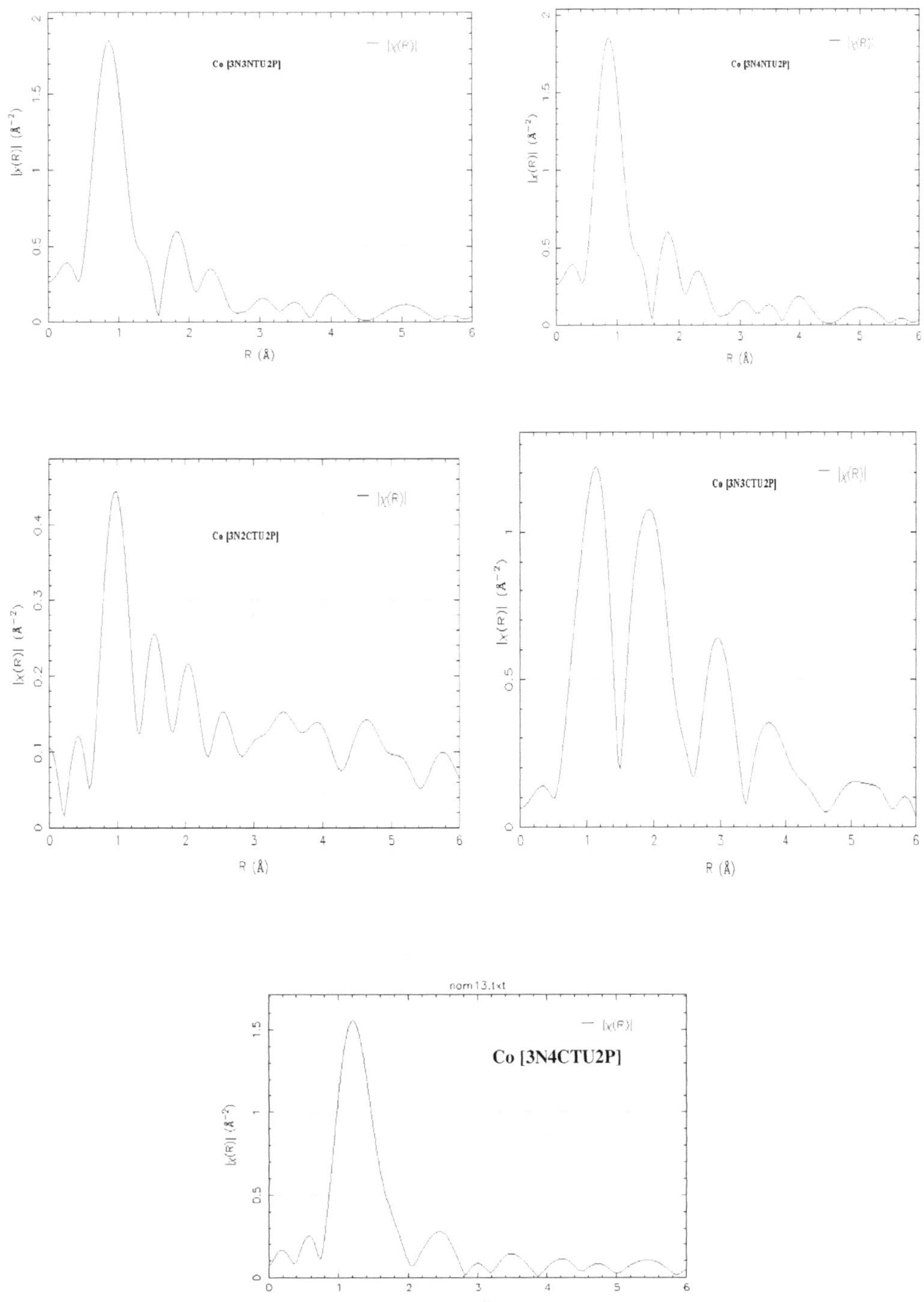


**Figure 2. Magnitude of F.T of χ(k) for cobalt complexes.**

Where k is the electron wave vector, $N_j$ is the number of atoms in the $j^{th}$ coordination shell, $R_j$ is the average radial distance to the jth atom, $t_j(2k)$ is the back-scattering matrix element encountered by the electrons, $\lambda$ is the mean free path of the electron, the $2^{nd}$ exponential containing $\sigma_j^2$ is a Debye-Waller-type term where $\sigma_j$ is the rms fluctuations of the atom about $R_j$, and $\delta_j(k)$ is a phase shift. The form of this equation is a sine like scattering from each shell of atoms at $R_j$ with the EXAFS signal proportional to the number of atoms surrounding the absorbing atom and inversely proportional to $R_j^2$. Each coordination shell contributes a sine like term of period $2kR_j$. The total result is a summation over all the coordination shells within range of the effect. Several attempts have been made to extract structural information from EXAFS. The most general methods used are fitting procedures [3] and Fourier Transform methods. The Fourier transform method can be applied to structures that are more complicated. From $\chi(k)$ a radial structure function $|\phi_n(r)|$ can be derived given by

$$\phi_n(r) = \frac{1}{\sqrt{2\pi}} \int_{k_{min}}^{k_{max}} k^n \chi(k) e^{2ikr} dk \qquad (2)$$

where n is usually either 1 or 3, and $k_{min}$ and $k_{max}$ are the minimum and maximum values of k, respectively. The n = 3 transform weights less the low-energy portion of $\chi(k)$, where the undesirable uncertainties occur, while it weights most the high-energy portion of $\chi(k)$. For this reason, we will employ $\phi_3(r)$ as our standard transform in the case. Shells of scattering atoms surrounding the absorbing atom generate the maxima of this function. The positions of the peaks in $|\phi(r)|$ are shifted compared to true distances due to contribution of the scattering phases that depend on k.

The analysis of Fourier transformation of $k^n \chi(k)$ into r-space was carried out using the IFEFFIT. This will ensure that the FT gives the correct contribution due to all constituent frequencies in $\chi(k)$. This is important for accurate determination of local structure. The k range for Fourier transform was $2 < k < 9$ $(\text{Å})^{-1}$.

## 3. Results and discussion

The XAFS and magnitude of the Fourier transform of XAFS are shown in figure 1 and figure 2 respectively for five cobalt complexes. The first strong peak at 2.1 Å for [3N3NTU2P] complex in the magnitude of the Fourier transform is due to the sulfur nearest neighbors around the cobalt ions the peaks are shifted towards smaller distances because of the phase shifts. The different Co-S bond distances are described in Table 1. Comparison of bond length to LSS, Lytle, Levy's to the bond length calculated by MathCAD programming and IFEFFIT have been done. The inverse Fourier transform of this peak in k-space gives the contribution to EXAFS from the sulfur nearest neighbors. The method employed for theoretical analysis of EXAFS and verification of various structural parameters using MathCAD programming is simpler than commonly used IFEFFIT technique. [4] In contrast to conventional EXAFS fitting analysis, other kinds of EXAFS analysis computation originated from the Tikhonov regularization method [3-4] is the iterative solution projection method using numerical algorithm [5] of manipulating matrix inversion computation under the ill-posed mathematical problem. The EXAFS spectra obtained using equation (1) in MathCAD programming for cobalt (II) complexes of thiourea are shown in figure 3. The bond length is measured from magnitude of Fourier Transform shown in figure 4 for cobalt (II) complexes of thiourea series.

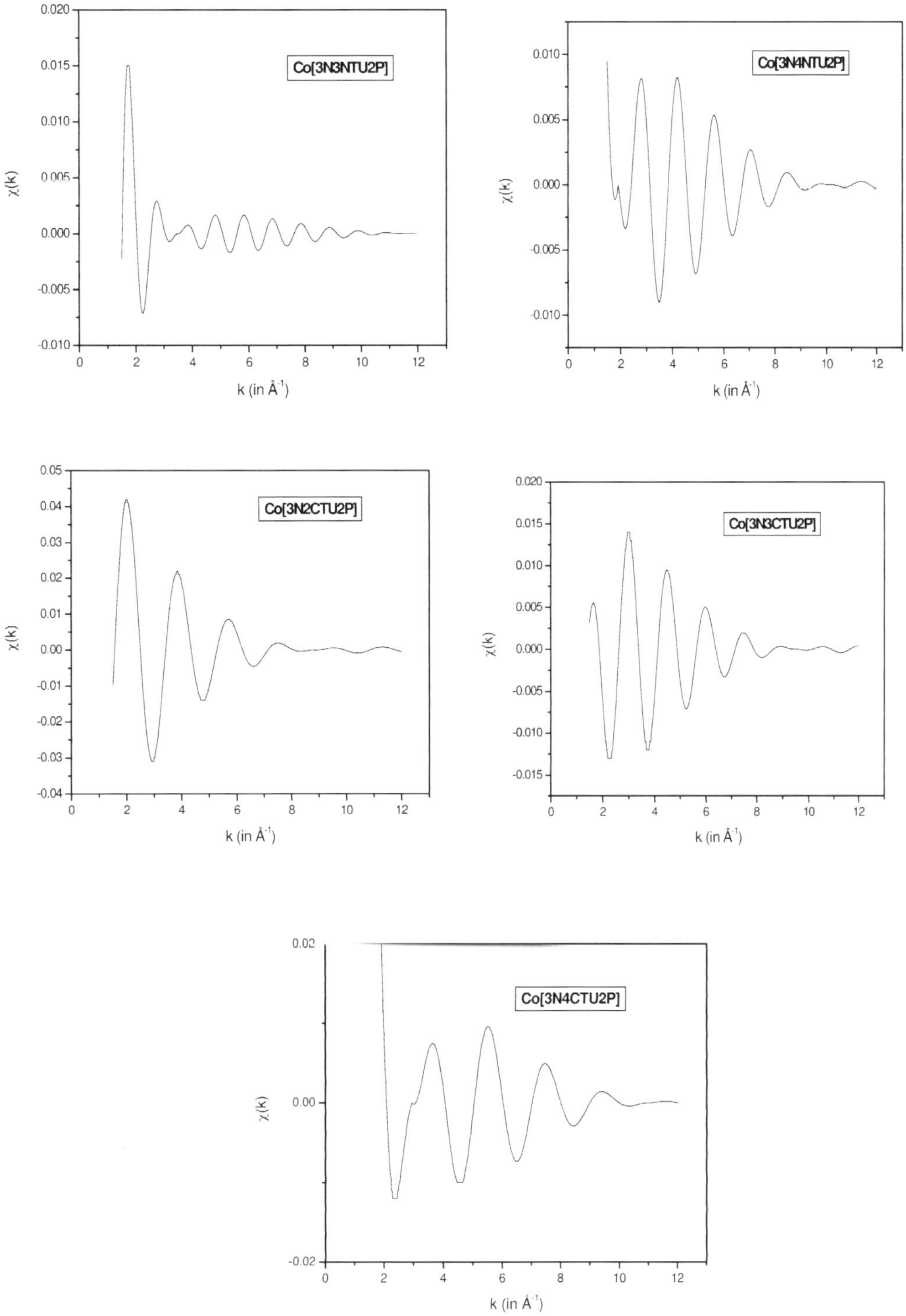


**Figure 3. Calculated $\chi(k)$ versus k curves for cobalt complexes.**

Journal of Physics: Conference Series **755** (2016) 012015 doi:10.1088/1742-6596/755/1/012015

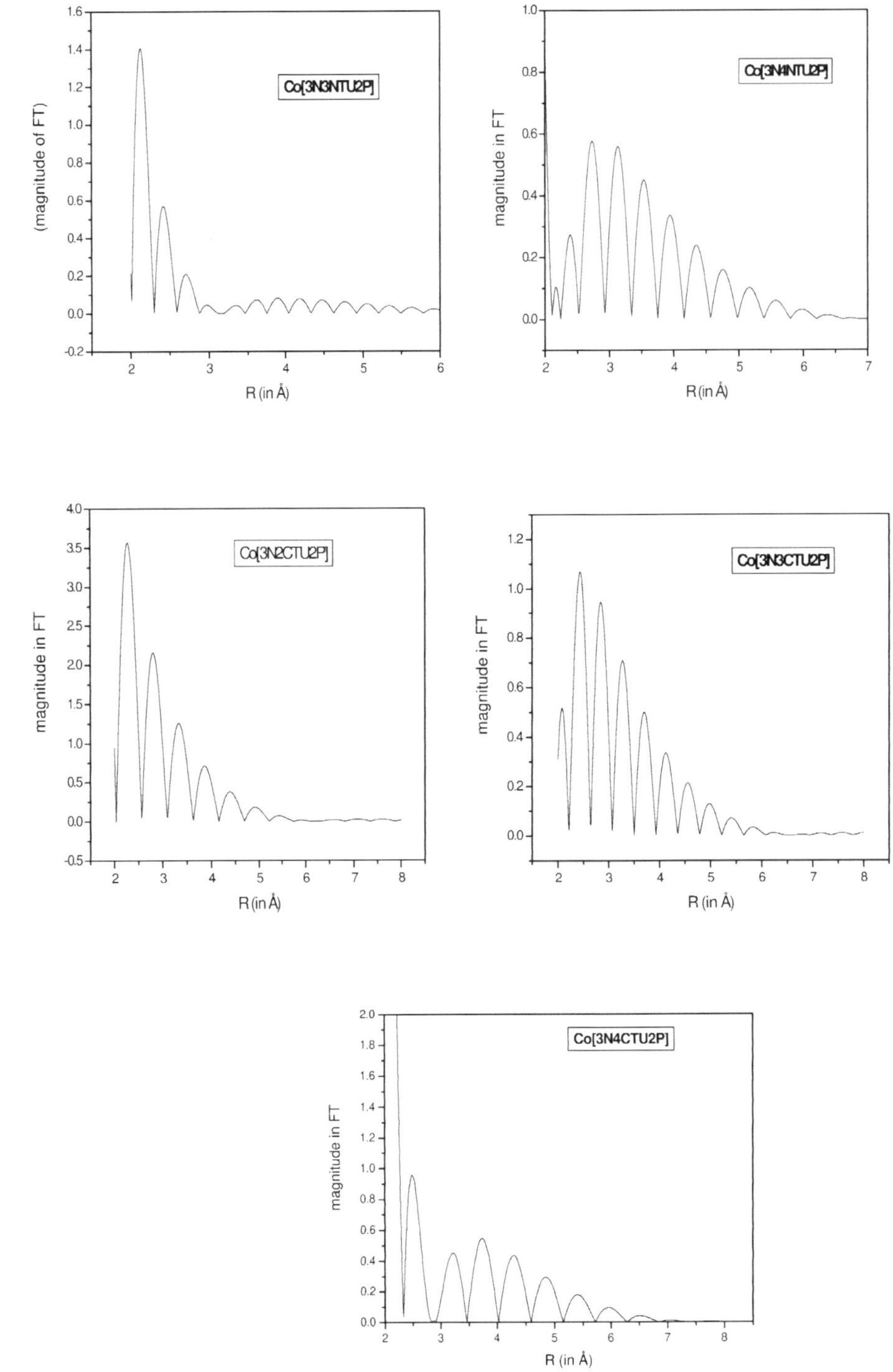


**Figure 4. Magnitude of Fourier transform of calculated χ(k) for cobalt complexes.**

## 4. Conclusions

Fourier transform technique provides results with great accuracy. This method is simple and straight forward, thereby providing a physical picture of the X-ray absorption process. This method appears to have a great potential in studying systems such as crystalline and amorphous materials. The atomic distance of central atom with the first nearest neighbour is not affected very much by changing the various substituents in the ligand. The reason is that the substituents are not directly linked to the central atom. Results indicate that bond lengths determined by Fourier transforming the calculated data resembles with the bond lengths obtained by LSS method, Levy method and Lytle method. As the imaginary part of Fourier Transform coincides with the amplitude of Fourier Transform, this verifies that experimentally obtained phase shift parameter is consistent with theoretically calculated data. Thus, the theoretical and experimental values agree well with each other. This means that the parameterized theoretical calculation of the EXAFS spectra describes in good agreement with physical reality. The calculated EXAFS curves for cobalt complexes of thiourea series are in good agreement with experimental EXAFS curves for all complexes studied within experimental error. A perusal of bond length values are obtained by magnitude of Fourier Transform of calculated EXAFS data indicates that the calculated values are in good agreement with the values obtained by LSS method, Levy method and Lytle method.

**Table 1. The average values of metal-ligand bond length in $(\acute{A})^{-1}$ for cobalt complexes.**

| *S.N* | *COMPLEXES* | $R^{a}_{LSS}$ | $R^{a}_{Lytle}$ | $R^{a}_{Levy}$ | $R_{FT}$ (MATHCAD) | $R_{FT}$ (ATHENA) |
|---|---|---|---|---|---|---|
| 1 | **Co [3N3NTU2P]** | **1.01** | **3.4** | **3.4** | **2.1** | **1.07** |
| 2 | **Co [3N4NTU2P]** | **1.04** | **2.5** | **2.5** | **2.7** | **0.88** |
| 3 | **Co [3N2CTU2P]** | **0.89** | **2.0** | **2.0** | **2.2** | **0.98** |
| 4 | **Co [3N3CTU2P]** | **0.90** | **1.9** | **2.4** | **2.4** | **1.13** |
| 5 | **Co [3N3CTU2P]** | **1.02** | **1.8** | **1.9** | **2.5** | **1.19** |

## 5. References

[1] Parsai N, Mishra A and Shrivastava B D 2012 *X-Ray Spectrom.* **41** 219.
[2] Mishra A, Parsai N, and Shrivastava B D 2011 *Indian Journal of Pure & Applied Physics* **49** 25.
[3] Teo B K 1986 *EXAFS: Basic Principles and Data Analysis* (Springer-Verlag: Berline).
[4] Koningsberger D C and Prins R 1988 *X-ray Absorption: Principles, Applications, Technique of EXAFS, SEXAFS and XANES* (Wiley-Interscience: NewYork).
[5] Vaarkamp M, Dring I, Oldman R J, Stern E A and Koningsberger D C 1994 *Phys Rev B* **50** 7872.

# Origin of the satellites L$\alpha_6$ and L$\alpha_7$ in the elements $_{40}$Zr to $_{50}$Sn

**Renuka Kendurkar and B D Shrivastava**
School of Studies in Physics, Vikram University, Ujjain -456010, India

E-mail: renukakendurkar@gmail.com; rashmibasant@gmail.com

**Abstract.** The origin of the satellites L$\alpha_6$ and L$\alpha_7$ have been explained in the elements from $_{40}$Zr to $_{50}$Sn. The energies and the intensities of the various transitions corresponding to the $L_3M_x$ - $M_xM_{4,5}$ (where x = 1-5) transition array, which may give rise to these satellites, have been calculated theoretically. The transition energies have been calculated using the available Hartree-Fock-Slater data for K-LM and L-MM Auger transition energies. The intensities of the various transitions have been estimated by considering cross sections $L_1$-$L_3M_X$ for Coster-Kronig transitions as well as for M-shell shake-off process occurring simultaneous to a $L_3$ hole creation. The total cross sections for initial two-hole states have been then distributed statistically amongst various allowed transitions from the initial states $L_3M_x$ to the final states $M_XM_{4,5}$. Each transition has been assumed to give rise to a Gaussian line and then the theoretical satellite spectrum has been computed by summing up these Gaussian lines. The calculated energies have been found to be comparable with the measured energies of the satellites L$\alpha_6$ and L$\alpha_7$. Consequently, the transitions which give rise to these satellites have been identified. The satellite L$\alpha_6$ has been assigned the transition $^3D_2L_3M_3$ - $^3F_3M_3M_5$ and the satellite L$\alpha_7$ the transition $^3P_0L_3M_3$ - $^3P_1M_3M_5$ in all the elements from $_{40}$Zr to $_{50}$Sn.

## 1. Introduction

L$\alpha$ satellites L$\alpha_6$ and L$\alpha_7$ have been observed [1] in the elements from $_{40}$Zr to $_{50}$Sn. Poonia and Soni [2] have theoretically studied the L$\alpha$ satellites in these elements but they have not studied L$\alpha_6$ and L$\alpha_7$ satellites. Hence, in the present study, we have theoretically studied these satellites using the multiple ionization theory. The origin of these satellites has been explained by computing the theoretical satellite spectra and then by assigning the transitions to these satellites. Earlier, we have theoretically investigated the satellites L$\alpha_3$, L$\alpha_4$ and L$\alpha_5$ in these elements and have assigned transitions to the satellites [3]. Hence, to complete the investigation of L$\alpha$ satellites in these elements, we have now studied L$\alpha_6$ and L$\alpha_7$ satellites.

## 2. Theoretical considerations

Poonia and Soni [2] have assigned the origin of L$\alpha$ satellites to the transition array $2p^{-1}3x^{-1}$ - $3x^{-1}$ $3d_{3/2,5/2}^{-1}$ (x = s, p, d). Hence, we have also assumed that the origin of the L$\alpha$ satellites observed near the dipole lines L$\alpha_1$($L_3$-$M_5$) and L$\alpha_2$ ($L_3$-$M_4$) can be ascribed to the transition scheme $2p^{-1}3x^{-1}$ - $3x^{-1}$ $3d_{3/2,5/2}^{-1}$ (x = s, p, d), i.e., $L_3M_x$ - $M_xM_{4,5}$, (X = 1,...5). This means that if an atom happens to be in a doubly ionized state with an electron missing from each of the two states $L_3$ and $M_x$, then a similar transition from $M_{4,5}$ to $L_3$ state will now leave the atom in the doubly ionized state $M_xM_{4,5}$. This hole in $M_x$ subshell remains a spectator and is called spectator vacancy. $L_3M_x$ is initial state and $M_xM_{4,5}$ is final state. All the possible initial $2p^{-1}\ 3x^{-1}$ ($L_3M_x$) states and final $3x^{-1}\ 3d_{3/2,5/2}^{-1}$ ($M_xM_{4,5}$) states are calculated in LS coupling scheme. Considering the dipole selection rules, 49 transitions have been found out corresponding to the transition scheme $L_3M_x$ - $M_xM_{4,5}$ [3]. In the earlier work [3], the energies and intensities of these 49 transitions have been calculated and then theoretical satellite spectra have been synthesized for all the elements under consideration. In the present investigation, we have utilised the same theoretical calculations and the same theoretical satellite spectra as reported in ref. [3]. However, in the present paper we are reproducing the method of calculations and the spectra, reported in ref. [3], for completeness.

## 3. **Calculations**

### *3.1. Calculation of energies of transitions in doubly ionized atoms*

The energy of a satellite has been calculated, on the basis of multiple ionization theory by using the energies of Auger transitions and by taking the energy of the diagram line, to which the satellite belongs. The formula for calculation of transition energy of say, $^3D_2L_3M_3$ - $^3F_3M_3M_5$ transition, has been written as

$$E\,(^3D_2L_3M_3 - {}^3F_3M_3M_5) = \quad E\,(K\alpha_1) + E\,(L_3 - {}^3F_3M_3M_5) - E\,(K - {}^3D_2L_3M_3)$$

The energy of $K\alpha_1$ line, i.e., E (K - $L_3$) has been taken from the tables of Cauchois and Senemaud [1]. The energies of the Auger transitions $L_3$ - $^3F_3M_3M_5$ and K - $^3D_2L_3M_3$ have been taken from the tables of Larkins [4].

### *3.2. Calculation of intensities of transitions in doubly ionized atoms*

After the creation of a single-hole state, it is the probability of a particular subsequent process that leads to the formation of initial two-hole state, required for the emission of a satellite. The single-hole state can get converted into the two-hole states by two processes, i.e., Auger transition and shake-off process. To calculate the probability of creation of the initial double vacancy state by the process of Coster-Kronig transition, the cross section for creation of single vacancy state is first calculated by the formulae given by Moores *et al.* [5] and then this cross section is multiplied by the transition rates for the Coster-Kronig transition as given in the tables of Chen *et al.* [6] The singly ionized state may get converted to doubly ionized state also by an associated shake-off of an electron. Hence, to calculate the probability of creation of the initial double vacancy state by the shake-off process, the cross section for creation of single vacancy state, as calculated above, is multiplied by the shake-off probability as calculated from the reports of Carlson *et al.* [7, 8] and Slater [9]. Finally, the total probability of creation of doubly ionized initial state can be determined by adding both the cross sections calculated above, i.e., by Coster-Kronig transition and by shake-off process. This probability can then be distributed statistically among all the allowed transitions using the tables of White and Eliason as given by Condon and Shortley [10]. In this method, all the multiplets of supermultiplets from various levels of the set are considered.

### *3.3. Synthesis of satellite spectra*

A composite spectrum formed by spectral lines emitted by the 49 transitions has been computed by taking each as a Gaussian line. For this we have taken energy on X-axis and transition probability on Y-axis. The peak height of each line has been taken equal to transition probability and peak position has been taken at the energy of transition. The widths of all the lines in one element have been assumed equal. A representative theoretical satellite spectrum for the element $_{48}$Cd is shown in Fig. 1 (reproduced from ref. [3]). The two satellites under investigations, i.e., $L\alpha_6$ and $L\alpha_7$ along with the transitions giving rise to them are clearly visible in this figure. The observed and theoretical positions are also marked in the figure. The theoretical spectra have been compared with the available experimental results for energies of the satellites. The intense peaks have been identified as the observed satellite lines and the transition assignment to the satellites, based on the identification of the peaks to the particular transitions, has then been done.

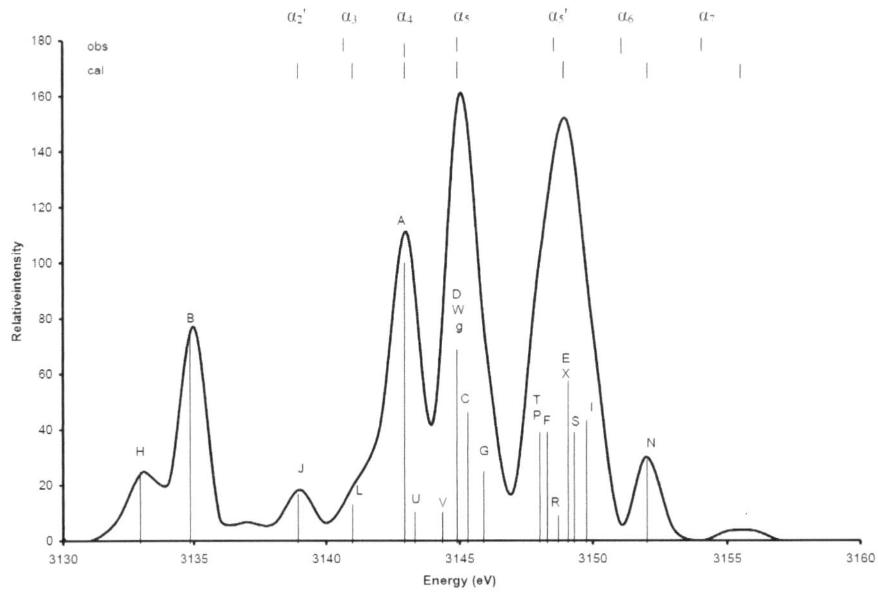


**Fig.1 The computed $2p_{3/2}^{-1}\ 3x^{-1} - 3x^{-1}\ 3d^{-1}$ ($L_3M_x$ - $M_x\ M_{4,5}$) spectrum of $_{48}$Cd. The satellite positions experimentally observed and theoretically calculated in the region of Lα satellites are shown at the top (reproduced from ref. [3]).**

**Table 1 Assignments of transitions to Lα satellites Lα$_6$ and Lα$_7$ in the elements from $_{40}$Zr to $_{50}$Sn.**

| S.No | Z | Theoretically computed data for satellite Lα$_6$ Transition $^3D_2$ - $^3F_3$ (3p) | | Measured energy (eV) | Theoretically computed data for satellite Lα$_7$ Transition $^3P_0$ - $^3P_1$ (3p) | | Measured energy (eV) |
|---|---|---|---|---|---|---|---|
| | | Energy (eV) | Normalized relative intensity | | Energy (eV) | Normalized relative intensity | |
| 1 | 40 | 2057.9 | 30.61 | 2056.6 | 2058.4 | 7.32 | 2059.2 |
| 2 | 41 | 2181.8 | 22.40 | 2181.3 | 2182.7 | 5.21 | 2183.3 |
| 3 | 42 | 2309.1 | 27.77 | 2303.5 | 2310.5 | 5.10 | 2307.1 |
| 4 | 43 | 2441.8 | 28.01 | - | 2443.4 | 5.23 | - |
| 5 | 44 | 2575.0 | 27.07 | 2575.1 | 2576.9 | 5.24 | 2578.7 |
| 6 | 45 | 2713.9 | 26.99 | 2713.8 | 2716.0 | 5.24 | 2717.9 |
| 7 | 46 | 2856.1 | 32.57 | 2856.7 | 2858.4 | 5.29 | 2861.4 |
| 8. | 47 | 3002.2 | 29.00 | 3003.3 | 3004.9 | 4.68 | 3008.7 |
| 9 | 48 | 3152.0 | 30.13 | 3153.9 | 3155.5 | 5.80 | 3158.0 |
| 10 | 49 | 3306.5 | 31.50 | 3308.7 | 3308.1 | 5.36 | 3315.2 |
| 11 | 50 | 3463.0 | 34.30 | 3469.4 | 3466.0 | 6.34 | 3473.7 |

**Measured energies are from Cauchois and Senemaud [1].**

## 4. Results and discussions

### *4.1. Assignments of transitions to satellite $L\alpha_6$*

$L\alpha_6$ satellite has been observed in most of the elements from $_{40}$Zr to $_{50}$Sn. The peak corresponding to this satellite should occur at the higher energy side of the $L\alpha_5'$ satellite and very near to it, according to the value of its energy. In our theoretical $L\alpha$ satellite spectra of the elements with Z = 40 to 50, there appears a peak on the higher energy side of $L\alpha_5'$ satellite and we have designated it as $L\alpha_6$ satellite. This peak arises due to transition $^3D_2$-$^3F_3$ (3p). The computed energy of this peak is in agreement with the energy of $L\alpha_6$ satellite in all the elements from $_{40}$Zr to $_{50}$Sn, except for $_{42}$Mo, $_{49}$In and $_{50}$Sn. On the basis of the present study, it is suggested that in $_{42}$Mo the $L\alpha_6$ satellite reported in the tables [1] at 2303.54 eV should be redesignated as $L\alpha_5'$ satellite and the satellite observed at 2307.1 eV and designated as $L\alpha_7$ satellite should be redesignated as $L\alpha_6$ satellite. Similarly, in $_{49}$In the satellite reported at 3308.69 eV and designated as $L\alpha_6$ satellite in the table [1] should be redesignated as $L\alpha_7$ satellite. Further, we suggest that $L\alpha_6$ satellite has not been observed in $_{49}$In. In $_{50}$Sn the satellite observed at 3464.43 eV and designated as $L\alpha_5'$ in the table [1] should be redesignated as $L\alpha_6$ satellite. After doing the redesignations as suggested above in the elements $_{42}$Mo, $_{49}$In and $_{50}$Sn and also after accepting that $L\alpha_6$ satellite has not been observed in $_{49}$In, the situation regarding $L\alpha_6$ satellite becomes clear that it arises due to the transition $^3D_2$-$^3F_3$ (3p) in all the elements from $_{40}$Zr to $_{50}$Sn. The table [1] gives the energies and intensities of the peaks identified as the two satellites $L\alpha_6$ and $L\alpha_7$. The measured energies [1] are also given in this table. It can be seen that the theoretical and measured energies agree well with each other.

### *4.2. Assignments of transitions to satellite $L\alpha_7$*

This satellite has been reported [1] in all the elements from $_{40}$Zr to $_{50}$Sn. In our theoretical $L\alpha$ satellite spectra there is a very weak peak on the higher energy side of $L\alpha_6$ satellite. This weak peak is the seventh peak in our theoretical satellite spectra and arises out of the transition $^3P_0$ - $^3P_1$ (3p), which has a small relative intensity. The energy of this peak agrees fairly well with the energy of the satellite $L\alpha_7$ in the elements from $_{40}$Zr to $_{50}$Sn, in which it has been observed. Thus, as a result of the present study, the $L\alpha_7$ satellite arises due to transition $^3P_0$ - $^3P_1$ (3p) in the middle-Z elements from $_{40}$Zr to $_{50}$Sn.

## 5. Conclusions

As a result of the present study, it has been found that the satellite $L\alpha_6$ arises due to the transition $^3D_2L_3M_3$ - $^3F_3M_3M_5$ and the $L\alpha_7$ satellite arises due to transition $^3P_0L_3M_3$ - $^3P_1M_3M_5$ in all the elements from $_{40}$Zr to $_{50}$Sn.

## References

[1] Cauchois Y and Senemaud C 1978 *Tables Internationnal de constantes selectionnnees. 18.Longeurs d'onde des emissions X et des discontinuites d'absorption X* (London: Pergamon Press).
[2] Poonia S and Soni S N 2002 *Ind. J. Pure & Appl. Phys.* **40** 786.
[3] Kendurkar R and Shrivastava B D 2014 *J. Phys. Conf. Series* **534** 012037.
[4] Larkins F P 1977 *At. Data Nucl. Data Tables* **20** 311.
[5] Moores D L, Golden L B and Sampson D H 1980 *J. Phys.* B **13** 385.
[6] Chen M H, Crasmann B and Mark H 1979 *At. Data Nucl. Data Tables* **24** 13.
[7] Carlson T A and Nestor C W Jr. Tucker T C and Malik F B 1968 *Phys. Rev.* **169** 27.
[8] Carlson T A and Nestor C W Jr. 1973 *Phys. Rev.* A **8** 2887.
[9] Slater J C 1930 *Phys. Rev.* **36** 57.
[10] Condon E U and Shortley G H 1967 *The theory of atomic spectra* (Cambridge: Cambridge University Press) p 241.

Journal of Physics: Conference Series **755** (2016) 012017 doi:10.1088/1742-6596/755/1/012017

# XRD and XANES study of some Cu-doped MnBi materials

**Ashutosh Mishra[a], Harsha Patil[a]**

[a]School of Physics, Devi Ahilya University, Indore-452001, India

E-mail:harshapatil9@gmail.com

**Abstract.** High purity MnBi low temperature phase has been prepared and analyzed using X-ray diffraction (XRD) and X-ray absorption near edge structure (XANES) measurements. The X-ray diffraction measurements were carried out using Bruker D8 Advance X-ray diffractometer. The X-rays were produced using a sealed tube and the wavelength of X-ray was 0.154 nm (Cu K-alpha). and X-rays were detected using a fast counting detector based on Silicon strip technology (Bruker LynxEye detector)[1]. and the X-ray absorption spectra has emerged as a powerful technique for local structure determination, which can be applied to any type of material. The X-ray absorption measurements of two Cu-doped MnBi alloys have been performed at the recently developed BL-8 Dispersive EXAFS beam line at 2.5 GeV Indus-2 synchrotron at RRCAT, Indore, India[2]. The X-ray absorption near edge structure (XANES) data obtained has been processed using data analysis program Athena. The energies of the K absorption edge, chemical shifts, edge-widths, shifts of the principal absorption maximum in the alloys have been determined.

## 1. Introduction

MnBi is a ferromagnetic inter-metallic Compound with hexagonal crystal structure, at present the MnBi binary compound is in low temperature phase (LTP) [1]. The structural properties of these MnBi compounds have been studied by using X-ray diffraction and XANES measurement of LTP[2]. In the present work, the X-ray diffract meter records of the alloys have been obtained from which the particle size and lattice parameter have been determined. and X-ray k-absorption spectra of Cu in two doped MnBi by Cu alloys , where 31% and 32% Cu doped MnBi are denoted MnCu(31)Bi and MnCu(32)Bi.

## 2. Experimental

Both the alloys have been prepared by solid state root method. The X-ray diffraction measurements were carried out using Bruker D8 Advance X-ray diffractometer. The X-rays were produced using a sealed tube and the wavelength of x-ray was 0.154nm (Cu K-alpha).and x-rays were detected using a fast counting detector based on Silicon strip technology (Bruker LynxEye detector)[3]. Analysis of X-ray diffraction gives information about the different planes present in the specimen, lattice parameters and phase purity.

The X-ray absorption spectra have been recorded using synchrotron radiation. The X-ray spectroscopic setup is available at Raja Ramanna Center for Advanced Technology (RRCAT), Indore, India and is called dispersive EXAFS beamline BL-8. This beamline has been recently commissioned at the 2.5 GeV Indus-2 synchrotron radiation source [4]. Using the relation, $I_t = I_0 e^{-\mu x}$, where μ is the absorption coefficient and x is the thickness of the absorber, the absorption μ(E) corresponding to the photon energy (E) are obtained. Te experimental data has been analyzed using the available computer software package Athena.

**Table 1.** Lattice Parameter and Particle size data

| Compound Samples | Lattice Parameter **a(A$^0$)** | **c(A$^0$)** | **Particle Size (nm)** |
|---|---|---|---|
| MnCu(31)Bi | 4.61 | 6.23 | 28.98 |
| MnCu(32)Bi | 4.62 | 6.23 | 28.99 |

**Table 2.** XANES data for the copper K-absorption edge in Cu doped MnBi alloys.

| Compound sample | $E_k$ (eV) | $E_A$ (eV) | Chemical Shift (eV) | Shift of the Principle ab. max. (eV) | Edge width ($E_A$-$E_k$) (eV) |
|---|---|---|---|---|---|
| MnCu(31)Bi | 8990.31 | 9020.13 | 11.31 | 41.13 | 29.82 |
| MnCu(32)Bi | 8989.78 | 9020.64 | 10.78 | 41.64 | 30.86 |

## 3. Results and discussion

### *3.1. X-ray diffraction studies:*

The X-ray diffractometer records of the compounds are given in figure 1. Particle size was calculated using Scherer's formula given by: $t=0.9\lambda/B\cos\theta$

where t is the crystal thickness (in the same units as $\lambda$), B is full width half maxima (in radians) of diffraction line, $\theta$ is the Bragg angle and $\lambda$ is the wavelength. The values of particle size and the lattice constant calculated from the diffractometer records are given in table 1.

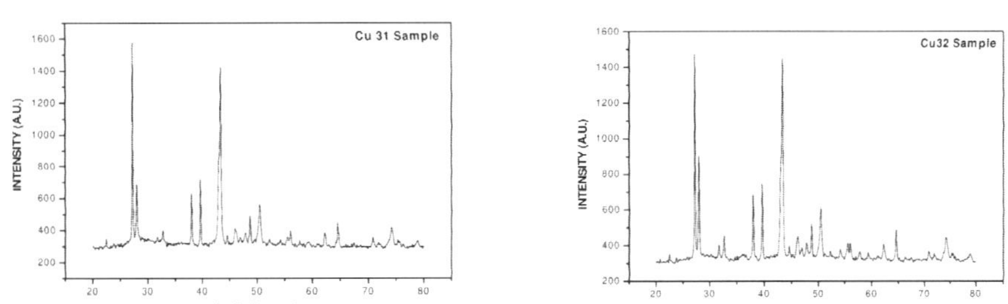


Figure1. X-ray diffractometer records of compounds

### *3.2 X-ray absorption near edge structure:*

The normalized XANES spectra in the region -30 < E < 50 eV, indicating positions of the absorption edges K and principal absorption maximum A are shown in figure 2. The first derivatives of the spectra in this region, indicating positions of the absorption edges K and principal absorption maximum A are shown in figure 3. First peak in the derivative spectra gives the position of first inflection point, i.e., the K-absorption edge (Ek). The position where the derivative is zero, gives the position of principal absorption maxima (EA). The results of the energies of the K-absorption edges (Ek) and the energies of principal absorption maximum A (EA) of copper in its two compounds are presented in table 2. The chemical shifts (in eV) of the K-absorption edge of copper in the compounds are also given in table 2. For all the compound the distance (in eV) of the principal absorption maximum A with respect to the respective K-absorption edge have also been computed and are collected in the same table. It can be readily seen from table that copper K-edge is found to be shifted towards the high energy side in all the two compounds, as compared to the copper metal K-absorption edge.

(i) Chemical shift:

The shift of X-ray absorption edge i (i=K, L, M...) of an element in a compound with respect to that of the pure metal is written as:

$$\Delta E = Ei\ (complex) - Ei\ (metal)$$

The results are given in table 2. For computing the chemical shift the value of EK (Cu metal) has been taken as 8979 eV. The values of chemical shifts are 11.31 and 10.78eV for compound 1 and 2 respectively

The chemical shift is largest for MnCu(31)Bi compound and the order for other compound is as follows:

MnCu(31)Bi > MnCu(32)Bi

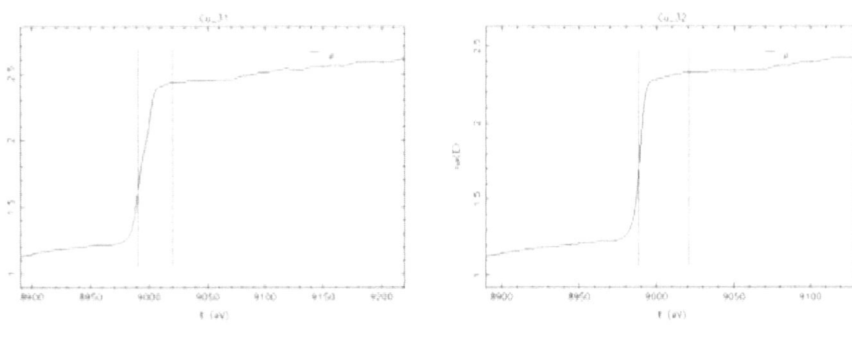

Figure 2. XANES spectrum for Copper doped MnBi compounds indicating position of absorption edges K and principal absorption maxima A.

(ii) Shift of principal absorption maxima:

In table 2, the data for the principal absorption maximum EA for the compound have also been included. It has been observed that with respect to copper metal, the value of EA is shifted towards the higher energy side. For the Cu doped MnBi compounds mentioned in table 2, the energy range of chemical shift in these compound is between 11.31 and 10.78 eV while the range for shift of principal absorption maximum is between 41.13 and 41.64 eV. The shift of the principal absorption maximum for Cu doped MnBi alloys are in the order

MnCu(31)Bi < MnCu(32)Bi

The order of the shift of the principal absorption maximum is in the reverse order of the chemical shift of the compounds.

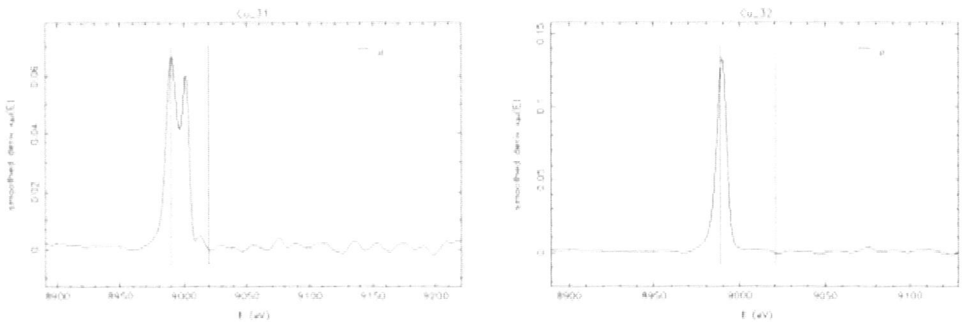

Figure 3. Derivative of the XANES spectrum for copper Copper doped MnBi compounds indicating position of absorption edge K and principal absorption maxima A.

(iii) Edge-width:

In table 2, the values of the edge-width (EA-EK) have been reported. The experimental data of edge width of both of compounds show that the edge-width increases as follows:

MnCu(31)Bi < MnCu(32)Bi

The relative order of the values of edge- width are not supported by the values of the chemical shift or the shift in principal absorption maxima in the present work. The reason may be that the values of edge-width not only depend on the factors responsible for the chemical shift but also on the geometry of the compounds[5].

## 4. Conclusions:

By the analysis of the X-ray diffraction pattern, one can conclude that both the samples are in crystalline form. Particle size is increasing as increase Cu % doped MnBi at same temperature.
The copper K-edge has been found to split into one components, i.e., K in all of these compounds. The energies of K edge (EK) and principal absorption maxima (EA) have been reported. From these, the shift of the K-edge (chemical shift), shift of the principal absorption maximum and edge-width has been obtained.

## Acknowledgement

Thanks are due to Dr. M. Gupta, UGC CSR, Indore for XRD measurements and Drs. S. N. Jha, D. Bhattacharya, A. Poswal, Applied Spectroscopy Division, BARC, Mumbai and S. K. Deb, RRCAT, Indore for their help in recording the spectra.

## References

[1] Wohlfarth E P 1980 *Ferromagnetic Materials* (North-Holland).
[2] "Heusler alloy," *Encyclopedia Britannica Online*, retrieved Jan. 23, 2005.
[3] X-Ray diffraction: a practical approach By C. Suryanarayana, M. Grant Norton (1998).
[4] Bhattacharya D, Poswal A K, Jha S N, Sangeeta and Sabharwal S C 2009 *Bull. Mater. Sci.* **32**(1) 103
[5] Kumar A, Nigam A N and Shrivastava B D 1981 *X-ray Spectrom.* **10** 25

# Structure Evolution of $BaTiO_3$ on Co Doping: X-ray diffraction and Raman study

**Amantulla Mansuri* and Ashutosh Mishra**

School of Physics, Devi Ahilya University, Khandwa Road Campus, Indore-452001, India

E-mail: a.mansuri14@gmail.com

**Abstract.** In the present study, we have synthesize polycrystalline samples of $BaTi_{1-x}Co_xO_3$ (x = 0, 0.05 and 0.1) with standard solid state reaction technique. The obtained samples are characterized by X-ray diffraction (XRD) and Raman spectroscopy. The detail structural analysis has been performed by Rietveld refinement using Fullprof program. The structural analysis reveal the samples are chemical pure and crystallize in tetragonal phase with space group *Pm3m*. We observe an increase in lattice parameters which results due to substitution of $Co^{2+}$ with large ionic radii (0.9 Å) for smaller ionic radii (0.6 Å) $Ti^{4+}$. Moreover peak at 45.5° shift to 45° on Co doping, which is due to structure phase transition from tetragonal to cubic. Raman study infers that the intensity of characteristic peaks decreases and linewidth increases with Co doping. The bands linked with the tetragonal structure (307 $cm^{1}$) decreased due to the tetragonal-towards-cubic phase transition with Co doping. Our structural study reveals the expansion of BTO unit cell and tetragonal-to-cubic phase transformation takes place, results from different characterization techniques are conclusive and show structural evolution with Co doping.

## 1. Introduction

Barium titanate ($BaTiO_3$) received intensive attention of researchers due to its ferroelectric, piezoelectric and dielectric properties.[1-2] $BaTiO_3$ has a typical perovskite structure with a variety of crystal structure modifications and known for wide range of applications as electric ceramics. Structure plays a key role in determining the physical properties of this ceramics material thus the relationship between structure and physical properties are widely explored.[1-5] $BaTiO_3$ crystal undergoes three successive phase transitions with decreasing temperature, from a cubic phase (Pm3m) to a tetragonal phase (P4mm) at 130 °C, then to an orthorhombic phase (Amm2) at 5 °C, finally to a rhombohedral phase (R3m) at -90 °C. It is observed that suitable doping at Ti site also induce structural phase transformation.[4-6]

Interesting and exotic properties are theoretically expected in doped $BaTiO_3$ like ferromagnetism at room temperature enhanced dielectric properties etc.[2-3] Moreover synthesis technique also influences the level of doping and physical properties. Based on this fact, several studies are made adopting different synthesis techniques such as sol-gel, solid state ceramic technique, laser ablation and chemical process etc.[6-7] In response to these studies different type of doping has been attempted for $BaTiO_3$ such as Fe, Mn, Co etc[1-7]

In the present work, we studied the structural phase transition of $BaTiO_3$ on cobalt (Co) doping. The detail structure analysis is made with using XRD and Raman spectroscopy techniques. We observed growth of lattice and structural phase transformation with increasing cobalt ($Co^{2+}$) doping concentration.

## 2. Experimental Detail

Polycrystalline samples with composition $BaTi_{1-x}Co_xO_3$ (x = 0, 0.05 and 0.1) have been prepared by solid state reaction technique. The ingredients $BaCO_3$, $TiO_2$ and CoO were taken in stoichiometric ratio and grounded well for homogenous mixing. The powder is then heated twice at 1250 °C with intermediate grinding. The powder samples are used for XRD and sample in pallet form is used for Raman spectroscopy. The XRD data is analyzed using Fullprof program.

## 3. Result and Discussion

### *3.1. X-ray diffraction*

Fig.1 shows XRD pattern of all samples taken at room temperature with monochromatized Cu-$K_\alpha$ radiation in 2θ range of 20-80 degree with step size $\Delta\theta = 0.02^o$. All samples crystallize in single phase. The comparative XRD plots of all samples in fig 1 shows a minor shift of Bragg peaks with Co doping. The shift can be explained due to structural expansion attributed to substitution of $Co^{2+}$ larger ionic radii for $Ti^{4+}$ smaller ion.

Fig.2 shows XRD pattern along with rietveld refinement of BTO sample, braggs reflections and difference of experimental and calculated pattern is also shown. Rietveld analysis of XRD shows the sample is single phase and crystallizes in tetragonal structure with space group *P4mm*. The goodness of fit ( $R_p/R_{wp}$) and $chi^2$ of rietveld refinement are 1.37 and 1.27 respectively shown reasonable refinement. Refinement for all samples is performed and the results are analyzed and are quite intriguing

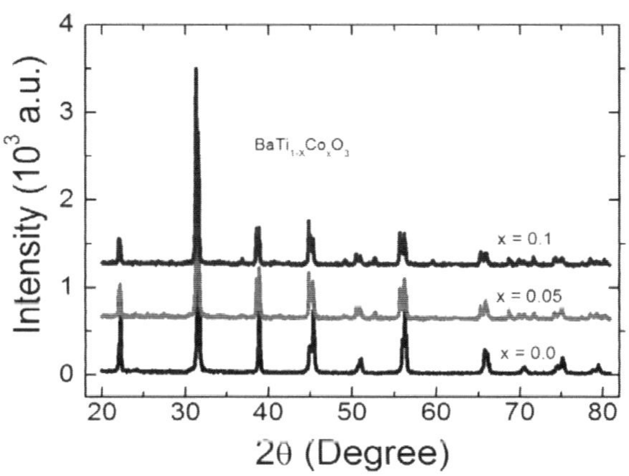


**Figure 1.** XRD pattern of $BaTi_{1-x}Co_xO_3$ for x= 0.0, 0.05, 0.1 recorded at room temperature.

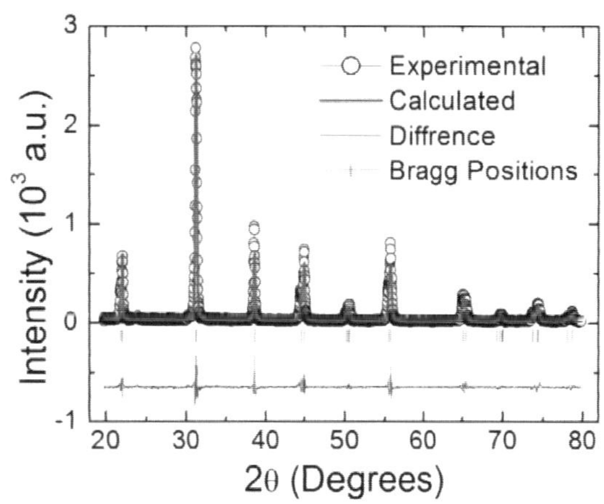


**Figure 2.** Rietveld refinement of XRD pattern for $BaTiO_3$ sample.

The structural study of all samples shows that with doping the lattice parameter *a* and *c* grows from 3.99406 Å to 4.052481Å and 4.03603Å to 4.066313Å respectively from x =0 to x = 0.1. The detail structure study reveal that on Co doping the phase transformation from tetragonal to cubic is observed which is confirmed by decrease in *c/a* from 1.036 for x =0 to 1.003 for x = 0.1. Ardent observation of XRD pattern reveal shifting of peaks which would be explained for tetragonal distortion of structure due to substitution of larger $Co^{2+}$ ion for smaller $Ti^{4+}$ ion.

### *3.2. Raman study*

Journal of Physics: Conference Series **755** (2016) 012018 doi:10.1088/1742-6596/755/1/012018

Raman spectroscopy provides an effective way of understanding the structure symmetry of any material. Since, Raman spectroscopy has better resolution of local microstructure so it is suitable technique to study structural phase transformation. Modification of Raman mode in terms of intensity, line broadening and even disappearance of mode is characteristics of phase transformation.

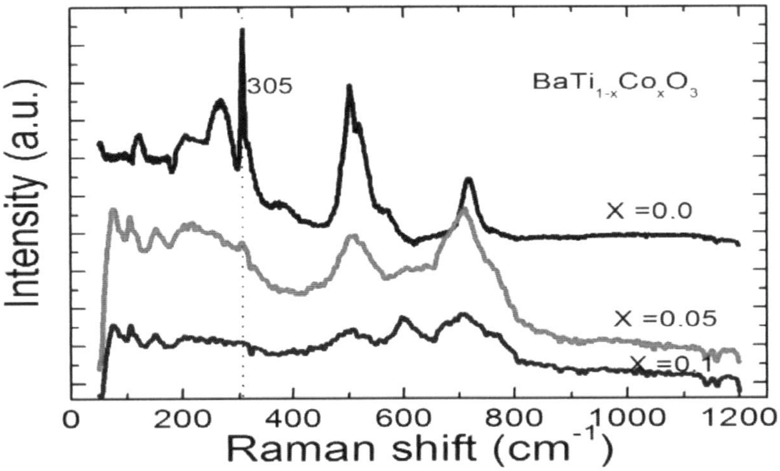


**Figure 3.** Raman Spectra recorded a roomtemperature for $BaTi_{1-x}Co_xO_3$ for x= 0.0, 0.05, 0.1 samples.

Fig.3 shows Raman spectral of all three samples taken in the energy range of 100 $cm^{-1}$ to 1200 $cm^{-1}$ at room temperature. The main feature of spectra is a sharp band at 305 $cm^{-1}$ assigned to [$B_1$, E(TO+LO)] Raman mode, However two other bands one intense broad band at 526 $cm^{-1}$ [$A_1$(TO)] and another around 717 $cm^{-1}$ [$A_1$(LO)] are also observed. Detail analysis of the spectra shows that the sharp intense peak at 305 $cm^{-1}$ becomes fader and border with Co doping which is result of appearance of more symmetric cubic.

### *3.3. Discussion*

Fig.4 reveal the whole story of our investigation where Fig 4(a) shows the enlarge XRD pattern between 2θ ~ 44° - 46° since the Bragg peak at 2θ = 45.5° is characteristic of tetragonal phase whereas

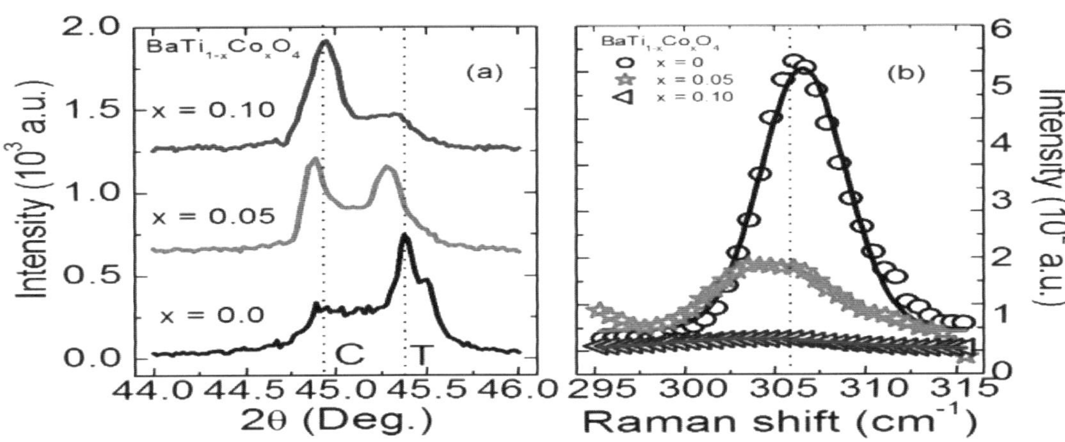


**Figure 4.** Evidences of structural phase transition observed from XRD and Raman Study (a)extended XRD between 2θ = 44°-46°. (b) Raman mode at 305 $cm^{-1}$Study.

the peak at 45$^{o}$ is due to cubic structure. It is quite obvious in fig 4(a) that the structure phase transformation occur from tetragonal to cubic on Co doping. Fig. 4(b) shows the Raman mode at 305 cm$^{-1}$ for all samples solid line is Gaussian fit to data. It is observed that the intensity of peak decreases and linewidth increases with doping concentration. The disappearance of peak at 305 cm$^{-1}$ and 715 cm$^{-1}$ is indicator of transformation of tetragonal phase to cubic symmetry.

In conclusion, our detail structural study on Co doped $BaTiO_3$ shows expansion of lattice parameters with increase Co concentration. XRD analysis shows increase of lattice parameter and shows phase transformation from tetragonal to cubic. Raman study also indicates the phase transformation. Our results from both XRD and Raman study are correlated and conclusive. For further understanding more samples with mediate composition will be prepare and studied for structural and physical properties.

## 4. Acknowledgmet

We acknowledge UGC DAE-CSR Indore for Raman spectroscopy and XRD data.

## 5. References

[1] Campos P. L, Diaz-Droguett D. E, Lavín R and Fuentes S 2015 *J. Molec. Struc.* **1099** 5029
[2] Li G. B etal., 2004 *Solid State Chem.* **177** 1695-1703
[3] Xu J, Zhai J, and Yao X J 2009 *J. Alloys Compd.* **467** 567-571
[4] Yang L, Qiu H, Pan L, Guo Z, Xu M, Yin J and Zhao X 2014 *J. Mag. Mag. Mat.* **350** 1-5
[5] Apostolova I. N, Apostolov A. T, and Bahoosh S. G, 2013 J. Appl. Phys. **113** 203904

Journal of Physics: Conference Series **755** (2016) 012019 doi:10.1088/1742-6596/755/1/012019

# Synthesis and Characterization of Transition Metal complexes with pyrimidine based ligand derivative

**Ruchita Awate[1], Ashutosh Mishra[2], A Mansuri[2]**
[1]Bangalore Technological Institute, Bangalore, 560 035
[2]School of Physics, Devi Ahillya University, Khandwa road, Indore 452 001

E-mail: ruchita.awate@gmail.com

**Abstract.** The article deals with a study of Synthesis Transition Metal Complexes Like copper and iron with Pyrimidine based ligand derivatives. The synthesized complex were characterized by XRD, SEM, FTIR. Mossbauer Spectra of Iron complex has also taken out to find oxidation state of iron after complexation.The aim of this study is to preparation and characterization with Transition Metal complexes by different physical and chemical characterization techniques.

## 1. Introduction

Pyrimidines are probably the most widely distributed heterocyclic compounds in nature. They show the immense commercial applications in medicine as antioxidant, antibacterial, antifungal, antiviral and hepatoprotective agents. The heterocyclic complexes are significant due to their use as models in naturally occurring systems; they also have greater biological importance. The coordination properties of pyrimidine are important in understanding the role of metal ions in biological system and its physicochemical importance creates great interest in complexation.[1] In view of the broad-spectrum importance associated with these metal complexes, it was considered worthwhile to prepare interesting complexes derived from hitherto unsubstantiated pyrimidine as the complexing agent. These complexes were characterized by XRD, SEM, FTIR, and Mossbauer Spectroscopy measurements corroborated towards the structure of both the synthesized compounds.

## 2 Experimental Techniques

### *2.1* Materials and methods

Elemental analyses (C, H, N, S) were performed by using a Carlo Erba 1160 elemental analyzer. Melting points were recorded in open capillaries and are uncorrected. Grain size of the powdered sample was measured by X-ray diffractometer (Philips X-pert pro diffractometer, PW 1830) at room temperature with CuK$\alpha$ radiation. Surface morphological study was observed by scanning electron microscopy (SEM, Philips XL 30). IR spectra were recorded using a Perkin-Elmer instrument (using KBr disks, 4000-400cm$^{-1}$). Mossbauer Measurement were carried out using a standard PC-based spectrometer equipped with a Weissel velocity drive operating in the constant acceleration mode.The data was fitted with the NORMOS-SITE program, and the obtained parameters are with respect to natural iron.

*2.2* Synthesis of the complexes

Both the complexes were prepared in a similar way by the usual diazotization process. [2] 6-Methyl-5-arylhydrazono-2thio-4-oxo-pyrimidine (ligand) were prepared by dissolving (0.01 mol) of Aniline in a mixture of concentrated hydrochloric acid (3ml) and water (4ml) and cooled to 0-5°C in ice bath. To it a cold aqueous solution of sodium nitrite (0.01mol) was then added. The Diazonium salt so obtained was filtered into a cold mixture of sodium acetate (7gms) and ethylacitoacitate (0.01mol) in ethanol (25ml) the resulting solid is 1-ethoxy-2-arylhydrezono-butene-1, 3dione was washed with water and dry it. Now 1-ethoxy-2-arylhydrezono-butene-1, 3dione of (0.01mol) and thiourea (0.01mol) dissolved in (10ml) Sodium ethoxide mixture (freshly prepared) in a 1 liter round bottomed flask. Then fit a reflux condensers to the flask, introduced a few fragments of broken porcelain pieces for boiling the mixture for 30mins, few crystals are appear after 15 minutes and gradually increase in amount as the refluxing is continued. Cool the flask in ice and filter the solution and dry it.

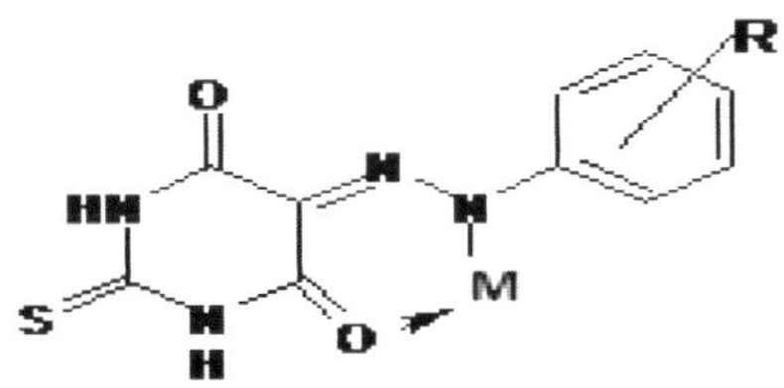


Synthetic path way for and Metal Complex here R is O Nitro

Formation of the metal complexes can be represented by the general equations. $MX_2 + 2L \rightarrow [M(L)_2X] + X$. In this equation, M is Cu (II) and Fe (II), X is Cl and L is [6-Methyl-5-arylhydrazono-2-thio-4-oxo-Pyrimidine]. Both the complexes are stable at room temperature

## 3. Result and Discussion

### *3.1 XRD*

The sample was characterized at room temperature by X-ray diffraction using Cu Kα radiation. The X-ray diffraction patterns of the complexes are indicative of crystalline nature. The diffraction pattern of complexes recorded between 2θ ranging from 10° to 75°. The crystalline size of the samples is estimated using the Scherrer's formula $D=k\lambda/\beta_{2\theta}\cos\theta$, here k is a constant taken to be 0.94, λ the wavelength of X-ray used (λ=0.154 nm) and $\beta_{2\theta}$ full width at half maxima of all peaks of the XRD patterns θ is Bragg angle. The diffraction patterns have been successfully indexed.[3] The crystalline size was found for copper and iron complex 48.22 nm and 33.89 nm respectively. It is observed that crystalline size is different for both the complexes, due to change in the R position. The XRD patterns are shown in Figure 1.

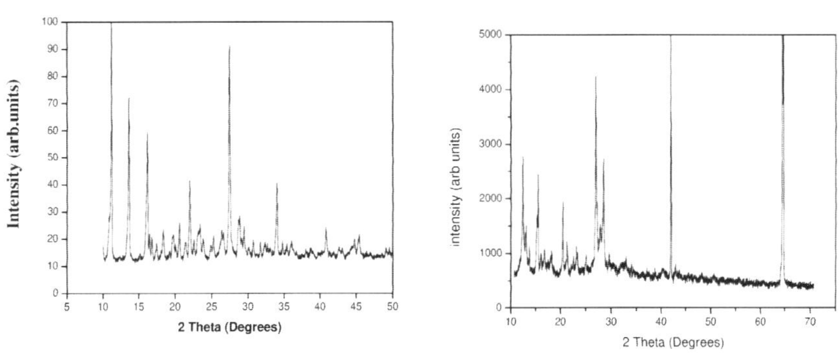


**Figure 1.** X-ray diffraction patterns of copper and Iron Metal complexe

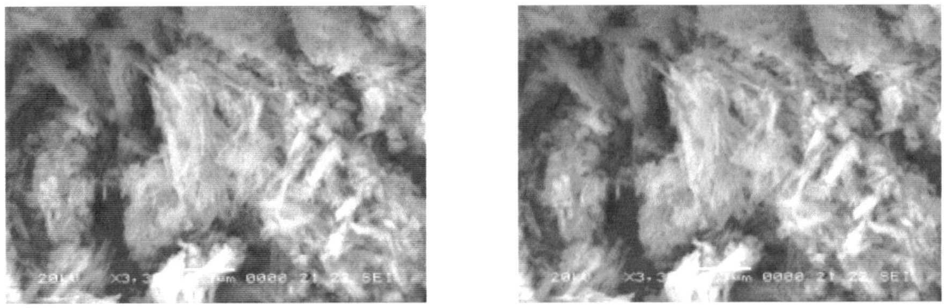

**Figure 2.** SEM images for copper and Iron complexes

*3.2 SEM*

The SEM images of the samples are shown in the figure 2. From the figure it can be seen that the average length of the crystals for the copper and Iron Metal Complex are 5 and 5.8μm respectively. Also, it can be seen that there are some void spaces. The surface morphology changes with change in substituted anilines, both the images having large number of irregular shaped and some having regular crystals associated with the images shown in figure.2 It is quite clear from the results that the average grain size estimated by SEM is quite larger than the average grain size measured by XRD.

*3.3 IR SPECTRA*

The IR spectra for both complexes were performed at field between 4000–400 $cm^{-1}$, along with tentative assignments. The presence of a single medium band in the region 3,250–3,350 $cm^{-1}$ in the spectra may be assigned due to N–H stretch. [4,5] The bands at 1580-1590$cm^{-1}$ due to the azomethine group of the ligand underwent a shift to lower frequency (1581-1495) $cm^{-1}$ after complexation, indicating the coordination of hydrazono nitrogen to metal atom and this can be explained by the donation of electrons from nitrogen to the empty d-orbital of the metal atom.[6,7] In addition, the ν(C=N) mode may be coupled with ν(C=S) to give the intense band observed at 1270-1190 $cm^{-1}$. [8] The band corresponding to (C=O) at 1680-1690 $cm^{-1}$ is shifted to lower frequency in the spectra for both the metal complexes. This is clear evidence for coordination by carbonyl oxygen. The vibrations absorption peaks of the ν(M-N),ν(M-Cl) and ν(M-O), which should be at 559-300$cm^{-1}$, as reported by other workers [9,10] could not be observed since it was beyond the extent of our measurement.

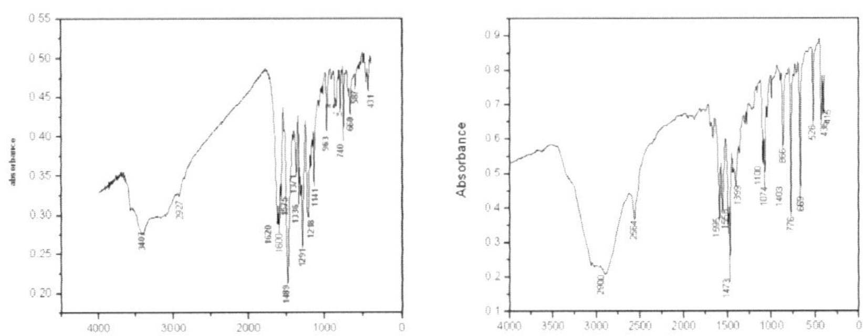


**Figure 3.** IR Spectra of Copper and Iron Metal Complexes

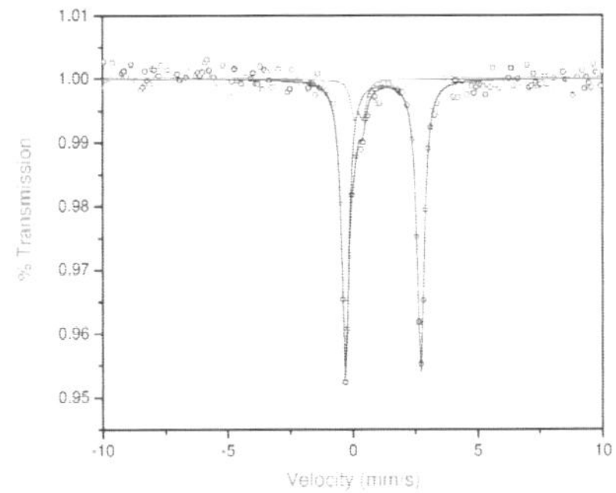


**Figure 4.** Mossbauer Spectra of Iron Metal Complex

*3.4 MOSSBAUER SPECTRUM*

The Mossbauer spectrum for iron complex is shown in figure 4. This spectrum was least-square fit with a superposition of two doublets and one singlet, all with Lorentzian line shape. According to the collected data the percentage of Fe (II) 88.9%. The isomer shift of 1.217 ± 0.005mm/s (35.6%), 1.213 ± 0.006 (53.3%) corresponds to the expected value for octahedral Fe (II) complex, as previously described complex.[11] which exhibits distorted octahedral symmetry with an isomer shift of 1.212mm/s and quadrapole splinting of 3.01mm/s. From the obtained isomer shift values one can conclude that Fe exists in $Fe^{2+}$ states with a relative fraction of 89.7 respectively.

**Acknowledgment**

The authors are sincerely thankful to UGC DAE Consortium for Scientific Research centre Indore.

**References**

[1] Mishra. A, Awate Ruchita, 2009 *Taylor & Francis* **184** 2624.
[2] Ramakrishnan S and Palaniandavar M, 2005 *J. Chem. Sci.* **117** 179.
[3] B. D. Cullity Elements of X-ray Diffraction, Second Edition, Addision-Wesley Publishing Company, 1978.
[4] Khan T A, Rather M A, Jahan N, Varkey S P and Shakir M,1998 *Transition Met. Chem.* **23** 283.
[5] Singh A K, Panwar A, Singh R and Beniwal S,2003 *Transition Met.Chem* **28** 160.
[6] Chandra S and Sharma S D 2002 *Transition Met. Chem.* **27** 732.
[7] Lodeiro C, Basitida R, Bertolo E, Macias A and Rodriguez R 2003 *Transition Met.Chem.* **28** 388.
[8] Offiong O E 1995 *Transition Met. Chem*, **20** 126.
[9] Satish M, Annigeri M P, Sathisha M P, and Revankar V K, 2007 *Transition Met.Chem.* **31** 81.
[10] Chandra S and Kumar R, 2004 *Transition Met. Chem.* **29** 269.
[11] Mishra A, Mishra N, Malviya P, Awate R, 2014 *international journal of Scientific Research in Physics and Applied Science* **02** 2348.

# Structural and Optical properties of poly-crystalline $BaTiO_3$ and $SrTiO_3$ prepared via solid state route

**Kanaka M. Jarabana, Ashutosh Mishra, Supriya Bisen**
School of Physics, Devi Ahilya University, Indore-452001, India

E-mail: jkanaka.hvp@gmail.com

**Abstract.** Polycrystalline $BaTiO_3$ (BTO) and $SrTiO_3$ (STO) were synthesized by solid state route method and properties of made polycrystalline were characterized by X-Ray diffraction (XRD), Raman Spectroscopy & FTIR Spectroscopy. XRD analysis shows that samples are crystalline in nature. In Raman Spectroscopy measurement, the experiment has been done with the help of JOBIN-YOVN HORIBA LABRAM HR800 single monochromator, which is coupled with a "peltier cooled" charge coupled device (CCD). Raman Spectroscopy at low temperature measurement shows the phase transition above & below the curie temperature in samples. Fourier transform Infrared spectroscopy was used to determine the Ti-O bond length position.

## 1. Introduction

Barium titanate ($BaTiO_3$) is one of the best known Perovskite ferroelectric compounds (A2+B4+O3) that have been extensively studied [1, 2] due to the simplicity of its crystal structure, which can accommodate different types of dopant. Barium titanate ($BaTiO_3$) is another type of $ABO_3$ perovskite, which has not only large ferroelectric response, but also very large nonlinear optical and electro-optic coefficients, which is attractive to designing nonlinear optical devices [3]. These properties can be dramatically enhanced when they are doped with transition metals. It is well known that $BaTiO_3$ has a paraelectric to ferroelectric transition at about 120 $^0$C with a very high dielectric constant. Thus, considerable attentions were paid to the doped $BaTiO_3$ [4-5] to modify the performance of the materials. Strontium titanate ($SrTiO_3$) with a perovskite-type structure is an important ceramic material having wide uses in the catalysis, sensors, actuators, electrooptical devices, random access memory devices, and multilayer capacitor. Because $SrTiO_3$ have a lot of physical and chemical properties, such as, high thermal and chemical stability, low coefficient of thermal expansion, large dielectric constant, low dielectric loss, high nonlinear optical coefficient, it is widely accepted that the practical performances of product are strongly influenced by its phase, morphology, particle size, crystal defects, surface properties, etc., which ultimately depend on its preparation method and condition.

## 2. Synthesis of Iron & cobalt doped barium titanate

Both samples barium titanate and strontium titanate were synthesized via solid state route. We started from highly pure fine powdered samples of $BaCO_3$, $SrCO_3$ and $TiO_2$ for bulk samples of $BaTiO_3$ and $SrTiO_3$. All these were mixed in the calculated percentage ratio. Samples were grinded for 8-hours and calcinations were done at 900 $^0$C and than sintered at 1100 $^0$C & 1350$^0$C for 24 hours.

## 3. Experimental Technique

### *3.1. X-ray Diffraction*

The synthesized powders were characterized by X-ray diffraction using a D8 Advanced X-ray diffractometer. X-ray diffraction (XRD) is a versatile, non-destructive technique that reveals the detailed information about the chemical composition and crystallographic structure of natural and manufactured materials. XRD pattern for polycrystalline $BaTiO_3$ and $SrTiO_3$ respectively was carried out using CuKα radiation in the 2Θ range of 10 to $85^0$ shown in figure 1(a) and 1(b).The XRD patterns of samples do not show any impurity peaks which indicates that the samples are single phase in nature. From the diffraction pattern, the lattice parameter for both the samples has been calculated. The lattice parameters calculated for $BaTiO_3$ are (a=0.3994nm, b=03994nm, c=0.4022nm) which are in accordance with the tetragonal structure of the $BaTiO_3$.The lattice parameters calculated for $SrTiO_3$ is a=0.3910nm.

### *3.2. Raman Spectroscopy Measurement*

In Raman Spectroscopy measurement, the experiment has been done with the help of JOBIN-YOVN HORIBA LABRAM HR800 single monochromator, which is coupled with a "peltier cooled" charge coupled device (CCD). The spectra have been recorded in unpolarized mode in the full range starting from 200 to 850 $cm^{-1}$. The temperatures have been recorded in unpolarized mode in the range from 80K to 350K. Spectra have the broad features characteristics of titanium disorder in the unit cell at all temperatures.During the continuous heating from 80 to 350 K, three phase transitions take place: R→O, O→T and T→C is shown in fig 2(a). At low temperature in phase rhombohedra (80 K), the spectrum shows the following main lines: a broad peak near 262.6 $cm^{-1}$ [$A_1$(TO)], a sharp peak at 311.1 $cm^{-1}$ [$B_1$, E(TO+LO)], a weak peak at 486.4 $cm^{-1}$, [E(LO), $A_1$(LO), E(TO)], asymmetric and

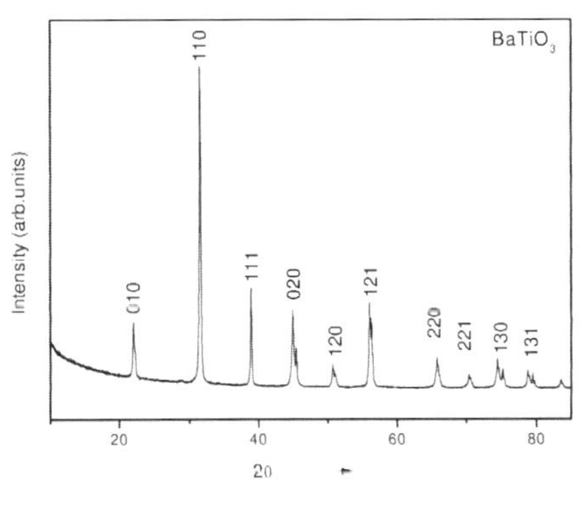


**Figure 1(a).** XRD pattern of $BaTiO_3$ at $900^0$C.

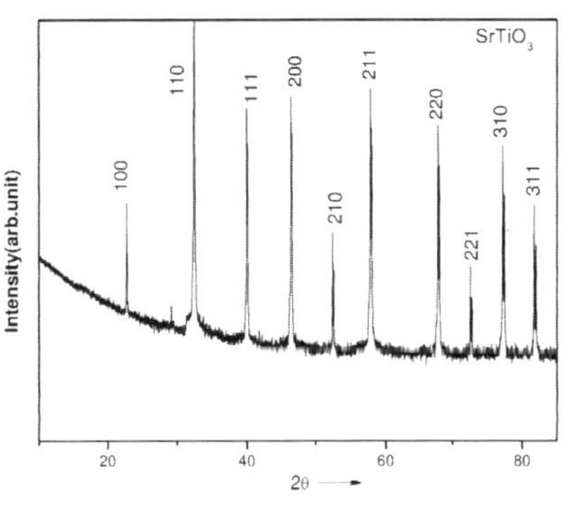


**Figure 1(b).** XRD Pattern of $SrTiO_3$ at $900^0$C.

**Table 1.** Particle size of pure $BaTiO_3$ & $SrTiO_3$ at $900^0$C.

| Sample | Particle size(nm) |
|---|---|
| $BaTiO_3$ | 26.4 |
| $SrTiO_3$ | 87.9 |

broad peak near 530.8cm$^{-1}$ [E(TO), $A_1$(TO)], and a weak peak at 714.8 cm$^{-1}$ [E(LO, $A_1$(LO)], where the phonon assignment is given inside square brackets. When the temperature is increased until 175 K, the Raman peaks disappeared or were shifted. The peak intensity at 486.4 cm$^{-1}$ decreases when the temperature increases, this behavior is due to the rhombohedral to orthorhombic phase transition. The next phase transition, orthorhombic to tetragonal, is carried out at 290K, the mode [E (LO), $A_1$ (LO), E (TO)] disappears and modes [$B_1$, E (TO+LO)] and [E (TO), $A_1$ (TO)] are shifted to 307.1 and 517.3 cm$^{-1}$, respectively. For phase transition T → C, at 350K, the several Raman modes have disappeared or diminished considerably.

A systematic low temperature Raman spectroscopy study has been done on polycrystalline $SrTiO_3$ is shown in fig 2(b). On cooling below 300K, no drastic change occurs till 90K. As the sample temperature comes closer to the transition temperature (105K) the band at ~250K starts to be very sharp. Since there is difference in the actual sample temperature and the temperature displayed on temperature sensor, thus the real phase transition occurs 2-3K lower than the actual $T_c$

As we further lower the temperature from 105K one new sharp peak starts developing at 90K with the peak positions at ~144cm$^{-1}$ . This gives a clear evidence of the occurrence of phase transition, as we don't expect sharp peaks in cubic structure. The strength of the peak increases with further lowering of the temperature. The peak at ~144cm$^{-1}$ is assigned as $E_g$, at 173.8cm$^{-1}$ as $B_{2g}$.

### *3.2. Fourier Transform Infrared Spectroscopy (FTIR)*

Fourier Transform Infrared Spectroscopy (FTIR) is a powerful tool for identifying types of chemical bonds in a molecule by producing an infrared absorption spectrum that is like a molecular "fingerprint". FTIR is simply a technical variant of a common infrared spectrometer, which yields an intensity signal as a function of wavelength or 'spectral color'. The FTIR spectrum was measured at room temperature. The measurement of FTIR spectra was carried out by Perkin-Elemer FTIR spectrometer. Figure 3(a) shows the FTIR spectra of BaTiO3.

The spectra of the powder calcined at 900$^0$C showed a wave number range from 400 - 4000 cm$^{-1}$ attributed to the Ti-O vibration, the carbonate peaks ($CO^{-2}$) at 857 cm$^{-1}$, 1745 cm$^{-1}$ and 2441.8 cm$^{-1}$ and hydroxyl (O-H) group at 1058.9cm$^{-1}$, 1634.6cm$^{-1}$. Bands corresponding to the carbonate and hydroxyl group were noted due to the intermeate $BaCO_3$ phase and water. Peaks corresponding to carbonate ($CO_3^{2-}$) bands have been noted 857 cm$^{-1}$ (out of plane deformation), 1745cm$^{-1}$ (symmetric stretching) and 2441.8cm$^{-1}$ (combined symmetry).
The FTIR spectrum of $SrTiO_3$ calcined at 900$^0$C is shown in fig 3(b), which exhibits a peak at 577.6 cm$^{-1}$ corresponding to Ti-$O_1$ stretching.

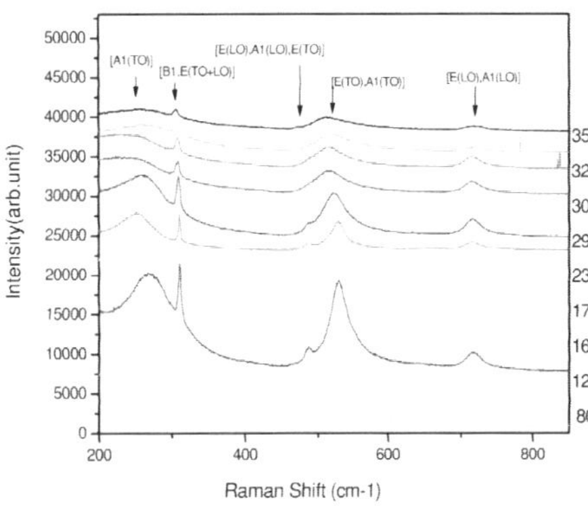


**Figure 2(a).** Unpolarized Raman spectra of polycrystalline $BaTiO_3$ recorded at selected temperatures.

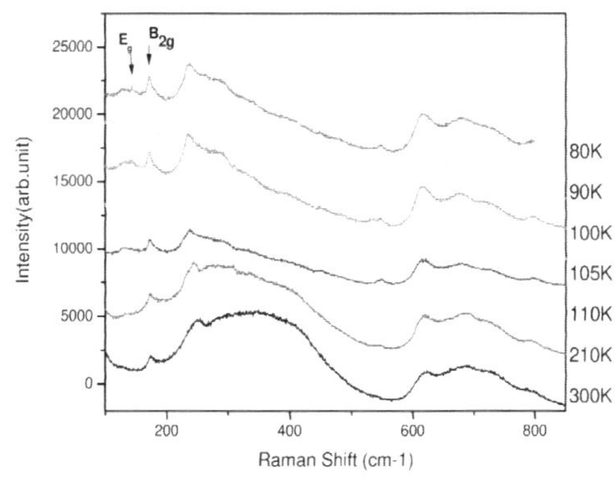


**Figure 2(b).** Unpolarized Raman spectra of polycrystalline $SrTiO_3$ recorded at selected temperatures.

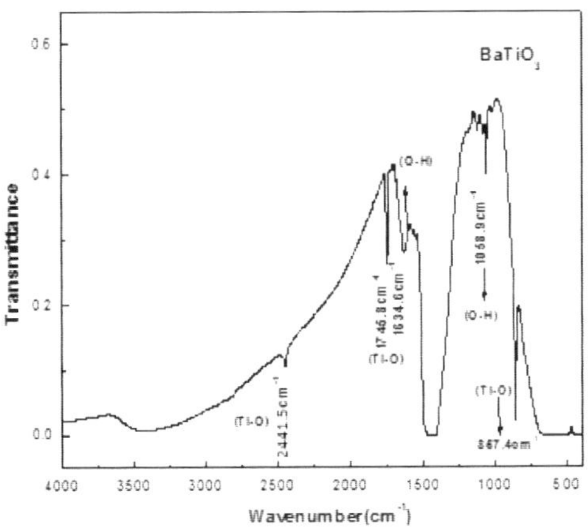


**Figure 3(a).** FTIR spectrum of pure $BaTiO_3$.

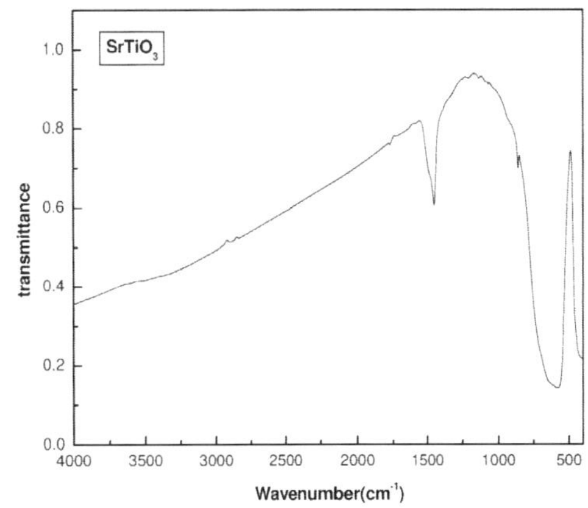


**Figure 3(b).** FTIR spectrum of pure $SrTiO_3$

## 4. Conclusion

Samples of polycrystalline $BaTiO_3$ and $SrTiO_3$ were synthesized by solid state method. The XRD patterns of samples do not show any impurity peaks which indicates that the samples are single phase in nature. We have carried out a systematic low temperature spectroscopy study of $SrTiO_3$ & $BaTiO_3$ single crystal using a micro-Raman setup. In $SrTiO_3$ we observed only broad humps in Cubic phase that is above 105K. Sharp peaks corresponding to normal modes in tetragonal symmetry are observed below this transition temperature. In $BaTiO_3$ we observed in two phase transition R→O, O→T when we increase the temperature Raman peaks disappeared or were shifted to upper frequency and Raman peak intensity decreases. But For phase transition T → C, at 350 K, the several Raman modes have disappeared or diminished considerably. The FTIR spectra reveal that they affect the vibration of crystal lattices. When a smaller ion replaces Ti4+, the cell becomes smaller; resulting in the distance between Ti and O becomes shorter. This enhances the interaction between these ions.

## Acknowledgments

The authors are thankful to Prof. Ajay Gupta, Dr. Mukul Gupta, UGC-DAE CSR for XRD, Dr. Vasant Sathe, UGC-DAE CSR for Raman measurement and Dr. Uday Deshpandey for FTIR measurement, Indore for providing the experimental facilities and for the precious guidance. One of the authors (Supriya Bisen) is thankful to UGC-DAE CSR Indore for providing the research fellowship.

## References

[1] Makovec D, Samadmija Z and Drofenik M J Am 2004 *Ceram. Soc.* **87** 1324
[2] Maga D, Igor P and Sergei M Mater J. 2000 *Chem.* **10**
[3] Jana A, Kundu T.K , Pradhan S.K and Chakravorty D J 2005 *Appl. Phys.* **97**
[4] Smolenski G.A, Isupov A.V, Dokl. Akad. Nauk. 1954 *SSSR* **9**
[5] Ismailzade I.H, Ismailov R.M 1980 *Phys. Stat. Sol.* **59** K

# X-ray diffraction and X-ray K-absorption near edge studies of Copper (II) Micro cyclic Carbamide complexes.

**P K Malviya[1], P Sharma[2], A Mishra[3]and D Bhalse[4]**

[1] Govt.College, Nagda (M.P), malviyapramod35@yahoo.co.in
[2] Govt.Holkar Science College Indore (M.P).
[3] School of physics, DAVV, Indore (M.P).
[4] Govt.Girls College, Khargon

Email: psharma29762@yahoo.co.in, amishra1960@ yahoo.co.in

**Abstract**:-Synthesis of metal complexes [Cu (Carbamide)] (X = Br, Cl, NO3, SO4,CH3COO) by the chemical root method. The XRD data have been recorded at DAE, IUC Indore.XANES spectra have been recorded at the K-edge of Cu using the dispersive beam line at 2.5GeV Indus-2 synchrotron radiation source RRCAT (Raja Ramanna Center for Advance Technology), Indore, India. XRD and XANES data have been analysed using the computer software Origin 8.0 professional and Athena. X-ray diffraction studies of all the complexes are indicative of their crystalline nature. The crystalline size of the samples is estimated using the Scherer's formula. The values of the chemical shifts suggest that copper is in oxidation state +2 in all of the complexes.
Key words – Cu (II) Carbamide complex, XRD, XANES, etc.

## Introduction:-

The present paper includes XRD and the X-ray absorption of copper (II) micro cyclic carbamide complexes. There has been a lot of interest in synthesis structure and properties of nitrogen donor ligands particularly carbamide, thiocarbamide and its derivatives due to their wide application as pharmaceuticals [1] and in wood protection [2-5].Metal complex with ligand systems containing oxygen and nitrogen donor atoms are very important class of coordination compound and complexes. It is known that the carbamide complex of copper plays important role [6-9] .X-ray studies of the following five copper (II) Carbamide complexes-[Cu(NH2CONH2)]Br2 , [Cu(NH2CONH2)]Cl2 , [Cu(NH2CONH2)](NO3)2 , [Cu(NH2CONH2)](SO4)2 and [Cu(NH2CONH2)](CH3COO)2

## Experimental details:-

(a)Synthesis of copper (II) carbamide with malonic acid

Reflexed solution of Carbamide (0.002 moles) and Cu metal salt (0.002 moles) in methanol (10 mL) was added drop wise to malonic acid (0.002 moles) in methanol (40 mL). After the addition was completed; the refluxing was continued for 12 h. The precipitate was filtered and washed with methanol, then dried in air.

For Carbamide

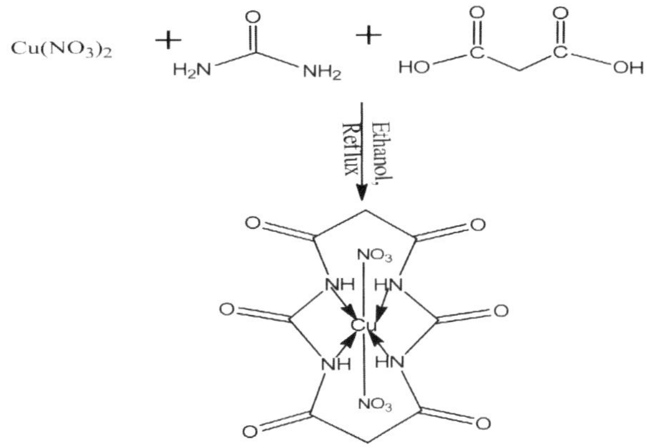


Fig 1- Chemical Scheme for copper (II) (Carbamide) Complexes.

In the present investigation, the preparation of complexes by chemical root methods, the X-ray diffraction have been recorded using bucker D8 DAE, IUC Indore, absorption spectra have been recorded using synchrotron radiation. The X-ray spectroscopic setup is available at Raja Ramanna Center for Advaced Technology (RRCAT) and is called BL-8 beamline. This beamline BL-8 has been recently commissioned at the 2.5 GeV Indus-2 synchrotron radiation sources.

## Results and Discussion:-

The sample was characterized at room temperature by X-ray diffraction using CuKα radiation. X-ray diffraction studies of all the complexes are indicative of their crystalline nature. The diffraction pattern of complexes recorded between 2θ ranging from 10° to 80°.The particle size and lattice parameter shown in the table 1 and XANES Parameter shown in table 2.

The table 2 presents the results for the K-absorption (EK) and the energy of the principal absorption maximum (EA) of copper metal and its complexes .The chemical shifts (eV) of the K-absorption edge of copper in the complexes are also given in Table 2 For all the complexes the distances (in eV)of principal absorption maximum EA with respect to the respective K –absorption edge have been computed and are collected in Table 2 .It can be readily seen from Table 2 that copper K-edge is found to be shifted towards the high energy side in all the five complexes as compared to the copper metal K- absorption edge .

Table1: Value of particle size and lattice parameter by XRD for copper (II) (Carbamide) complexes

| Complexes | 2θ | hkl | Particle size(nm) | Lattice parameter in Å |
|---|---|---|---|---|
| Copper(II)Carbamide Bromide | 27.04 | 1,1,1 | 9.21 | 5.70 |
| Copper(II)Carbamide Chloride | 12.56 | 2,2,2 | 8.46 | 12.15 |
| Copper(II)Carbamide nitrate | 12.12 | 1,1,1 | 9.87 | 12.70 |
| Copper(II)Carbamide sulphate | 18.77 | 1,1,1 | 11.04 | 11.94 |
| Copper(II)Carbamide acetate | 13.76 | 1,1,1 | 10.61 | 11.03 |

Table-2 Chemical shift of K-absorption edge of Copper (II) Carbamide complexes

| Complexes | $E_K$-Edge (eV) | $E_A$(eV) | Chemical Shift (eV) | Shift of principal absorption maxima (eV) | Edge width (eV) | ENC Electron/ atom | Percentage covalency (%) |
|---|---|---|---|---|---|---|---|
| Copper metal | 8980.5 | 9003.7 | - | - | 23.2 | - | - |
| [Cu(II)U]($NO_3$) | 8992.8 | 8999.2 | 12 | 18 | 6 | 1.53 | 36.34 |
| [Cu(II)U)] (Br) | 8990.8 | 8996.8 | 10 | 16 | 5 | 1.03 | 42.75 |
| [Cu(II)U)](Cl) | 8989.8 | 9002.5 | 9 | 22 | 12 | 0.96 | 45.00 |
| [Cu(II)U)] ($CH_3COO$) | 8988.8 | 8999.7 | 8 | 19 | 10 | 0.93 | 47.25 |
| [Cu(II)U)]($SO_4$) | 8988.5 | 8998.5 | 8 | 18 | 9 | 0.90 | 49.29 |

Energy of copper K-edge (Practical $E_K$) present study =8980.5eV

**Chemical Shift:-**The shift of the X-ray absorption edge (i=K, L, M…..) of an element in a compound/complex with respect to that pure element is written as:

$$\Delta E_k = E_{k\,(complex)} - E_{k\,(metal)}$$

The values of chemical shifts obtained for all the complexes are tabulated in Table 1.for the present complexes, the chemical shifts values lie in the **range 8 to 12 eV.** Hence on the basis of the values of the chemical shifts, all our complexes are found to have copper in oxidation state +2.

**Shift of the Principal Absorption Maxima:-**In the Table 2, we have also included the data for the principal absorption maximum $E_A$ in the complexes and metal .it has been observed that for copper metal, the value of $E_A$ is 9003.7 eV.

The shift of principal absorption maximum depends upon the type of overlap between metal atom and the ligand orbital's. The greater the overlaps of the metal and the ligand orbital, the more stable are bonding molecular orbitals and hence the corresponding antibonding molecular orbitals are more unstable. Since the principal absorption maxima occurs due to the transitions from 1s orbital to the unoccupied antibonding molecular ($1s \rightarrow 5t_{1u^*}$ in octahedral), the principal absorption maxima, therefore, shifts to the higher energy side is towards the high energy side of the edge [9].

**Edge Width:-**The edge widths are shown in Table 1 and related to the electro negativity differences between the central metal ion and its surrounding neighbours in the co-ordination sphere according to a semi-empirical correlation proposed by Nigam and Shrivastava [9] Equation [ $E_w\Sigma(X_M-X_L)]^{1/2}$= constant for given metal in given region}.Table 1 shown that the measured edge widths for the all complexes are similar. In the present work edge width of Cu (II) complexes in Table 1 are ranging from **5 to 12 eV**

**Effective Nuclear Charge:-**Effective charge can be defined as a total charge within a specified volume around the nucleus. Various theoretical and phenomenological methods have been proposed for the estimation of the effective nuclear charge .In the present work effective nuclear charge has been obtained from the measured chemical shift by using the semi- experimental methods by employing the procedure suggested by Nigam and Gutpa [10].ENC on the copper in the complexes under present study varies **between 0.90 to 1.53 electrons /atom.**

**Percentage Covalency:-**. The percentage covalency of metal ligand bonding in copper complexes is the ranging from **36.34 to 49.29 and they are reported in table 2.**

**Conclusion:-**

All the copper (II) complexes are crystalline in nature and the analysis of X-ray diffraction pattern shows that the samples exhibit simple cubic phase. The chemical shift values between 8 to 12 eV, shift of the principal maximum values between 16 to 22 eV and edge width values between 5 to12eV .The effective nuclear charge ranging from 0.90 to 1.53 electron/atom. The percentage covalency values ranging from **36.34 to49.29 %**.The values of the chemical shifts suggest that copper is in oxidation state +2 in all of the complexes.

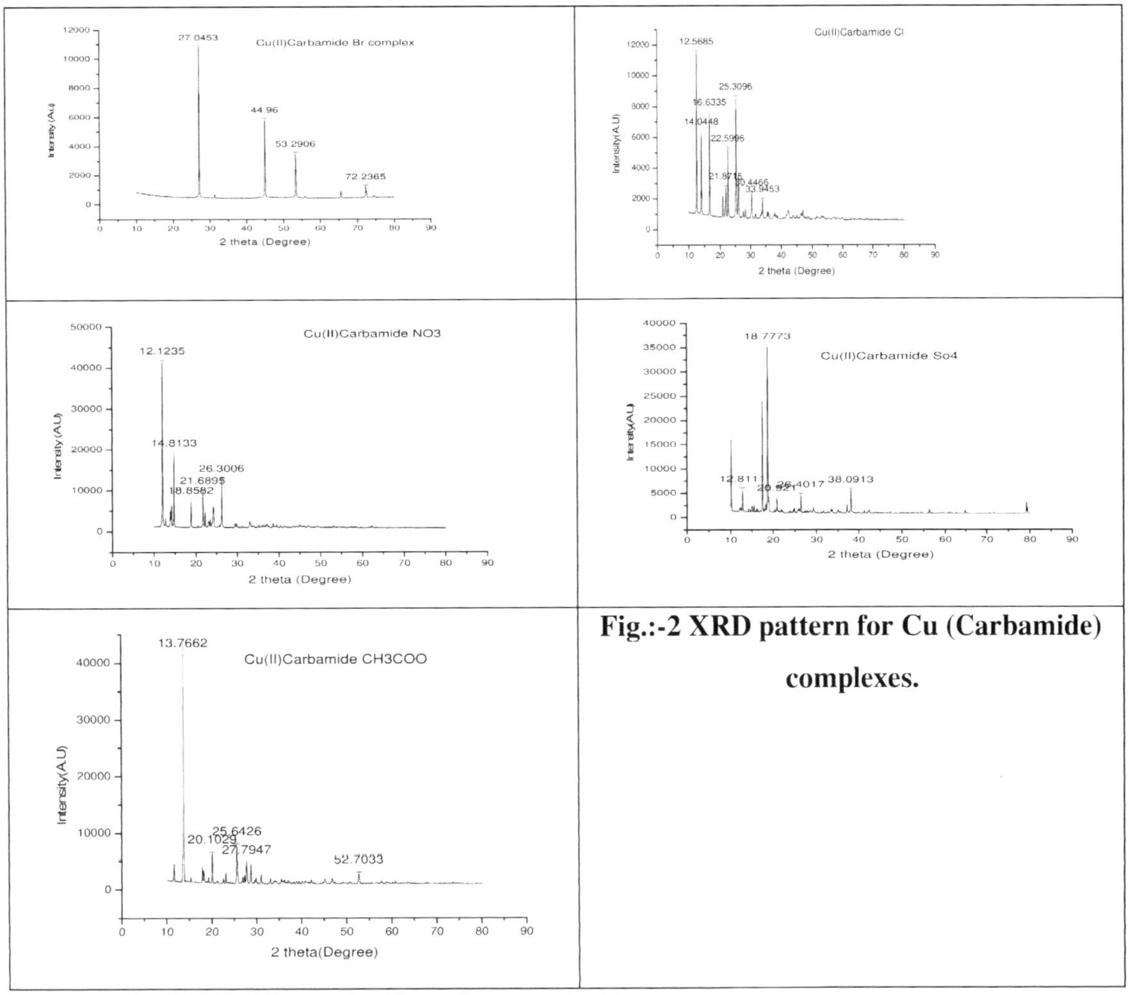


**Fig.:-2 XRD pattern for Cu (Carbamide) complexes.**

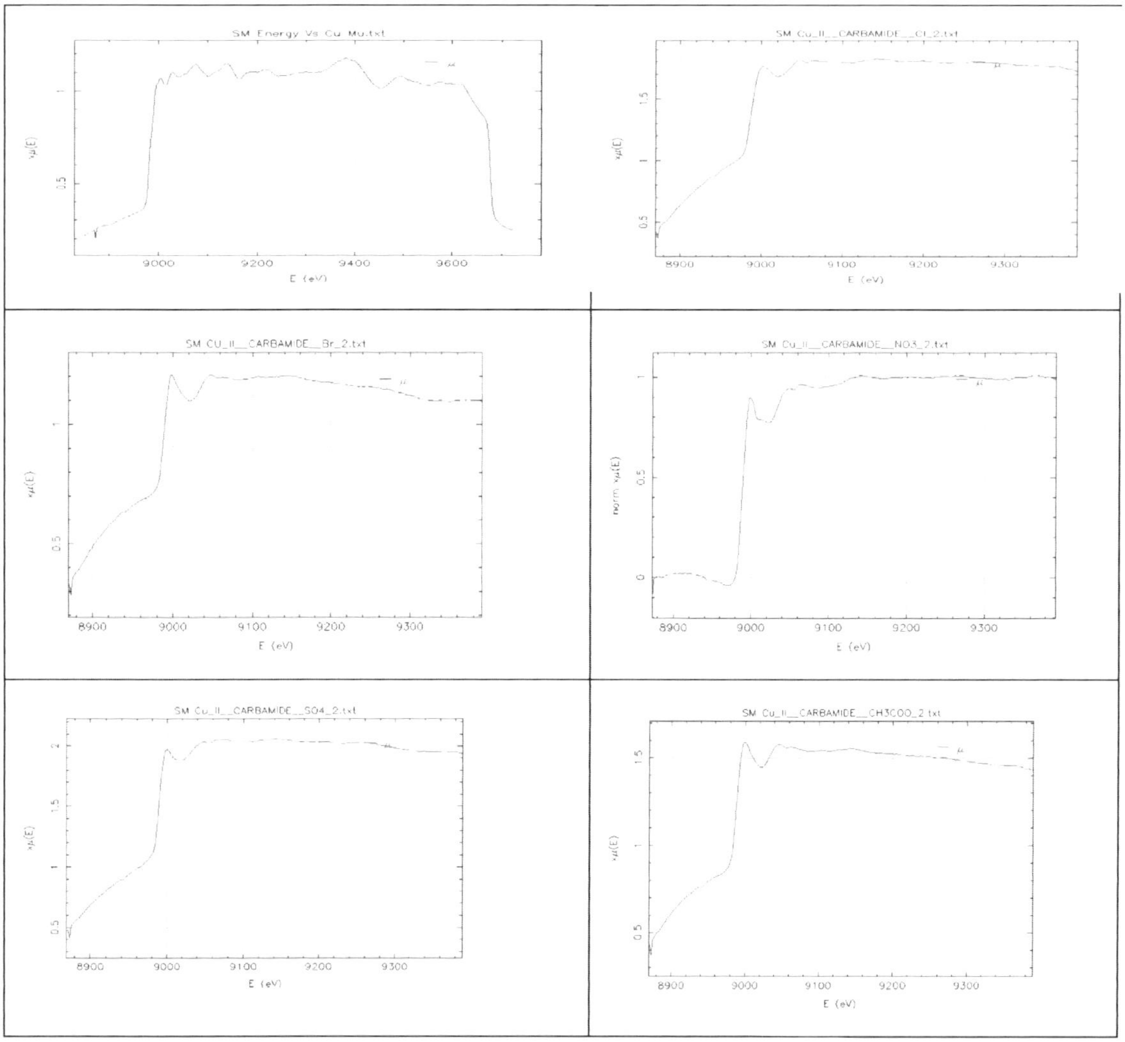


**Fig2. X-ray absorption near edge structure of copper II complexes and metal foil.**

## References

[1] Mixed Ligand Complex of Copper (II) Containing ON Donor Ligands Ashish Kumar Sahoo, Department of Chemistry, National Institute of Technology, Rourkela, Orissa 2010 page no.01

[2] Dalen M B., Mamza P A P 2009 *science world journal* **4** 112.

[3] Laboratory module 01, indexing X-ray diffraction patterns, scottA speakman Ph.D 13-4009A

[4] Huang Y B, Yi W B, and Cai C 2012 *Top Curr Chem* **308** 191.

[5] AL-Obaidi O H S and A-Hiti A R H 2012 *American Chemical Science Journal* **2(1)** 1.

[6] Scherrer P 1918 *Math.Phys.K* **1.2** 96.

[7] Deshmukh J H and Deshpande M N 2011 *J. Chem. Pharm. Res.* **3** 706

[8] R.L.Snyder, X-ray diffraction. In: X-ray characterization of mater (Eds.E.Lifshin, and Weinheim), Wiley-VCH New York (1999).

[9] C.Suryanarayana and M.G.Norton.X-ray diffraction, a practical approach. Plenum press (1998).

[10] J.R.J Sorenson. 1982. In Metal Ions in Biological Systems. Edited by H. Sigel and Marcel Dekker. New York. 14: 77-124.

[11] Ninalahi B, Makuc S, Segedin P. 2000 *Acta Chim Slov.* **47** 421.

[12] B.A Richardson. 1993. Wood Preservation. Second Edition; E & FN Spon an imprint of Chapman & Hall London.
[13] Prisakar V I, Tsapkor V I, Buracheva S A, Bryke M S and Gulya A P. 2006 *Pharmaceutical Chemistry Journal*. **39** 6.
[14] Premkumar, Tand and Govindarajan S 2005 *World Journal of Microbiology* **21** 479.
[15] StojcevaRadovanovic B C and Premovic P I 1992 *Journal of Thermal analysis* **38** 1992 715.
[16] Agarwal B K and Verma L P 1970 *J.Phys.C:Solid state Phys* **3** 533.
[17] Hinge V K, Joshi S K and Shrivastava B D 2012 *IOP Publishing* 365.
[18] Rajput O P, Nigam A N and Shrivastava B D 1984 *X-ray Spectron* **13** 156.
[19]. Nigam A K and Gupta M K, 1974 *J.Phys. F: Metal Physics* **4** 1084.

# High Temperature Multiferroicity in Cupric Oxide

**Suchita Pandey and A M Awasthi***
Thermodynamics Laboratory,
UGC-DAE Consortium for Scientific Research, Khandwa Road, Indore- 452 001, India

E-mail: amawasthi@csr.res.in

**Abstract.** Multiferroic materials where magnetic and electrical orders coexist have been subject of extensive studies during last few years. Non-collinear/spiral magnetic structures are expected to exhibit concurrent ferroelectricity. CuO which adopts the structure as mineral Tenorite, with monoclinic structure (space group C2/c), has been reported to possess two high temperature antiferromagnetic (AFM) transitions, $T_{N1}$ ~ 213 K &$T_{N2}$ ~ 230 K. Below the first AFM transition, magnetic structure is reported to be collinear while below later one magnetic structure is non-collinear . In the present study observation of strong anomalies in dielectric and heat capacity measurements around both the $T_N$'sconfirm magnetically induced ferroelectricity in CuO.

## 1. Introduction

Materials which exhibit co-existence of magnetic and electrical orderings foresee tremendous technological applications and have attracted considerable attention in recent years [1]. Magnetic order and ferroelectric polarization in spiral magnets are expected to exhibit strong magneto-electric (ME) coupling effects; recent observation of large magneto-electric and magneto-capacitive effects in $TbMnO_3$ that provides a novel approach to the mutual control of magnetization and electric polarization in magnetic ferroelectrics (FE) materials [2]. However, the multiferroicity in these materials typically occurs at low temperatures, at ~ 40K. Recently, significant progress has been achieved in some hexaferrites, where spiral magnetic order (SMO) induced multiferroics and ME effects were discovered at or near room temperature [3]. In addition to the complex compounds mentioned above, a prominent exception was recently discovered in the very well-known oxide, cupric oxide CuO, where SMO-induced polarization is observed in the temperature range of 215-230K [4-5]. Such discoveries extend the family of high-temperature multiferroics. Copper (II) oxide, occurs in nature as the mineral Tenorite, and has a monoclinic structure with the space group C2/c [6].

There are two well-established magnetic transitions in CuO, one at $T_{N1}$ ~ 213 K and another at$T_{N2}$ ~ 230K [4-5]. A commensurate (CM) collinear magnetic structure develops below $T_{N1}$. In this phase, magnetic moments are aligned collinear along the *b*-axis, such as parallel and anti-parallel zigzag chains of Cu-O along the [101] and $[10\bar{1}]$, respectively. With increasing temperature, an incommensurate (ICM) spiral magnetic structure develops, exists between $T_{N1}$ and $T_{N2}$, due to the competition between antiferromagnetic (AFM) and ferromagnetic (FM) interactions. This noncollinear structure results in a net electric polarization through the inverseDzyaloshinskii-Moriya (D-M) interaction and transverse spin-lattice coupling [7]. Above $T_{N2}$, the magnetic susceptibility of CuO undergoes an anomalous continuous rise, instead of a peak at its Nèel temperature, and rises to a broad maximum at about 540K [5]. Even in the strongest magnetic field, the transition temperatures and the associated dielectric anomalies were found to be unaffected by Wang et.al.[8].Considering the SMO as very robust, these authors concluded that the ME effects are nominal in CuO at both $T_{N1}$ and $T_{N2}$. In the present work we have performed structural, dielectric spectroscopy and heat capacity, measurements and investigated the above mentioned transitions.

## Experimental Details

High-purity CuO (99.995%) powder was pelletized and then heat treated at 800°C for few hours. Phase purity of sintered material was checked by powder X-ray diffraction (XRD) at room temperature using Cu $K_\alpha$ radiation by Bruker D-8 advanced diffractometer. The dielectric measurements were performed using Novo control Alpha-A High Frequency Analyzer, with a homemade test-cell, across 10Hz-10MHz, from room temperature down to 10K. Heat capacity was measured using Modulated Differential Scanning Calorimeter (MDSC) from TA-2910 instruments with temperature range 150-300K in Nitrogen atmosphere, with the scanning rate of 10K/min.

## Results and Discussions

Figure 1shows the room temperature Rietveld fitted XRD pattern of CuO, confirming the sample having monoclinic structure with the space group C2/c. A smallunavoidable secondaryphase of cubic $Cu_2O$ was detected (a small peak at ~ 36.4deg.) during Rietveld fitting. Wang et al. [8] also observed the small quantity of same secondary phase. Lattice parameters of synthesized CuO,obtained from fitting, are as follows; $a$ = 4.6878 Å, $b$ = 3.4228 Å, $c$ = 5.1316 Å, $\alpha$ = 90°, $\beta$ = 99.51°, $\gamma$ = 90°.These are similar to those reported elsewhere [8].

Further, we explored the dielectric behaviour of this material in wide frequency range, across magnetic transitions reported in CuO. Zhenget.al. [9] reported both the AFM transitions ($T_{N1}$&$T_{N2}$), and unlike other antiferromagnets, they found that its magnetization increases with temperature above $T_{N2}$, and transformation to proper paramagnetic state occurs only at higher temperatures (630K >>$T_{N2}$ = 230K).Fig.2 (a) shows the temperature dependence of real part of dielectric permittivity in the temperature region 200-250K, at two different high frequencies (0.6 &0.8 MHz).

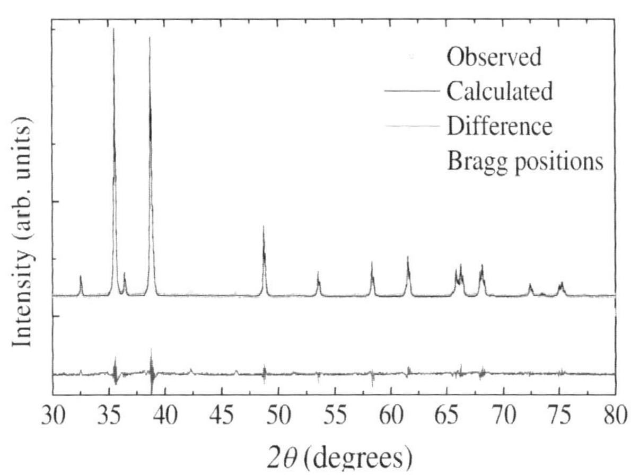


Fig. 1: Room temperature Rietveld fitted XRD pattern of CuO.

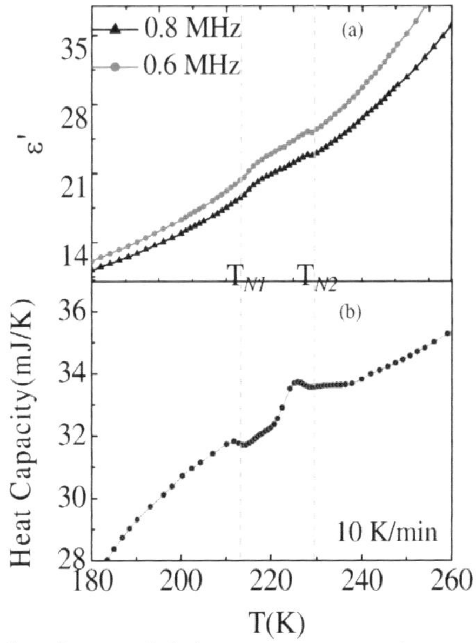


Fig. 2: (a) Real part of dielectric permittivity across magnetic transition temperatures in CuO, at two selected high frequencies. (b) Heat capacity measurement showing peaks at magnetic transition temperatures.

The dielectric constant shows small FE-like peak at $T_{N2}$ and a decreasing step-like behaviour at$T_{N1}$. It shows no substantial ferroelectric polarization (***P***) exists below $T_{N1}$ and above $T_{N2}$. Observation of a finite ***P*** between $T_{N1}$ and $T_{N2}$ is a signature of magnetically induced ferroelectricity.It is to be highlighted here that in an earlier study on CuO single crystal, though the existence of magnetically induced polarization was confirmed in between $T_{N1}$ and $T_{N2}$through pyroelectric current measurement [4], but the anomaly in dielectric constant around $T_{N1}$ was not as evident as here.

The manifestation of these transitionsis also clear in our heat capacity measurement and it show sharp peak at both the commensurate-incommensurate antiferromagnetic transition temperatures (fig.2b).The anomalies observed in the heat capacity are reported here for the first time. Also the anomalies observed here in the dielectric data are much clearer than the previous reports [4, 9].

**Conclusions**

Dielectric anomalies across the incommensurate magnetic phase boundaries, at$T_{N2}$ ~ 230K and $T_{N1}$ ~ 213Kin CuO evidence magnetically-induced ferroelectricity. Anomalies in the heat capacity are observed for the first time, confirming the AFM transitions.

**Acknowedgments**

Authors are thankful to the Director,UGC-DAE CSR and the Centre-Director, Indore Centre, for their help and support.

**References**

[1] Cheong S-WandMostovoy M 2007 *Nat.mater.* **6** 13.
[2] Kimura T, Goto T, Shintani H, Ishizaka K, Arima T, and Tokura Y 2003 *Nature* **55** 426.
[3] KimuraT, Lawes G, and Ramirez A P 2005 *Phys. Rev. Lett.* **94** 137201.
[4] Kimura T, Sekio Y, Nakamura H, Siegrist T and Ramirez A P 2008 *Nat. mat.***7** 291.
[5] KeeffeM O and Stone F S 1962 *J. Phys. Chem. Solids* **23** 261.
[6] Asbrink S and Norrby L J 1970*ActaCrystallogr. B***26** 8.
[7] Giovannetti G, Kumar S, Stroppa A, Brink J van den, Picozzi S, and Lorenzana J 2011 *Phys. Rev. Lett.* **106** 026401.
[8] Wang F, Zou T, Liu Y, Yan L Q, and Sun Y 2011 *J. Appl. Phy.* **110** 054106.
[9] Zheng X G, Xu C N, Tanaka E, Tomokiyo Y, Yamada H, Soejima Y, Yamamura Y, and Tsuji T 2001 *J.Phys. Soc.Jpn* .**70** 1054.
[10] Huang Z J, Cao Y, Sun Y Y, Xue Y Y, and Chu C W 1997 *Phys. Rev. B* **56** 2623.

# Randic and Schultz molecular topological indices and their correlation with some X-ray absorption parameters

**Sunil Khatri[1], Pravin A Kekre[2], Ashutosh Mishra[3]**

[1] Shri R.G.P.G. Professional Institute, Indore, India
[2] P.M.B. Gujarati Science College, Indore, India
[3] School of Physics Devi Ahilya University, Indore, India

Email:drpravinkekre@yahoo.com

**Abstract.** The properties of a molecular system are affected by the topology of molecule. Therefore many studies have been made where the various physic-chemical properties are correlated with the topological indices. These studies have shown a very good correlation demonstrating the utility of the graph theoretical approach. It is, therefore, very natural to expect that the various physical properties obtained by the X-ray absorption spectra may also show correlation with the topological indices. Some complexes were used to establish correlation between topological indices and some X-ray absorption parameters like chemical shift. The chemical shift is on the higher energy side of the metal edge in these complexes. The result obtained in these studies shows that the topological indices of organic molecule acting as a legands can be used for estimating edge shift theoretically.

## 1. Introduction

Earlier, we have established a correlation between topological Index (Winer and Randic) and X-ray parameters.1, 2, 3, In present work we have now established a correlation between X-ray absorption parameter like chemical shift with Randic and Schultz index respectively.

## 2. Generation of Randic and Schultz Index

The Randic index[4] makes use of vertices present in the chemical graph and therefore, it is sensitive to the shape also. Hence this index is also known as molecular connectivity index $x$. It is equal to the some of the values of all molecular edges.

$$\chi = \sum_{\text{all bonds}} \left( \frac{1}{\text{m.n}} \right)^{1/2}$$

The variables m and n are the valencies of the bond end points of each edge. Multiple bonds and hetro atoms have been included in the Randic index by assigning valencies value ($\partial$) to the concerned atoms using the expression:

$$\partial = z^{v} - h_{i}$$

and/or the expression

Where z is the total number of electrons. $z^v$ is the number of valence electrons of atom i and $h_i$ is the number of hydrogen atom attached to it.

The Schultz[5] molecular topological index of G denoted by MTI, is defined as

$$\mathrm{MTI} = \sum_{i=1}^{n} e_i$$

where $e_i$'s (1, 2, 3, ….. n) represents the elements of the following row matrix of order n.

$$V(G)\,[A(G) + D(G)] = [e_1, e_2, e_3 \ldots\ldots\ldots e_n]$$

where V(G) is the valence row matrix, A(G) and D(G) are adjacency and distance matrix respectively.

## 3. Results and Discussion

Table 1 contains name of complexes and their molecular graph. Table 2 contains chemical shift, Randic and Schultz Index.

Table 1 Name of the complexes and their molecular graphs

| No. | Name of complexes | Molecular graph |
|---|---|---|
| 1. | N-Phenyl-3-methyl-4- [4' – (2" – (5, 6-dimethyl oxazolyl) Sulphonamoyl] Phenylazo-Pyrazolin-5-one Cobalt (II) (CoPzS) | 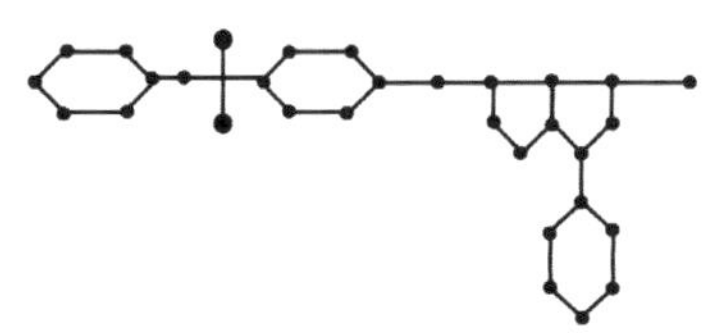 |
| 2. | N-Phenyl-3-Methyl-4- [4' – (2" – Pyrimidimyl sulphonamoly)] Phenylazo-Pyrazolin-5-one cobalt (II) (CoPZ) | 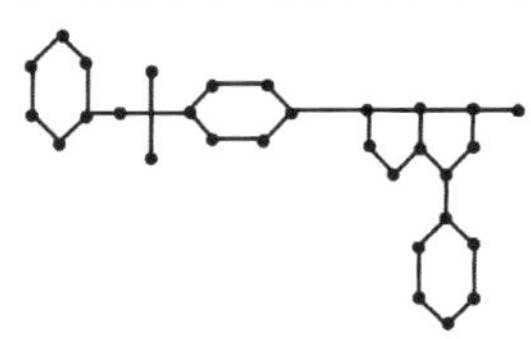 |
| 3. | Co-P Complex using Sulphadiazine (CoP5) | 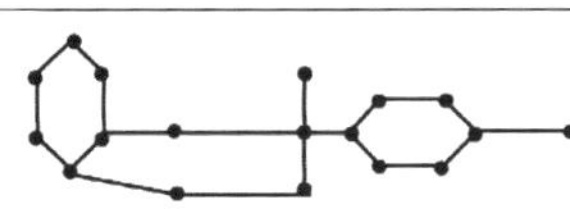 |
| 4. | N-Phenyl-3-Methyl-4- (4' – nitro) Phenylhydrozone Pyrozolone-5-one (CoPN) | 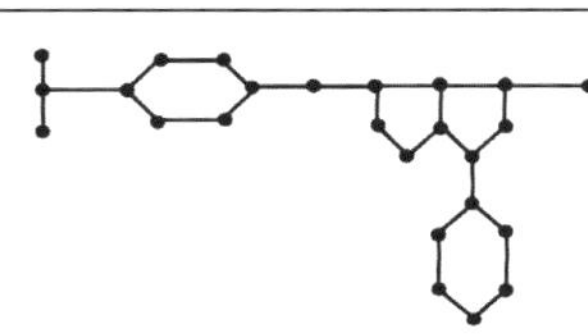 |
| 5. | Co complex of P-toluidine [CoPT] | 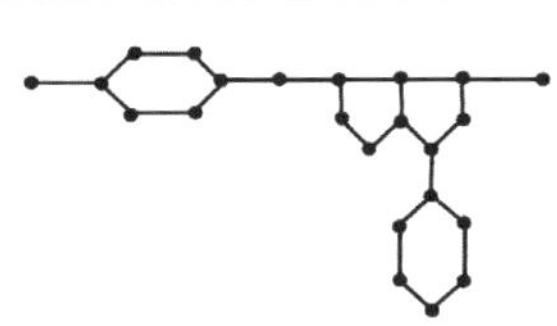 |

Table 2 Chemical Shift ( $\Delta$ E), Randic and Schultz Index for Co-substituted Pyrazolm-5-one complex

| **No.** | **Complex** | **Chemical Shift** $\Delta E \pm 0.3$ ev | **Randic Index** | **Schultz Index** |
|---|---|---|---|---|
| 1. | CoPZ5 | 10.0 | 10.60 | 26433 |
| 2. | CoPZ | 11.3 | 9.82 | 26240 |
| 3. | CoPS | 8.5 | 5.34 | 4569 |
| 4. | CoPN | 12.1 | 7.48 | 14000 |
| 5. | CoPT | 10.7 | 7.39 | 9246 |

The position of X-ray absorption edge depends upon the valence of the absorbing ion the effective charge on the central atom and also upon the geometry of the complex[6, 7] In the present work the K-absorption edge for all Co(TI) complexes is found to be on the higher energy side. The order of chemical shifts for complexes as indicated by their values has been found to be as follows:

CoPN > CoPZ > CoPT > CoPZ5 > CoP5

The order is due to fact that –NH and –C = 0 stretching frequencies are observed at 3380 and 1750 $cm^{-1}$ respectively. However, upon complexation the bond due to the –NH stretching is again retained at 3380 $cm^{-1}$ because of the presence of another –NH group.

In sulphonomoyl moiety large values of chemical shift in all the complexes show that the complexes are more ionic in character. As the shift is taken to be proportional to ionic character[7] the role of various bonds attached to the central metal ion is prominent in determining the overall chemical shift in a metal complex. Further not only ligation of a particular type, e.g. Co–O, Co–N is responsible in imparting ionic character to the complex, but their position in the planar ring / axial position also contributes to the ionic character.

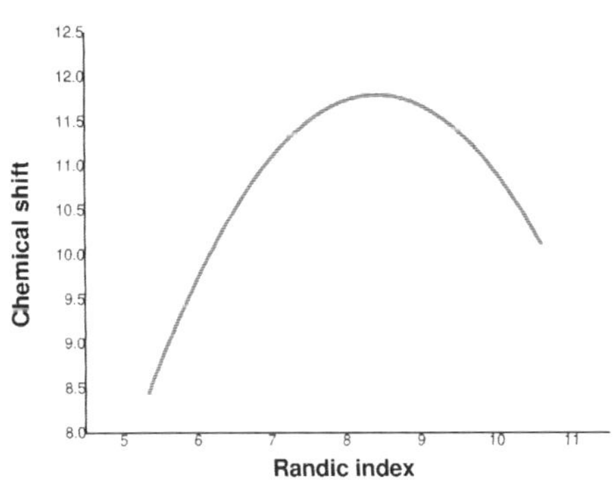


Randic Index
Figure 1 Correlation of Randic Index with chemical shift for Co (II) complexes

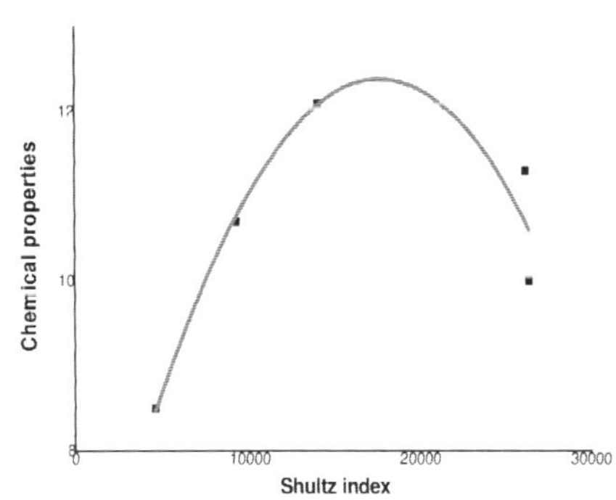


Schultz Index
Figure 2 Correlation of Shultz Index with chemical shift for Co (II) complexes

Figure 1 and 2 shows correlation between chemical shift with Randic Index and Schultz Index respectively.

The correlation assumes the general form

$$(TI)^2 = -4a\,(\Delta E)$$

Where TI stands for a particular topological index and a is constant.

The Figure 1 represents the correlation for chemical shift and Randic Index.

While Figure 2 demonstrates the correlation for Schultz indices and chemical shift. Both the plots yield excellent parabola (second order polynomial)

It is gratifying that the correlations with the Schultz index are better than that with Randic Index. As the chemical shift not only depends on ligation of particular type e.g. Co–$O^-$ Co–N is responsible in imparting ionic character to the complex, but their position in the planar ring / axial position also contributes to the ionic character. Thus result obtained in this study show that the correlation with Schultz indices is very effective.

## 4. Conclusion

Randic and Schultz molecular topological indices contain important structural information. One can say that the X-ray absorption parameter, thus estimated, indicate that topological indices of the organic molecules acting as legands can be used for estimating X-ray absorption parameter theoretically. Thus the topological understanding of molecular properties can lead to the development of new area of the present and future interest i.e. tracking the effect of pollutants in the environment, designing of drugs and prediction of carcinogenicity of molecule.

## 5. References

[1] Kekre P A, Joshi K P, Khatri S and Mishra A 1996 *Bull Soc Belg.* **105** 169
[2] Mishra A, Khatri S, Joshi K P and Kekre P A 2000 *Asian Journal of Physics* **9** 345
[3] Kekre P A, Khatri S, Mishra A and Joshi K P 2012 *Journal of Physics: Conference Series* **365** 012011
[4] Randic M 1975 *J Am. Chem. Soc.* **97**i6609
[5] Schultz H P 1989 *J. Chem. Inf. Computer Sci.* **29** 227
[6] Agrawal B K & Verma L P 1970 *J. Phy* **C3** 535
[7] Shrivastava U G, Nigam H L & Vishnoi A N 1971 *Ind. J Pure and applied Phy.* 963

# Study of structural, electronic and magneto transport properties of $La_{0.7}Ca_{0.2-x}Sr_xAg_{0.1}MnO_3$

**P Subhashini[1], B Munirathinum[2], M Krishnaiah[1], R Venkatesh[3], D Venkateswarlu[3] and V Ganesan[3]**

[1] Department of Physics, Sri Venkateswara University, Tirupati, India
[2] NDT / SPP, SDSC SHAR, Sriharikota, India
[3] Low Temperature Laboratory, UGC-DAE Consortium for Scientific Research, University Campus, Khandwa road, Indore-452001, (MP) India

E-mail: subhashinisvu@gmail.com

**Abstract.** Structural, electrical and magneto transport properties of Lanthanum based manganites $La_{0.7}Ca_{0.2-x}Sr_xAg_{0.1}MnO_3$ (x=0 & 0.1) synthesized by low temperature nitrate route is studied systematically. The X-ray Diffraction patterns confirm the presence of orthorhombic structure with Pnma space group. The temperature dependence of MR (-35%) from 233-272K for x=0 and an MR (-26%) from 281-309K for x=0.1composition with an overall variation of 1% is very much advantageous for device application. Interestingly, in low temperature regime, the MR value of -47% obtained in x=0.1 composition at 10T around 5K is 20% higher than the MR obtained at 10T around the metal insulator transition. Significant changes happening in the low temperature MR measurements is discussed in the light of electron-electron interactions and weak localization mechanisms while the additional broad hump responsible for flat MR is attributed to the intrinsic electronic in homogeneity driven phase competition created due to the presence of mono valent Ag ions. The complex localization mechanism associated with insulating regime is in accordance with Variable range hopping of small polarons.

## 1. Introduction

Research and development of highdensity magnetic storage devices requires materials with large magnetoresistivity (MR) at room temperature. In this context, manganites appear to be promising materials due to their colossal magnetoresistivity (CMR) property. The attempts of synthesis and characterization of manganite systems of different compositions are accordingly on the rise in recent years to produce large MR at room temperature. It is known that the physics of manganites is challenging due to the strong interplay between charge, spin, and lattice degrees of freedom [1]. Specifically the electronic transport properties and the CMR effect are understood in terms of Zener's double exchange (DE) mechanism, strong electron-phonon interaction, charge and orbital ordering, average size of A-site cations ($<r_A>$), oxygen stoichiometry and so on[2]. One of the main challenges in CMR property is that the MR reaches a peak at metal to insulator transition temperature and hence shows a strong dependence of temperature. A temperature independent MR that too around the room temperatue is very much preferable for device applications. Among the different types of manganites, Lanthanum based manganites of type $La_{1-x}A_xMnO_3$ , ($A = Ca^{2+}, Sr^{2+}, Ba^{2+}$, etc.) have been extensively studied and the ground state is found to be ferromagnetic metallic due to the creation of mixed valence of manganese ions ($Mn^{3+}$ & $Mn^{4+}$). A metal–insulator transition is also found to occur close to Curie temperature ($T_C$) accompanied with a very large magneto-resistivity. Similarly, the substitution of mono valent cations such as $Na^+$, $Ag^+$, etc., partly in the place of trivalent lanthanum $La^{3+}$ ions also creates mobile carriers to form a ferromagnetic metallic ground state. Among the monovalent manganites, the silver (Ag) doped lanthanum manganites exhibit significantly lower resistivity and 25% MR is reported at room temperature [3]. In the present paper, the structural, electronic and magneto resistive transport properties of silver doped manganites $La_{0.7}Ca_{0.2-x}Sr_xAg_{0.1}MnO_3$ (LCSAMO) are investigated systematically. Creation of intrinsic chemical inhomogeneity is utilized as a tool to realize the temperature independent MR in these manganite systems.

## 2. Experimental

The synthesis of $La_{0.7}Ca_{0.2-x}Sr_xAg_{0.1}MnO_3$ (x=0 & 0.1) manganites is carried out through the low temperature nitrate route [4]. The phase purity of the samples is checked using Philips X-ray diffractometer (XRD) with Cu-$k_\alpha$ radiation ($\lambda$=1.5406 Å) and XRD patterns are recorded in angular steps of 0.02°. The temperature dependent electrical resistivity and magneto resistivity of the samples are measured using Quantum design Physical Property Measurement System (PPMS) down to a temperature of 2K and upto a magnetic field of 10T.

## 3. Results and analysis

### 3.1 Structural properties

The XRD patterns of all the samples are recorded at room temperature in the angular range of 10-80°. The XRD patterns are analyzed with Rietveld's method using FULLPROF program which confirms orthorhombic crystal structure with Pnma space group as shown in figure 1 for x=0. The average crystallite size is calculated using Scherrer's formula and is found to be in the range of 22 to 27 nm.

### 3.2 Electronic Transport

The electrical resistivity of samples is measured as the function of temperature from 2 to 380 K and the magnetic field is varied up to 10 T. The zero field resistivity as a function of temperature shown in figure 2 reveals two transition peaks at $T_{p1}$ and $T_{p2}$ as the temperature is lowered from 380 K. The peak appearing at $T_{p1}$ (290K for x=0) is found to decrease as the percentage of Sr is increased (313 K for x=0.1) at the expense of Calcium and systematic shifting of these two peaks due to Sr doping is in accordance with the literature [7].

The temperature dependent electrical resistivity of samples for different magnetic field strengths is shown in figure 3 and figure 4. The first peak at $T_{p1}$ is found to disappear for all the samples as the magnetic field is increased. The hump observed at $T_{p2}$ of the samples however is observed to shift to higher temperatures with the increase of magnetic field. With the increase of magnetic field strength, ρ (T) is found to decrease significantly as $e_g$ carriers suffer less scattering during hopping process due to favourable alignment of spins at $t_{2g}$ orbitals of Mn sites.

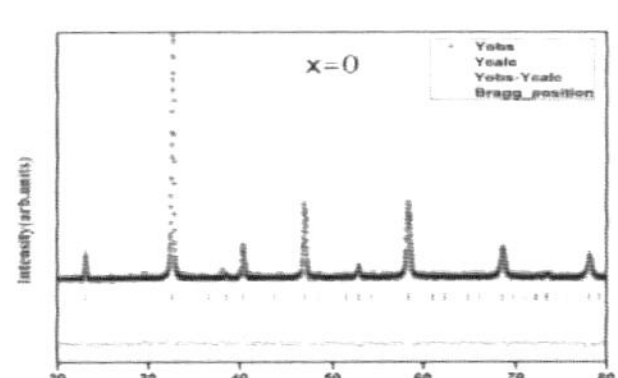


**Figure 1.** Rietveld fits for XRD patterns of $La_{0.7}Ca_{0.2-x}Sr_xAg_{0.1}MnO_3$ (x=0).

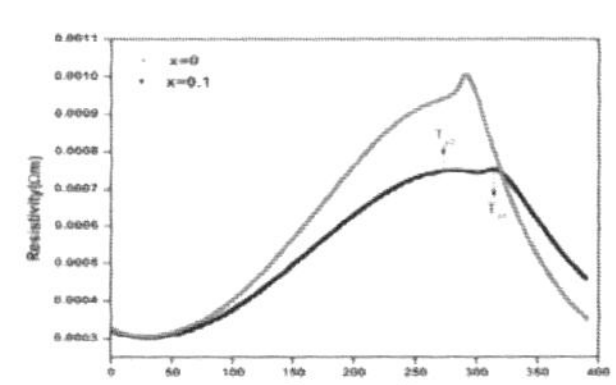


**Figure 2.** Temperature dependent resistivity of LCSAMO (x=0 & 0.1).

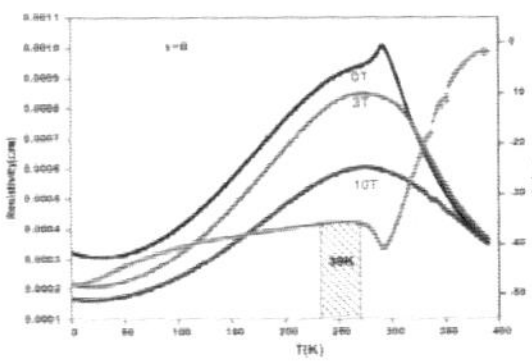

**Figure 3.** Temperature dependent ρ (T) and MR of LCSAMO (x=0) for different magnetic field strengths.

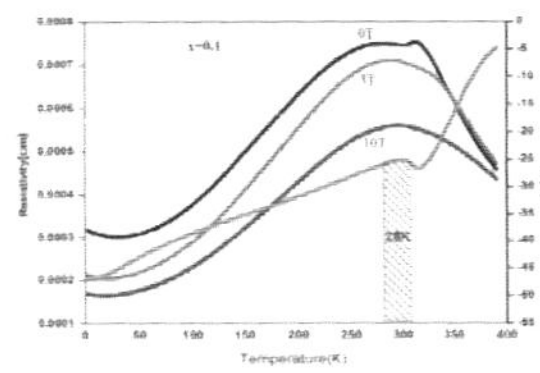

**Figure 4.** Temperature dependent ρ (T) and MR of LCSAMO (x=0.1) for different magnetic field strengths.

Decrease of electrical resistivity is observed with the increase of magnetic field strength. This negative magneto resistivity is usually calculated by the following formula.

$$MR = \frac{\rho(T,H) - \rho(T,H=0)}{\rho(T,H=0)} \times 100\%$$

Where ρ (T, H) and ρ (T, H=0) are the resistivity of samples measured in applied and zero magnetic field respectively. MR as a function of temperature for different magnetic fields is shown in right axis of figure 3 and figure 4. It is clear from the figure that the MR increases significantly with the increase of magnetic field around MIT. MR around the MIT reaches a maximum value of 40% for x=0 and 26% for x=0.1 composition in an applied magnetic field of 10T. Interestingly a weak temperature dependence of MR (-35%) from 233-272K for x=0 and an MR (-26%) from 281-309K for x=0.1 composition is observed respectively with an overall variation of 1% as given by highlighted region in figure 3 and figure 4. This is advantageous for memory device applications. Only few reports using either hetero structures or by composites are available in literature, which addresses the issue of broadening the temperature dependence of MR [8]. In lower temperature regime a maximum MR of 48% for x=0 and 47% for x=0.1 is observed respectively in an applied magnetic field of 10T. In order to understand the interesting features observed in MR, different models are used to fit the temperature dependent resistivity measurements.

The electronic transport behaviour specifically in the low temperature range is usually explained using various models [9-11] that combine various mechanisms including weak localization and electron–electron scattering. The resistivity of the sample is fitted using the following equation (1) in the temperature $T<T_P$.

$$\rho(T) = \rho_0 - \rho_1 T^{1/2} + \rho_2 T^2 \quad (1)$$

The fitting graphs of ρ (T) for zero field and magnetic field strength of 10 T are shown in figure 5 and figure 6. The fitting data shows reduction in $\rho_0$ from 3.38 x $10^{-4}$ to 1.62 x $10^{-4}$ for x=0 and 3.28 x $10^{-4}$ to 1.66 x $10^{-4}$ for x=0.1 sample. The behaviour of $\rho_2$ for the increase of magnetic field strength is observed to be similar [12].

Attempt is made to fit the electrical resistivity data in the high temperature range to popular models such as small polaron model (SPM) and Mott's variable range hopping (VRH) model. In the adiabatic regime where the charge carrier motion is faster than the lattice vibrations, the small polaron hopping is described by the following equation.

$$\rho = A\, T \exp\left(E_a/k_B T\right) \qquad (2)$$

Where the constant A is linked to scattering by impurities, defects, grain boundaries and domain walls, $E_a$ is the activation energy and $k_B$ is the Boltzmann's constant. The fitting of resistivity data of samples with the SPM is shown in figure 5 and figure 6. The mechanism of electron hopping between states of nearly equal energy is described by the VRH model as given below.

$$\rho(T) = \rho_\infty \exp\left[(T_o/T)^{1/4}\right] \qquad (3)$$

Where $\rho_\infty$ is the temperature independent parameter and $T_o$ is the characteristic temperature [13]. The fit of resistivity data with VRH model is shown in figure 5 and figure 6. The characteristic temperature is found to decrease with the increase of x indicating a decrease in effective mass of carriers and

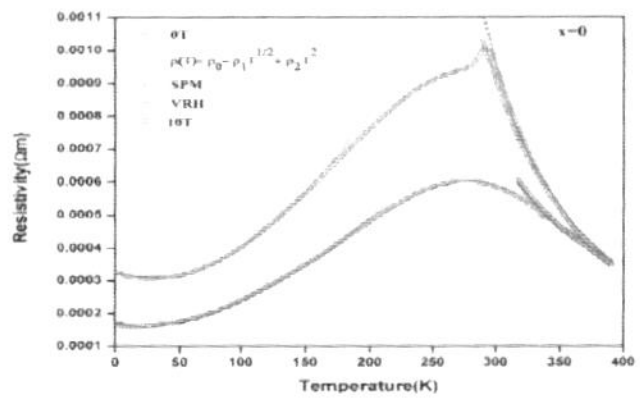


**Figure 5.** Fitting of electrical resistivity data of LCSAMO (x=0) at zero field and 10T in the low temperature and high temperature regions.

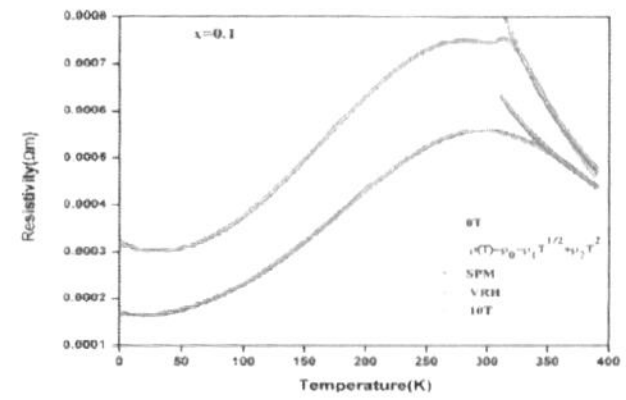


**Figure 6.** Fitting of electrical resistivity data of LCSAMO (x=0.1) at zero field and 10T in the low temperature and high temperature regions.

broadening of $e_g$ electron band width [9]. The further decrease of $T_0$ with increase of magnetic field shows reduced bending of Mn-O-Mn bond and further broadening of $e_g$ electron bandwidth [14]. Comparing the performances of the above models in fitting the resistivity data, the VRH model is found to be marginally better than the small polaron model in the high temperature paramagnetic domain. In manganites systems a strong electron phonon coupling due to $\sigma^2$(variation in A site ionic radius) variation is reported to show small polaron conduction mechanism in paramagnetic regime [15]. Localization mechanism which is weak in nature such as disorder or magnetic localization indicates the presence of large spin polarons and the transport is in accordance with VRH model [16]. In our present investigation even though VRH appears more applicable in paramagnetic domain, the resistivity can also be accounted by small polaron model. A recent report in manganites predicts VRH in small polarons especially above the MIT [17].

## 4.Discussion

The occurrence of MR close to metal to insulator transition ($T_{MI}$) reveals the intrinsic contribution of samples. The second broad peak or the hump noticed at $T_{p2}$=280 K for x=0.1 is indicating the phase separation induced by intrinsic chemical inhomogenity [18]. Substitution of non magnetic $Ag^+$ ions is known to give rise to a random distribution of A-site ions and consequently the $e_g$ electrons are believed to experience site to site local distortion as they move through the crystal grains. Moreover for every mono valent $Ag^+$ ion substituted for the trivalent $La^{3+}$ ion, two $Mn^{3+}$ ions oxidized to $Mn^{4+}$ ions giving rise to individual regions rich in $Mn^{4+}$ and $Mn^{3+}$ respectively. This in turn is reported to induce electron inhomogenity leading to phase separation in ferromagnetic metallic (FMM) phases in the hump region. A strong competition between enhanced double exchange mechanism in FMM region promoted by magnetic field and phase separation created by enhanced $\sigma^2$ has led to the weakly temperature dependent magneto resistance in the investigated LCSAMO samples around the MIT transition. This has also reflected in complex localization mechanisms observed as very low temperature as well as at high temperature regime which demand more than a single model to describe the nature of electrical transport.

## 5. Conclusion

The structural, electrical and magneto transport properties of manganite system $La_{0.7}Ca_{0.2-x}Sr_xAg_{0.1}MnO_3$ (x=0 & 0.1) synthesized through nitrate route are investigated. The temperature dependent electrical resistivity ρ(T) of samples shows two transition peaks which signifies a significant competition between double exchange mechanism and magnetic phase separation below MIT. The complex insulating nature at the high temperature regime is attributed to the VRH of small polarons, The anomalous low temperature behavior is explained in terms of a model which combines electron-electron scattering and weak localization. The study of magneto transport behavior reveals broad weakly temperature dependent magneto resistance at room temperature which can be exploited for magnetic sensing applications.

**Acknowledgments**
The first author is greatful thank to University Grant Commission (UGC) for providing Rajiv Gandhi national fellowship to pursue doctoral degree research program at Sri Venkateswara University, Tirupati. The authors are thankful to UGC-DAE CSR, Indore, India, for providing MR facilities.

**References**

[1] Pekala M, Drozd V, Mucha J 2005 *J. Magn. Magn. Mater.* **290–291** 928–932
[2] Kalyana Lakshmi Y, Venugopal Reddy P 2009 *J. Magn. Magn. Mater.* **321** 1240–1245
[3] Battabyal M, Dey T K 2004 *J. Phy. Chem. Sol.* **65** 1895–1900
[4] Pai M R, Wani B N, Bharadwaj S R 2006 *J. Indian. Chem. Soc.* **83** 336-341
[5] Tang T, Gu K M, Cao Q Q, Wang D H, Zhang S Y and Du Y W 2000 *J. Magn. Magn. Mater.* **222** 110–114
[6] Gencer H, Pektas M et al 2012 *J. Magn.* **17** 176-184
[7] Xiao-Bo Yuan, Yi-Hua Liu, Bao-Xin Huang, Cheng-Jian Wang and Liang-Mo Mei 2005 *Solid State Commun.* **135** 170–173
[8] Pattabiraman M, Adepu R, Singh N P, Venkatesh R, Angappane S, Rama N and Rangarajan G 2008 *J. Alloys Compd.* **452** 230–233
[9] Venkataiah G, Venugopal Reddy P 2005 *Solid State Commun.* **136** 114–119
[10] Lalitha G, Venugopal Reddy P 2010 *J. Alloys Compd.* **494** 476 – 482
[11] Pekala M, Drozd V, Fagnard J F, Vanderbemden P H and Ausloos M 2009 J. Appl. Phys. **105** 013923
[12] Srivastava S K, Manoranjan Kar and Ravi S 2008 *Mater. Sci. Eng., B* **147** 84–89
[13] Munirathinam B, Krishnaiah M, Devarajan U, Esakki Muthu S and Arumugam S 2012 *J. Phys. Chem. Sol.* **73** 925–930
[14] Subhashini P, Munirathinam B, Krishnaiah M, Venkatesh R, Venkateshwarlu D and Ganesan V 2015 *Bull. Mater. Sci.* **38** 1–7
[15] Millis A J 1998 *Phil. Trans. R. Soc. Lond.* A **356** 1473
[16] Coey J M D, Viret M, Ranno L and Ounadjela K 1995 *Phys. Rev. Lett.* **75** 3910
[17] Young S, Xu X, and Zhang Y 2000 *J. Phy. Cond. Matt.*10475
[18] Battabyal M, Dey T K 2004 *Solid State Commun.* **131** 337–342

# Temperature-time induced changes in magnetic properties of multi-component $Fe_{78}Si_{3.6}C_{13.4}Mn_{0.65}B_{4.35}$ alloy

**S N Kane[a], S S Modak[b], M Shah[a], M Satalkar[a], K Gehlot[a], N Ghodke[c], J P Araujo[d] and L K Varga[e]**

[a]Magnetic Materials Laboratory, School of Physics, Devi Ahilya University, Khandwa Road Campus, Indore-452001, India
[b]Department of Physics, Jaypee University of Engineering and Technology, Guna-473226, India
[c]UGC-DAE-CSR, University campus, Khandwa Road, Indore-452001, India
[d]IFIMUP, Departmento de Fisica, Universidade de Porto, 4169-007 Porto, Portugal
[e]RISSPO, Hungarian Academy of Sciences, P.O. Box, 49-125 Budapest, Hungary

Corresponding author: kane_sn@yahoo.com

**Abstract**. Influence of thermal annealing as a function of temperature and time, on magnetic and structural properties of $Fe_{78}Si_{3.6}C_{13.4}Mn_{0.65}B_{4.35}$ ($Fe_{78}Si_{3.6}C_{13.4}Mn_{0.65}$ = Ci) alloy has been reported. Information on structure, formed nano-crystalline phases and their correlation with magnetic properties has been studied using, magnetic measurements, differential scanning calorimetery (DSC) and x-ray diffraction (XRD). Thermal annealing treatment leads to rather coarse grain microstructure (~31 nm) accompanied by surface crystallization that appears to deteriorate the magnetic properties with annealing temperature.

## 1. Introduction

Fabrication of Fe-based bulk metallic glasses using high purity raw elements and the strict processing under high vacuum results in high production cost. That eventually restricts their industrial mass production. In this respect cast iron (Ci) based multi component bulk-type Fe-based amorphous alloys emerged as suitable cheap magnetic alloys prepared with industrially available raw materials [1, 2]. These alloys exhibit large glass forming region and good soft magnetic properties. Depending upon the specific application requirement these cast iron based alloys can be prepared in the shape of ribbons, cylinders with transversal dimensions up to ~ 0.5 mm [3] and, their magnetic properties should also be optimized.

In the present work we report the investigation of the effect of isochronal and isothermal annealing treatment on the structural and magnetic properties of cast iron based multi component $Fe_{78}Si_{3.6}C_{13.4}Mn_{0.65}B_{4.35}$ which is also represented as $Ci_{95.65}B_{4.35}$ alloy.

## 2. Experimental Details

Ribbons (about 16 μm thick and 3.3 mm wide) of nominal composition $Fe_{78}Si_{3.6}C_{13.4}Mn_{0.65}B_{4.35}$ were prepared using planar flow casting technique. In order to determine the crystallization peak temperature differential scanning calirimetery (DSC) measurements were done at heating rate of 10

$^{o}$C/min. Utilizing DSC results, isochronal thermal annealing was done at 350, 380, 400, 425 and 450 $^{o}$C for 60 minutes. Samples were also isothermally annealed at 380 $^{0}$C for 15, 30 and 45 minutes. X-ray diffraction (XRD) measurements on as-cast and annealed samples were carried out using Bruker D8 Advance X-ray diffractometer with Cu-$K_{\alpha}$ radiation (wavelength: $\lambda$ = 0.154 nm). XRD patterns were analyzed by fitting a crystalline and an amorphous component using pseudo-Voigt line profiles. Crystalline volume fraction ($V_x$) was obtained by integrating the main diffraction peak. The average grain size $D$ is obtained by Scherrer's formula using integral width of the (110) line and the lattice parameter $a$ is calculated using Nelson-Taylor-Sinclair correction in order to take into account the peak shift due to sample offset. For the amorphous phase, the first nearest neighbour distance $X_m$ is calculated using the expression: $X_m = 1.227\ \lambda / 2\mathrm{Sin}\theta$. Hysteresis loops of as cast and annealed samples were measured using a loop tracer at 50Hz using maximum applied magnetic field of ± 3000 A/m. Measured hysteresis loops were used to get information on coercive field ($H_c$) and, saturation induction ($B_{3000}$) values.

## 3. Results and Discussion

Figure 1 depicts the DSC curve of as cast sample measured at 10 $^{o}$C/min. It can be seen that the studied alloy system exhibit a two step crystallization process with onset of the first peak at 435$^{0}$C. The crystallization temperatures corresponding to the two peaks are 463 and 524 $^{o}$C respectively. XRD patterns of as cast and annealed samples at various temperatures for 60 min. are presented in figure 2. Perusal of figure 2 shows that the crystallization starts after annealing at 380 $^{o}$C. X-ray diffraction data was analyzed using a MATLAB® based program that fits the data using pseudo Voigt line profiles. Table 1 represents the annealing temperature evolution of various structural parameters obtained by analyzing the XRD data. Effect of isothermal annealing at 380 $^{o}$C for different time period on the structural properties of the studied alloy samples is shown in table 2. Perusal of tables 1 and 2 shows that with increasing annealing temperature, although $D$ increases but $V_x$ does not increase much, exhibiting that increase in annealing temperature results in grain coarsening. Change of annealing time does not have appreciable effect on, $D$ and $V_x$. Obtained $a$ values suggest that the precipitated crystalline phase is bcc Fe ($\alpha$–Fe). Within the experimental errors the first nearest-neighbour distance in the residual amorphous matrix (0.250 nm ± 0.001) almost remains constant for the studied specimens, showing that the studied samples have comparable mass densities.

Effect of annealing temperature and time on the magnetic response of the studied alloy composition is represented in the figure 3. Perusal of figure 3 (left panel) indicates that the flattening of hysteresis loops occur for the annealing temperature of 400 $^{o}$C and above as also observed in case of other cast iron based samples ascribable to surface crystallization [4,5].

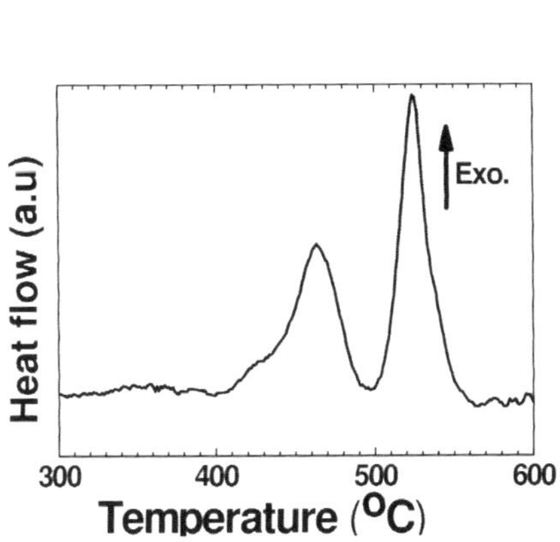


**Figure 1:** DSC curve of as cast $Fe_{78}Si_{3.6}C_{13.4}Mn_{0.65}B_{4.35}$ sample measured at 10 $^{o}$C/minute.

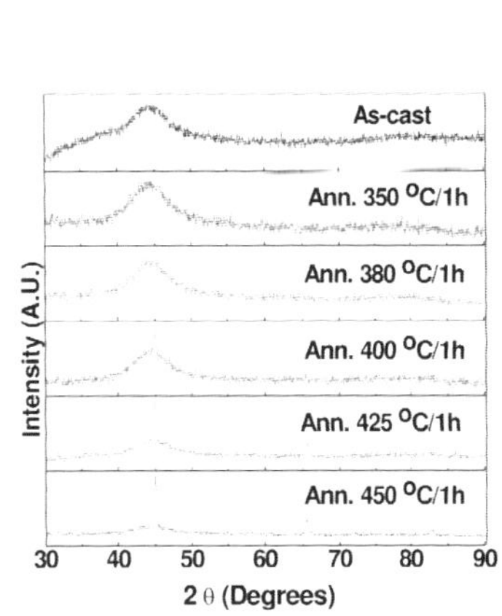


**Figure 2:** XRD patterns of as cast and annealed $Fe_{78}Si_{3.6}C_{13.4}Mn_{0.65}B_{4.35}$ samples at various temperatures for 60 minutes.

**Table1:** Temperature transformation of structural parameters for samples annealed for 60 minutes.

| Sample details | <D> (nm) (±1) | *a* (nm) (±.0001) | $V_x$ (%) (±1) |
|---|---|---|---|
| As-cast | – | – | – |
| 380 °C | 11.5 | 0.2849 | 5.5 |
| 400 °C | 17.5 | 0.2848 | 10.5 |
| 425 °C | 25 | 0.2849 | 19 |
| 450 °C | 31.5 | 0.2848 | 30.5 |

**Table2:** Time transformation of structural parameters for samples annealed at 380 °C.

| Sample details | <D> (nm) (±1) | *a* (nm) (±0.0001) | $V_x$ (%) (±1) |
|---|---|---|---|
| As-cast | – | – | – |
| 15 min | 11.0 | 0.2868 | 3.2 |
| 30 min. | 11.0 | 0.2869 | 3.9 |
| 45 min. | 11.5 | 0.2845 | 6.0 |
| 60 min. | 11.5 | 0.2849 | 5.5 |

In order to further optimize the magnetic properties of the studied alloy system, the annealing temperature of 380 °C was chosen, which shows partial crystallization with average grain diameter of ≈ 11 nm. It is worth noting that XRD pattern of the specimen annealed at 380 °C, shows presence of small crystalline fraction ( ≈ 5 %) after annealing for 1 hour. By choosing different annealing time would lead to various levels of structural relaxation and/or crystallization having different average grain diameter of the nano-grains, giving rise to the variation in magnetic properties. Keeping the above mentioned reason in account, specimens were also annealed at 380 °C for different times e.g. 15, 30, 45 and 60 min. Time evolution of the hysteresis loops corresponding to annealing temperature of 380 °C is presented in figure 3 (right panel).

Figure 4 represents the annealing temperature and time evolution of coercive field behaviour of the studied samples. It shold be noted that insets in the figure 4 (both in left and right panel) exhibit the variation of saturation induction ($B_{3000}$) values as a function of isochronal and isothermal annealing treatment respectively. It can be seen in the figure 4 (left panel) that there is an initial decrement in the coercvity till annealing temperature of 350 °C, which may be due to relaxation of quenched in stresses. Further coercive field values increases rather sharply with annealing temperature.

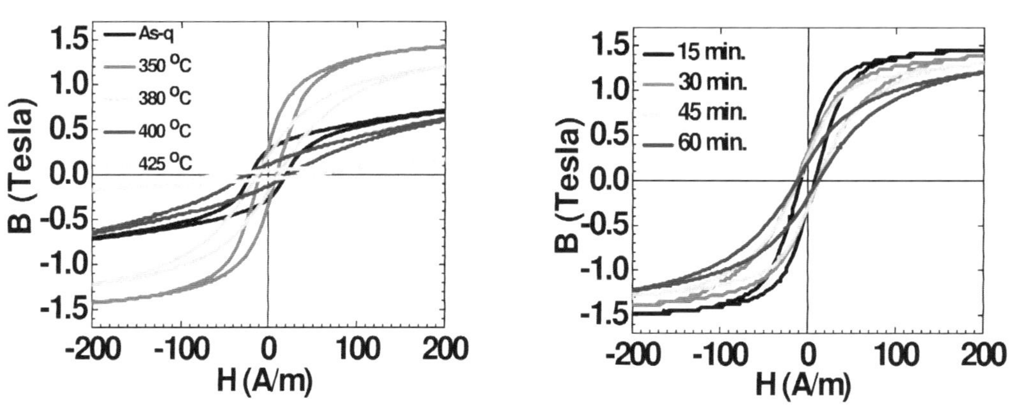


**Figure 3:** Annealing temperature evolution of representative hysteresis curves of $Fe_{78}Si_{3.6}C_{13.4}Mn_{0.65}B_{4.35}$ alloy (left panel). Annealing time evolution of $Fe_{78}Si_{3.6}C_{13.4}Mn_{0.65}B_{4.35}$ alloy samples annealed at 380 °C (right panel).

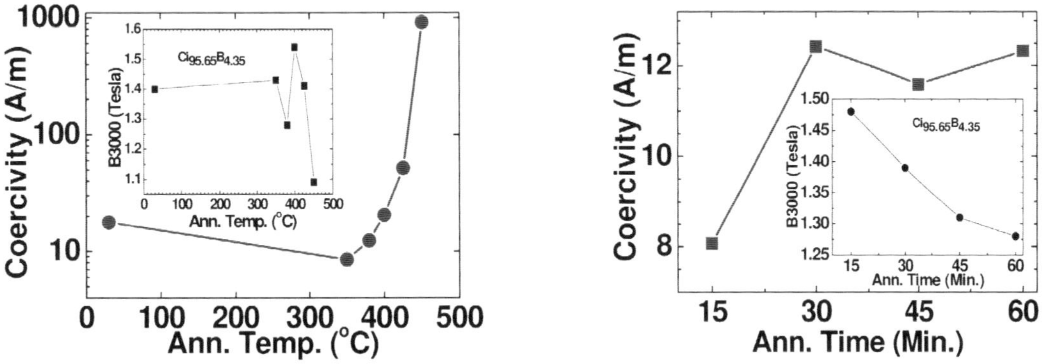


**Figure 4:** Annealing temperature evolution of coercivity and saturation induction values (inset) of $Fe_{78}Si_{3.6}C_{13.4}Mn_{0.65}B_{4.35}$ alloy (left panel). Annealing time evolution of coercivity and saturation induction (inset) of $Fe_{78}Si_{3.6}C_{13.4}Mn_{0.65}B_{4.35}$ alloy samples annealed at 380 °C (right panel).

This behaviour indicates the presence of non-exchange coupled nanograins. It can be seen in table 1 that grain coarsening (average grain diameter~31 nm and $V_x$ ~ 30%) is responsible for large distance between the grains and hence deteriorating the magnetic properties. Coercive field values of the studied alloy system varies between 8 to 914 A/m. The lowest value of coercivity ($H_c$ ~ 8 A/m) is exhibited for the sample annealed at 380$^0$C for 15 min. It should also be noted that the saturation induction ($B_{3000}$) values ranges between 1.28 to 1.54 Tesla. Highest value (1.54 T) of $B_{3000}$ corresponds to the annealing treatment at the temperature of 400 °C for 60 min.

## 4. Summary

Changes in the structural and magnetic properties as a function of thermal annealing temperature and time for cast iron based multi component $Fe_{78}Si_{3.6}C_{13.4}Mn_{0.65}B_{4.35}$ alloy in the form of ribbons have been studied using DSC, XRD and hysteresis measurements. Studied alloy system exhibit two-step crystallization, exhibiting a pre-peak, before the main peak. For the samples isochronally annealed at various temperatures, coercive field values range between 8 – 914 A/m,whereas $B_{3000}$ values vary between 1.28 - 1.54 Tesla. The lowest $H_c$ value of 8 A/m was obtained for the sample annealed at 380 °C for 15 min. Best $B_{3000}$ value of 1.54 Tesla was obtained for the one annealed at 400 °C for 1 hr. Obtained lattice parameter values suggest that the precipitated crystalline phase is bcc Fe (α–Fe). Thermal annealing leads to rather coarse grains ~ 31 nm, but low crystalline fraction ~ 30%. This appears to reduce the ferromagnetic exchange interaction between α–Fe nano-crystals, thereby deteriorating the magnetic properties. Thermal annealing leads to the flattening of the hysteresis loop that ascribed to surface crystallization.

## Acknowledgement

S. N. Kane acknowledges gratefully the financial support from Indo Portuguese joint project No.IND/Portugal/P-07/2013 and, also by the project CSR-IC-IC-BL-25/CRS-122-2014-15/218.

## References

[1] Inoue A, Shen B L, Koshiba H, Kato H and Yavari A R 2004 *Acta Mater.* **52** 1631
[2] Qin C,. Zhang W, Asami K, Ohtsu N and Inoue A 2005 *Acta Mater.* **53** 3903
[3] Inoue A and Wang X M 2000 *Acta Mater.* **48** 1383
[4] Kane S N, Lee H J, Kim S B, Jeong Y H, Hyun S W, Kim C S and Varga L K 2009 *J. Physics: Conference series* **144** 012040
[5] Herzer G and Hilzinger H R 1986 *J. Magn. Magn. Mater.* **62** 143

# Crystal structure and electrical transport properties of single layered perovskite $LaSrCoO_4$

**Abdul Ahad[1†], D K Shukla[2], F Rahman[1], S Majid[1], Tarachand[2], G S Okram[2] and D M Phase[2]**
[1]Department of Physics, Aligarh Muslim University, Aligarh 202002, India
[2]UGC-DAE Consortium for Scientific Research, Indore 452001, India

[†]Corresponding author, E-mail: matrixabdulahad@gmail.com

**Abstract.** We present here investigations on the influence of structure on electrical transport properties of polycrystalline $LaSrCoO_4$ that is single layered perovskite with $K_2NiF_4$ type structure synthesized using solid state reaction route. Using Reitveld refinement of X-ray diffraction (XRD) data, it is found that the sample is in single phase with tetragonal structure (space group *I4/mmm*). Electrical resistivity performed in the temperature range 140-300K shows semiconducting character of the sample. Considerable contrasts in the Co-O bond length is associated with the intermediate spin (IS) state of Co ion that correlates the structural and transport properties. Detailed analysis indicates that the temperature dependent electrical resistivity follows the three-dimensional variable range hopping (VRH) model in low temperature region below 225K. The high temperature (225-300K) resistivity data has been found to follow the thermally activated behaviour.

## 1. Introduction

Compounds which crystallize in the tetragonal $K_2NiF_4$ type structure are the members of Ruddlesden-Popper series $(ABO)_3(AO)_n$ with n=1 known as single layered perovskites. Since the discovery of high temperature superconductivity in cuprate [1], these types of structures have been the subject of intense research. In single layered perovskite $La_2CoO_4$ structure, $CoO_6$ octahedron are connected by corner-sharing oxygen atoms to form a two dimensional $CoO_2$ layers which are separated by the rock salt structure type LaO slabs [2].

An extra degree of freedom that comes on top of the orbital, spin (up or down) and charge degree of freedoms (that makes manganite and cuprate systems so exciting), is the spin state degree of freedom that distinguish the cobaltate from the other similar systems. Cobalt ion can exist in the three spin states, i.e. low spin (LS), intermediate spin (IS) and high spin (HS) depending on the competition between crystal field splitting and the on-site Coulomb exchange or Hund's coupling [2]. Considerable research efforts have been made on cobalt containing layered structures $La_{2-x}Sr_xCoO_4$ because of interesting properties similar to manganese and copper containing layered oxides like colossal magnetoresistance [3] and superconductivity [4]. The single layered perovskite $La_{2-x}Sr_xCoO_4$ contains mixed valent state of cobalt ion i.e. $Co^{2+}$ and $Co^{3+}$ in (1-x):x ratio. For $Co^{3+}$ in $LaSrCoO_4$ compound (x=1) there is controversy in literature regarding spin state, it has been proposed to be in either intermediate spin state [5] or in a mixture of low spin and high spin states [6].

Consider the information about spin states of $LaSrCoO_4$ compound we have planned to understand the structure and electrical transport behaviour of this single layered perovskite. We present here the detailed analysis and study of $LaSrCoO_4$ by means of room temperature X-ray diffraction and temperature dependent resistivity measurement in the range 140-300K. A possible correlation between structure and transport based on the already known spin state of the Co ions is proposed.

## 2. Experimental details

Polycrystalline sample $LaSrCoO_4$ has been synthesized by employing solid state route. Stoichiometric mixture of high purity $La_2O_3$, $SrCO_3$ and $Co_3O_4$ was grinded and calcined for 8 hours at 1000°C. After calcination the obtained powder was re-grinded, palletized in 10 mm disc and sintered for 12 hours at 1150°C.

The room temperature X-ray diffraction have been done using Bruker D2 Phaser in 2θ range 20°-80°, step size 0.02° with Cu Kα radiation (λ=1.54 Å). The diffraction pattern was analysed by Reitveld technique [7] using FULLPROF code. Temperature dependent resistivity was done using standard four probe method in the temperature range 140-300K.

## 3. Results and Discussion

Figure 1 shows the Reitveld refined XRD pattern of $LaSrCoO_4$, which reveals that the sample is in single phase having space group *I4/mmm* (space group no. 139). From figure 1, it is clear that the calculated pattern is in excellent agreement with the experimentally observed pattern. The estimated lattice parameters $a$ and $c$ from the refinement are 3.8055 Å and 12.5010 Å, respectively. In the crystal structure of $LaSrCoO_4$ the Co ions are surrounded by oxygen in the octahedral symmetry with four O1 oxygen in the equatorial plane and two O2 oxygen in the apical direction along the $c$ axis of the unit cell resulting two different Co-O bond lengths one along the $c$-axis Co-O2 and other in the $ab$ plane Co-O1. From the refinement results we have found Co-O2 and Co-O1 are 2.0085 Å and 1.9027 Å, respectively. The ratio of these bond lengths i.e. Co-O2/Co-O1 is ~1.055 indicating the elongation of $CoO_6$ octahedron. We may correlate this distortion with the occupation of $e_g$ orbital [8]. Elongation of octahedron along the $c$-axis indicate about the IS of $Co^{3+}$ ion containing non-equivalent electrons in $e_g$ orbital, yielding Jahn-Teller distortion [8].

Electrical resistivity of sample in the range 140-300K is shown in figure 2, which shows the semiconducting nature of the sample. Inset (Figure 2) shows the derivative of resistivity curve. Anomaly at around 225K indicates a possibility of change in conductivity mechanism. For further detail analysis, resistivity data has been plotted in such way so that typical conductivity behaviour can be inferred. From the plot of lnρ against 1/T in Figure 3, the curve deviates from linearity. It is clear

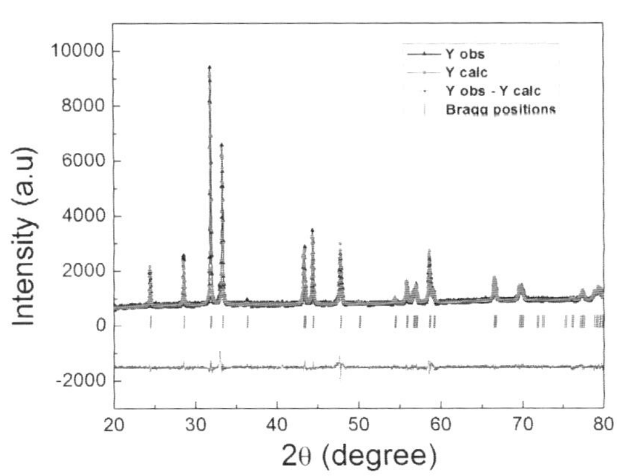


**Figure 1.** Room temperature Reitveld refined XRD pattern of $LaSrCoO_4$.

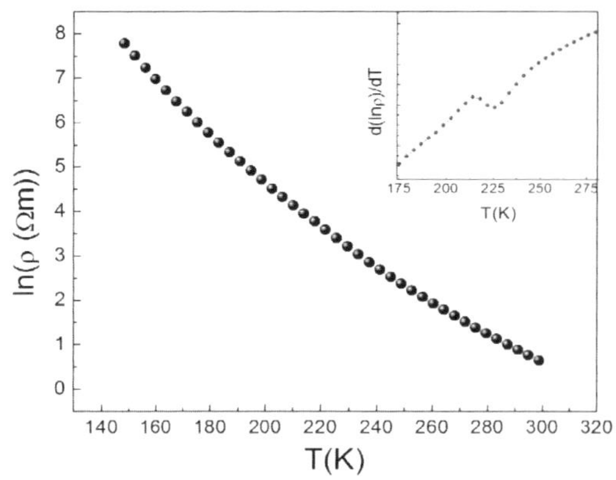


**Figure 2.** Temperature dependent electrical resistivity

that the high temperature (above 225K) resistivity data can be well fitted by the thermally activated conduction law, which follow the following form, equation (1).

$$\rho(T) = \rho_o \exp(E_a / k_B T) \tag{1}$$

Where $E_a$ represent the activation energy and $k_B$ is the Boltzmann constant. The estimated value of activation energy $E_a$ is 211meV.

Figure 4 shows the plot of lnρ vs $T^{-1/4}$ reveals that the low temperature (below 225K) resistivity follow the Mott's three dimensional variable range hopping model which have following form, equation (2).

$$\rho(T) = \rho_o \exp(U / T^{1/4}) \tag{2}$$

Where U represent the fitting parameter.

The high value of electrical resistivity is the signature of absence of mobile charge carriers. For tetragonally elongated $CoO_6$ octahedron, we may expect that there is less repulsion between Co ion and oxygen ligand along the *z* coordinate so to attain stability $d_{xz}$, $d_{yz}$ and $d_{3z^2-r^2}$ may shifted to lower energy. Simply we can say that degeneracy of $e_g$ orbital is lost and it has been reported previously that, if electrons in orbitals other than $3d_{x^2-y^2}$ tend to be localized, one may expect the semiconducting behaviour [9].

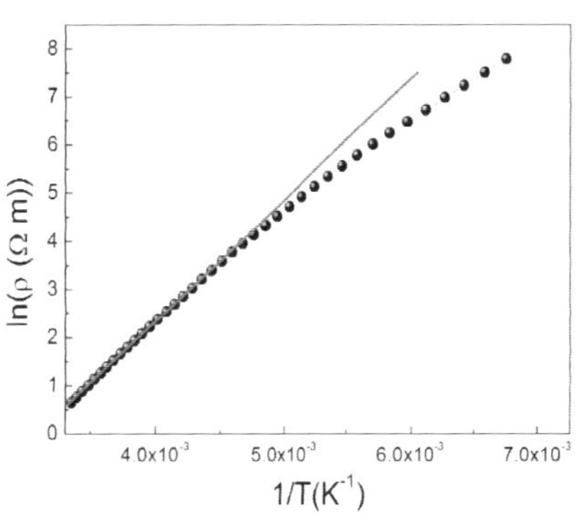


**Figure 3.** The plot of lnρ as a function of 1/T, thermally activated behaviour

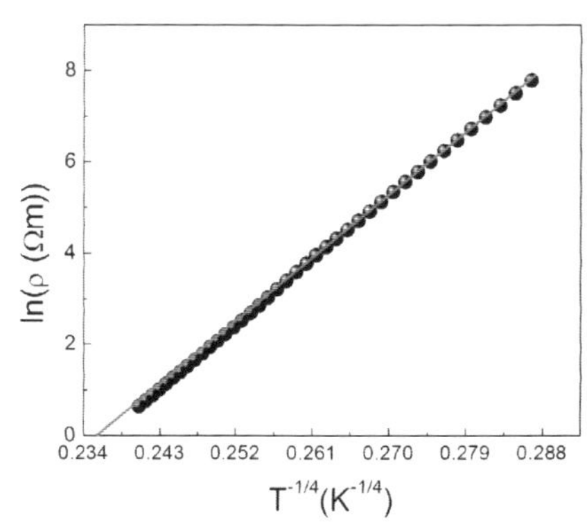


**Figure 4.** lnρ vs $T^{-1/4}$ plot, VRH model.

## 4. Conclusion

We have successfully synthesized polycrystalline $LaSrCoO_4$ sample using solid state reaction route. Reitveld refinement confirms that the sample is in single phase with tetragonal $K_2NiF_4$ type structure. The lattice parameters and bond lengths have been obtained from refinement. The difference between two Co-O bond lengths shows the distortion of $CoO_6$ octahedron, which gives an idea about the presence of IS state of $Co^{3+}$ ion in our sample. Temperature dependent resistivity shows the semiconducting behaviour of the sample. Detailed analysis revealed that it follow the three dimensional Mott's variable range hopping formula below 225K and thermally activated conduction law with activation energy 211meV in high temperature region (above 225K).

**Acknowledgments**

One of the authors (A Ahad) gratefully acknowledge to UGC-DAE CSR, Indore for providing necessary experimental facilities regarding this work.

## 5. References

[1] Chu C W, Hor P H, Meng R L, Gao L and Huang Z J 1987 *Science* **235** 567
[2] Tealdi C, Ferrara C, Malavasi L, Mustarelli P, Ritter C, Chiodelli G and Diaz-Fernandez Y A 2010 *Phys. Rev. B* **82** 174118
[3] Briceno G, Chang H, Sun X, Schultz P G and Xiang X D 1995 *Science* **270** 273
[4] Takada K, Sakurai H, Takayama-Muromachi E, Izumi F, Dilarian R A and Sasaki T 2003 *Nature* **422** 53
[5] Chichev A V *et al.* 2006 *Phys. Rev. B* **74** 134414
[6] Wang J, Zhang W and Xing D Y 2000 *Phys. Rev. B* **62** 14140
[7] Rietveld H M 1967 *Acta Crystallogr.* **22** 151
[8] Ang R, Sun Y P, Luo X, Hao C Y and Song W H 2008 *J. Phys. D: Appl. Phys.* **41** 045404
[9] Wang X L and Takayama-Muromachi E 2005 *Phys. Rev. B* **72** 064401

# Characterization of pulsed laser deposition grown $V_2O_3$ converted $VO_2$

**Suhail Majid[1†], D K Shukla[2], F Rahman[1], Kamini Gautam[2], V G Sathe[2], R J Choudhary[2] and D M Phase[2]**
[1]Department of Physics, Aligarh Muslim University, Aligarh 202002, India
[2]UGC-DAE Consortium for Scientific Research, Indore-452001, India

[†]Corresponding author, E-mail: suhailphy276@gmail.com

**Abstract.** Controllable tuning of Metal-insulator transition in $V_xO_y$ thin film has been a field of extensive research. However controlled synthesis of desired Vanadium oxide phase is a challenging task. We have successfully achieved $VO_2$ phase on Silicon substrate after post deposition annealing treatment to the PLD grown as deposited $V_2O_3$ thin films. The annealed thin film was characterized by x-ray diffraction (XRD), resistivity, Raman spectroscopy, X-ray absorption spectroscopy (XAS) and X-ray photoelectron spectroscopy (XPS) measurements. XRD confirms the crystalline nature and growth of $VO_2$ phase in thin film. The characteristic MIT was observed from resistivity measurements and transition temperature appeared at lower value around 336 K, compared to bulk $VO_2$. The structural transition accompanied with MIT from lower temperature monoclinic phase to higher temperature Rutile phase became evident from temperature dependent Raman measurements. Chemical state of vanadium was examined using XAS and XPS measurements which confirm the presence of +4 oxidation state of vanadium in thin film.

## 1. Introduction

Vanadium exhibits variable oxidation states towards oxygen, hence results in the formation of different number compound with oxygen. Among all the vanadium oxide phases, Vanadium Dioxide ($VO_2$) has been extensively studied because of its near room temperature metal-insulator (MIT) transition ($T_t$ = 341 K), transition from high temperature metallic phase to low temperature insulating phase. The MIT in $VO_2$ is accompanied by the structural transition from high-temperature tetragonal (Rutile) structure to low temperature monoclinic (M1) structure [1]. The MIT in $VO_2$ is of both fundamental and technical interest, the former due to the important questions about its origin while the latter due to the possible applications in electronic devices such as electrical switches and field effect transistors [2]. The mechanism of the metal-insulator transition in $VO_2$ is considered to be of Peierls type or Mott-Hubbard type involving the electron-phonon and electron-electron interactions respectively [2].

The tendency of Vanadium to form different compounds with oxygen poses a stiff challenge on synthesis of high purity single phase of vanadium oxides. Various deposition methods including molecular beam epitaxy, chemical vapour deposition, R-F sputter deposition, sol-gel, ion implantation and pulsed laser deposition have been utilised for the deposition of vanadium dioxide thin films [3].

In this paper, we report the attempt of the growth of $VO_2$ thin film using pulsed laser deposition technique on the [0 0 1] Silicon substrate using $V_2O_5$ as the target material. The $VO_2$ phase was obtained after the annealing treatment to the as grown Vanadium sesquioxide ($V_2O_3$) thin film. The annealed thin film was using X-ray diffraction (XRD), Raman, Resistivity, X-ray photoelectron spectroscopy (XPS) and X-ray photoelectron spectroscopy (XAS) measurements.

2. Experimental details

KrF excimer laser ($\lambda$ = 248 nm, repetition rate of 5Hz and pulse laser energy of 210mJ) was focussed onto a target (pressed $V_2O_5$ with purity > 99.9%) with a fluence of 3 J/cm$^2$. The angle between the incident laser beam and normal to the target surface was 45°. The base pressure during the deposition was $3.6 \times 10^{-6}$ Torr. During ablation the target was rotated at the rate of 10 rotations per minute to avoid the depletion of the material at the same spot continuously and a uniform thin film $V_2O_3$ was deposited on ultrasonically cleaned Si (001) substrate. The substrate temperature was maintained at 650°C during the deposition of 50 minutes. The $V_2O_3$ phase was confirmed through X-ray diffraction measurements (not shown here). The thickness of the film as measured by the stylus profilometer comes out to be 70 nm. The as grown thin film was annealed at 650$^0$C for 2 hours in presence of 100 mTorr oxygen partial pressure. X-ray diffraction (XRD) was carried out by Bruker D8 X-ray diffractometer with Cu K$\alpha$ radiation ($\lambda$ = 1.54 Å). Temperature dependent resistivity measurements were performed using four-probe resistivity setup. Temperature dependent Raman spectra were collected in backscattering geometry using a 10 mW Ar (488 nm) laser as an excitation source coupled with a Labram-HR800 micro-Raman spectrometer equipped with a 50X objective. All the spectra were corrected in frequency by using a silicon substrate. The X-ray photoelectron spectroscopy (XPS) measurement was performed using Omicron energy analyzer (EA-125) with Al $K_\alpha$ (1486.6eV) X-ray lab source. Room temperature soft X-ray absorption spectroscopy across V $L_{3,2}$ and O K edges were carried out at the beam line BL-01 (soft X-ray absorption spectroscopy beamline) at Indus-2 Raja Ramana centre for Advanced Technologies Indore. SXAS measurements were performed in total electron yield mode (TEY). Energy resolution during SXAS measurements at oxygen Kedge energy was better than ~ 250 meV energy.

**3. Results and discussion**

Figure 1(a) shows the $\theta$-2$\theta$ XRD patterns of annealed thin film grown on silicon substrate along with XRD of Si substrate. The XRD pattern of thin film shows peaks at 2$\theta$ = 29.15, 39.85, 41.50 and 85.88. The peak at 2$\theta$= 29.15 correspond to the reflection from the (0 0 2) plane of metastable monoclinic $VO_2$ (B) phase which is one of the polymorphic structures of $VO_2$ with space group C2/m [4], while as the peak at 2$\theta$=39.85 can be indexed to the reflection from the planes (0 0 2) of insulating monoclinic phase of $VO_2$ (M) with the space group $P2_1/c$ [JCPDS card no 82-0661], which is one of the stable phases of $VO_2$. The third peak at 2$\theta$= 41.50 can be assigned to the reflection from the (1 1 3) plane that belongs to the rhombohedral phase of $V_2O_3$ with space group $R\overline{3}C$ [JCPDS card no 85-1411], while all the other remaining peaks are from the substrate Si. Fig 1(b) represents the characteristic metal-insulator transition in $VO_2$ thin film obtained from the temperature dependent resistivity measurements. The observation of thermal hysteresis confirms the first order nature of this transition. From the graph, it is clear that the change in the magnitude of resistivity across the metal-insulator transition is not large. The reason behind this may be the presence of $V_2O_3$ phase which does not undergo change in the resistivity [5]. More over the presence of $VO_2$ (B) polymorphous also affect the MIT [4]. The phase transition temperature defined as $T_t = (T_{c.h} + T_{c.c}) /2$ (Where $T_{c.h}$ and $T_{c.c}$ correspond to the phase transition temperature of heating and cooling branch respectively) has been calculated as 336.1 K. The transition temperature (~ 336.1 K) is appearing at the lower temperature compared to bulk $VO_2$ (~ 341 K). This may be because of substrate induced stress present in the thin film due to substrate [ref]. The metal-insulator transition of the compound is characterized by transition temperature ($T_t$), hysteresis width ($\Delta$H = 7.07 K ) and transition sharpness ($\Delta$T = 5.04

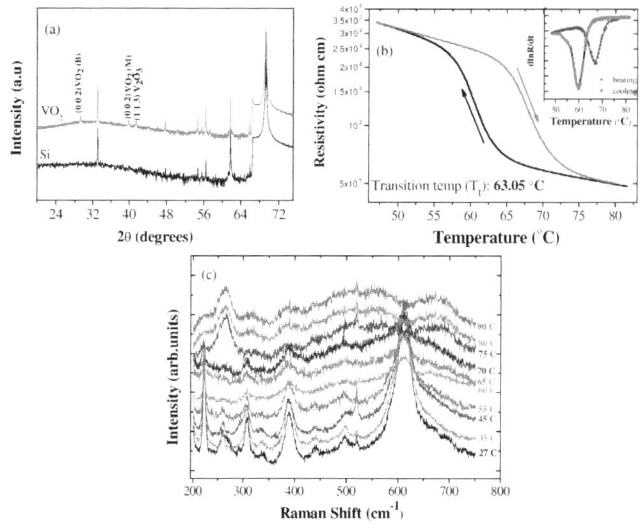

**Figure 1.** (a) X-ray diffraction pattern of annealed thin film. (b) Temperature dependent resistivity of annealed thin film. Inset: Gaussian fitting of the differential curves of resistivity Vs temperature. (c)Temperature dependent Raman measurements of annealed thin film in heating cycle only.

(heating) and 5.42 K (cooling) ).All these parameters are calculated from the Gaussian fittings of the differential curves of lnR-T plots shown in the inset of Fig 1b.

To investigate the structural transition accompanying with MIT in the annealed thin film, temperature dependent Raman scattering measurements were performed from room temperature up to 363 K, shown in Figure 1(c). The structural transition from the low temperature monoclinic structure to higher temperature tetragonal structure is quite visible and occurs around the MIT of $VO_2$ thin film signifying the importance of periels nature to the transition. In our spectrum we have observed 8 Raman modes in low-temperature monoclinic insulating phase at 222.1, 259.9, 308.8, 334.9, 388.8, 439.9, 497.5, 611.8 $cm^{-1}$ with $A_g$ symmetry [6]. The low intensity Raman mode around 520 $cm^{-1}$ is from the Si substrate. The peaks at 194 and 223$cm^{-1}$ were assigned to V–V vibration modes, whereas those in the high-frequency region 300–700 $cm^{-1}$ were assigned to V–O vibration modes [7]. The Ag phonon mode at 221 $cm^{-1}$ appears to be at lower wave number compared to that observed in single crystal $VO_2$ [8] which signifies the decrease in V-V distance hence an increase in the direct overlapping of V- 3d orbitals that results in the strain induced decrease of MIT in annealed thin film.

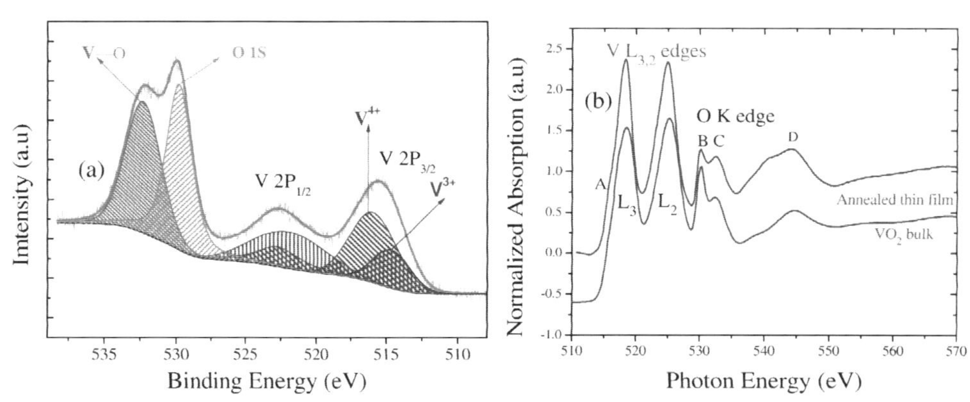

**Figure 2.** (a) XAS of annealed thin film along with pure bulk $VO_2$. (b) XPS of annealed thin film

The correct chemical composition of thin film was confirmed from XPS measurements as shown in Fig 2(a). The graph shows that annealed thin film is composed of both +3 and +4 oxidation states of Vanadium [9]. X-ray absorption spectroscopy XAS has proven to be a valuable tool to study the unoccupied conduction bands of $VO_2$ and improve the understanding of this system. We have performed the XAS measurements across the V $L_{3,2}$ and O-K edges of annealed thin film and pure $VO_2$ bulk (99.9% purity) shown in Fig 2(b). V $L_{2,3}$ edges correspond to the transition from V $2p_{1/2}$ ($L_2$) and $2p_{3/2}$ to empty V $3d$ states while O K-edge involves the transition from O $1s$ to O $2p$ state hybridised with V $3d$ orbitals. The fine feature A for V L portion corresponds to the transition to $t_{2g}$ band obtained because of the octahedral crystal field splitting of V $3d$ orbitals. In the O-K edge the features B and C emerge from the the transition to $\pi$ and sigma bands which result from the hybridization of O 2p with $t_{2g}$ and $e_g$ orbitals of V 3d band respectively, The broad feature D around ~544.1 eV is attributed to V4sp bands formed by V4sp-O2p antibonding interactions [10]. The magnitude of relative intensity of O K edge features reflects the amount of hybridization among metal ligand orbitals [ref]. The magnitude of the relative intensity (B/C = 1.08) in thin film is found to be smaller compared to bulk $VO_2$ (~ 1.54). This magnitude difference is attributed to the presence of $V_2O_3$ impurity phase which has different strength of hybridization with O 2p orbitals compared to $VO_2$.

## 4. Conclusion

We have been successful in obtaining a significant $VO_2$ phase from as deposited V2O3 thin film. XRD confirm the presence of VO2 polymorphic phase along with small secondary phase of V2O3in annealed thin film. The effect of strain and impurity phase on the MIT was observed from resistivity measurements. Raman measurements confirm structural transition accompanying with MIT dominated by $VO_2$. XPS reveal the +3 and +4 oxidation states of Vanadium. Presence of +3 Vanadium observed here from XPS in annealed film is in agreement to XRD data. The effect of impurity phase on the magnitude of hybridization among V $3d$ and O 2p orbitals was observed from XAS measurements.

## Acknowledgement

S Majid greatly acknowledges UGC-DAE CSR, Indore for providing the experimental facilities to carry out this research work.

## References

[1] Tan X, Yao T, Long R, Sun Z, Feng Y, Cheng H, Yuan X, Zhang W, Liu Q, Wu C, Xie Y and Wei S 2012 *Scientific Reports* **2** 466
[2] Laverock J, Preston A R H, Newby Jr. D, Smith K.E, Sallis S, Piper L F J, Kittiwatanakul S, Lu J W, Wolf S A, Leandtrsson M, and Balasubramanian T 2012 *Phys Rev B* **86** 195124
[3] Fu D, Liu K, Tao T, Lo K, Cheng C, Liu B, Zhang R, Bechtel H, and Wu J 2013 *Journal of Applied Physics* **113** 043707
[4] Wang Y L, Li M C, Zhao, L C 2006 *Science Direct* **201** 6772–6776
[5] Yang Z, Ko C and Ramanathan S 2010 *Journal of applied physics* **108** 073708
[6] Petrov G I, Yakovleva V V, Squier J 2002 *Applied Physics Letters* **81** 1023
[7] Shibuya K, Tsutsumi J, Hasegawa T and Sawa A 2013 *Applied Physics Letters* **103** 021604
[8] P. Schilbe 2002 *Physica B* 316, 600
[9] Zimmerman R, Claessen R, Reinert F, Steiner P, and Hfner S 1998 *J. Phys. Condens. Matter* **10** 5697.
[10] Abbate M, Pen H, Czyzyk M T, de Groot F M F, Fuggle J C, Ma Y J, Chen C T, Sette F, Fujimori A, Ueda Y, and Kosuge K 1993 *J. Electron Spectrosc. Relat. Phenom.* **62** 185

# Magnetic properties of thin films of samarium-cobalt alloy prepared by magnetron sputtering

**Gyanendra Panchal, Mukul Gupta, R J Choudhary, D M Phase**
UGC-DAE Consortium For Scientific Research, Khandwa Road,
Indore, Madhya Pradesh 452001

Email: gyanendra.panchal@gmail.com

**Abstract.**We examine the magnetic properties of samarium-cobalt thin films on quartz and Si(111) substrates grown by dc magnetron sputtering. Both films are deposited on Cr buffer layer and subsequently a capping layer of Cr was also deposited. Secondary ion mass spectroscopy results reveal that Cr diffused in to Sm-Co layer.This lead to local change in magnetocrystalline anisotropy. As the result of this we observed the two coercive behaviors in magnetization of thin film.

## 1. Introduction

For the application of magnetic recording, materials with high magnetic anisotropy are required with smaller grain size to increase the areal density and thus the reduction in the recording bit. Perpendicular magnetic anisotropy (PMA) in thin ferromagnetic films has attracted a great deal of attention due to interesting physics and promising application in recording media and spintronic devices [1].Therefore $SmCo_5$ with perpendicular magnetic anisotropy is a potential candidate for high-density magnetic recording. It has the highest magnetic anisotropy energy density. Recently magnetic anisotropy constant reached $4.0\times10^7$ erg/cm$^3$ which was achieved by introducing a Cu/Ti dual underlayer and controlling the substrate temperature during sputter deposition [2,3]. Singh et al [4] prepared epitaxial thin film with a single in-plane texture on Cr buffered MgO(110) substrates. They observed that epitaxial growth of the $SmCo^5$ phase is highly anisotropic and its remanent magnetization is improved by 38% as compared to $Sm_2Co_7$. J.S.Chen et al [5] reported that the magnetic properties of $SmCo_5$ hugely depended on the growth conditions and very much sensitive to the crystallinity of the film. They observed that the crystallinity, and magnetic properties were sensitive to deposition temperature, the texture of the Cu (111) underlayer, roughness of the Cu underlayer, and formation of SmCoCu alloy through the Cu diffusion. L. N. Zhang [6] et al showed that the crystallinity of the SmCo5 layer plays an important role in coercivity and in plane anisotropy, which is strongly dependent on the deposition temperature and texture of the Cr underlayer. Y. K. Takahashi [7] et al showed that due to composition fluctuation within the SmCo5 layer because of Cu diffusion and hence formation of $Sm(Co,Cu)_5$ phase, caused the fluctuation of the magnetocrystalline anisotropy within the $Sm(Co,Cu)_5$ layer. It turns out that the stoichiometric ratio of Sm and Co in samarium cobalt alloy thin films hugely depend on the technique employed for deposition and plays a pivotal role in shaping its magnetic property. In the present study we used magnetron sputtering to prepare Sm-Co alloy and study its structural and magnetic properties.

## 2. Experimental

The samarium cobalt thin films were grown by dc magnetron sputtering on Si(111) and quartz substrates. Before starting deposition, the chamber was evacuated to the base pressure of $2\times10^{-6}$ Torr. The deposition was performed at the argon partial pressure of 2 mTorr. A 10 nm buffer layer of chromium was deposited to increase the adhesion on quartz and Si(111) substrate. Samarium cobalt alloy ($SmCo_5$) thin film was grown by cosputtering from elemental target of Sm(99.9% pure) and Co(99.95% pure). Finally the 10nm capping layer of Cr was deposited to protect the oxidation of samarium cobalt thin film. Both films are grown at room temperature. The atomic composition of Sm-Co alloy thin film was controlled by adjusting the sputtering power of Sm and Co targets. Energy Dispersive X-ray analysis (EDX) is used for quantitative elemental analysis of the composition of the Sm-Co alloy thin films. Magnetic behavior of the thin film was studied by the SQUID-vibrating sample magnetometer (SVSM; Quantum design Inc, USA). The crystallographic structure of the films were studied by table top Bruker D2 PHASER X-ray diffractometer with Cu Kα radiation. To understand the phenomena of inter diffusion, secondary ion mass spectroscopy (SIMS) measurements was carried out on a Hiden Analytical SIMS Workstation using $O^{+2}$ as the primary source with 5 keV energy and a 400-nA beam current.

## 3. Results and discussion

Fig 1 (a) shows the X-ray diffraction pattern of Sm-Co alloy thin films with Cr used as under layer as well as capping layer on quartz and Si(111) substrates. The broad peak at 2θ =44.8° reveals that the film is nanocrystalline in nature.The EDX is used for quantitative elemental analysis of the composition of the Sm-Co alloy. In fig 1(b) the EDX spectrum shows signals due to their component elements Sm, Co and Cr, whereas no additional impurity peak is observed suggesting the thin film is free from contamination or impurities. The detailed analysis of thin film shows the atomic weight ratio of Sm and Co is 1:6.7. Therefore, the composition of our grown Sm-Co alloy thin film is SmCo6.7.

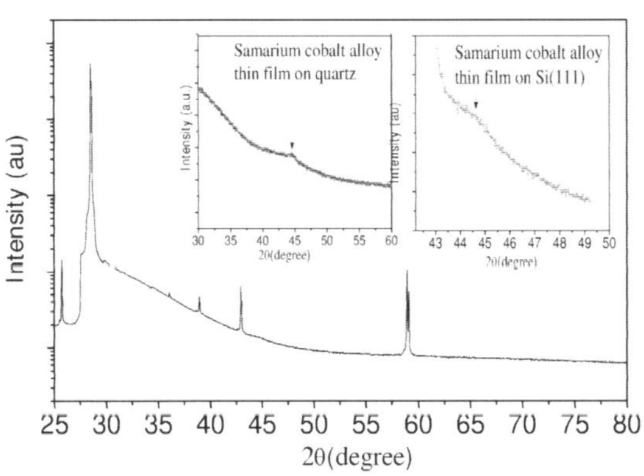


**Figure 1(a).** XRD pattern of samarium cobalt alloy thin film on quartz and Si(111) substrate

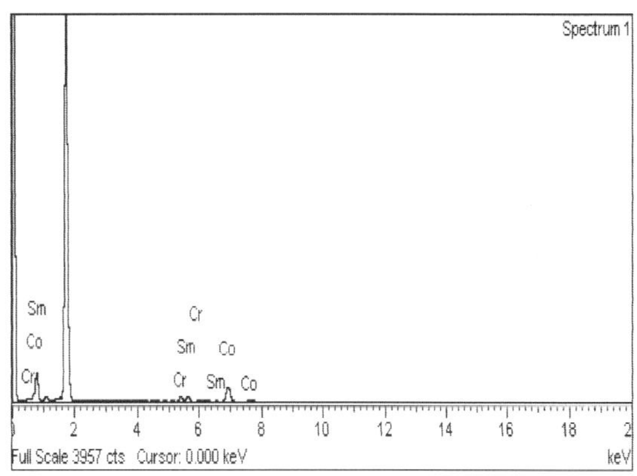


**Figure 1(b).** EDX of Sm-Co thin film under Chromium layer

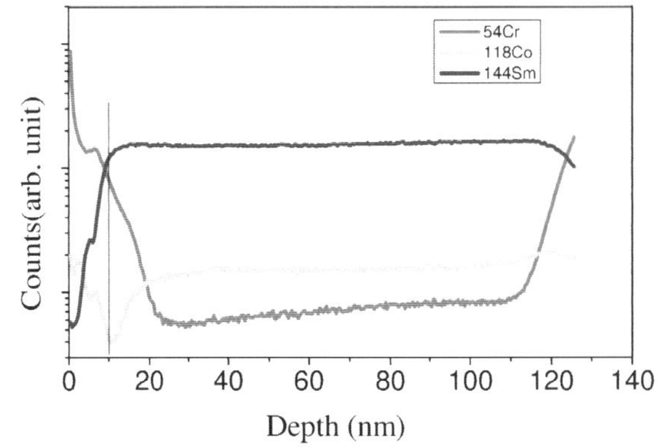


**Figure 2.** SIMS depth profile of various elements of annealed Sm-Co alloy thin film on Si(111) substrate

In order to probe the stability of the structure of the grown film, and interdiffusion process among Sm-Co alloy film and capping/ buffer layer of Cr, we annealed the film at 500 °C for 2 hour in vacuum and performed XRD and SIMS measurements. The XRD pattern of the annealed film remains similar to the as grown film, suggesting a good structural stability. SIMS depth profile of $Sm^{144}$, $Co^{188}$(Dimer) and $Cr^{54}$ are shown in fig 2.

In this figure we see that there is a gradual fall in concentration of $Cr^{54}$ and $Co^{118}$ up to few nanometer of depth. After 10 nm the concentration of $Sm^{144}$ and $Co^{118}$ increase up to near about 20 nm and then become constant , while concentration of $Cr^{54}$ decrease gradually up to approximately 20 nm , which clearly indicate that there is diffusion of Cr and intermixing of Cr with Co takes place at the top interface.

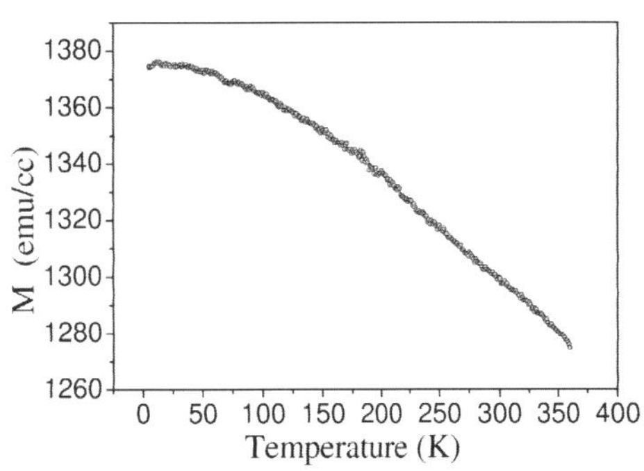


**Figure3(a).**Magnetization versus temperature curve of annealed Sm-Co alloy thin film on Si(111) at 100 Oe.

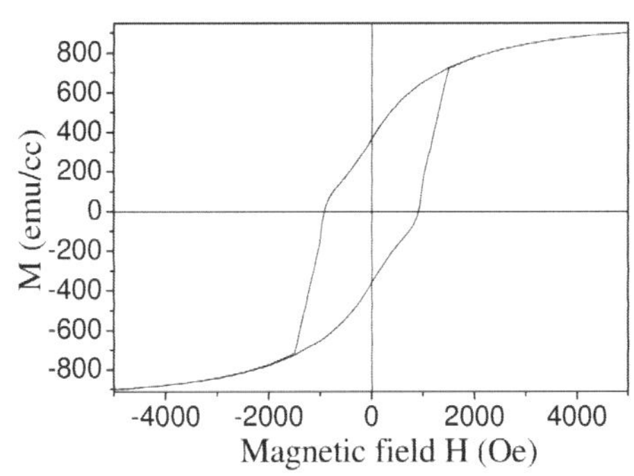


**Figure 3(b)** In planemagnetization of annealedSm-Co alloy film on Si(111) substrate .

In Fig. 3(a) we show the result of zero field cooled magnetization versus temperature behavior measured at 100 Oe. For this the sample was cooled in zero field down to 5 K, after that 100 Oe field was applied and measurement was done in warming cycle. It is clear from the M-T behavior that the sample shows magnetic ordering, with its Curie temperature larger than 360 K. This is consistent with the expected ferromagnetic nature of Sm-Co alloy. In Fig. 3(b) we show the results of magnetic hysteresis (M-H) measurements performed at room temperature. It clearly shows an open hysteresis loop with coercivity ~ 1000 Oe. Interestingly we also notice a double loop feature in M-H behavior, suggesting towards the possibility of two magnetic phases in the system. In this context it is worth recalling the results obtained from SIMS measurements, where we observed diffusion of capping layer of Cr in SmCo alloy at top 10 nm of the film. In such situation our film can be treated as combination of Cr-SmCo alloy and SmCo alloy. Cr from top layer diffused in Sm-Co layer which caused the local change in magneticrystaline anisotropy. Indeed, the presence of these two different magnetic phases is also reflected in the two coercive phase hysteresis behavior.

## 4. Conclusion

In conclusion Sm-Co alloy thin films were depositedby cosputteringSm and Co elemental targets on Cr under layer with Cr capping layer. Effect of Cr diffusion on magnetic properties of Sm-Co alloy thin film has been investigated. It is found that the Cr is diffused into Sm-Co layer and intermixed with Co. Magnetization versus temperature behavior suggested its ferromagnetic nature with Curie temperature larger than 360 K. The interdiffusion of Cr in SmCo alloy would cause the local change in magnetocrystalline anisotropy. As a result of this we observed the two coercive behaviors in magnetization of thin film.

## Acknowledgments

The authors express their gratitude to Dr. A.K. Sinha, Director UGC-DAE CSR & Dr. V. Ganesan Centre Director UGC-DAE CSR Indore for the encouragement.

## References

[1] Wolf S A, Awschalom D D, Buhrman R A, Daughton J M, Von Molna S Roukes M L, Chtchelkanova A Y and Treger D M 2001 *Science* **294**.
[2] Zhang L N, Hu J F, Chen J S, Ding J 2009 *J. Appl. Phys.* **105** 07A743.
[3] Sayama J, Asahi T, Mizutani K, and Osaka T 2004 *Appl. Phys. Lett.* **85** 23.
[4] Singh A, Neu V, Tamm R, SubbaRao K, Fähler S, Skrotzki W, Schultz L and Holzapfel B 2005 *Appl. Phys. Lett.* **87** 072505.
[5] Chen J S, Zhang L N, Hu J F and Ding J 2008 *J. Appl. Phys.* **104** 093905.
[6] Zhang L N, Chen J S, Ding J, Hu J F 2008*J. Appl. Phys.* **103** 113908.
[7] Takahashi Y K, Ohkubo T and Hono K 2006 *J. Appl. Phys.* **100** 053913.

# Synthesis and Mossbauer spectroscopy of macrocyclic complexes of iron(III)

**A Mishra[1], Kamaljeet S. Sura[1,2] and P Sharma[3]**

[1]School of physics , vigyan bhavan, Devi Ahilya University, Indore [M.P.] 452001,India
[2]Acropolis Institute of Technology and Research, Indore [M.P.] India
[3]School of chemical science, vigyan bhavan, Devi Ahilya University, Indore [M.P.] 452001,India

E-mail: kamaljeetsura@gmail.com

**Abstract.** The article deals with a fresh series of the complexes of the type: [Fe(III)(TML)Cl]Cl2; where TML is a tetra-dentate macrocyclic ligand; has been synthesized by condensation of o-phenylenediamine, diethyl malonate and diazonium ion in the ethanolic medium, through refluxing with $FeCl_3$.The synthesized metal complexes were characterized by Mossbauer spectroscopy. Mossbauer measurements were carried out using standard PC-based spectrometer equipped with Weissel velocity drive operating in the constant acceleration mode. Mossbauer study interprets paramagnetic nature of complexes. Mossbauer measurement of complex 1 and 2 has been taken to find out the value of isomer shift and quadrapole splitting and oxidation state after complaxsation.

## 1. Introduction

In all metals, iron has a relevant position due to its indispensability in life, since it takes part in processes such as oxygen transport and electron transfer and DNA syntheses. Iron metal complexes possess a variety of structures and are of current interest not only in bioinorganic chemistry [1] but also in material science, since they can find technological applications due to their promising magnetic, electronic and catalytic properties [2, 3], that are related to their structural characterization. It has been reported that metal complexation alters the various physiological properties especially the cytotoxic and antitumor activities of many naturally occurring compounds [4] It also shows antifungal potential against Alternaria alternate,Fusarium oxysporum and Myrothecium roridum by inhibiting spore germination [5]. It is also demonstrated that the coordination of metal ions with bioactive ligands can actually improve the pharmaceutical activity of drugs [6, 7].

## 2. Experimental Detail:

### *2.1. Chemicals and Solvent.*

All chemicals (sodium nitrite, sodium acetate, O-phenylenediamine, diethyl malonate, ferric chloride and anilines) are of Loba Chemie and solvents (absolute ethyl alcohol) are of Loba Chemie all are laboratory reagent grades and were used without further purifications.

### *2.2. Experimental analysis.*

Mössbauer spectroscopy is a spectroscopic technique based on the Mössbauer effect which was discovered by Rudolf Mössbauer in 1957, observed in recoil-free, resonant absorption and emission of gamma rays in solids. Mössbauer spectroscopy probes tiny changes in the energy levels of an atomic nucleus in response to its environment. Typically, three types of nuclear interaction may be observed: an isomer shift, quadrupole splitting and magnetic splitting. Due to high energy and extremely narrow line widths of gamma rays, Mössbauer spectroscopy is one of the most sensitive techniques in terms of energy resolution, capable of detecting change in just a few parts per $10^{11}$. In its most common form, Mössbauer absorption spectroscopy a solid sample is exposed to a beam of gamma rays and detector measures the intensity of the beam transmitted through the sample. The atom in the source emitting the gamma rays must be of the same isotope as the atoms in the sample absorbing them. In the resulting spectra, gamma ray intensity is plotted as a function of the source velocity. At velocities corresponding to the resonant energy levels of the sample, a fraction of the gamma rays is absorbed, resulting in a drop in the measured intensity and a corresponding dip in the spectrum. The number, position, and intensity of the dips (also called peaks; dips in transmitted intensity are peaks in absorbance) provide information about the chemical environment of the absorbing nuclei and can be used to characterize the sample. Mössbauer spectroscopy is a versatile technique that can be used to provide information in many areas of science such as physics, chemistry, biology and Metallurgy. It can give very precise information about chemical, structural and magnetic properties of a material [8]. Mössbauer measurements of iron complex were carried out using standard PC-based spectrometer equipped with Weissel velocity drive operating in the constant acceleration mode. The data was fitted with the NORMOS-SITE program, and the obtained parameters are with respect to natural iron.

### *2.3.1. Synthesis of complex 1.*

The ligand was prepared by diazotization process. A freshly prepared diazonium ion from sodium nitrite, 2-Chloroaniline and hydrochloric acid in equimolar proportion, is added drop wise with diethyl malonate in continuous stirring .Keep remember that the temperature of reaction should not be more than $10^0$C other wise diazonium salt will be decomposed. After keeping the mixture in refrigerator for one day, precipitated it and dry in air for few hours, TML1 is formed. Place O-phenylenediamine, TML 1 and Ferric chloride (equimolar proportion) in round bottom flask containing 25 ml of ethyl alcohol. Reflux it for 2-3 hours (Figure 1). Remove the flask from the heat source and allow for cooling until just warm to the touch. Separate the precipitate from the solution and allow it for drying in air for 1 hour. Complex 1 is formed.

### 2.3.2. Synthesis of complex 2.

The ligand was prepared by same analogy as we discussed above only difference is that 2-Chloroaniline is replaced by O-Toluidine. Now place O-phenylenediamine, TML 2 and Ferric chloride (equimolar proportion) in round bottom flask containing 25 ml of ethyl alcohol. Place this flask for reflux for 2-3 hours (Figure 1). Remove the flask from the heat source and allow for cooling until just

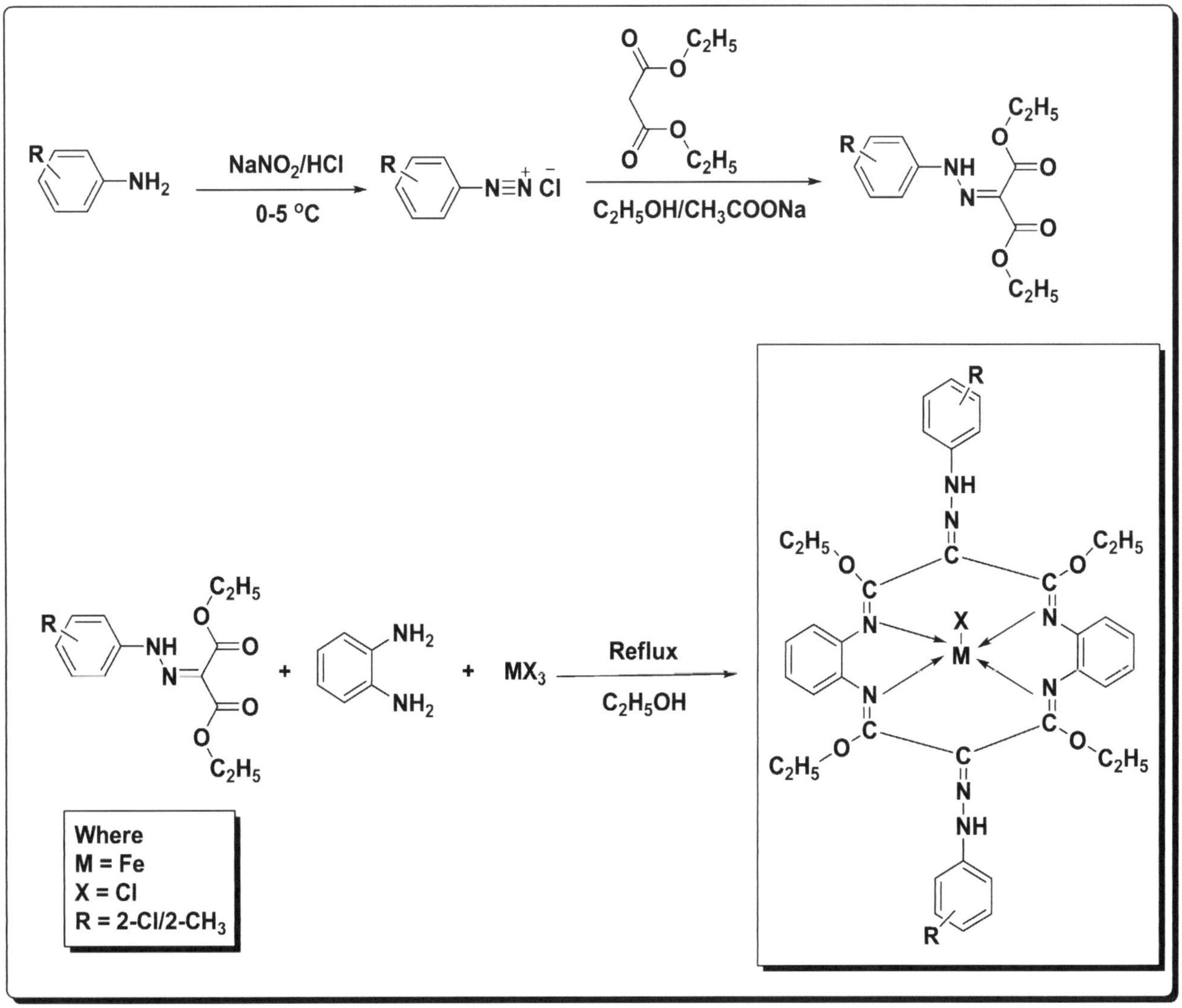


**Figure 1 .** Schematic overview of synthetic strategy for ligand and metal complexes; R=Cl & $CH_3$ for complex 1 and complex 2 respectively.

warm to the touch. Separate the precipitate from the solution and allow it for drying in air for 1 hour. Complex 2 is formed.

## 3. Result and Discussions.

The Mossbauer effect provides a powerful probe to chemical state and the environment of iron atoms. The Mossbauer spectrum for iron complexes at room temperature is shown in Figure 2. The various Mössbauer parameters obtained from the computations of spectral data are presented in Table 1.These spectrums were least-square fit with a superposition of two doublets with Lorentzian line shape. The fitted parameters are given in the table 1. According to the collected data presented in table, least square fitting and Lorenzian line shape were used for the spectral plot, which indicates two doublets corresponding to low and high spin states. Values for the isomer shift yield information on the degree of covalency in the metal ligand bond. $Fe^{2+}$ has higher isomer shift values than $Fe^{3+}$ because a decrease in number of d electron with increasing iron oxidation state increases s electron density at the nucleus. This in turn, tends to decrease the value of isomer shift for $Fe^{3+}$ state of iron [9]. The spectra of quadrupole splitting at room temperature suggest that the presence of electric field gradient at the metal ion is due to the presence of ligand around the ion. This may be due to the distorted octahedral geometry. According to collected data the isomer shift of complex 1 is 0.32 ± 0.020 and 0.46±0.05 mm/s and for complex 2 has isomer shift of 0.22 ± 0.06 and 0.37± 0.03 mm/s and quadrupole splinting of complex 1 is 0.50 ± 0.05 and 1.15±0.16 mm/s and for complex 2 is 0.43 and 0.71 mm/s.

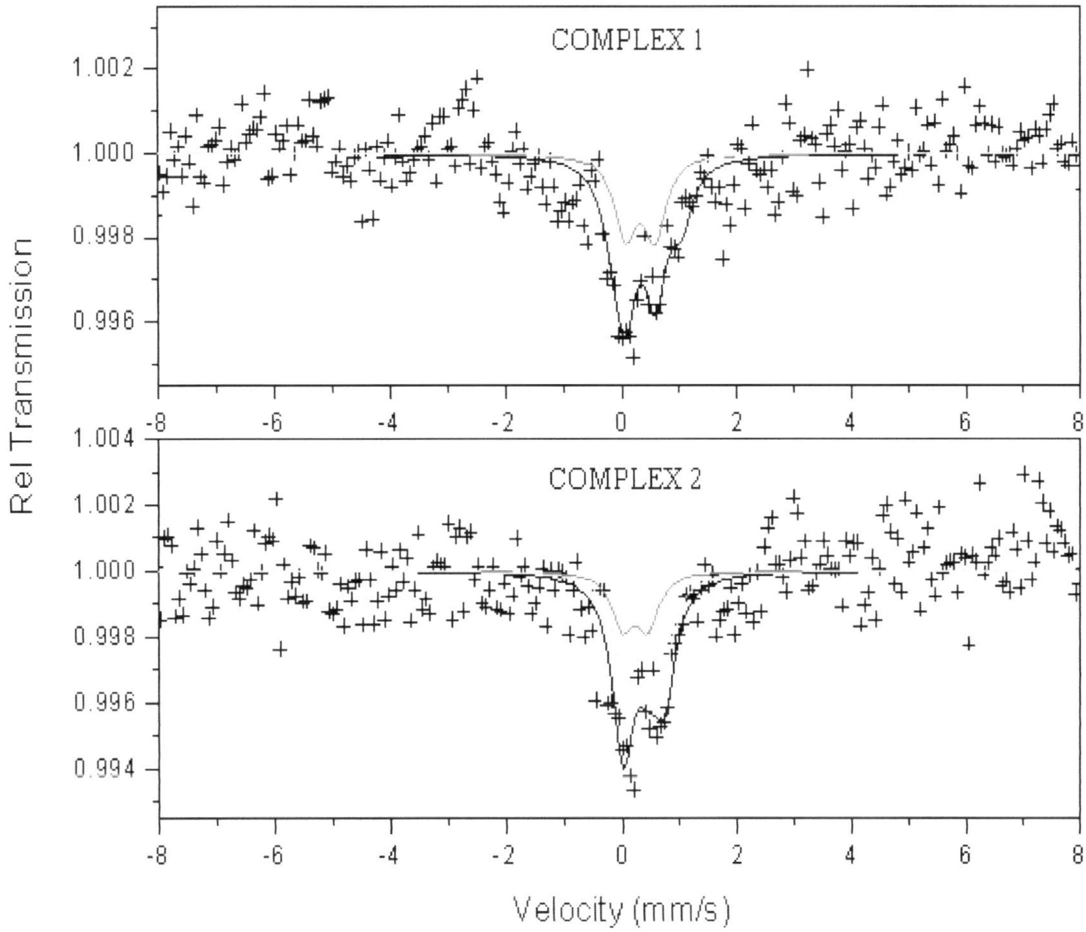


**Figure 2**. Mössbauer spectra of Complex 1 and Complex 2.

**Table1.** Parameters from Mossbauer spectra of iron complexes

| Sample | Isomer shift (mm/s) | FWHM (mm/s) | Quadrupole splitting (mm/s) | Area |
|---|---|---|---|---|
| Complex 1 | 0.32 ± 0.02 | 0.45 | 0.50 ± 0.05 | 69.7 |
| | 0.46±0.05 | 0.45 | 1.15±0.16 | 30.3 |
| | 0.22 ± 0.06 | 0.45 | 0.43 | 36.6 |
| Complex 2 | 0.37 ± 0.03 | 0.45 | 0.71 | 63.4 |

## 4. Conclusion

Complaxsation of Fe with tetra-dentate macrocyclic ligand has been successfully done via chemical route method. Mössbauer fitting reveals the value of isomer shift and quadrupole splitting of the complexes. It exhibits that the complexes possess distorted octahedral geometry.

## Acknowledgments

Authors are thankful to Dr. V. Reddy (UGC-DAE Consortium Indore) for his support in experimental work, and our colleagues who encourage us.

## References

[1] Tanase S, Bouwman E, Long G J, Shahin A M, Gelder R de, Mills A M, Spek L A, Reedijk J 2005 *Polyhedron* **24** 41 b) Xu X, Persson H L, Richardson D R 2005 *Mol. Pharmacol.* **68** 261

[2] Chaudhuri P, 2003 *Coord. Chem. Rev.* **243** 143; Soler M, Wernsdorfer W, Folting K, Pink M, Christou G *2004 J. Am. Chem. Soc.* **126** 2156.

[3] Be´nisvy L, Chottard J - C, Marrot J, Li Y 2005 Eur. *J. Inorg. Chem.* **999**.

[4] Thompson K H, B¨ohmerle K, Polishchuk E, Martins C, Toleikis P, Tse J, Yuen V, Mcneill J H and Orvig C *2004 Journal of Inorganic Biochemistry* **98** 2063

[5] Hothi H S, Makkar A, Sharma J R & Manrao M R 2006 *Eur J Med Chem* **41** 253

[6] Zhang S, Tu C, and Wang X 2004 *European Journal of Chemistry* **98** 4028

[7] Dutta S, Murugkar A, Gandhe N, and Padhye S 2001 *Metal-Based Drugs* **8** 183

[8] Mishra A, Mishra N, Malviya P, Awate R 2014 *International Journal of Scientific Research in Physics and Applied Sciences* **2** 2348

[9] Mishra A, Sharma R, & Shrivastava B D 2011 *Indian Journal of pure & Applied physics* **49** 740

International Conference on Recent Trends in Physics 2016 (ICRTP2016) IOP Publishing
Journal of Physics: Conference Series **755** (2016) 012030 doi:10.1088/1742-6596/755/1/012030

# Resistivity study of $Dy_{0.93}Y_{0.07}$ alloy

**Rudra Prasad Jena and Archana Lakhani**[1]
UGC-DAE Consortium for Scientific Research,
University Campus, Khandwa Road, Indore-452001, India.

E-mail: archnalakhani@gmail.com

**Abstract.** The electrical resistivity measurements of the polycrystalline rare earth alloy $Dy_{0.93}Y_{0.07}$ in the temperature range 2-300K at different magnetic fields are carried out. Two transitions from paramagnetic (PM) to spiral antiferromagnetic (S-AFM) and S-AFM to ferromagnetic (FM) states are clearly seen in the zero field resistivity data. The various contributions to the electrical resistivity of Dy and $Dy_{0.93}Y_{0.07}$ in three different magnetic states are studied. In the ferromagnetic region, the spin wave excitation energy term increases on decreasing the temperature due the enhanced magnetic anisotropy at lower temperatures.

## 1. Introduction

The rare earth element Dysprosium shows interesting magnetic properties with change in temperature and magnetic field. It is paramagnetic (PM) at room temperature. On reducing the temperature, it undergoes a transition from the paramagnetic state to the spiral antiferromagnetic (S-AFM) state at the Neel temperature ($T_N$). On lowering the temperature further, it transforms from the S-AFM state to the ferromagnetic (FM) state at the Curie temperature ($T_C$) [1,2]. The spiral antiferromagnetic arrangement in the temperature range of 180-90 K is formed due to the competition between antiferromagnetic and ferromagnetic exchange interactions. Substituting Dysprosium with Yttrium changes the strength of exchange interactions and magnetoelastic coupling in the Dy lattice [3,4]. As a consequence, both the transition temperatures are affected and shift towards the lower temperature region on adding the Y in Dy.

The electrical resistivity study of rare earth metals is of significant importance due to changing dynamics at magnetic phase transitions. The electrical resistivity is assumed to be a combination of various scattering phenomena in different magnetic states at different temperature regions. The resistivity of Dysprosium is addition of mainly three contributions viz: (i) residual resistivity due to impurities and disorder present in the system, (ii) phonon scattering and (iii) magnon scattering.

$\rho(T) = \rho_{res} + \rho_{mag}(T) + \rho_{ph}(T)$.

The resistivity due to impurities and imperfection in the lattice is independent of temperature whereas the scattering of electrons due to phonons is expected to be a linear dependence of temperature at T > 0.5θ, where θ is the Debye temperature. The magnetic scattering contribution increases with magnetic entropy [5].

[1]Email: archnalakhani@gmail.com
UGC-DAE Consortium for Scientific Research
University Campus, Khandwa Road, Indore-452001, India.

Journal of Physics: Conference Series **755** (2016) 012030 doi:10.1088/1742-6596/755/1/012030

It is interesting to study the electrical resistivity in changing magnetic states of polycrystalline Dy in different temperature regions. This manuscript describes various contributions to the resistivity in the PM, S-AFM and FM regions of Dy and $Dy_{0.93}Y_{0.07}$ alloy.

## 2. Experimental

For preparing the polycrystalline rare earth alloy, $Dy_{0.093}Y_{0.07}$, high purity Dy and Y elements having purity 99.9% are taken in stoichiometric amounts and are melted in pure argon atmosphere by arc melting method. Melting is done several times for making the sample homogenous. The spherical ball obtained after melting is cut into a flat piece. X-ray diffraction (XRD) was carried on Bruker D8 Advance Diffractometer (Cu K$\alpha$ = 1.54 Å) for initial characterization of the sample. A rectangular bar shaped piece was cut for electrical resistivity measurements. Electrical resistivity measurements have been carried out using standard four probe method on the 9T-ACT PPMS system i.e. AC-Transport Physical Property Measurement System from Quantum Design, USA in the temperature range of 2 - 300 K and magnetic field up to 1.5 T.

## 3. Results and Discussions

Figure 1 shows the XRD pattern obtained at room temperature, the Rietveld refinement fit, Bragg peak positions and the difference between the observed and calculated pattern for $Dy_{0.93}Y_{0.07}$. The XRD data confirm the single phase formation of the prepared sample and no impurity phase is detected. Rietveld refinement of the XRD data using Fullprof software confirms the hexagonal structure of the sample having space group $P6_3/mmc$ and the lattice parameters are found to be a = b = 3.61Å and c = 5.67Å.

The temperature dependence of electrical resistivity in the temperature range of 2-300 K has been measured at various constant magnetic fields up to 1.5T. For in-field resistivity measurements, magnetic field was applied at 300K and then resistivity was measured during cooling and heating cycles. Figure 2(a) and 2(b) shows the temperature dependence of resistivity of $Dy_{0.93}Y_{0.07}$ in zero and 1.5T magnetic field respectively. Pure Dysprosium goes from PM to S-AFM state at the Neel temperature ($T_N$) =180K via second order phase transition and from S-AFM to FM state at the Curie temperature ($T_C$) =90K via a first order phase transition [6,7]. On substituting 7% Y in Dy, transformation from PM to S-AFM state takes place at ~170K, whereas the S-AFM state transforms to FM state at ~56K at zero field. Both the transition temperatures are shifted towards low temperature region due to the changes in exchange interactions and magnetoelastic coupling with the incorporation of non-magnetic Y in Dy lattice. Figure 2(b) shows that the application of magnetic field ~1.5T on $Dy_{0.93}Y_{0.07}$ alloy suppresses the hysteresis observed at zero field, indicating a transformation from paramagnetic to ferromagnetic state, overcoming the antiferromagnetic interactions as seen by earlier magnetization measurements in $Dy_{0.965}Y_{0.035}$ alloy [3].

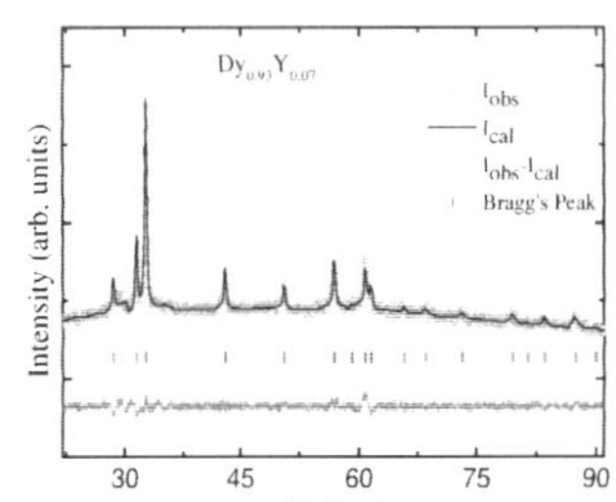


**Figure 1.** Room temperature XRD pattern of $Dy_{0.93}Y_{0.07}$ with Rietveld refinement fit, difference between the observed data and the fit along with Bragg peak positions.

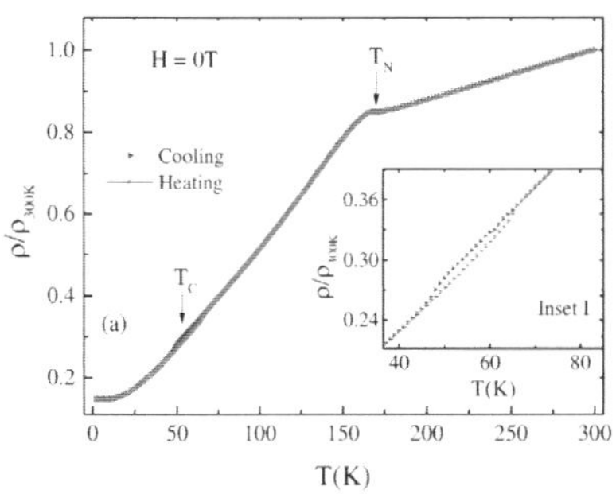


**Figure 2(a).** Electrical resistivity of $Dy_{0.93}Y_{0.07}$ as a function of temperature in zero field. Inset shows the thermal hysteresis across $T_C$ in cooling and heating cycles.

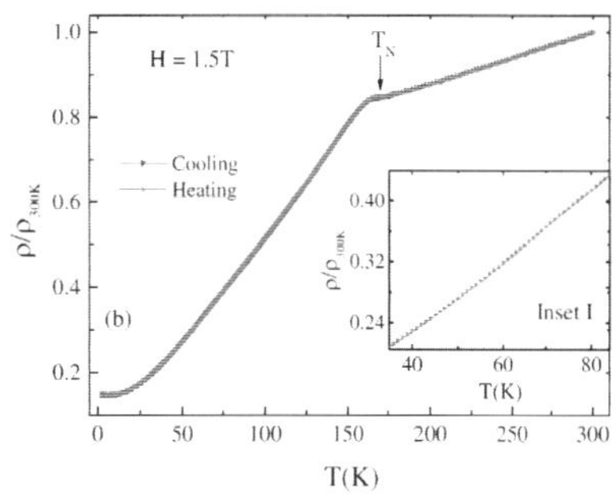


**Figure 2(b).** Electrical resistivity of $Dy_{0.93}Y_{0.07}$ as a function of temperature at 1.5T. Inset shows the overlapping of cooling and heating resistivity curves at 1.5T.

The zero field resistivity data of pure Dy and $Dy_{0.93}Y_{0.07}$ has been fitted in the different magnetic states. Figure 3(a) and 3(b) show the fitting of the resistivity data for Dy and $Dy_{0.93}Y_{0.07}$ respectively. We have subtracted the residual resistivity from the temperature dependent resistivity data and fitted different scattering contributions to $\Delta\rho = \rho(T) - \rho_0$. The resistivity in the paramagnetic state above the Neel temperature follows a linear T dependence in the temperature range of 300-180K on account of electron - phonon scattering i.e. $\Delta\rho(T_{300\text{-}180K}) \propto T$. Below the Neel temperature, in the spiral antiferromagnetic state, temperature dependent resistivity fits well to the relation $\Delta\rho = A*T^{1.5} + B*T$. The linear T dependence in this expression corresponds to the electron - phonon scattering whereas the $T^{1.5}$ dependence is the spin contribution.

Below the Curie temperature these systems transform to the ferromagnetic state where the major contribution in the resistivity is by magnetic scattering. Kasuya [8] has shown that such scattering gives a $T^2$ dependence to the resistivity at low temperatures. Mackintosh [9] considered the role of spin wave excitations in the resistivity for the anisotropic ferromagnetic materials like Dysprosium and Terbium. At low temperatures the magnetic contribution to the resistivity for anisotropic Dy and Tb is expected to follow a $T^2\exp(-\Delta/ k_BT)$ behaviour, where $\Delta$ is the energy gap, which represents the energy required for exciting the spin wave against the magnetic anisotropy [9]. Lodge et al. [10] also found the $T^2\exp(-\Delta/ k_BT)$ fit more satisfactory than simple $T^2$ term for the thin films of Dy samples below 60K. We obtained the best fit of the resistivity data in FM state to the following equation,

$$\Delta\rho = C*(T^2)*\exp(-D/T) + ET^5 \qquad (1)$$

where $D = \Delta/ k_B$. $ET^5$ term represents the phonon contribution at low temperatures. We found that the D values vary significantly with temperature, therefore we divided the FM region in three temperature ranges below $T_C$ and fitted equation (1) in these regions. Values of D obtained in different temperature regions for Dy and $Dy_{0.93}Y_{0.07}$ are shown in table 1. The large values of D in the lower temperature region indicate the large spin excitation energy required due to higher anisotropy in the lower temperature region. Further, D values for $Dy_{0.93}Y_{0.07}$ alloy are also found to be large in comparison to Dy, indicating the large spin excitation energy possibly due to enhanced anisotropy in comparison to Dy.

**Table 1.** Values of D (K) in different temperature regions.

| Sample/ T range | (2 - 20)K | (20 - 40)K | (40 - 70)K |
|---|---|---|---|
| **Dy** | 30 | 25 | 0.5 |
| **$Dy_{0.93}Y_{0.07}$** | 39 | 27 | - |

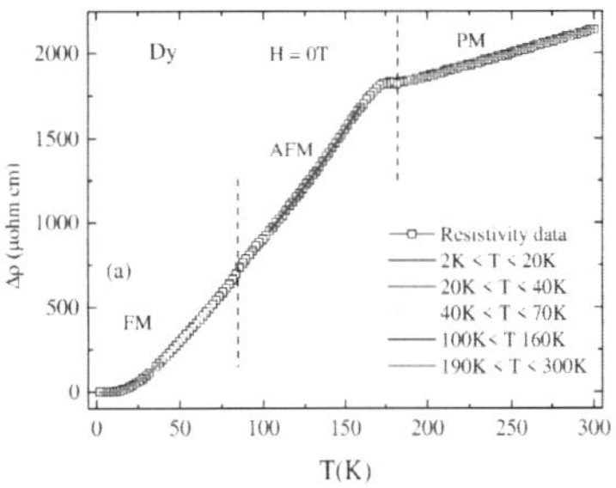


**Figure 3(a) (Colour lines).** Fitting of temperature dependent resistivity data of Dy in the PM state ($\Delta\rho \propto T$), AFM state ($\Delta\rho = A*T^{3/2} + B*T$)and FM state ($\Delta\rho = C*T^2\exp(-D/T)+E*T^5$) in different temperature regions.

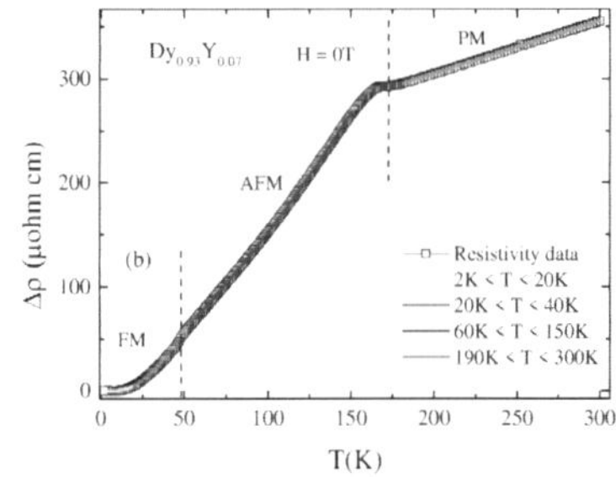


**Figure 3(b) (Colour lines).** Fitting of temperature dependent resistivity data of $Dy_{0.93}Y_{0.07}$ in the PM state ($\Delta\rho \propto T$), AFM state ($\Delta\rho = A*T^{3/2} + B*T$) and FM state ($\Delta\rho = C*T^2\exp(-D/T)+ E*T^5$) in different temperature regions.

## 4. Conclusion

The transition temperatures vary significantly with 7% Y substitution in Dy as well as with the application of magnetic field on $Dy_{0.93}Y_{0.07}$ alloy. The various contributions to the resistivity in different temperature ranges, corresponding to the different magnetic states in Dy and $Dy_{0.93}Y_{0.07}$ alloy have been fitted. A $T^2\exp(-\Delta/k_BT)$ behaviour fits well in the ferromagnetic state, indicating the presence of energy gap corresponding to the energy required to excite the spins against the anisotropy of Dy and $Dy_{0.93}Y_{0.07}$ alloy. The larger values of energy gap in the low temperature range and with Y substitution in Dy are noticed on account of higher anisotropy.

**Acknowledgments**

We are thankful to D. Kumar for scientific discussions. We acknowledge M. Gupta and L. Behera for XRD measurements, V. Ganesan (Centre Director) and A.K. Sinha (Director) for the support.

**References**

[1] Jensen J and Mackintosh A R, Rare Earth Magnetism: Structure and Excitations, Clarendon Press, Oxford, 1991d.
[2] RhyneJ J, Thesis, Iowa State University (1965)
[3] Lakhani A 2014 *Physica B* **448** 13
[4] Benito L, Ward R C C and Blamire M G 2013 *Phys. Rev. B* **88** 224407
[5] Hall P M, Legvold S and SpeddingF H 1960 *Phys. Rev. B* **117** 971
[6] Elliott J F, Legvold S and Spedding F H 1954 *Phys. Rev.* **94** 5
[7] Behrendt D R, Legvold S and Spedding F H 1958 *Phys. Rev.* **109** 5
[8] Kasuya T, Progr. Theoret. 1956 *Phys.* (Kyoto) **16** 58
[9] Mackintosh A R 1963 *Phys. Lett.* **4**, 2
[10] Lodge F M K and Taylo K N R 1969 *Phys. Lett.* **30A** 6

Journal of Physics: Conference Series **755** (2016) 012031 doi:10.1088/1742-6596/755/1/012031

# Co (II) Complexes with Schiff Base Ligands: Synthesis and EXAFS Study

**Ashutosh Mishra[1], Amantulla Mansuri[1], Samrath Ninama[2], Apurva Trivedi[3], Sushma Patidar[1], Mahesh Jamod[1], Ruchita Awate[1]**

[1] School of Physics, Devi Ahilya University, Khandwa Road Campus, Indore-452001, India
[2]Kasturbagram Rural Institute, Kasturbagram Indore-452001, India
[3]SICA College Indore-452003, India

Email :a.mansuri14@gmail.com, monujamod7@gmail.com

**Abstract.**Thesynthesis of transition metal Schiff base complexes of Co(II) are prepared by chemical root method. Obtained by the condensation of O-phenylenediamine, salicylaldehyde and isatin / 2-hydroxy- 1 Naphthaldehyde is presented. The complexes were characterized by Co- K- edge EXAFS measurements using the dispersive beam line at 2.5GeV energy of Indus-2 synchrotron radiation source RRCAT Indore. The recorded EXAFS data were analyzed using the computer software Athena for determine the nearest neighboring distances (bond lengths) of these complexes with conventional methods and were compared with Fourier transform (FT) analysis.

## 1. Introduction:

Schiffbase have been playing an important part in the development of Co-ordination chemistry. Schiff base metal complexes have been studied extensively because of their attractive chemical and physical properties and their wide range of application in numerous scientific areas. They play in important role in both synthetic and structural research, because of their preparative accessibility and structural diversity. Schiff base of o-phenylediamine and its complexes have a variety of application including biological, clinical and analytical [1-2].

The nature of a complex sample is best revealed by the application of several different experimentaltechniques, with each individual measurement providing both unique and complementaryinformation.EXAFS have been used extensively in the investigation of local atomic structures such asthe number and type of neighbouring atoms, inter-atomic distances and disorder. Since the applicationof EXAFS does not require the materials to have long-range order. It is well suited for determining thelocal structures of both non-crystalline materials. The well-markedEXAFS feature on the high energyside of the K-absorption edge up to several hundred eV has been observed in the cobalt complexes.From the knowledge of EXAFS, we have calculated the bond length for the following cobalt complexeswith the help of Levy's, LSS, Lytle and Fourier transform methods.

## 2. Experimental Technique:

### *2.1. Preparation of Schiff bas* $L_1/L_2$:

A solution of o-phenylenediamine (0.005 mol) in alcohol was added to a mixture of isatin/ 2-hydroxy 1naphthaldehyde (0.005mol) and salicylaldehyde(0.005mol) in 20 ml alcohol. The mixture was refluxed for about 30 minutes. The mixture was cooled in ice.The resulting precipitate was then filtered, washed with ethanol and dried.

**2.1.** ***Preparation of complex*:** To anethanolic solution of the Schiff base Ligand $L_1/L_2$ an ethanolic solution of the metal (cobalt chloride) was added in molar ratio (1:1).The mixture was refluxed for about 30 minutes.The mixture was cooled in ice.The resulting precipitate was then filtered, washed with ethanol and dried.

**EXAFS Analysis**

The X-ray absorption spectra have been recorded using synchrotron radiation.The X- ray spectroscopy setup is available at Raja Ramanna Centre for Advanced Technology (RRCAT) and is called beam line .This beam line BL-8 has been recently commissioned at the 2.5 GeV Indus-2 synchrotron radiation sources.

## 3. Result and discussion:

**EXAFS Bond determination using Different methods**

**(i) Levy's method**

In Levy's method, the bond lengths are given by $R1 = [151/\Delta E]^{1/2}$ Å,Where, ΔE is the difference in eV of the energies of the EXAFS maximum B and minimumand R1is the radius of the first coordination sphere [1].

**(ii) Lytle's method**

The energy values (E) of the EXAFS maxima, according to Lytle for p symmetry, i.e., Q = 2.04, 6.0,12.0, and 20.0. To evaluate the radius Rsof equivalent polyhedron through the relation [2]$Rs= [37.60 / M]^{1/2}$.

**(iii) Lytle, Sayers and Stern's (LSS) method**

In the LSS method for determination of the nearest neighbour distances, gives the value of $2(R1 - \alpha 1) / \pi$ where R1 is the bond length [3-4].

**(iv) Fouriertransforms method**

Bond length has also been determined by the Fourier transformation method for the cobalt complexesstudied however, determined only the phase uncorrected bond lengths by this method. No attempt hasbeen made to employ the fitting procedures by which phase corrected bond length can be determined,because the required crystallographic data is not available for any of the complexes studied.

The bond lengths are calculated using Levy's, LSS and Lytle methods. The values of bond length R is tabulated in table 1. The bond lengths obtained in Co(II) Complexes by LSS, Levy's, and Lytle method with F.T. methodare comparable each other.

## 4. Conclusion

As pointed out above, the LSS method, Levy's, Lytle and the Fourier transformation method give the value of bond length R, are comparable to each other. This distance is called the phase uncorrected bond length. It can be seen from table.1 that the phase uncorrected bond lengths obtained from these two methods agree with each other within the limits experimental error.

**Table 1:** Average values of the bond length in (Å) for the Co (II) Complexes.

| Complexes | RLSS | RLevy | RLytle | RF.T. |
|---|---|---|---|---|
| Co foil | 1.45 | 1.55 | 1.35 | 1.38 |
| Co1 | 1.38 | 1.48 | 1.37 | 1.37 |
| Co2 | 1.39 | 1.44 | 1.32 | 1.35 |

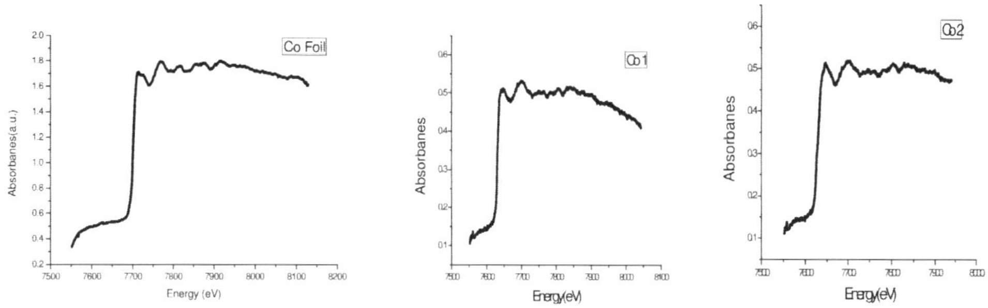


**Figure 1** Energy spectra of Co foil, Co complex1 and Co complex2.

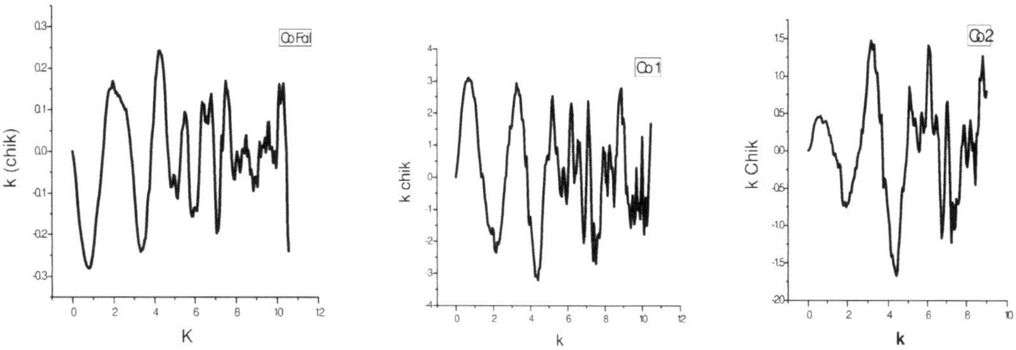


**Figure 2** k vs chi (k) spectra of Co foil, Co complex 1and Co complex2.

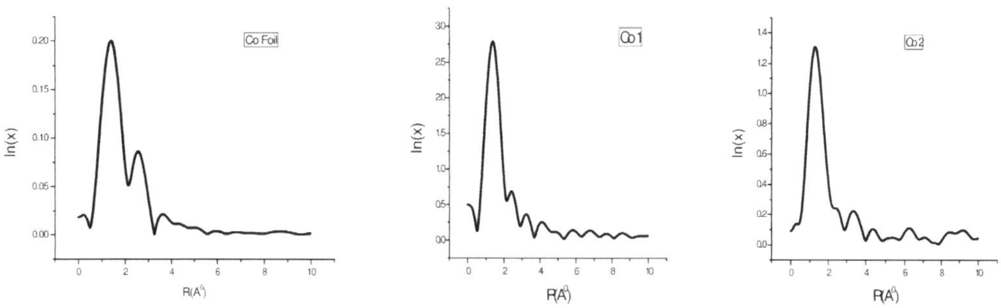


**Figure 3** bond length by Fourier Transformation magnitude of Co foil, Co complex1 and Co complex2.

**Acknowledgments**

The author is thankful to Dr.Pratibha Sharma School of Chemical Science DAVV Indore for sample preparation andDr. S N Jha RRCAT Indore India for providing EXAFS facility of synchrotron Radiation Source.

**References**

[1] kumar B, Rai B K and Ambastha N 2011 *J Chem.* **27** 1173
[2] Raman N, Pitchaikani Raja Y and Kulandaisamy A 2001 *ProcIndAcad Sci.* **113** 183
[3] Levy R M 1965 *J. Chem. Phys.* **43** 1846
[4] Lytle F W, Sayers D E and Stern E A 1975 *Phys. Rev. B* **11** 4825
[4] Lee J D 1991 *Concise of Inorganic Chemistry* ***4th*** *ed.* (Chapman & Hall, London)
[5] Sandler S R and Karo W 1971 *Organic Functionally Group Preparations II*, (New York: Academic press).

# Synthesis of Lead Sulfide Nanoparticles by Chemical Precipitation Method

**L S Chongad[1,3], A Sharma[2,3], M Banerjee[3] and A Jain [3,4]**
[1]Government Arts and Science College, Ratlam, India
[2]Shri Dadaji Institute of Technology and Science, Indore Road, Khandwa, India.
[3]Nanoscience and Nanotechnology laboratory,School of Physics, Devi Ahilya University, Vigyan Bhawan, Khandwa Road, Indore, India.
[4]Indore Institute of Science and Technology, Pithampur Road, Rau, Indore, India.

E-mail: mandira_bm@rediffmail.com

**Abstract.** Lead sulfide (PbS) nanoparticles were prepared by chemical precipitation method (CPM) with the assistance of $H_2S$ gas. The microstructure and morphology of the synthesized nanoparticles have been investigated using X-ray diffraction (XRD) and transmission electron microscopy (TEM). The XRD patterns of the PbS nanoparticles reveal formation of cubic phase. To investigate the quality of prepared nanoparticles, the particles size, lattice constant, strain, dislocation density etc. have been determined using XRD. TEM images reveal formation of cubic nanoparticles and the particle size determined from TEM images agree well with those from XRD.

## 1. Introducation

Lead sulphide(PbS) is an important IV-VI group semiconductor and bulk lead sulfide has cubic crystal structure and narrow direct band gap 0.41 eV at room temperature (300k) has a large excitonic Bohr radius about 18nm [1-2]. It has high dielectric constant and very high carrier mobility [3]. Hence PbS nanoparticles (NPs) show strong quantum size effects for relatively large size. Also, their absorption edge can be tuned to anywhere between red to violet covering the entire visible spectrum. PbS nanoparticles have been prepared by various techniques but they require sophisticated instruments. Keeping in mind the importance of PbS nanoparticles development of a simple and economic method of preparation is highly desirable. In the present work we have prepared PbS nanoparticles by a simple, time and cost effective method. To the best of knowledge of authors this method has not been reported earlier. The prepared nanoparticles have been characterized using x-ray diffraction (XRD), transmission electron microscopy (TEM) and selected area electron diffraction (SAED).

## 2. Experimental details

### *2.1 Synthesis*

Lead nitrate $Pb(NO)_3.5H_2O$ procured from Merck India was used for synthesis of PbS nanoparticles. Lead nitrate was dissolved in double distilled water to make N/20 aqueous solution. $NH_4OH$ solution was slowly added with constant stirring to obtain required pH value. Then $H_2S$ gas was bubbled through the solution for a few seconds. The resulting black precipitate was collected and washed with distilled water many times and then centrifuged for fifteen minutes. The nanoparticles were dried by heating for half an hour at 373 K and then allowed to cool at room temperature. Finally we get black powder which is PbS. The PbS nanoparticles were obtained with pH value = 8. The samples were prepared at room temperature (308K). The as-prepared samples without any further processing were used for all the experiments.

### *2.2. Characterization*

XRD patterns were recorded using Burker D8 Advance X-ray diffractometer with Cu Kα radiation (λ =1.54 Å) operated at 40 kV, 100 mA with a step size of 0.1792°. TEM images were recorded with Philips CM 200 microscope operated at 200 kV.

## 3. Results and discussion

XRD patterns of PbS nanoparticles are presented in Figure 1. All the patterns show well defined but broad diffraction peaks indicating the formation of nanocrystalline materials. It can be seen that in case of PbS (Figure1) all the major crystalline planes of PbS are present. The interplanar distance d were determined by Bragg's law [4] given below

$$n\lambda = 2d \sin\theta$$

where n is the order of diffraction, λ is the wavelength of the incident x-ray, d is the interplanar distance, θ is the Bragg angle The peaks corresponding to (111), (200), (220), (311) and (222) planes of PbS with cubic phase are present. Further, no peaks other than those corresponding to PbS are observed, which indicates the high purity of synthesized nanoparticles. The lattice constants were estimated using the formula [4]

$$d = \frac{a}{\sqrt{h^2 + k^2 + l^2 +}}$$

Where a = lattice constant; h, k, l = miller indices

The interplanar distances as well as the lattice constants match well with the values reported in JCPDS Card No.65-0132 and can be attributed to cubic phase and confirm the formation of PbS nanoparticles.

The average crystallite size D was calculated by Scherrer formula [5] given below.

$$D= 0.9\lambda/ B \cos \theta$$

where λ is the wavelength of x-ray used, B is the full width at half maximum (FWHM), θ is Bragg angle. The average crystalline size of PbS nanoparticle is estimated were found to be 25.6 nm.

The origin of strain is related to lattice "misfit" which depends upon the growing condition of nanoparticles [6]. The strain **ε** value have been calculated the following formula

$$\varepsilon = \beta\cot\theta/4$$

Where θ = Bragg angle, β = Full width of half maxima

The calculated value of strain has been shown in Table 1. From this table the strain data indicate the strain values increases corresponding to pH value such type of change in the strain may be due to the predominant recrystallization process in nanoparticles.

The dislocation density is known as the length of dislocation lines per unit volume of the crystal. A dislocation is a crystallographic defect or irregularity in the crystal structure [7]. The presence of dislocation strongly influences many of the properties of materials.

**Table 1.** Particle size and lattice constant, dislocation density and strain of PbS nanoparticles prepared by chemical precipitation method with assistance of $H_2S$ gas.

| Sample name | 2θ (degree) | d-Spacing (Å) | | Lattice constant **a** (Å) | | hkl | Dislocation density Lines/m$^2$ | Strain |
|---|---|---|---|---|---|---|---|---|
| | | Present study | Reported* | Present study | Reported* | | | |
| PbS | 26.05 | 3.41 | 3.41 | 0.5921 | 0.59 | 111 | 0.0012 | 0.0039 |
| | 30.16 | 2.96 | 2.95 | 0.5924 | 0.59 | 200 | 0.0008 | |
| | 43.12 | 2.09 | 2.09 | 0.5931 | 0.59 | 220 | 0.0012 | |
| | 51.06 | 1.78 | 1.78 | 0.5930 | 0.59 | 311 | 0.0016 | |
| | 53.50 | 1.71 | 1.70 | 0.5930 | 0.59 | 222 | 0.0020 | |

* JCPDS Card No.65-0132

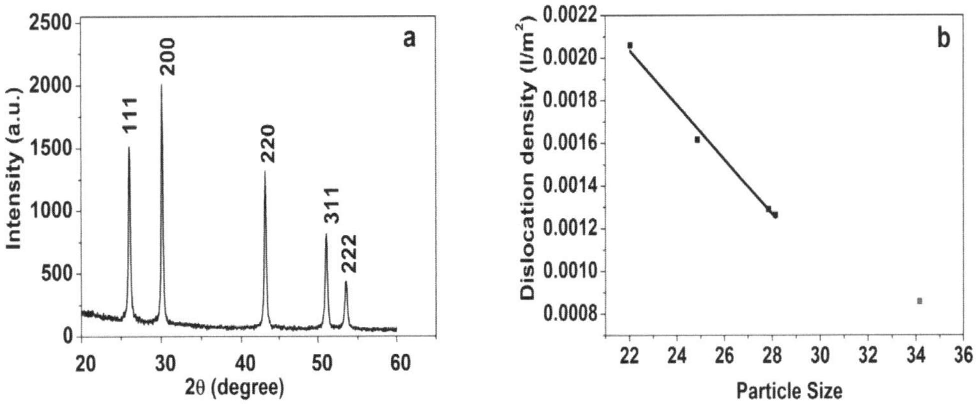


**Figure 1.**(a) XRD pattern and (b) Particles size Vs dislocation density of Lead sulfide nanoparticles prepared by chemical precipitation method

The X-ray line profile analysis has been used to determine the intrinsic stress and dislocation density [8]. The dislocation density (δ) in the sample has been determined using following expressions

$$\delta = 1 / D^2$$

where D is the size of the nanoparticle. Figure 1b shows the relation between particle size as obtained from each XRD peak and its corresponding dislocation density.

TEM image of PbS nanoparticles is presented in Figure 3. The images of PbS nanoparticles reveal formation of cubic structure. The size of PbS nanoparticles is found to between 29-49 nm. Furthermore, the selected area electron diffraction (SAED) patterns are also included in Figure 3. These SAED patterns index well to the cubic rock salt PbS crystal structure and diffraction rings match well with corresponding XRD patterns of the PbS nanoparticles. The SAED pattern shows concentric rings with bright spots, which indicates the nanocrystalline nature and good crystallinity of PbS NPs.

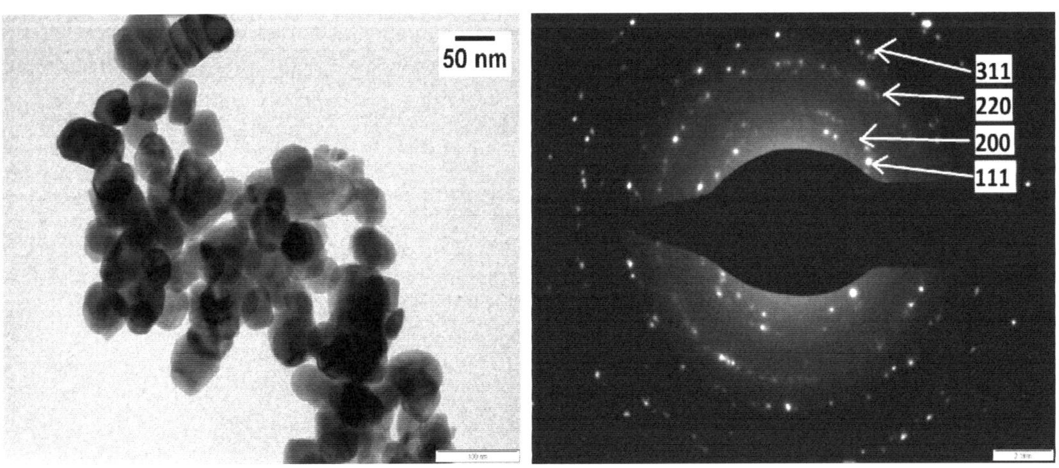


**Figure 2.** TEM image and SAED pattern of Lead sulfide nanoparticles

## 4. Conclusion

The present technique is a simple, cost effective as well as time effective method for producing good quality lead sulfide nanoparticles which is confirmed by XRD, TEM and SAED study.

## Acknowledgement

The authors would like to thank Dr. V. Ganesan Director, and Dr. Mukul Gupta Scientist UGC- DAE – CSR, Indore, India for their help in recording XRD pattern.

## References

[1] Machol J L, Wise F, Patel R C and Tanner D B 1993 *Phys. Rev.* **B 48** 2819
[2] Zhou S and Feng Y 2003 *J. Mater. Res.* 18 1188
[3] Scalon W W 1958 *Phys. Rev.* **109** 47
[4] Cullity B D 2001 *Elements of X-ray Diffraction* (New Jersey: 3rd edn. Prentice Hall)
[5] Scherrer P and Nachr Ges 1918 *Wiss. Göttingen* **2** 98
[6] Sharma K, Sharma R and Das H L 2008 *Chalcogenide letter* **5** 153
[7] Nehru L C, Swaminathan V and Sanjeeviraja C 2012 *American.J.Mat.Sci.* **2(2)** 6
[8] Chinh N Q, Gubicza J and Langdon T G 2007 *J.Mater.Sci.* **42**(5) 1594

# Electrical and Mössbauer study of polycrystalline $PrFeO_3$

**Anjali Panchwanee[1], V Raghavendra Reddy[1], Ajay Gupta[2]**

[1]UGC DAE CSR, University Campus, Khandwa Road, Indore-452001, India.
[2]Amity Center for Spintronic Materials, Amity University, Noida 201303

Email: vrreddy@csr.res.in

**Abstract** Preparation and study of polycrystalline $PrFeO_3$ is presented in the present work. The crystal structure and phase purity was determined through x-ray diffraction and Rietveld analysis.Magnetic ordering, $Fe^{3+}$ ionic state of iron and existence of only one type of crystallographic site for iron ion in $PrFeO_3$ is confirmed by the Mössbauer spectra at room temperature. The temperature dependent leakage current density (J-E) measurments are carried out over a temperature range of 150 – 250 K in order to sutdy the conduction mechanism and activation energy.

## 1. Introduction

Rare-earth orthoferrites ($RFeO_3$) are drawing increasing interest because of potential applications such as ultrafast magneto-optic recording and electric field control of magnetismi.e, multiferroic propertyetc[1]. In orthoferrites, $PrFeO_3$ (PFO) appears as a potential candidate in microelectronic industry as it enables device miniaturization due to the high dielectric constant and low dielectric loss.$PrFeO_3$ crystallizes in orthorhombic distorted ($ABO_3$) perovskite structure with centro-symmetric space group *Pbnm*. $PrFeO_3$ exhibits $Fe^{3+}$-$Fe^{3+}$, $Pr^{3+}$-$Fe^{3+}$ and $Pr^{3+}$-$Pr^{3+}$magnetic exchange interactions and as a result exhibits antiferromagnetic (AFM) to weak ferromagnetic transition corresponding to $Fe^{3+}$ at $T_N^{Fe}$= 705 K, spin reorientation of $Fe^{3+}$ spins ($T_{SR}^{Fe}$) at about 140 K [2]. Though several studies are reported on structural and magnetic properties of $PrFeO_3$but very few studies are focused on electrical properties of this important compound. In view of the multiferroic nature of the sample, it is necessary to study the electrical properties such as different conduction mechanisms, electrical break down of this compound. Therefore, in the present work a systematic study is undertaken to measure the leakage current density of the polycrystalline $PrFeO_3$ in order to understand the different conduction mechanisms present and estimate the activation energy.

## 2. Experimental details

### 2.1. Synthesis

Polycrystalline sample of $PrFeO_3$ was prepared using conventional solid state route. The starting materials $Pr_6O_{11}$ and $Fe_2O_3$of high purity (99.9%) taken in stoichiometric amounts were thoroughly mixed and ground in pestle & mortal for several hours. The ground powder was calcinated at 1100°C for 24 hrs with intermediate grinding. For final sintering the calcinated powder is pressed into pellets and sintered at 1300°C for 24 hrs.

2.2. Characterization

The structural study of the prepared sample was carried out by performing X-ray Diffraction measurement using D8-Discover system of M/s Brucker equipped with Cu-Kα radiation. In order to know the phase purity of the prepared sample along with unit cell parameters, Reitveld refinement was done using FullProfsoftware[3]. $^{57}$Fe Mössbauer measurements were performed in transmission geometry with $Co^{57}$ (Rh matrix) radioactive source, using standard PC-based spectrometer calibrated with natural iron at room temperature. TheMössbauerspectra is fitted with NORMAS-SITE software. The sample morphology was checked with a JEOL JSM 5600 scanning electron microscope (SEM).The temperature dependent leakage current density (J-E) measurements were carried out with ferroelectric loop tracer of Radiant technologies USA and homemade sample holder. The sample was kept in Janis make CCR for temperature variation.

**3. Results and discussions**

Figure 1(a) shows the X-ray diffraction pattern along with the Reitveld refinement [2]. All the diffraction peaks are well matched with a single phase orthorhombic perovskite structure having space group *Pbnm*, with no impurity peak present. The refined lattice parameters are a=5.48 (1) Å, b=5.58(1) Å c=7.79(1)Å match well with literature values [2].Figure-1(b) shows the SEM image of the sample at room temperature and figure 1(c)shows particle size distribution which indicates non-uniform size distribution having average particle size approximately7.2 ± 2.5 μm. The particle size is estimated using ImageJ software[4]. The average grain size is comparable to those for leaky samples [5].

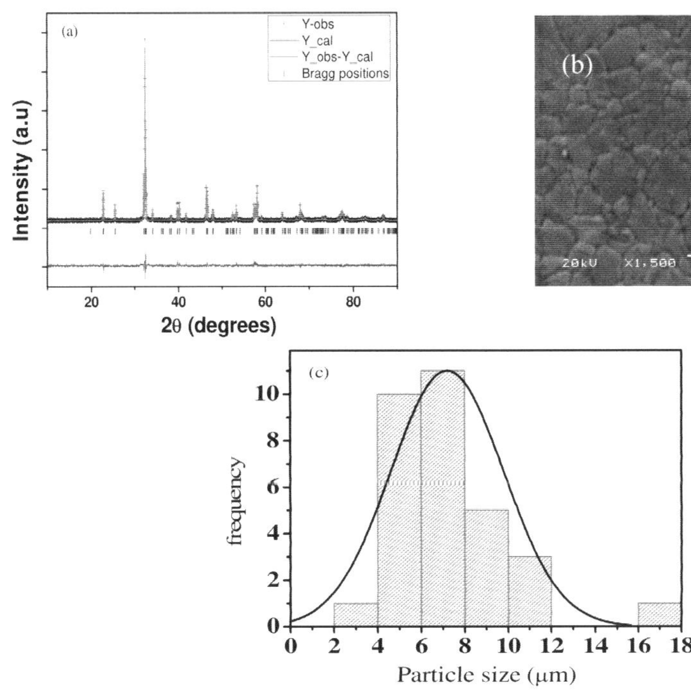


**Figure 1.**(a) X-ray diffraction pattern of polycrystalline $PrFeO_3$along with Reitveld refinement (b)SEM image of $PrFeO_3$ at room temperature (c) Particle size distribution.

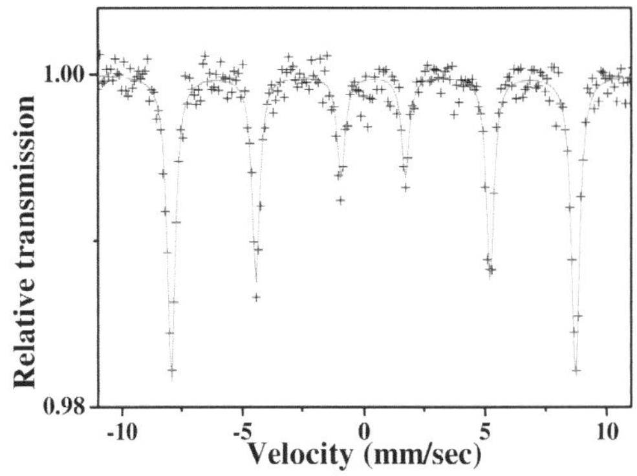


**Figure 2.** $^{57}$FeMössbauer spectra of $PrFeO_3$measured at room temperature. Symbols represent the experimental data and the solid line is the best fit to the data.

Figure-2 shows the Mössbauer spectra of $PrFeO_3$ recorded at room temperature in transmission geometry. The observed sextet indicates the presence of magnetic ordering in the prepared sample consistent with literature [6]. The data is fitted with one sextet considering only one site as the lines are relatively sharper consistent with the literature of these types of compounds. TheMössbauer data covering wide range of temperatures i.e., below and above magnetic transition ($T_N$)for $RFeO_3$ compounds is fitted in literature by considering only one site as orthoferrites are expected to have only one type crystallographic iron [6]. The obtained isomer shift 0.392 ± .003 mm/sec corresponds to the high spin state of $Fe^{3+}$ ion and internal magnetic hyperfine field ($B_{HF}$) is 51.5± 0.1 Tesla matches with literature[6]. Therefore, the x-ray and Mössbauerdata indicate structural and magnetic phase purity, high spin state of $Fe^{3+}$, presence of only one type of crystallographic site around iron ion, the magnetic ordering of Fe-sub lattices.

Figure 3(a) shows leakage currentdensity (J) versus applied electrical field (E) at different temperatures.One can clearly see that as the temperature is increasing the leakage current density is also increasing indicating the thermally activated process, phenomena typically observed in leaky dielectric samples. The leakage current as a function of electrical field E and temperature represents a temperature activated process in a dielectric material [7].The activation energy $E_a$ has been calculated using following equation

$$J(E,T) \propto E e^{\left(\frac{-E_a}{KT}\right)} \quad ...... (1)$$

where K is the Boltzmann constant.Inset of figure 3(a)shows Arrhenius plot for leakage current density according to equation -(1). From the data the activation energy $E_a$ is found to be approximately 0.329 ± 0.015 eV whichis comparable to the literature value[8] and agrees well with 0.29 eV for the polaron relaxation due to electrons hopping between $Fe^{2+}$ and $Fe^{3+}$ ions [9]. Further, the J-E data is used to study the dominant mechanisms that contribute to the measured leakage current in polycrystalline $PrFeO_3$ as discussed below. According to the law $J \propto E^m$,a plot between log(J) and log(E) is shown in Figure 3(b). As one can see, the data can be divided into two regions (corresponding to low and high applied electric fields) and can be fitted by two straight lines of different slopes. The slope *m* of these plots give information about various leakage mechanisms contributing in the leakage behavior of the sample[10].The value of m is generally less than one for grain boundary limited conduction[5], equal or greater than one for Ohmic conductionand equal or more than two space charge limited conduction[11]. For 225K the slope of low electric field region is ~1.15 and of high electric field region is ~2.28 which indicates the dominance of Ohmic and Space charge limited conduction mechanism however for 250K it shows space charge limited

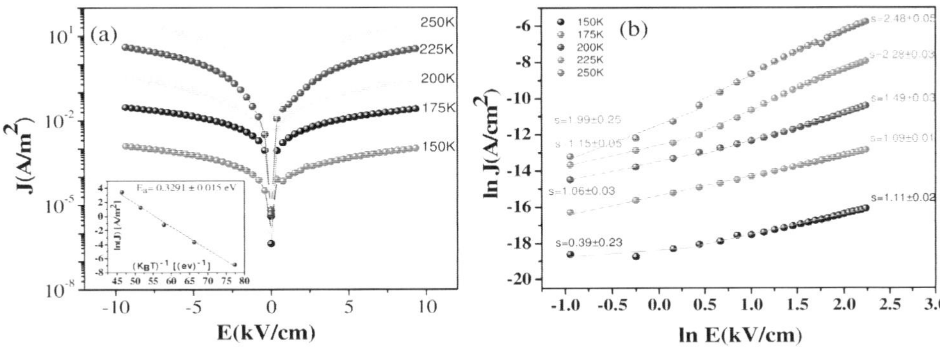

**Figure 3.** (a)Current density J versus applied electrical field E at different temperatures of $PrFeO_3$ with inset of Arrhenius plot(b) log(J) vs. log(E) plot measured at different temperatures for $PrFeO_3$

conduction in low as well as high electric field region as the value of slope is greater than two. For 200K and 175 K the theOhmic conduction is dominant over both electric field region as the value of slopes are between one & two. For 150 K the slope of low electric field region is ~0.39 and that of high electric field region is ~1.1 which indicates that grainboundary limited conduction is dominant in low electric field and Ohmic conduction mechanism is dominant in high electric field region.

## 4.Conclusions

Single phase polycrystalline $PrFeO_3$ was successfully synthesized via solid state reaction. Different leakage mechanisms are obtained at different temperatures along with the estimation of activation energy. Temperature sensitivity was seen in leakage current Magnetic ordering, no magnetic impurity and $Fe^{3+}$ state of iron ion is confirmed by Mössbauerspectroscopy.

## Acknowledgments

We acknowledge Dr. Mukul Gupta / Mr. L.Behrafor XRD and Dr.D.M.Phase / Mr.V.Ahire for SEM measurements. Dr. A.K.Sinha, Director UGC-DAE CSR & Dr.V.Ganesan, Centre Director UGC-DAE CSR Indore for the encouragement.

## References

[1] Tokunaga Y, Furukawa N, Sakai H, Taguchi Y, Arima T-H & Tokura Y 2009 *Nat. Mater.* 8558
[2] Zhou Z, Guo L, Yang H, Liu Q, Ye F 2014 *Journal of Alloys and Compounds* 58321
[3] Carvajal R 1993 *J Physica B* **192** 55
[4] ( http://imagej.nih.gov/ij/index.html
[5] Makhdoom A R, Akhtar M J, Rafiq M A, Hassan M M 2012 *Ceramics International* **38** 3829
[6] Eibschutz M, Shtrikman S and Treves D 1967 *Phys. Rev.* 156562
[7] Hassine N B, Mercier D, Renaux P, Parat G, Basrour S, Waltz P, Chappaz C, Ancey P, and Blonkowski S 2009 *Journal of Applied Physics* **105** 044111
[8] Hong F, Cheng Z, Zhang S and Wang X *J.Appl Phys.***111** 034104
[9] Ikeda N, Ohsumi H, Ohwada K, Ishii K, Inami T, Kakurai K, Murakami Y, Yoshii K, Mori S, Horibe Y & Kito H 2005 *Nature* **436** 1136
[10] Qi X, Dho J, Tomov R, Blamire M G, and MacManus-Driscoll J L 2005 *Appl. Phys. Lett.* **86** 062903
[11] Wu J, Wang J, Xiao D, and Zhu J 2011 *AIP Adv.* **1** 022138

# Microstructure and optical properties of Cu doped CdS nanostructured thin Films

**A Sharma[1, 2], L S Chongad[1, 3], M Banerjee[1]**

[1]Nanoscience and Nanotechnology laboratory, School of Physics, Devi Ahilya University, Vigyan Bhawan, Khandwa Road, Indore, INDIA
[2]Shri Dadaji Institute of Technology and Science, Indore Road, Khandwa, INDIA
[3]Government Arts and Science College, Ratlam, INDIA

E-mail: mandira_bm@rediffmail.com

**Abstract:** In the present work, we have synthesized pure CdS and Cu doped CdS thin films using chemical bath deposition technique. The crystal structure and the grain size of the particles were studied using XRD. The optical absorption spectra of the samples were recorded using UV-VIS-NIR spectrophotometer .The optical properties have been discussed in light of the microstructure of nanostructured thin films and the role of dopant.

## 1. Introduction

II–VI semiconductor nanoparticles and thin films are presently of great interest for their applications in optoelectronics and photonics [1,2]. The optical properties get modified dramatically due to the confinement of charge carriers within the nanoparticales and deposited thin films. Doping with different suitable dopants is also an easy treatment for tailoring the grain size and energy band gap which is one of the important factors of nanostuctured semiconducting materials.[3-5]. The chemical bath deposition (CBD) method is a relatively simple and low cost method to prepare a homogenous films with controlled composition. CdS is a wide band gap semiconductor with energy band gap equal to 2.4 eV [6] is used in solar cells. Keeping in view the imporatnce of CdS thin films, development of preparation techniques which cut down both the cost as well as preparation time is highly desirable. In the present work , we have prepared pure CdS and Cu doped CdS thin films by CBD method at room temperature in very short time. The microstructure and optical properties of these films have been studied and the results are discussed in this paper.

## 2. Experimental details

The CdS films were prepared by CBD technique . The bath solution comprised of 2M solutions of cadmium chloride [$CdCl_2.H_2O$] (Qualigens Fine Chemicals Mumbai) and 1.5 M solution of thiourea ($((NH_2)_2CN)$) (Merck , India) prepared in double distilled water. The pH of the solutions play major role in reaction. The pH of the bath solution during deposition was maintained at 10 by adding ammonia solution. Cleaned glass substrates were mounted vertically in the solution at room temperature for 10 minutes After the formations of the films the substrates were removed from the bath and rinsed several times in distilled water to remove any adherent particles and unreacted material. Samples were dried at room temperature. Adequate precaution was taken to protect the films from dust. For preparing Cu doped CdS films Copper sulphate [Cu(SO)$_4$ .7 $H_2O$ ] (Qualigens

Fine Chemicals Mumbai)was also added to $CdCl_2$ solution in appropriate amount and film was prepared by the same procedure as that for CdS thin film. The thickness of the films were estimated gravimetrically and were found to be 154 nm and 168 nm for pure and Cu doped CdS thin films. XRD patterns of the as-prepared films were recorded using Burker D8 Advance X-ray diffractometer with Cu Kα radiation (λ =1.54 Å) operated at 40 kV, 100 mA with a step size of 0.1792°. Optical spectra were recorded using Shimadzu 2450 UV-VIS-NIR spectrophotometer .

## 3. Results and Discussion

3.1 *XRD studies*: The X-ray diffraction patterns of the prepared CdS and Cu doped CdS thin films are presented in Figure 1. It can be seen that CdS nanostructures are polycrystalline in nature and exhibit hexagonal(H) and cubic (C) phases. CdS can exist in three crystalline phases, hexagonal, cubic (zincblend) and cubic (rock salt). The first two phases have been reported for CdS thin films prepared by CBD method. The crystal structure of CdS films in the present study are in agreement with the observation of earlier researchers. The interplanar distance d were calculated using Bragg's law [7]

$$n\lambda = 2d\sin\theta$$

where n is the order of diffraction, λ is the wavelength of the incident x-ray, d is the interplanar distance, θ is the Bragg angle.The d values corresponding to prominent peaks are presented inTable 1. The interplanar spacing d of the CdS film under study match well with those reported by JCPDS files (JCPDF#892944 hexagonal,890440 cubic).The slight shift in the peak positions in case of Cu doped CdS thin film can be attributed to doping of Cu in the CdS structure. X-ray diffraction peak profiles as seen in figure 1a and 1b are broad which can be attributed to small crystallite size. The crystallite size of the nanoparticles is estimated by using Scherrer's formula [7]

$$D= 0.9\lambda/ \beta\cos\theta$$

where λ is the wavelength of x-ray used, β is the full width at half maximum (FWHM) , θ is Bragg angle. The calculated crystallite size estimated from the main peak $(111)_C$ is presented in Table 1.

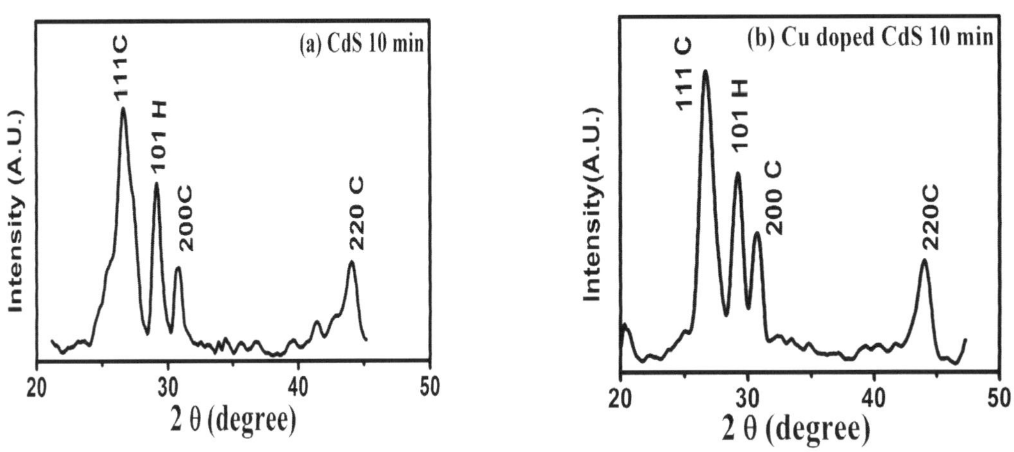

Figure1;XRD patterns of as-prepared CdS(a) and Cu doped CdS (b) thin films prepared by CBD method.

**Table1:** Particle size and lattice constant of CdS and Cu doped CdS thin film

| Sample | hkl | Interplaner Spacing d (Å) | | Lattice Constant **a** (Å) | | Particle Size D (nm) By using | |
|---|---|---|---|---|---|---|---|
| | | Observed | Reported* | Observed | Reported* | XRD studies | Absorption studies |
| Pure CdS | $(111)_C$ | 3.34 | 3.36 | 5.76 | 5.81 | 5 nm | 6nm |
| | $(200)_C$ | 2.9 | 2.91 | | | | |
| | $(220)_C$ | 2.05 | 2.06 | | | | |
| Cu:CdS | $(111)_C$ | 3.36 | | 5.78 | | 7 nm | 7.7 nm |
| | $(200)_C$ | 2.8 | | | | | |
| | $(220)_C$ | 2.05 | | | | | |

* (JCPDF# 890440).

3.2 *Absorption studies:* The absorption spectra of pure CdS and Cu doped CdS thin films were recorded at room temperature. The optical band gap $E_g$ of the film was estimated from the absorption data. The photon energy (hυ) and absorption coefficient (α)for direct optical transition are related by the following equation [7]

$$(\alpha h\upsilon) = B\,(h\upsilon - E_g)^{1/2}$$

where hυ is photon energy,α is the absorption coefficient , $E_g$ is the optical band gap and B is a constant .For calculation of optical band gap of the prepared films the graphs of $(\alpha h\upsilon)^2$ Vs hυ were plotted .The $E_g$ values of these film were determined from Fig 2(a),(b) and are found to be 2.68 eV for CdS thin film.The blue shift in case of CdS thin film confirms the formation of nanocrystalline CdS. The band gap of Cu doped CdS thin film is found to be 2.56 eV. The decrease of band gap in case of Cu doped CdS thin film shows that Cu is incorporated into CdS .

Particle size has been calculated by using effective mass approximation [8]

$$(E_g)\text{nano} - (E_g)\text{bulk} = \hbar^2\pi^2 / 2M^*R^2$$

where $\hbar = h/2\pi$ ,$M^*$ is effective mass of specimen ,h is plank's constant,$(E_g)$ nano is the band gap of the sample, $(E_g)$ bulk is the bulk band gap The particle size estimated by this method are presented in Table 1 and are in agreement with the XRD results.

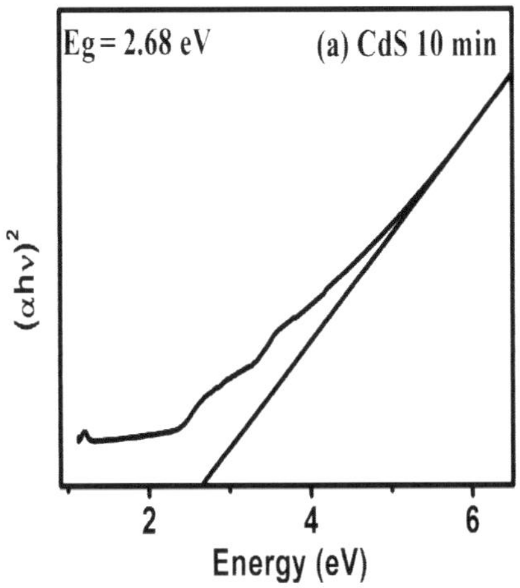


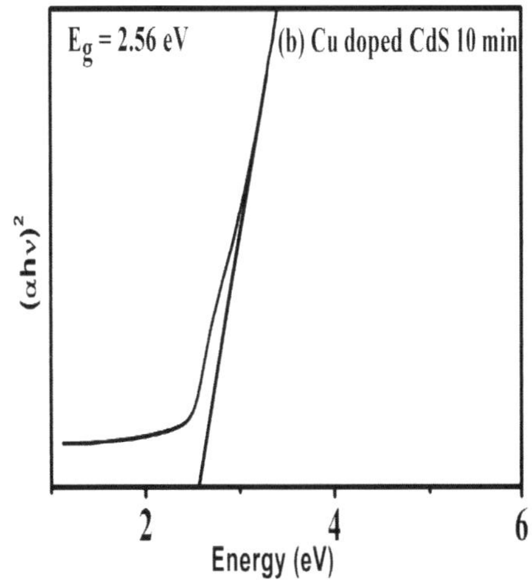


Figure 2; $(\alpha \; h \; \upsilon)^2$ Vs h$\upsilon$ graph of as –prepared pure CdS(a) and Cu doped CdS (b) thin films prepared by CBD technique.

## 4. Conclusions

Using the preparation conditions of the present work, good quality CdS and Cu doped CdS thin films were obtained using CBD technique in very short deposition time. The band gap of CdS nanoparticles show blue shift as compared to its bulk counterpart and doping by Cu shows shift in its bandgap. Grain size increases with the copper doping.

**Acknowledgment**
Authors would like to thank UGC-DAE-CSR Indore, specially Dr. Mukul Gupta for their help in recording XRD Patterns.

**References:**

[1] Oladejil I O, Chow L and Chu R 2000 *Thin solid Films* **359** 154
[2] Yang C C and Li S 2000 *J. Phys.Chem.B.* **112** 14193
[3] Kumar P, Sexsena N, Chandra R 2012 *Nanoscale Research Letters* **7** 584
[4] Sreelekha N ,Subramanyam K, Murali G, Giribabu G,Vijayalakhmi R,and Madhusudhana N 2014 *International Journal Of Chem. Tech Research* **6** 2113
[5] Reyes P and Velumani S 2012 *Mater. Sci. Eng. B* **177** 1452
[6] Das R, Pandey S 2011 *International Journal of material sciences* **1** 35
[7] Cullity B D 1956 *Elements of X-Ray Diffraction* **1** 149
[8] Brus L 1991 *Applied Physics A.* **53** 465

# Thermal annealing induced modification of structural and soft magnetic properties of $Fe_{33.8}Co_{50.7}Nb_5B_{8.5}P_2$ alloy

**M Shah[a], S S Modak[b], N Ghodke[c], J P Araujo[d], L K Varga[e] and S N Kane[a]**

[a] Magnetic Materials Laboratory, School of Physics, Devi Ahilya University, Khandwa Road Campus, Indore-452001, India
[b]Department of Physics, Jaypee University of Engineering and Technology, Guna-473226, India
[c]UGC-DAE-CSR University Campus, Khandwa Road, Indore-452017, India
[d]IFIMUP, Departmento de Fisica, Universidade de Porto, 4169-007 Porto, Portugal
[e]RISSPO, Hungarian Academy of Sciences, P.O. Box, 49-125 Budapest, Hungary

Corresponding author: kane_sn@yahoo.com

**Abstract**. Effect of thermal annealing treatment aimed to optimize the soft magnetic properties of $Fe_{33.8}Co_{50.7}Nb_5B_{8.5}P_2$ alloy system has been investigated. Information on the correlation between micro-structure and magnetic properties have been obtained using differential scanning calorimetry (DSC), x-ray diffraction (XRD), and hysteresis measurements. Annealing treatment enhances the saturation induction in studied alloy composition. Whereas there is a moderate increment in the coercive behaviour with annealing temperature that may be ascribed to weakly exchange coupled nano grains.

## 1. Introduction

Owing to their excellent soft magnetic properties Fe-based nanocrystalline alloys obtained from partially crystallized amorphous precursors have been studied widely. Microstructure of these alloys consists of ferromagnetic nanocrystallites embedded in amorphous matrix. Coercivity of nanocrystalline alloys can be significantly reduced provided the magnetic anisotropy gets averaged over many randomly oriented grains. This happens if the average grain size is smaller than the exchange coherence length [1, 2]. So, a suitable alloy composition and post-preparation thermal annealing treatments play a very important role in order to achieve a uniform nano-granular structure.

Soft magnetic nanocrystalline alloys have been extensively studied specially the compositions based on FeSiNbBCu, FeZrNbCu and FeCoZrBCu under the patented trade names of FINEMET [3], NANOPERM [4] and HITPERM [5] respectively. Fe-Co based HITPERM alloy family was developed specially for high temperature applications that can be cast in air [6, 7]. Addition of Co to Fe-based amorphous precursor enhances the Curie temperature of the amorphous as well as that of the precipitated crystalline phase [8].

In the present work we report the investigations on the effects of thermal annealing treatment on the structural and soft magnetic properties of $Fe_{33.8}Co_{50.7}Nb_5B_{8.5}P_2$ alloy in the form of ribbons.

**Table1:** Annealing temperature evolution of structural parameters obtained by XRD.

| Sample details | <*D*> (nm) | *a* (nm) | $V_x$ (%) | $X_m$ (nm) | Co % in Fe-Co Phase |
|---|---|---|---|---|---|
| As-cast | 20 | 0.2852 | 25 | 0.248 | 55 |
| Ann. 400 °C /1h | 12 | 0.2847 | 50 | 0.249 | 62 |
| Ann. 450 °C/1h | 9.5 | 0.2848 | 76 | 0.248 | 61 |
| Ann. 500 °C /1h | 10.5 | 0.2851 | 81 | 0.251 | 56 |

## 2. Experimental details

Ribbons (4mm wide and 20μm thick) of nominal composition $Fe_{33.8}Co_{50.7}Nb_5B_{8.5}P_2$ were prepared in air by planar flow casting method using single Cu wheel. Differential scanning calorimetry (DSC) at a heating rate of 20 °C/min was done. Following the DSC results thermal annealing of the as cast ribbons was done at 400, 450 and 500 °C for 1 hr. In order to investigate the micro-structural parameters: Scherrer's grain diameter (***D***), lattice parameter (***a***),Crystalline volume fraction ($V_x$) and nearest neighbor distance ($X_m$), X-ray diffraction (XRD) patterns were recorded on Bruker advance diffractometer equipped with a fast counting detector and making use of Cu-$K_\alpha$ (λ= 0.154 nm) radiation. Room temperature hysteresis measurements on as cast and annealed samples were done using conventional induction technique based loop tracer working at 50Hz with a maximum applied magnetic field of ± 1000 A/m to obtain coercivity ($H_c$) and saturation induction ($B_s$).

## 3. Results and Discussion

DSC results reveal that the onset of crystallization of the studied alloy composition occurs at 416 °C. X-ray diffraction data was analyzed using a MATLAB® program that fits the data using one crystalline and an amorphous component based on pseudo Voigt line profiles. Table 1 represents the annealing temperature evolution of various structural parameters obtained by analyzing the XRD data. Perusal of table 1 suggests the presence of crystalline phase in as quenched ribbons. As ribbons are prepared by rapid quenching of alloy melt on to a single copper wheel and, here the wheel side of the ribbon is rapidly cooled as compared to the shiny side. This may lead to growth of crystalline phase mainly in a thin layer near the surface region as is also observed earlier for metal-metalloid glass systems [9].

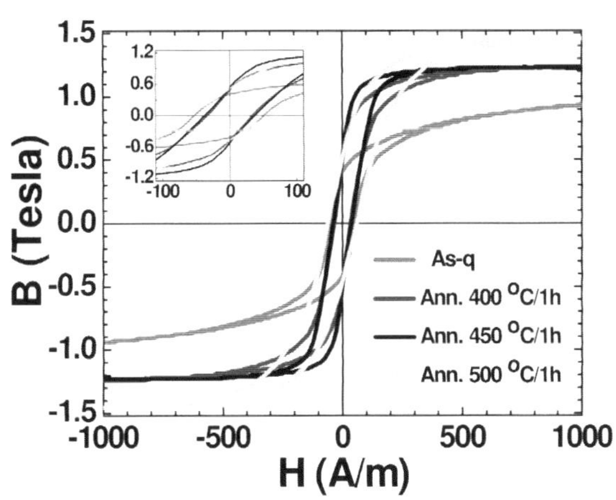


**Figure1:** Hysteresis loops of as-quenched and annealed ribbons measured at 50 Hz with a maximum applied field of 1 kA/m. Inset: Magnified region of the loops around low field.

**Table2:** Annealing temperature dependence of magnetic parameters.

| Sample details | $H_c$ (A/m) | $B_s$ (Tesla) | $B_r/B_s$ |
|---|---|---|---|
| As-cast | 52.4 | 0.94 | 0.43 |
| Ann.400 °C/1h | 33.8 | 1.23 | 0.39 |
| Ann. 450 °C/1h | 38.1 | 1.23 | 0.43 |
| Ann. 500 °C/1h | 60.1 | 1.46 | 0.35 |

Moreover presence of P reportedly favors the formation of high density nucleation sites by early precipitation of Fe-rich crystalline phase [10]. As can be seen from table 1, the lattice parameter values indicate the precipitation of Fe-Co crystalline phase. The change in the lattice parameter values is indicative of the diffusion of Co from amorphous matrix to the Fe-Co nanocrystalline phase. Table 1 also contains information on the amount of Co (%) present in the Fe-Co nanocrystalline phase. For the observed change in the lattice parameter values suggest that the amount of Co present in the crystalline phase corresponding to as cast and annealed samples range between 51 - 62%. Modest variation in the $X_m$ values suggests that the studied samples have similar mass densities.

It is interesting to note that average grain size is decreasing with increasing annealing temperature. Contrary to this the volume fraction of crystalline phase is increasing as a function of thermal treatment. It may be interpreted in terms of diffusion of Nb in the nanocrystalline phase. There can also be a possibility of Nb pile up at the interface of Fe-Co nanocrystals and amorphous matrix which restricts the grain growth. It is also worth noting that B and P may not be able to be solidified with nanocrystalline Fe-Co phase. Thus with increasing annealing temperature and hence the crystalline fraction, replacement of B by P in the amorphous matrix will increase as also observed in Fe-based nanocrystalline alloys [10]. This might also be a probable reason for the suppressed grain growth with higher annealing temperature.

Figure 1 represents the hysteresis curves corresponding to the as cast and the samples annealed at 400, 450 and 500 °C. Magnetic parameters obtained by hysteresis loop measurements are shown in table 2. It should be noted that lowest value of coercive field (~33 A/m) corresponds to the sample annealed at 400 °C and this sample is not characterized by the smallest grain size and highest volume fraction. Best value of saturation induction (~1.46 T) was obtained for the sample annealed at 500 °C, but this sample also exhibited highest coercivity (~60 A/m). As can be seen from table 2 the squareness ratio does not change much as is also reflected in terms of loops shape in figure 1. Initial decrease in coercivity till annealing temperature of 400 °C may be associated with the relaxation of stresses. Further it appears that this annealing treatment has resulted in optimized grain structure (~12 nm average grain diameter) where the nanograins are well exchange coupled through amorphous matrix.

According to random anisotropy model [1] the average crystalline anisotropy in soft nanocrytsalline alloys scale as $\propto 1/\sqrt{N}$ where N is the number of randomly oriented grains in exchange coupled volume. With increasing annealing temperature crystalline fraction increases. Consequently the number of exchange coupled grains should increase and eventually this would lead to lowering of magneto-crystalline anisotropy. As a result the coercivity should decrease. Contrary to this the coercive field is increasing with increasing annealing temperature for the studied alloy system. It may be associated with the possibility that at higher annealing temperatures of 450 and 500 °C grain size is decreasing which may result in increasing lattice strain presumably due to increasing number of structural defects. This may result in increasing magneto elastic anisotropy contributions that weakens the exchange coupling interaction among the Fe-Co nanograins despite of their reduced size and increased volume fraction.

## 4. Summary

Annealing temperature dependence of structural and magnetic properties of $Fe_{33.8}Co_{50.7}Nb_5B_{8.5}P_2$ alloy has been investigated using differential scanning calorimetry (DSC), x-ray diffraction (XRD), and magnetic measurements. Micro-structural evolution with annealing treatment is characterized by precipitation of single Fe-Co crystalline phase with average grain diameter ranging between 9-20 nm. Observed behaviour of coercive field is ascribed to the reduction of the internal stresses in the residual amorphous matrix during early stages of annealing. With increasing annealing temperature the growth of nanograins is suppressed ascribing to the diffusion of Nb and P through the amorphous matrix. Reduction in average size of nanograins appears to enhance the structural defects that lead to weaker exchange coupling among nano-grains. Annealing treatment appears to enhance the saturation induction values with moderate increment in coercivity.

## Acknowledgement

S. N. Kane acknowledges gratefully the financial support from Indo Portuguese joint project No.IND/Portugal/P-07/2013 and, by the project CSR-IC-IC-BL-25/CRS-122-2014-15/218.

## References

[1] Herzer G, 1990 *IEEE Trans. Magn.* **26** 1397.
[2] Hono K, Hiraga K, Wang Q, Inoue A, Sakurai T 1992 *Acta Metall. Mater.* **40** 2137.
[3] Yoshizawa Y, Oguma S, Yamauchi K 1988 *J. Appl. Phys.* **64** 6044.
[4] Suzuki K, Makino A, Inoue A, Masumoto T 1991 *J. Appl. Phys.* **70** 6232.
[5] Willard M A, Laughlin D E, McHenry M E, Thoma D, Sickafus K, Cross J O, Harris V G 1998 *J. Appl. Phys.* **84** 6773.
[6] Willard M A, Laughlin D E, McHenry M E 2000 *J. Appl.Phys.* **87** 7091.
[7] Gomez-Polo C, Marin P, Pascual L, Hernando A, Vazquez M 2002 *Phys. Rev. B.* **65** 024433.
[8] Gomez-Polo C, Pérez-Landazabal J I, Recarte V, Campo J, Marín P, López M, Hernando A, Vazquez M 2002 *Phys. Rev. B* **66** 12401
[9] Životský O, Postava K, Kraus L, Jirásková Y, Juraszekd J, Teilletd J, Barčová K, Švec P, Janičkovič D and J. Pištora 2008 *J. Magn. Magn. Mater.* **320** 1535
[10] Wang A D, Men H, Shen B ., Xie G Q, Makino A, Inoue A 2011 *Thin Solid Films* **519** 8283

# Electronic and optical studies of pulse laser deposited ZnO/NiO bilayer film

**P Baraskar[a], T A Dar*[b], R J Choudhary[c], P K Sen[b] and P Sen[a]**

[a]School of Physics, Devi Ahilya University, Khandwa Road Indore-452001, India.
[b]Shri G. S. Institute of Technology and Science, Park Road, Indore-452003, India.
[c]UGC-DAE Consortium for Scientific Research, University campus Khandwa Road, Indore-452001, India

E-mail: tanveerphysics@gmail.com

**Abstract.** We report the structural, optical and electronic properties of polycrystalline ZnO and NiO thin films and amorphous ZnO/NiO bilayer film, prepared by pulsed laser deposition technique. Despite of the presence of both Zn and Ni in +2 state in the bilayer film, the grown bilayer shows no reflections (in XRD) corresponding to ZnO or NiO. The difference in crystal structure of ZnO and NiO leads to the strain in the grown bilayer film. An increase in the band gap has been observed in bilayer film which can be attributed to the amorphous nature of the structure.

## 1. Introduction

Wide band gap transition metal oxides have established themselves as the important candidate materials in electronic industry. These oxides are suggested as alternative to the conventional electronics based on silicon [1]. ZnO is a transparent conducting oxide and it has unique optical and electrical properties due to the high electron mobility, high thermal conductivity, wide and direct band gap (3.3eV) and large excition binding energy (60meV). ZnO is an important and natural n-type semiconductor material having applications in photodetectors, light emitting diodes, laser diodes and transparent thin film transistors etc., NiO is a p-type semiconductor and has wide, direct band gap of 3.6 to 4.0 eV at room temperature. Due to its chemical stability it can be used as catalysts, in fuel cells and gas sensors. The study of the ZnO/NiO bilayer (ZnO/NiO) will result in the formation of stable p-n junction. However these structures are quite sensitive to the structural and optical properties of the individual semiconductors. The n-type nature of the ZnO is well understood and is attributed to the presence of oxygen vacancies or zinc interstitials [2]. Similarly the p-type conductivity in the NiO is due to holes generated by Ni vacancies in the lattice [3]. In the present manuscript we have deposited ZnO, NiO thin films and ZnO/NiO bilayer film. The grown samples have been characterized by XRD, XPS and UV-Vis techniques. The amorphous nature of the ZnO/NiO bilayer film is discussed.

## 2. Experimental details

In the present we report preparation and characterization of pulsed laser deposited thin film of ZnO, NiO (single layer) and ZnO/NiO (bilayer). The samples were grown by the laser ablation of the 25mm

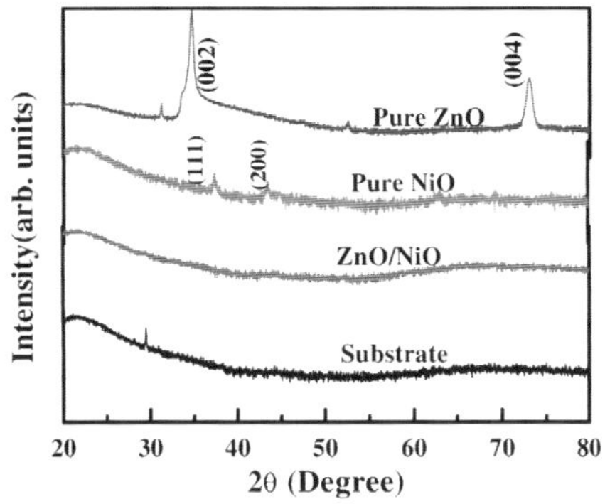


**Figure 1.** XRD patterns of the grown samples.

circular pellets which were prepared by solid state reaction route. After carefully weighing the powder materials, they are mixed together in pestle mortar. The powder was grinded for about 8 hours followed by heat treatment at 1000°C for 12 hours to bring about the thermal decomposition, phase transformation, removal of moisture and evaporation of volatile materials. After the calcinations the powder was again grinded for 4 hours and then mixed in a binder (polyvinyl alcohol PVA) for allowing the particles to remain bound to each other. The pellet shaping is done at room temperature by applying 150KN pressure. The prepared pellets were sintered at 1200°C for 20 hours for achieving the densification and strong bonding. This deposition was carried out on quartz substrate at 500°C. The base pressure was $2\times 10^{-5}$ Torr and oxygen partial pressure was $1\times 10^{-4}$ Torr. The distance between the target and substrate is 5 cm and the laser energy density was set to 2J/cm$^2$ (KrF laser with wavelength 248 nm). Repetition rate was

## 3. Result and discussion

Fig.1 shows the XRD pattern of the grown films in logarithmic scale. The pattern confirms wurtzite structure of ZnO in polycrystalline form and is highly c axis oriented. NiO film possesses rocksalt structure while the bilayer film of ZnO/NiO is amorphous in nature. We have calculated the grain size of pure ZnO and pure NiO as calculated from Scherer's formula was found to be 36.16nm and 19.57nm respectively. Lattice parameter ***a*** of both the films were calculated were found to be 2.98 Å and 4.16 Å for ZnO and NiO respectively. We also calculated the strain which is 28 % between ZnO and NiO layers in bilayer film by using the formula

$$\text{Strain}\ (\%) = \frac{a_{\text{substrate}} - a_{\text{film}}}{a_{\text{substrate}}}$$

Where, NiO has been considered as a substrate for the calculation.

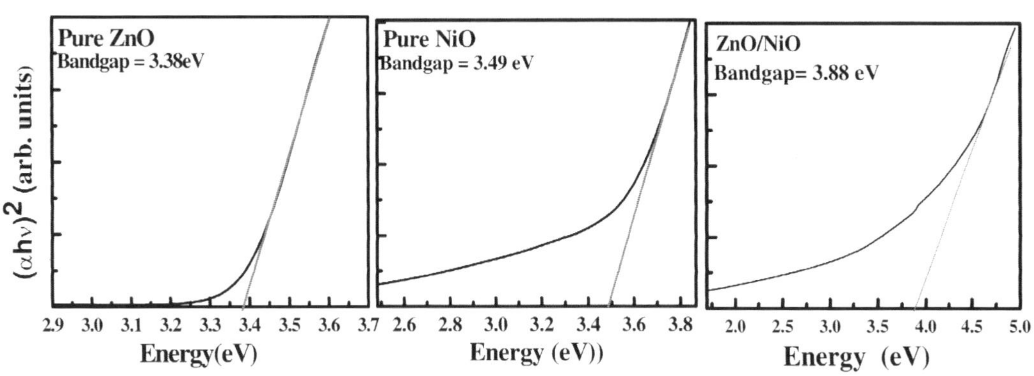


**Figure 2.** $(\alpha h\nu)^2$ verses hn of the grown films for calculating the band gap.

Fig. 2 represents the optical band gap of grown film which was obtained from $(\alpha h\upsilon)^2$ versus $h\upsilon$, where $\alpha$ is absorption coefficient and $h\upsilon$ is the incident photon energy [4]. The band gap for ZnO, NiO and ZnO/NiO bilayer was found to be 3.38eV, 3.49eV and 3.88eV, respectively. These values clearly indicate that band gap of bilayer is increased. The increased value of the band gap in the amorphous bilayer can be attributed to the amorphous nature. Similar results have been observed in as prepared amorphous chalcogenide films [5].

After studying the structural and optical properties of the grown samples, we will now focus on the electronic core level studies by analyzing the XPS data obtained using Al ka, X-ray source. The spectra were recorded using Specs XPS instrument at UGC-DAE CSR Indore in the vacuum better than $10^{-9}$ Torr. Prior to the measurements, all the samples were subjected to a surface clean procedure by Ar+ sputtering for 15 min at 200 V. The data taken before and after the Ar+ sputtering resulted in the reduction of C content accumulated on the film surface due to the exposure of the samples in the atmosphere. All the core levels were fitted using Touguard background and Voigt profile and were carefully corrected by a correction factor arising from the C 1s (284.6 eV) CL shift to compensate the charging effect. The XPS core level spectra of Zn2p in ZnO and Ni2p in NiO and its bilayer films is shown in fig.3. From the given figure it can be seen that $Zn2p_{3/2}$ peak (1021.7eV) in ZnO film shifts to higher values (1022.8eV) in bilayer film while $Ni2p_{3/2}$ peak (854.6eV) shifts to lower values (853.2eV) in the bilayer film. In all the samples we found that both Zn and Ni are in +2 states [6].

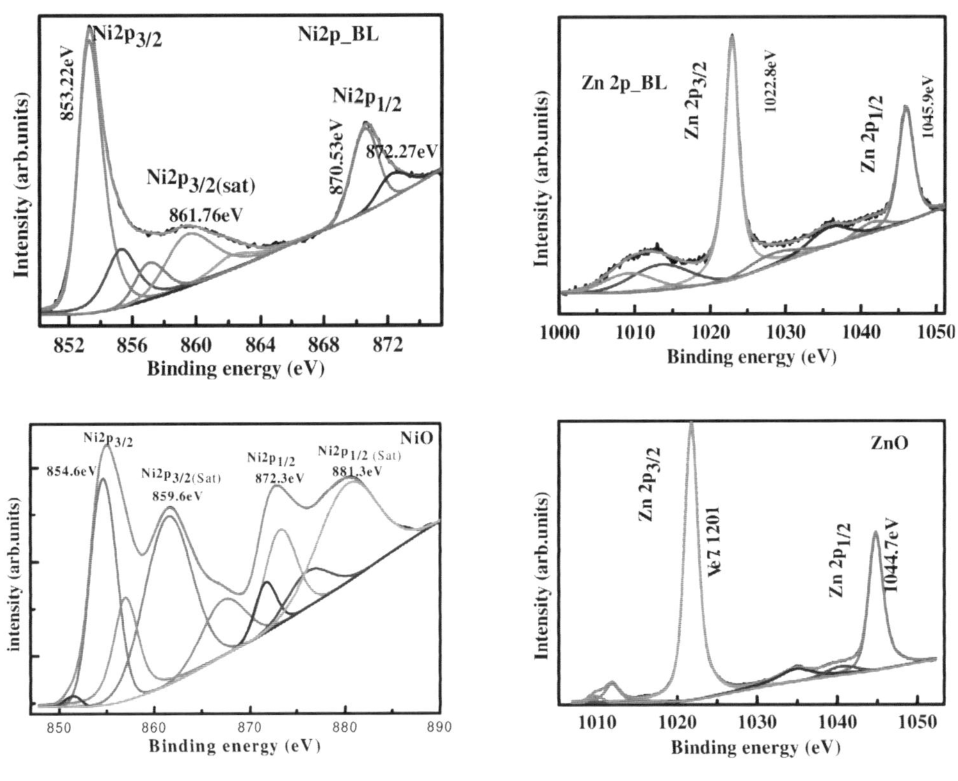


**Figure 3.** Core level spectra of Ni2p and Zn2p in bilayer (ZnO/NiO) and single layer of NiO and ZnO thin films.

## 4.Conclusion

We deposited polycrystalline wurtzite phase ZnO and rock salt NiO thin films by pulse laser deposition. The bilayer film prepared by same technique was amorphous in nature as was evident from its XRD pattern. The amorphous nature occurs due to large strain (28%) arising from different crystal structure of NiO and ZnO. XPS data confirms the presence of Ni2+ and Zn2+ in all films. A shift of $Zn2p_{3/2}$ and $Ni2p_{3/2}$ peaks in bilayer film was observed.

## Acknowledgements

The authors are thankful to Dr. D.M. Phase, Dr. T.Shripathi and Dr. U.P. Deshpandey, of UGC-DAE CSR Indore for providing XRD, XPS and UV-Vis measurements. We are also thankful to MPCST, Bhopal and UGC-DAE CSR, Indore, for providing financial support.

## References

[1] Imada M, Fujimori A and Tokura Y 1998 *Rev. Mod. Phys* **70** 1039–1263.
[2] Janotti A and Vande Walle C G 2009 *Rep. Prog. Phys.* **72** 126501
[3] Fasaki I, Koutoulaki A, Kompitsas M and Charitidis C 2010 *Appl. Surf. Sci.* **25** 429–433.
[4] Tauc J, Grigorovici R and Vancu A 1966 *phys. stat. sol.* **15** 627.
[5] Dwivedi D K, Pathak H P, Shukla N and Kumar A 2014 *Journal of Ovonic* **10** 15–22.
[6] Wagner C D, Riggs W M, Davis L E and Moulder J F 1979 *Hand Book of X-Ray Photo Electron Spectroscopy,* ed. G.E. Mulinberg (Minnesota: Perkin Elmer: Physical Electronics division) pp 80-85.

# Low Temperature Heat Capacity of polycrystalline MnSi

**Ashish Mishra, M Krishnan, Mohan Gangrade, Durgesh Singh, R Venkatesh and V Ganesan**
UGC-DAE Consortium for Scientific Research, Indore-452001, MP, India

E-mail: ashishkm@csr.res.in

**Abstract.** The itinerant helimagnet MnSi has been investigated by heat capacity using relaxation calorimetry in *H*=0. The paramagnetic to helimagnetic transition temperature $T_c$ is found to be 29.6 K. Low temperature heat capacity data in the range 2-18 K is fitted with a usual electronic and phonon contribution. However, an additional magnetic term $T^{3/2}$ needs to be added to witness the role of spin wave excitations in the magnetically ordered state below $T_c$. As well the energy gap on the order $\Delta$=1.29 meV in the spin wave spectrum is evidenced. Small value of magnetic coefficient $\delta$ = 1.66 $mJmol^{-1}K^{-5/2}$ hints the itinerant nature of the magnetism in MnSi. Obtained coefficients gives the Debye temperature $\theta_D$ = 534 K and the Sommerfeld coefficient of specific heat $\gamma$ = 0.0367 $Jmol^{-1}K^{-2}$. Noteworthy observation is the negative trend in the magnetic part of the entropy after $T_c$.

## 1. Introduction

The non centro-symmetric crystal MnSi crystallizes in B20 (space group $P2_13$), lacking inversion symmetry. Its properties are determined by the competing interaction between isotropic exchange, Dzyaloshinskii-Moriya (DM) interaction and a relatively weaker anisotropic exchange interaction. The interest in this system is due to the formation of Skyrmionic structures at low temperatures in a narrow interval of (T,H) phase diagram centred around (~30K, 0.1-0.25T) [1], confirmed by the small angle neutron scattering (SANS) [2] and the Topological Hall effect (THE) [3]. However, below $T_c$ the role of spin excitations is still not very clear. To shed some light on this, heat capacity in the absence of magnetic field, *H*=0 has been performed and the data in low temperatures has been fitted with the addition of magnetic term $T^{3/2}$. Further in this study spin fluctuations [4, 1] above $T_c$ has also been addressed using the magnetic contribution to the entropy.

## 2. Experimental Methods

Polycrystalline MnSi sample was prepared by arc melting technique in high purity argon atmosphere (99.999%) with high quality Mn (alfa aesar 99.9% pure) and Si (alfa aesar 99.9% pure). The stoichiometric compositions were arc melted and the ingot was remelted three to four times to ensure the homogeneity. The obtained ingot was vacuum sealed in a quartz tube and then annealed for 48 hours at 900° C. X-ray diffraction (XRD) study was done using Cu-Kα radiation in a Bruker D8 Advance X-ray Diffractometer for structural characterization. Heat Capacity measurement was performed down to 2 K in Zero field by relaxation calorimetry using 14T/2K QD-PPMS.

## 3. Results and Discussion

The room temperature XRD was performed on the crushed powder sample. Obtained pattern was Rietveld refined using Fullprof software and the results are consistent with B20 ($P2_13$ space group) with lattice parameter a= 4.558 Å [5] as shown in figure 1.

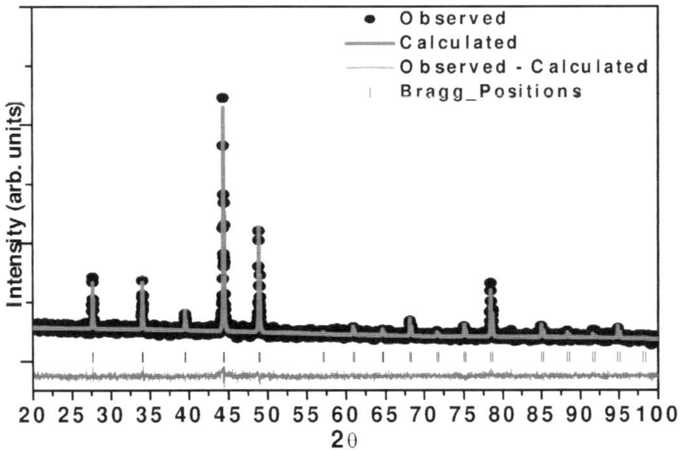


**Figure 1.** Rietveld refined room temperature XRD showing the single phase MnSi.

Figure 2, shows the heat capacity data from 2-45 K, at *H*=0. The transition temperature from paramagnet to helimagnet was found to be $T_c$ = 29.6 K [6, 1]. Observed shoulder after the transition is marked as $T_1$ onset of the feature and $T_2$ end of the shoulder [7].

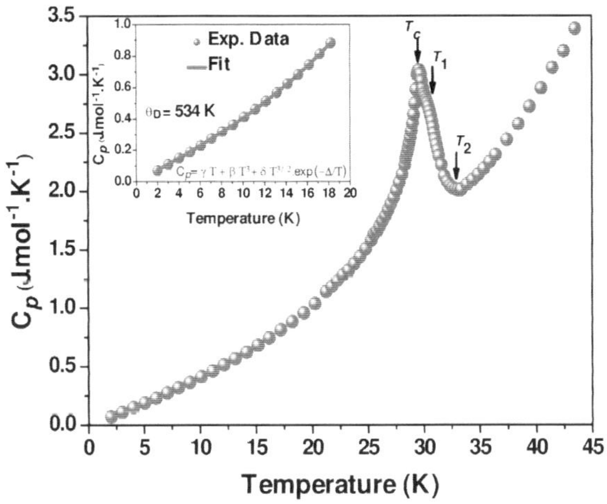


**Figure 2.** Heat Capacity from 2-45 K. $T_c$ is shown by an arrow. $T_1$ and $T_2$ being the beginning and end of the feature (shoulder). Inset: low temperature fit (including the magnetic part) to the heat capacity data in the range 2-18 K.

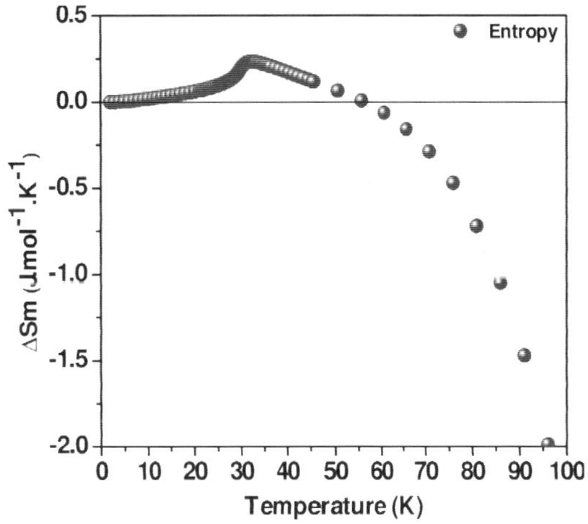


**Figure 3.** Magnetic contribution to the entropy, calculated using $\Delta C_m = C_p - C_{elec} - C_{phonon}$.

Top inset in figure 2, shows the low temperature specific heat data in the range 2-18 K and is fitted with the equation

$$C_p = \gamma T + \beta T^3 + \delta T^{3/2}\exp(-\Delta/T)$$

Where $\gamma$, $\beta$, and $\delta$ are the Sommerfeld, lattice and magnetic contribution coefficients respectively and $\Delta$ being the spin wave gap. A similar approach to fit resistivity at low temperatures using spin gap was incorporated in [8, 9, 10]. As the system enters a magnetically ordered state below $T_c$, a magnetic term $T^{3/2}$ was needed in addition to lattice and electronic contribution to the specific heat. Also the energy gap of the order $\Delta$=1.29 meV is present in the spin wave spectrum, which can be caused by the higher order anisotropic exchange term due to spin-orbit coupling term in the system. Small value of magnetic coefficient $\delta$ = 1.16 $mJmol^{-1}K^{-5/2}$ [11, 12] hints the itinerant nature of the magnetism in MnSi. The Sommerfeld coefficient of specific heat $\gamma$ = 0.0367 $Jmol^{-1}K^{-2}$ and the Debye temperature calculated from the relation $\theta_D = (5\beta/12R\pi^4)^{-1/3}$ found to be $\theta_D$ = 534 K [6], which matches well with the value obtained from the elastic constant [5]. Excluding the magnetic term in the heat capacity; $C_p = \gamma T + \beta T^3$, resulted in different value for Debye temperature $\theta_D$ = 485 K [1]. Figure 3, shows the magnetic contribution to the entropy calculated after subtracting the lattice and electronic part of the heat capacity from the relation, $\Delta C_m = C_p - C_{elec} - C_{phonon}$;

$$\Delta S_m = \int \frac{\Delta C_m(T)}{T} dT$$

It is evident from the figure 3 that entropy follows a down trend just after the transition temperature which further becomes negative after 60 K and is attributed to the spin fluctuations above $T_c$ [4]. The negative contribution to the entropy is predicted theoretically in [13] as a result of mode-mode driven transition. This implies that the system is in minimum energy state and preparing for some minimum energy ground state.

## 4. Conclusions

The present work focuses on the low temperature study of heat capacity to witness the role of spin excitations as well as the presence of a gap in the spectrum below $T_c$. The contribution of magnetic term in the total heat capacity at low temperatures is must. The negative magnetic contribution to the entropy before $T_c$ (from higher temperature side to the lower temperatures), further confirms that the system is preparing to be in minimum energy ground state in paramagnetic region itself, which means an indication of some sort of preparedness of ordering in this region. So, the work insights the role of spin excitations below $T_c$ and a complex spin fluctuation scenario above $T_c$.

**Acknowledgements**

Authors thank to Dr. A. K. Sinha, Director, UGC-DAE CSR, Indore for his kind encouragement, the staff of low temperature laboratory Ms. M M Patidar, Mr. V Gorli and cryogenics laboratory especially Mr. P Saravanan, for their assistance. DST India is thanked for their initial support while establishing the 14T PPMS at CSR Indore.

## References

[1] Samatham S S and Ganesan V 2013 *Phys. Status Solidi* RRL **7** 184–186
[2] Muhlbauer S, Binz B, Jonietz F, Pfleiderer C, Rosch A, Neubauer A, Georgii R and Böni P 2009 *Science* **323** 915
[3] Neubauer A, Pfleiderer C, Binz B, Rosch A, Ritz R, Niklowitz P G and Böni P 2009 *Phys. Rev. Lett.* **102** 186602
[4] Stishov S M, Petrova A E, Khasanov S, Panova G K, Shikov A A, Lashley J C , Wu D and Lograsso T A 2008 *J. Phys.: Condens. Matter* **20** 235222
[5] Bauer A, Garst M and Pfleiderer C 2013 *Phys. Rev. Lett.* **110** 177207

[6] Stishov S M, Petrova A E, Anatoly A S, Thomas A L, Isaev E I, Johansson B and Daemen L L 2010 *Phys. Rev. Lett.* **105** 236403
[7] Murata K K and Doniach S 1972 *Phys. Rev. Lett.* **29** 285
[8] Bhattacharyya A, Deepti J, Ganesan V, Giri S and Majumdar S 2011 *Phys. Rev.* B **84** 184414
[9] Ghimire N J, McGuire M A, Parker D S, Sales B C, Yan J Q, Keppens V, Koehler M, Latture R M and Mandrus D 2012 *Phys. Rev.* B **85** 224405
[10] Samatham S S and Ganesan V 2015 *Phys. Status Solidi* B **252** 1810-1816
[11] Lamago D, Georgii R, Pfleiderer C and Böni P 2006 *Physica B: Condensed Matter* **385** 385-387
[12] Bauer A, Regnat A, Blum C G F, Schonmeyer S G, Pedersen B, Meven M, Wurmehl S, Kunes J and Pfleiderer C 2014 *Phys. Rev.* B **90** 064414
[13] Sidorov V A, Bauer E D, Frederick N A, Jeffries J R, Nakatsuji S, Moreno N O, Thompson J D, Maple M B and Fisk Z 2003 *Phys. Rev.* B **67** 224419

Journal of Physics: Conference Series **755** (2016) 012038 doi:10.1088/1742-6596/755/1/012038

# Coexistence of Kondo effect and Ferromagnetism in $Ce_{1.5}Nd_{1.5}Al$

**Durgesh Singh, Manju Mishra Patidar, Mohan Gangrade and V Ganesan**

UGC DAE Consortium for Scientific Research, Indore- 452001, MP, India

E-mail: durgeshkushinagar@gmail.com

**Abstract.** The possibility for a coexistence of Kondo effect and ferromagnetism in $Ce_{1.5}Nd_{1.5}Al$ has been studied using heat capacity and resistivity. The sample is of polycrystalline in nature. Heat capacity data confirms the heavy fermion behavior of this compound with a Sommerfeld coefficient $\gamma$ = 190mJ/mol-K$^2$ and Debye temperature ~ 180K. An upturn in resistivity and heat capacity observed near ~ 24K is attributed to the paramagnetic to ferromagnetic transition as reported. Kondo like behavior is observed below 10K and vanishes for magnetic fields of the order of 2T. Around 60K a prominent jump or cusp like behaviour is seen and observed to be robust against magnetic fields up to 14T. Heat capacity below 20K is found to be consistent with equation $C_p(T) = \gamma \mathrm{T} + \beta \mathrm{T}^3 + \alpha \mathrm{T}^{3/2} \mathrm{e}^{-\Delta/\mathrm{T}}$ that signifies the presence of a gap in magnon excitation energy of the order of 70K.

## 1. Introduction

Cerium based intermetallic compounds show many properties like Fermi and non-Fermi liquid behavior, heavy fermion behavior, Kondo effect, valence fluctuation, quantum criticality, superconductivity, RKKY interaction and various types of magnetic ground states due to the hybridization between localized 4f electrons and conduction electron band. Quantum critical behavior appeared as a result of competition between RKKY interaction and Kondo Effect [1-5]. Conventionally Kondo effect opposes the magnetic order but there are many cerium and uranium based intermetallic systems where metallic ferromagnetism exists in Kondo Lattice system [5], examples are $CeCrGe_3$ [6], $CeTiGe_3$ [7], UTe [8] and $UCo_{0.5}Sb_2$ [9]. Here in all above mentioned compounds ferromagnetic ordering temperature ($T_C$) is less than Kondo minima temperature ($T_K$). Parent compound $Ce_3Al$ shows Kondo effect, heavy fermion behaviour, antiferromagnetism and first order structural phase transition from hexagonal to monoclinic with space group P $6_3$/m m c [10 -11]. Doping of Nd at the site of Ce in $Ce_3Al$ up to 35%, vanishes the Kondo effect as well as the structural transition [12]. But at 50% Nd doping again shows a Kondo behavior that consistent with the co-existence of ferromagnetic ordering.

## 2. Experimental Detail

Polycrystalline $Ce_{1.5}Nd_{1.5}Al$ sample was prepared by arc melting in high purity argon gas using constituent elements of purity > 99.99%. For homogeneity we have melted the ingots three times and then make rectangular rod for resistivity measurement with a special type of dye. Then rod sealed in a quartz tube at high vacuum with pressure (P) < $10^{-5}$ mbar followed by annealing at 500°C for one week and at 200°C for two weeks. X-ray diffraction has been carried out using Cu-Kα radiation with a Bruker D8 Advance X-ray Diffractometer for phase analysis. Resistivity measurements were carried out down to 2K in magnetic fields 0-14T with conventional four probe method. Heat capacity in 0T field is also reported in support of the ferromagnetic and heavy fermion behaviour and is carried out by a relaxation calorimetry. QD-PPMS system has been used for both of these measurements.

## 3. Results and Discussions

Resistivity of the sample up to 14T and temperatures down to 2K are shown in figure 1. The resistivity curve shows Kondo effect at low temperature as evident by the shallow upturn (inset of figure 1). Around 24K there is a drop in resistivity whose behaviour is attributed to the ferromagnetic nature of transition. An interesting behaviour has also been observed where Kondo effect appears below 10K in an otherwise ferromagnetic state. As in the nonmagnetic metal the action of an applied magnetic field on the resistance is associated with the curving of the electronic orbits, and results in an increase of electrical resistance. If there are magnetic impurities in a metal, then the applied magnetic field will polarize the spins of impurity atoms. Thus the direction of spins become fixed and the probability of spin-flip scattering declines i.e. breakdown of Kondo effect. As a result, the resistivity falls off in field. Since this compound is dense Kondo system hence Kondo singlet breakdown is happening here also and resistivity is decreases with magnetic field. A more interesting feature at 60K has been observed in 0T resistivity, which is robust against magnetic field. The dispersion in resistivity has also been visualized upon the application of magnetic fields. Above 60K all the curves coincide in the fields. Even though at this composition of Nd, the structural transition is ruled out, as evident from the nil hysteresis seen, this seems to be a transition of unknown origin.

Magnetoresistance (MR) vs temperature curve up to 14T is shown in figure 2. Maximum MR is 18% at 2K and 14T. There is a deep in MR at ~24K which vanishes after 6T indicated by a red arrow in figure 2. Above 60K all the MR curves are coincide means magnetic response of the sample is only below 60K. Therefore upturn at the 60K may be a magnetic in nature.

Heat capacity curve of the sample is shown in figure 3. Inset of figure 3 shows the close interval points measurement around 25K. Inset curve shows an upturn near 25K resulting in a magnetic transition from PM to FM as reported [12]. This compound has hexagonal structure up to lowest measured

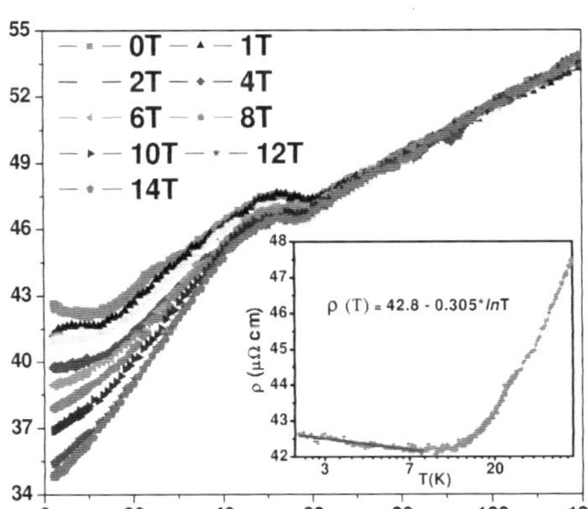


**Figure 1.** ρ vs. T in magnetic field up to 14T and temperature range 2-120K. Inset shows the 0T resistivity vs. lnT below 50K and blue line is fitted with equation ρ = a – b*lnT

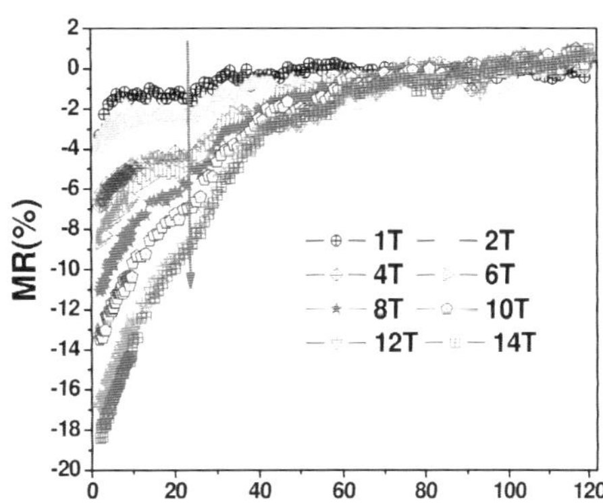


**Figure 2.** MR (%) vs. T up to 14T. Red arrow is pointing the dip in MR around 24K attributed to the FM ordering.

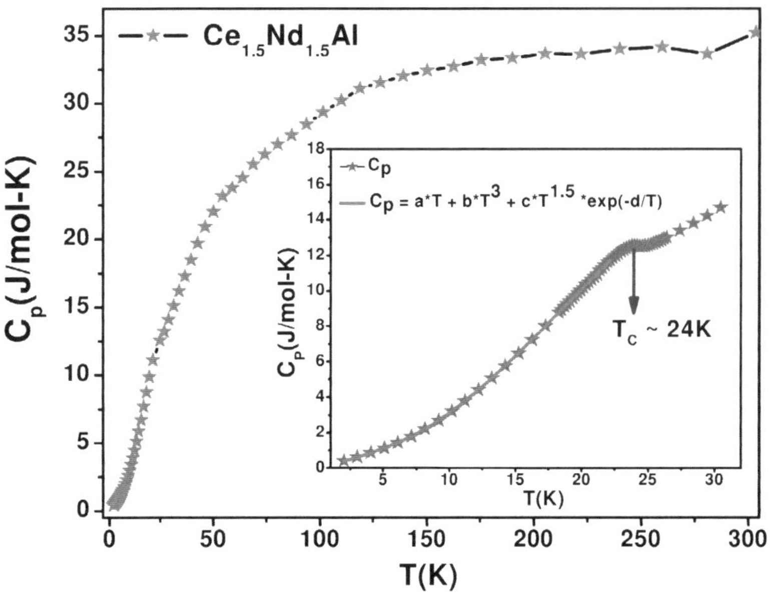


**Figure 3.** Heat capacity (Cp) vs. Temperature in 0T. Inset shows the $C_p$ curve below 30K and red line is fitting curve with equation (1).

temperature. For a ferromagnetic metal with a noncubic crystal structure; the heat capacity well below $T_C$ is generally expressed as

$$C_p(T) = \gamma T + \beta T^3 + \alpha T^{3/2} e^{-\Delta/T} \tag{1}$$

Where first and second terms represent the electronic and phonon contributions in heat capacity respectively. The third term represents the conventional form of a magnon contribution, where $\Delta$ is an anisotropy gap in magnon energy. We have fitted the heat capacity data up to 20K (inset of figure 3) with the above given equation and we have found $\gamma$ = 190.8mJ/mol-$K^2$, $\beta$ = 1.24 mJ/mol $K^4$, $\alpha = -1.4 \times 10^3$ mJ/mol –$K^{2.5}$ and $\Delta$ ~ 70K. From the fitting with equation (1) we conclude that system is moderate heavy fermion with Debye temperature ~ 180K.

A small and broad upturn in heat capacity curve observed near ~ 24K may represent a short range magnetic correlation. Ce has a strong Kondo nature while Nd has a ferromagnetic tendency. Hence for temperatures $T_K < T_C$ (= 25K) there may be a hybridization between conduction electron band of Ce and the uncompensated excess Nd magnetic moments, giving rise to a possible FM correlations.

## 4. Conclusions

A good quality polycrystalline sample of $Ce_{1.5}Nd_{1.5}Al$ was prepared using arc melting in Argon atmosphere. XRD pattern shows the sample has single phase. Resistivity and specific heat measurements have been done by the QD 14T/2K PPMS. Heat capacity curve is showing heavy fermion behavior with $\gamma$ = 190mJ/mol-$K^2$ with Debye temperature ($\Theta_D$) =180K. The Magnetic moment of Nd and of Ce in this compound is (3.62*1.5 = 5.43$\mu_B$) and (2.54*1.5= 3.81$\mu_B$) respectively. Therefore magnetic moment in the compound is not fully compensated and hence a small amount of Nd magnetic moment is available for magnetic ordering. MR vs T plot is showing the negative behavior below 40K. This is a clear signature of FM ordering. Kondo effect is vanishing in the magnetic field after 2T. The deep in MR around 25K is a direct evidence of ferromagnetic nature of the material.

**Acknowledgments**
Authors thank to Dr. A. K. Sinha, Director, UGC-DAE CSR, Indore for his kind encouragement. Authors would also like to give their gratitude towards the staff of low temperature especially Dr. R. Venkatesh and cryogenics laboratory especially Mr. P. Saravanan for technical support and supplying cryoliquids. DST India is thanked for their initial support while establishing the 14T PPMS at CSR Indore.

**References**

[1] Stewart G R 1984 *Rev. Mod. Phys.* **56** 755
[2] Hewson, A C 1997 *The Kondo Problem to Heavy Fermions*, Cambridge, UK, Cambridge University Press, ISBN 0521599474
[3] Schofield A J 1999 *Contemporary Physics* **40** 2 95
[4] Gegenwart P, Si Q and Steglich F 2008 *Nature Phys.* **4** 186
[5] Yamamotoa S J and Si Q 2010 *PNAS* **107** 36 115704
[6] Debarchan D, Gruner T, Pfau H, Paramanik U B, Burkardt U 2014 *J. Phys. Condens. Matter* **26** 106001
[7] W. Kittler, V. Fritsch, F. Weber, G. Fischer, D. Lamago, G. André, and H. v. Löhneysen 2013 *Phys. Rev. B* **88** 165123
[8] Schoenes J, Frick B. and Vogt O 1984 *Phys. Rev. B* **30** 6578
[9] Tran V H, Troc R, Bukowski Z, Badurski D and Sulkowski C 2005 *Phys. Rev. B* **71** 094428
[10] Sakurai J, Murashita Y, Aoki Y, Fujita T, Takabatake T and Fujii H 1989 *J. Phys. Soc. Jpn.* **58** 4078
[11] Chen Y Y, Lawrence J M, Thompson J D, and Willis J O 1989 *Phys. Rev. B* **40** 10766
[12] Lingwei L, Yamagata R, Nishimura K and Yamaoka H 2009 *Phys. Rev. B* **80** 134429

# Low temperature dielectric relaxation of poly (L-lactic acid) (PLLA) by Thermally Stimulated Depolarization Current

**Manju Mishra Patidar[1*], Deepti Jain[2], R Nath[1] and V Ganesan[1]**

[1] UGC-DAE Consortium for Scientific Research, Indore, MP - 452001, India
[2]MITS, Gole ka Mandir, Gwalior, MP-474005, India

[*]E-mail: manjumishra.iuc@gmail.com

**Abstract.** Poly (L-lactic acid) (PLLA) is a biodegradable and biocompatible polyester that can be produced by renewable resources, like corn. Being non-toxic to human body, PLLA is used in biomedical applications, like surgical sutures, bone fixation devices, or controlled drug delivery. Besides its application studies, very few experiments have been done to study its dielectric relaxation in the low temperature region. Keeping this in mind we have performed a low temperature thermally stimulated depolarization current (TSDC) studies over the temperature range of 80K-400K to understand the relaxation phenomena of PLLA. We could observe a multi modal broad relaxation of small but significant intensity at low temperatures while a sharp and high intense peak around glass transition temperature, $T_g$~ 333K, of PLLA has appeared. The fine structure of the low temperature TSDC peak may be attributed to the spherulites formation of crystallite regions inter twinned with the polymer as seen in AFM and appear to be produced due to an isothermal crystallization process. XRD analysis also confirms the semicrystalline nature of the PLLA film.

## 1. Introduction

Chemical structure of Poly (L-lactic acid) (PLLA) can be given by the formula $(C_3H_4O_2)_n$. Among the well known biodegradable polyesters, PLLA has attracted much attention in medical and industrial field due to its combined biocompatibility and degradation character. Being a prominent candidate of polymer industry, we could find only a few measurements on its low temperature dielectric relaxation [1]. Present paper attempt to probe it by employing thermally stimulated depolarization current (TSDC) technique. It measures the thermally stimulated release of the polarization frozen in during a previous application of an electric field during rapid cooling. During subsequent TSDC, the aligned dipoles randomly disorient at a rate proportional to the number of dipoles still aligned. TSDC spectra show peaks corresponding to different relaxations. Shape and location of peak are characteristic of mechanism by which polymers store their charges [2, 3].

## 2. Experimental Techniques

Solvent cast PLLA film of 15 mm diameter and 0.11 mm thickness is used as a sandwich between metallic copper electrodes for TSDC measurements using a homemade but versatile setup [4]. The PLLA film has been poled at temperature, $T_P$=350K. $T_p$ is chosen well above the glass transition temperature, $T_g$~ 333K of PLLA, so that maximum number of dipoles are available that aligned in the direction of poling field, $E_P$= 2.7MV/m. In the presence of $E_p$, sample has been cooled down to the freezing temperature, $T_f$ ~80K, which restrict the disorientation of aligned dipoles. Upon reaching the desired freezing temperature film has been short circuited for 10min to remove effect of stray surface charge. Global TSDC spectrum of PLLA has been recorded with the Keithley 617 electrometer over the temperature range of 80K-400K by maintaining the linear heating rate of 5 K/min using Lakeshore DRC93CA in the vacuum of $10^{-3}$ mbar. X-ray diffraction (XRD) and Atomic Force Microscopy (AFM) measurements have been used for the structural and morphological studies of the sample.

## 3. Result and Discussion

XRD pattern (Figure 1) of PLLA is showing the semi crystalline (~ 25% of crystallinity) nature of the film. Broad background is due to the amorphous contribution, while sharp peaks reflect the crystallized nature of the film. XRD analysis reveals that the film is in orthorhombic in nature and carries $P2_12_12_1$ space group symmetry. Lattice parameters calculated for the film are a=10.584 Å, b=6.247 Å and c=28.806 Å.

The morphology of freshly prepared PLLA film was studied by AFM Model Nanoscope E in contact mode (figure 2). AFM images clearly show emerged crystallites of PLLA from an amorphous background region due to iso-thermal crystallization and has a characteristic spherulite structures similar to those observed through a polarized optical microscope by Yasuniwa et al [5]. The AFM results support the XRD results, where the semicrystalline nature of the film has been depicted.

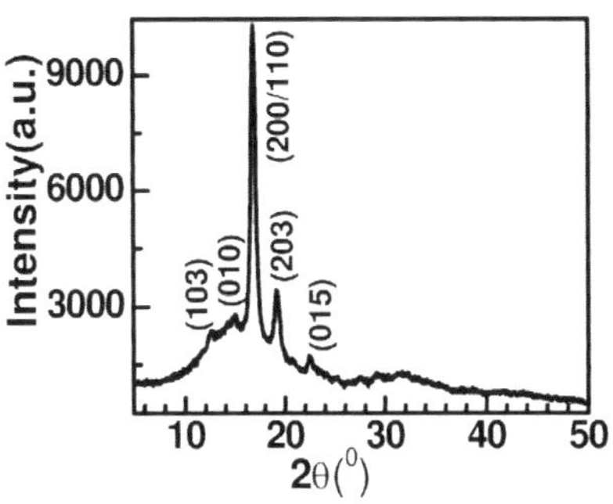


**Figure 1.** XRD shows the semicrystalline nature of PLLA film.

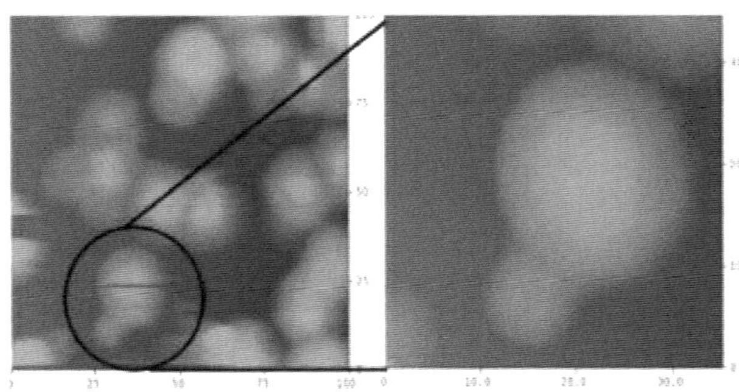

**Figure 2.** AFM scan shows the spherulite kind of crystallised region in the film.

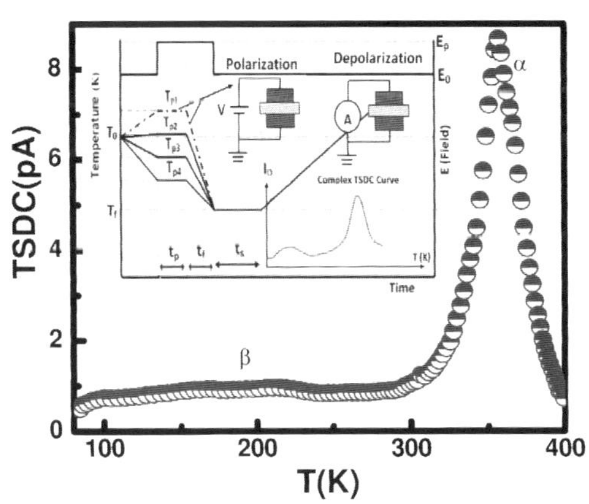


**Figure 3.** Global TSDC spectrum of PLLA poled at $T_p$=350K and $E_p$=2.7MV/m. Inset shows the protocol opted for the measurement.

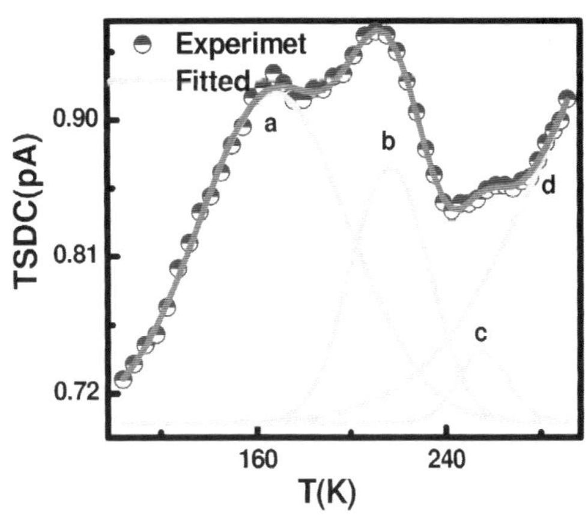


**Figure 4.** Low temperature β relaxation of the TSDC spectra of PLLA along with constituent peaks a, b, c and d.

Figure 3 shows the global TSDC spectrum of PLLA. Inset of the figure3 representing the protocol opted for the TSDC measurement followed by $T_{p1}$ poling route. One can use the same protocol but followed by different poling temperature route according to the systems demand and the range of relaxation desired to be studied. Two prominent features can easily be identified in the global TSDC curve. A broad and low intensity peak has been observed at low temperature which covers the temperature range of 100-290K called as β peak where as a sharp and high intense peak , named as α, appears in the range 300K-400K. α peak gives an insight into the relaxation around the glass transition temperature $T_g$~ 333K and has one order higher intensity than the low temperature β peak. The high temperature relaxation peak around $T_g$ is considered due to the segmental mode of relaxation. During the TSDC process, the aligned dipoles will randomly disorient at a rate proportional to the number of dipoles still aligned.

Therefore in the primitive form, the TSDC is expected to follow the Debye rate equation

$$dP(t)/dt + P(t)/\tau(T) = 0, \qquad (1)$$

Where $P(t)$ is the polarization at time t and $\tau(T)$ is the relaxation time.

Various parameters, as tabulated in table 1, like activation energy $E_g$ and relaxation time $\tau_0$ of the PLLA polymer have been calculated by adopting BFG (Bucci-Fieschi-Guidi) or initial rise method applicable for Debye like single relaxations of dipoles [2] and using single Arrhenius relaxation time equation

$$\tau = \tau_0 \exp(E_g/kT) \qquad (2)$$

Figure 4 shows the multiple constituent peaks fitted in low temperature part of TSDC spectra, which is helpful to understand more about the low temperature relaxation behavior of the macromolecules present in the PLLA polymer. One can see that a broad relaxation peak is composition of a number of small elementary peaks. These low temperature relaxations seem to be due to the small amplitude local motion in the amorphous phase. Parameters which are shown in table 2 are helpful to understand more about the low temperature relaxation behavior of the macromolecules present in the PLLA polymer. Discussion on broad peak'd' is beyond the scope the present work due to its unusually broad nature.

The fine structures of the low temperature TSDC β peak may be attributed to the spherulites formation of crystallite regions , already observed in XRD and AFM results , as they are inter twinned with the

**Table 1.** Parameters of high temperature α-peak

| $T_m$ (K) | $I_m$ (A) | $P_0$ ($Cm^{-2}$) | $J_0$ ($Am^{-2}$) | $E_g$ (eV) | $\tau_0$ (ns) |
|---|---|---|---|---|---|
| 358 | 8.7E-12 | 8.3E-10 | 1.2E-08 | 0.49 | 0.89 |

**Table 2.** Parameters of the low temperature β-relaxation peaks

| Peak | $T_m$ (K) | $I_m$ (A) | $P_0$ ($Cm^{-2}$) | $J_0$ ($Am^{-2}$) | $E_g$ (meV) | $\tau_0$ (ms) |
|---|---|---|---|---|---|---|
| a | 167.6 | 9.2E-13 | 6.6E-11 | 2E-09 | 7.58 | 34.2 |
| b | 216.6 | 8.7E-13 | 2.8E-11 | 4E-09 | 22.28 | 7.0 |
| c | 255.8 | 7.5E-13 | 5.1E-11 | 2E-09 | 14.33 | 2.6 |
| d | 288.4 | 9.0 E-13 | 7.1E-10 | 3E-09 | 22.55 | 235.7 |

polymer and are expected to yield broken segments of varying lengths and degrees of crystallization. Hence sharpness appears in an otherwise broad β relaxation, which is a slow process called 'Reptation' that resembles a reptile motion of side chains of the polymers [6].

## 4. Conclusion

Low temperature dielectric relaxation of PLLA film has been studied by TSDC technique. Two features named β (100-290K) and α (300K-400K) are clearly observed in global TSDC spectrum. Various parameters are also calculated for β as well as α peak. The prominent change in the parameters signifies the different origin responsible for the β and α relaxation. XRD and AFM analysis shows the semicrystalline nature of the film. The fine structures of the β peak may be recognized due to the crystallization of the film.

## Acknowledgment

The authors are thankful to Dr. A. K. Sinha, Director, UGC-DAE CSR, Indore for his encouragement and members of Low Temperature and Cryogenic laboratories for their help in cryogens. Authors thank Dr. M. Gupta for XRD measurements. The authors are also grateful to the DST, Govt. of India, New Delhi for providing research grant in the form of major research project SR/S2/CMP-005/2009 with fellowship to MMP and honorarium to RN.

## References

[1] Laredo E, Prutsky N, Bello A, Grimau M, Castillo R, Muller A J and Dubois Ph 2007 *Eur. Phys. J. E.* **23** 295
[2] Turnhout J V 1975 *Thermally Stimulated Discharges of Polymer Electrets* (Amsterdam: Elsevier)
[3] Sessler G M 1980 *Electrets* (Springer-Verlag Berlin Heidelberg New York)
[4] Jain D, Sharath Chandra L S, Nath R and Ganesan V 2012 *Meas. Sci. Technol.* **23** 025603
[5] Yasuniwa M, Tsubakihara S, Lura K, Ono Y, Dan Y and Takahashi K 2006 *Polymers* **47** 7554
[6] Das-Gupta DK, 2001 *J. Electrostat* **51-52** 159-66

# Structural optical and electronic properties of Fe and Ga doped ZnO thin films grown using pulsed laser deposition technique

**Karmvir Singh[1], D K Shukla[2], S Majid[3], R Dhar[1], R J Choudhary[2] and D M Phase[2]**
[1]Department of Applied Physics, Guru Jambheshwar University of Science and Technology, Hisar, Haryana, India
[2]UGC-DAE Consortium for Scientific Research, Indore-452001, India
[3]Department of Physics, Aligarh Muslim University, Aligarh 202002, India

[†]Corresponding author, E-mail: karamvirnehra@gmail.com

**Abstract.** Band gap engineering in ZnO thin films have been subject of intensive studies. The thin films of 2 wt % Fe and 2 wt % Ga doped ZnO and undoped ZnO were deposited on glass substrate by pulse laser deposition technique. Structural, optical and electronic structure properties of these thin films were investigated by X-Ray diffraction (XRD), UV-Vis spectroscopy and X-ray absorption spectroscopy (XAS), respectively. XRD studies show that all the thin films are highly oriented along the *c*-axis and maintain the wurtzite structure. Out of plane lattice parameter in Ga doped is smaller while in Fe doped is larger, compared to undoped ZnO. The band gaps of doped films have been found to increase due to doping of the Ga and Fe ions. XAS studies across O K edges of doped thin films show that the conduction band edge structure probed via oxygen 1*s* to 2*p* transitions have modified significantly in Ga doped sample.

## 1. Introduction

ZnO is a very promising material for semiconductor device applications and for the understanding of semiconductor Physics [1]. ZnO has large direct wide bandgap in the near-UV spectral region. With this property, ZnO is considered as potent applicant in optoelectronics in the blue/UV region, including light-emitting diodes, laser diodes and photodetectors [2]. Band gap engineering plays an important role in the modification of the optical properties of ZnO. Recently, doping has been found an effective method for bandgap regulation [3].
On the other hand, current technological applications of ZnO such as varistors, transparent conductive electrodes for solar cells, piezoelectric devices and gas sensors finds the use of ZnO polycrystalline thin films on glass substrate. So, the effect of doping in ZnO on glass substrate has become a topic of numerous scientific investigations.

Various techniques have been established to deposit the thin films such sol gel, chemical vapor deposition (CVD), molecular beam epitaxy (MBE), metal-organic chemical vapor deposition (MOCVD), R.F sputtering and PLD techniques [3]. Among all of these PLD technique is considered to be one of the best techniques to grow the thin film because of its excellent stoichiometric transfer of the material from target to substrate. We have successfully grown the pure and doped ZnO (2 wt % Fe (FZO) and 2 wt % Ga (GZO)) on glass substrate using pulsed laser deposition technique. Thin films were characterized by X-ray diffraction (XRD), UV visible and X-ray absorption (XAS) techniques to study the structural, optical and electronic properties of thin films respectively.

## 2. Experimental details

Pure ZnO and doped ZnO ($Zn_{0.98}Ga_{0.02}O$, $Zn_{0.98}Fe_{0.02}O$) thin films were deposited using by pulsed laser deposition technique on the glass substrate. Substrate was ultrasonically cleaned with acetone and methanol prior to deposition. Pure and doped ZnO were used as the target materials. KrF eximer laser ($\lambda$=248nm) with the repetition rate of 10 Hz and laser pulse energy of 220 mJ was focussed on highly compressed targets. During deposition, oxygen partial pressure was maintained at 1 mTorr and the substrate temperature was maintained at 450°C. Prior to deposition vaccum base pressure of $1\text{x}10^{-6}$ Torr was achieved.. The target to substrate distance was maintained at 5cm.The thickness of thin films as calculated by Stylus profilometer came out to be ~ 100 nm. X-ray diffraction (XRD) measurements were performed by Bruker D8 Advance diffractometer using Cu Ka radiation of wavelength 1.54 Å. The optical band gap energies of the samples were calculated using UV-vis-NIR spectrophotometer from JASCO V-550. Room temperature soft X-ray absorption spectroscopy across O K edge was carried out at the beam line BL-01 (soft X-ray absorption spectroscopy beamline) at Indus-2 Raja Ramana centre for Advanced Technologies Indore. SXAS measurements were performed in total electron yield mode (TEY). Energy resolution during SXAS measurements at oxygen K edge energy was better than ~ 250 meV energy.

## 3. Results and discussion

Figure 1 shows the XRD patterns of the ZnO, GZO and FZO thin films. The ZnO, FZO and GZO thin films shows a strong reflection from (0 0 2) plane corresponding to wurtzite structure of ZnO [4]. XRD patterns confirm the crystalline nature, single phase and highly oriented growth of thin films along the *c*-axis. The [0 0 2] reflection can be used to calculate the grain size of both pure and doped ZnO thin films using Deby Scherrer formula explained below:

$$D = \frac{0.9\lambda}{\beta\cos\theta} \quad (1)$$

Where $\lambda$ = 1.54 Å is the X-ray wavelength, $\beta$ is the full width half maximum (FWHM) in radian and $\theta$ is the Bragg's angle. The calculated values of grain size are 17, 29 and 8 nm for pure ZnO, GZO and FZO respectively. Ga doping has been found to increase the crystallinity, while Fe doping decreases the crystallinity. With Ga doping *c*-lattice parameter decreases slightly, while Fe doping causes increase in c-lattice parameter as observed from the position of (0 0 2) reflection.

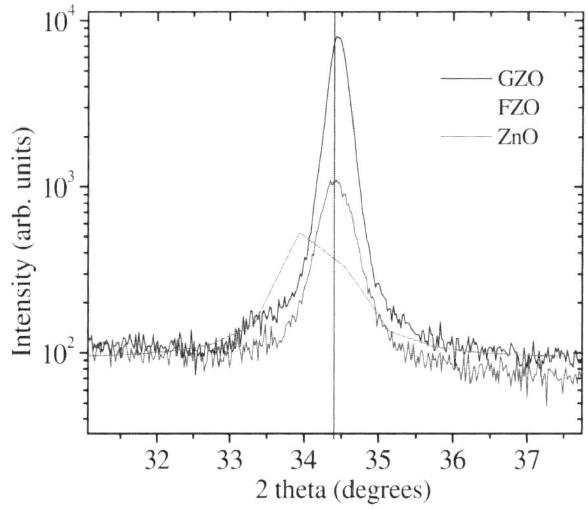


Figure 1. X-ray diffraction pattern of pure ZnO and doped ZnO (FZO and GZO) thin films.

Figure 2 (a) represents the optical transmittance spectra of PLD grown pure ZnO, FZO and GZO thin films. The transmittance is measured as a function of wavelength in the range 300–800 nm. The transmittance is more than 90% in the visible range and starts to decrease as the UV visible range (300-400 nm) approaches. The optical band gap energy was calculated with from Tauc plots (shown in Fig 2(b)) with the help of Tauc's equation [5] read as:

$$(\alpha h\nu)^2 = A(h\nu - E_g) \quad (2)$$

Band gap energies of pure ZnO, FZO and GZO are calculated as 3.24 eV, 3.27 eV and 3.5 eV, respectively. The band energy for FZO thin film does not change much from the pure ZnO, while there is a significant increase in the band gap of GZO thin film. This blue shift in the band gap can be explained in terms of an increase of the carrier concentration which blocks the lowest states in the conduction band, this phenomenon is called Burstein–Moss shift [5].

Figure 3 Figure shows the O K edge XAS in Fe doped and Ga doped ZnO samples, deposited on glass substrate. Different features of O K-edge XAS spectra are marked as **a** (530-535 eV), **b** (~537 eV), c & **d** (~540-550 eV), and **e** (~557 eV). The energy region of ~530-539 eV forms the bottom of the conduction band and is attributed to O 2p-Zn 4s/Fe 3d/Ga 4s hybridized states. The features in between 540 to 550 eV can be assigned to O 2p hybridization with Zn 4p/Fe 4sp/ Ga 4p states and features above 550 eV arise due to the hybridization of O 2p with extended Zn orbitals [6]. Ga doping

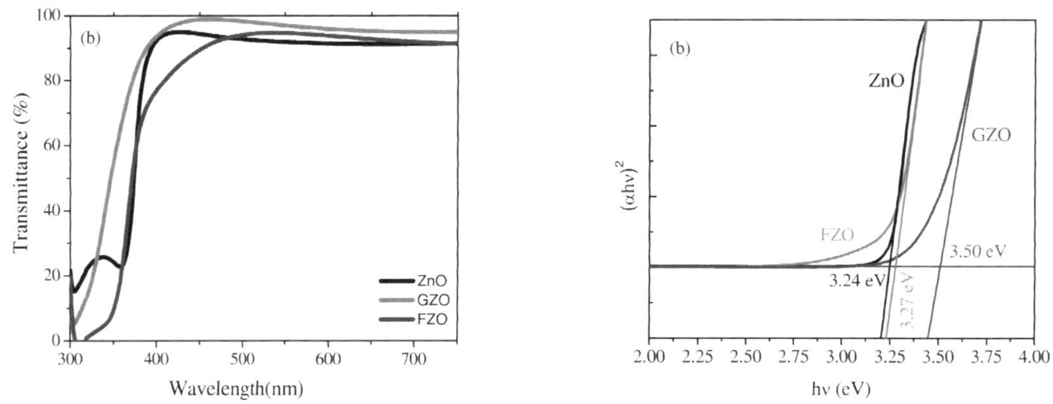


**Figure 2.** (a) UV visible spectra of pure ZnO, FZO and GZO thin films. (b) Tauc plot of Pure ZnO, FZO and GZO thin films.

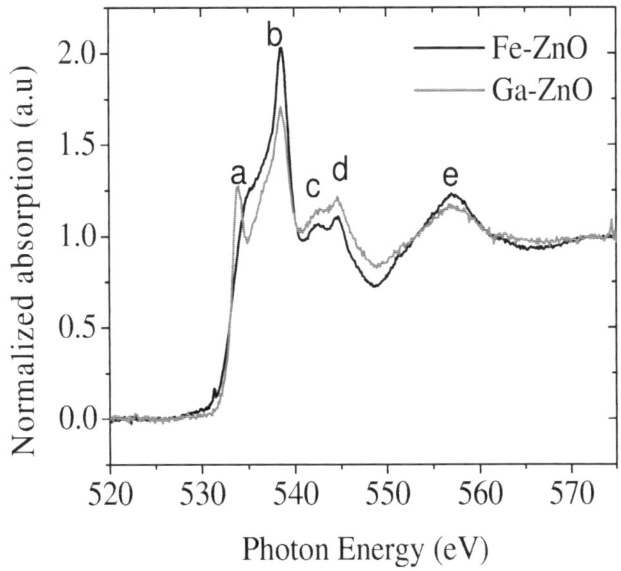


**Figure 3.** O K-edge X-ray absorption spectrum of pure and doped ZnO (FZO and GZO)

has significantly modified the bottom of conduction band in Ga doped ZnO sample. Additional localized feature (a) observed in Ga doped sample at the conduction band edge may be related to increased band gap/ Burstein–Moss shift.

## 4. Conclusion

XRD studies show that all the thin films are highly oriented along the *c*-axis and maintain the wurtzite structure. However, out of plane lattice parameter in Ga doped is smaller while in Fe doped is larger, compared to undoped ZnO. The band gaps of doped films have been found to increase due to doping of the Ga and Fe ions. XAS studies across O K edges of doped thin films show that the conduction band edge structure probed via oxygen 1*s* to 2*p* transitions has modified significantly in Ga doped sample.

**Acknowledgement**
K. Singh greatly acknowledges UGC-DAE CSR, Indore for providing the experimental facilities to carry out this research work.

## References

[1] Ozg ur U, Alivov Y I, Liu C, Teke A, Reshchikov M A, Dogan S, Avrutin V, Cho S J and Morkoc H 2005 *J. Appl. Phys.* **98** 041301
[2] Jagadish C and Pearton S J (ed) 2006 *Zinc Oxide Bulk, Thin Films, and Nanostructures* (New York: Elsevier)
[3] Janotti A and Walle C G V 2009 *Rep. Prog. Phys.* **72** 126501
[4] Gandhi V, Ganesan V R Syedahamed H H A and Thaiyan M 2014 *J. Phys. Chem. C* **118** 9715
[5] Zhou H et.al. 2015 *J. Mater. Chem. C* **3**, 5089
[6] Devi V, Kumar M, Kumar R, Singh A and Joshi B C 2015 *J. Phys. D: Appl. Phys.* **48** 335103

# Exafs and comparative studies of copper (II) complexes

**A Mishra[1], S Mishra[2]**
[1] School of physics, Devi Ahilya University, Khandwa Road, Indore 452001 (India)

Email: amishra1960@yahoo.co.in,

**Abstract.** The present paper deals with the synthesis of transition metal Schiff base complexes of copper by chemical root method. The synthesized metal complexes were characterized by Extended X-ray absorption fine structure (EXAFS) is a technique that has been used for determining the metal ligand bond length using conventional X-ray source and also by EXAFS analysis using IFEFFIT programming. Bond lengths determined from these data analysis methods are compared with the bond lengths obtained from LSS, Levy's and Lytle's methods. Bond lengths of the complexes are in good agreement obtained by theoretical and experimental method.

## 1. Introduction

The transition metal Schiff base complexes of copper have been playing an important role in the development of coordination chemistry. Schiff base metal complexes have been widely studied, because of their antifungal and biological applications. [1]. In the present studies we have taken Series of copper (II) complexes thiosemicarbazide as ligand have been taken.

The structure extending from 40 to 1000eV from the edge is called extended X-ray absorption fine structure (EXAFS) and it gives information about the local structure of the complexes. The aim of the present investigation is to measure the spectra of copper complex, and to compare first shell inter atomic distance using the programmed called IFEFFIT.

## 2. Material synthesis

### Preparation of schiff base ligand

Standard method was adopted for the preparation of the complexes. The series of the copper (II) complexes with thiosemicarbazide as ligand is shown below.

## 3. Experimental technique

In the present thesis, X-ray absorption spectra at the K-edge have been recorded for series of copper(II) complexes using the dispersive EXAFS beamline (BL-8) at Indus-2 synchrotron at Raja Ramanna Centre for Advanced Technology (RRCAT), Indore.

After this process, the scanning of the X-ray films was completed on Carl-Ziess microdensitometer coupled with computer to convert the data into IFEFFIT. The FEFF series of programs for the calculation of X-ray absorption spectra has had a transformative impact on EXAFS analysis because of its accuracy, flexibility, and portability. Its primary use has been in supporting a path by path analysis of experimental data using auxiliary programs such as IFEFFIT, Artemis, Six Pack, etc. In this paper alternative strategies for XAFS analysis that combine FEFF are described[2].

## 4. Results and discussion

### Exafs analysis

The bond lengths of copper complexes were calculated using IFEFFIT method and compared with LSS, Levy's, lytle's methods [3, 4, 5]. According to LSS method, the bond lengths can be determine using the slope of n vs k plot. The phase parameter $\alpha_1$and $\beta_1$, the metal ligand bond length $R_1$ have been estimated with expression

$$(1/2+n)\pi=2k\,(R_1-\alpha_1)+2\,\beta_1-\pi$$

The EXAFS curves are shown in figure1.1a Using the Levy's method, the bond length has been obtained from the expression

$$r=(151/\Delta E)^{1/2}$$

where, $\Delta E$ is the energy difference between first maxima and first minim [7]. Lytle method is modification of LSS method and the bond length is given by the expression

$$R=(37.60/M)^{1/2}$$

where M is slope between E vs Q plots. In FEFFIT programming the EXAFS data were analyzed by the standard procedure. The pre edge background absorption was subtracted to yield the atomic absorption spectrum of the atom of interest background above the edge was removed from the spectrum using spline fit. The data converted to energy space and then to k space. A Fourier transform is applied to these data to convert into the r-space . The bond lengths of copper complexes calculated by IFEFFIT programming is compared with LSS, Lytle and Levy's methods as depicted in table2.

The theoretical analysis of EXAFS of copper complexes having thiosemicarbazide as ligand has been done and compared with experimental values. The results of the average values of metal ligand bond length are reported in table 2.

**Table 1.** Series of copper (II) complexes thiosemicarbazide as ligand.

| S. N. | Name | Molecular Formula | Abbreviation |
|---|---|---|---|
| 1 | Cu[(2,5-dimethoxy-2-nitrobenzylidene)] thiosemicarbazide | C20H22Cl2 CuN8O8S2 | Cu[2,5-dimethoxy] |
| 2 | Cu[ (3,5-dimethoxy-2-nitrobenzylidene)] thiosemicarbazide | C20H22Cl2 CuN8O8S2 | Cu[3,5- dimethoxy] |
| 3 | Cu[ (4,5-dimethoxy-2-nitrobenzylidene) ]thiosemicarbazide | C20H22Cl2 CuN8O8S2 | Cu[4,5- dimethoxy] |

**Table 2.** The average values of metal-ligand bond length in ( $\overset{0}{A}$ )$^{-1}$ for copper (II) complexes, thiosemicarbazide as ligand.

| S.N. | Abbreviations | $R_{Lavy}$ | $R_{Lss}$ | $R_{Lytel}$ | $R_{IFEFFIT}$ |
|---|---|---|---|---|---|
| 1 | Cu[2,5-dimethoxy] | 1.2 | 1.2 | 1.4 | 1.5 |
| 2 | Cu[3,5- dimethoxy] | 1.8 | 1.7 | 1.9 | 1.1 |
| 3 | Cu[4,5- dimethoxy] | 1.5 | 1.2 | 1.4 | 1.6 |

$R_{Levy's}$ : *Bond Length by Levy's method*

$R_{Lss}$ : *Bond Length by LSS method*

$R_{Lytel}$ : *Bond Length by Lytel method*

$R_{IFEFFIT}$ : *Bond Length by IFEFFIT*

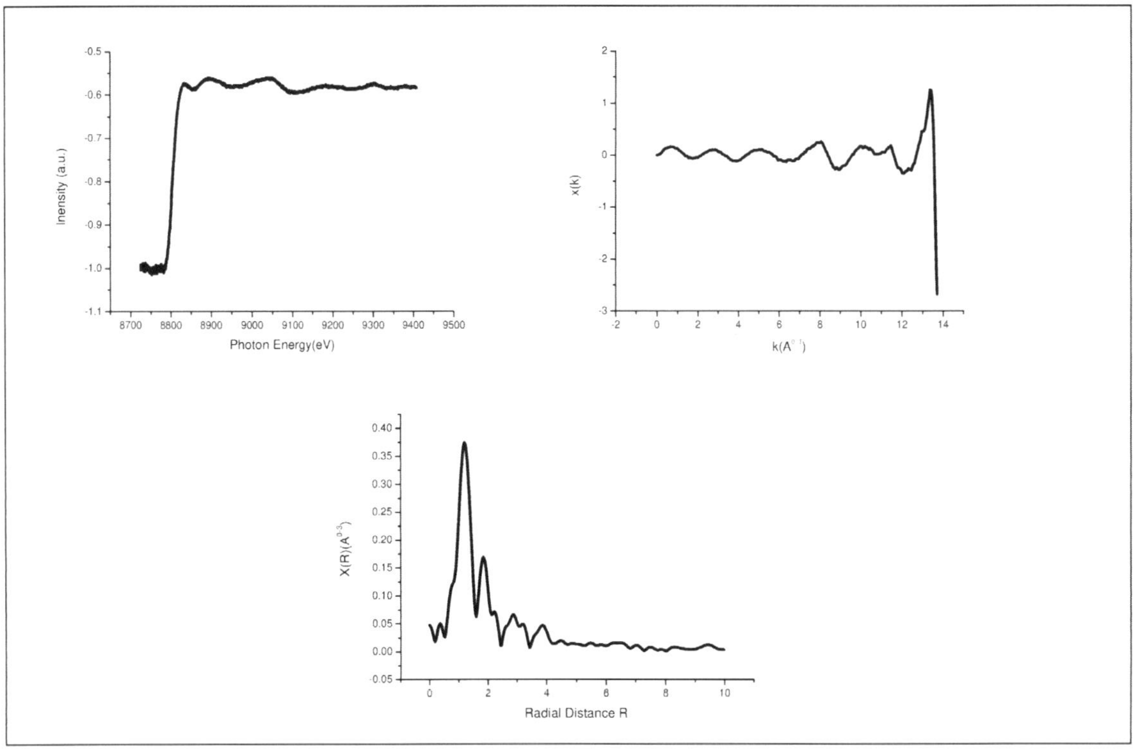


**Figure 1.** EXAFS curve of Cu[2,5-dimethoxy]

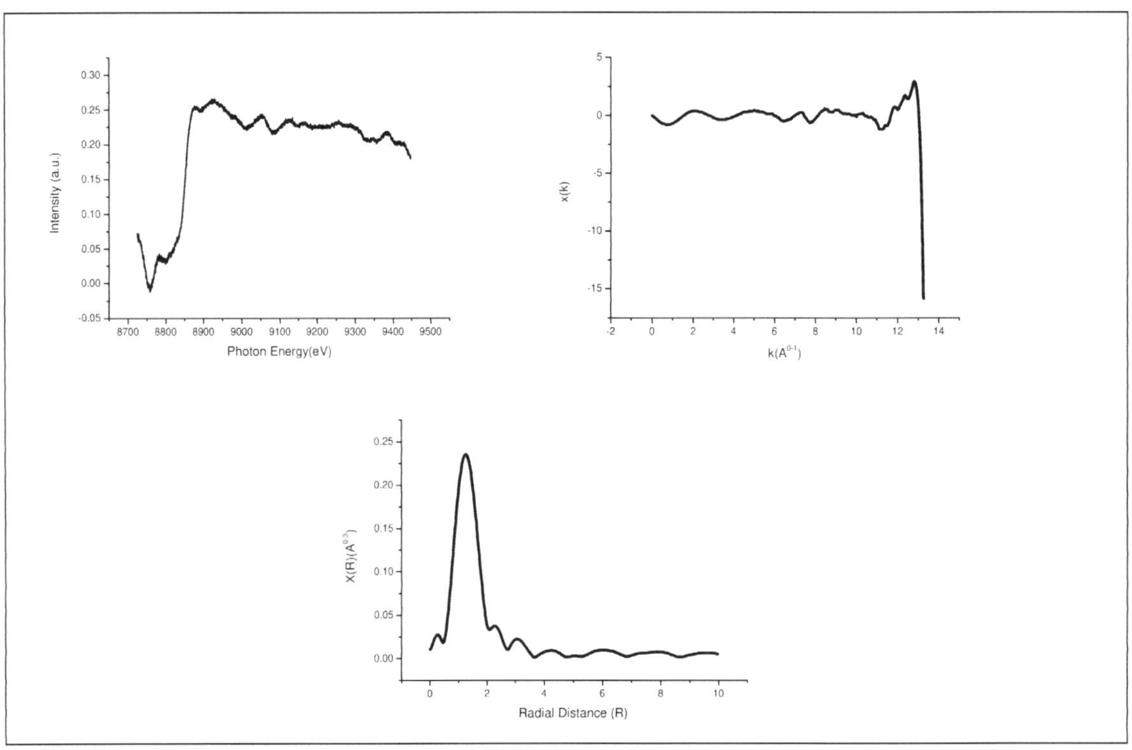


**Figure 2.** EXAFS curve of Cu[3,5-dimethoxy]

Journal of Physics: Conference Series **755** (2016) 012041 doi:10.1088/1742-6596/755/1/012041

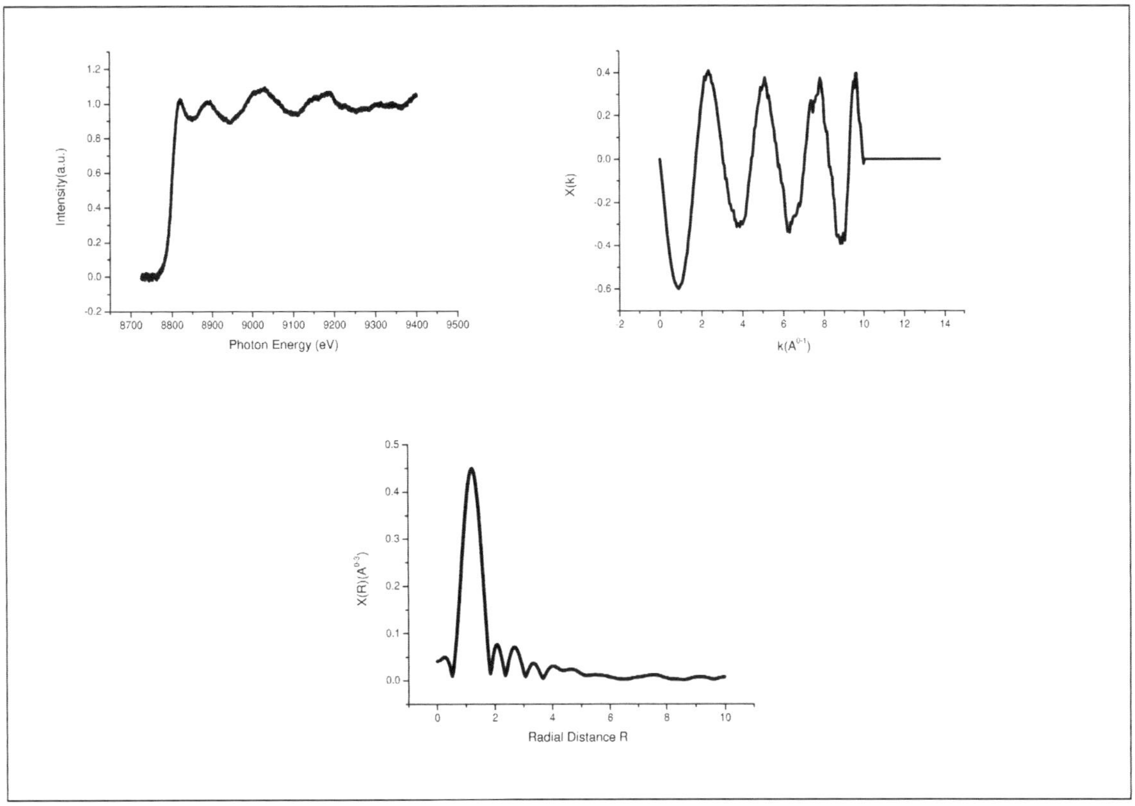


**Figure 3.** EXAFS curve of Cu[4,5-dimethoxy]

## 5. Conclusion

As it is evident from the analysis that the bond lengths determined by Fourier transforming the FEFFIT programming are comparable with the bond length obtained by LSS, Levy's and Lytle methods. The theoretical and experimental values agree well with each other. This means that the parameterized theoretical calculation of the EXAFS spectra of copper complexes described here is in good agreement with physical reality.

## References

[1] Kekre P A, Joshi K P and Mishra A, 1996 *Indian Journal of Pure and applied Physics* **34** 306
[2] Alternative Strategies forXafs Data Analysis using Feffand Mathematica 7JCPDSInternational Centre for Diffraction Data 2011 ISSN 1097-0002
[3] Konningsberger D C, R. Prins (Eds.), "X-ray Absorption Principle, Technique of EXAFS, SEXAFS, XANES" Wiley, New York, 1998
[4] Cullity B D, Element of X-ray Diffraction, ewville M 2001 *J Synchrotron Rad.***8** 322
[5] Stern E A, Sayers D E and Lytle F W 1975 *Physical Review B* **11** 4836
[6] Stern E A, 1974 *Physical Review B* **10** 3027.
[7] Stern E A, Sayers D E and Lytle F W 1975 *Physical Review B* **11** 4825

# Magnetic performance of DEVU25.20h undulator at IDDL, DAVV,India

**G. Mishra, , Mona Gehlot, Geetanjali Sharma#, Roma Khullar**
Insertion Device Development Laboratory, School of Physics, DAVV, Indore.
corresponding author: gmishra_dauniv@yahoo.co.in
#soleil,Paris,France

**Abstract:** In this paper, we describe the design ,development and magnetic performance of a prototype DEVU25.20h, variable gap hybrid undulator at IDD laboratory, DAVV, Indore, India. The undulator is NdFeB based hybrid undulator with twenty five period and *20 mm* each period length .Low cost *M35* grade cobalt steel is used as the pole material of the undulator. The abc coefficients of the new pole material is prescribed and the measured performance and characteristics of the undulator is evaluated using RADIA. The detailed design parameters and performance indicators are summarized in Table 1 and Table 2 respectively.

## I.Introduction

Pure Permanent magnet and hybrid undulators are in use in synchrotron radiation sources and free electron lasers over past few decades .The first NdFeB based pure permanent magnet (PPM) undulator was proposed by Halbach [1] . The hybrid scheme with high permeable poles was later proposed [2] to extract higher field on axis for given gap, undulator period ratio.Several reviews on PPM and Hybrid schemes have been reported in the last few decades [3-4]. In a PPM structure, four magnets (two vertical and two horizontal) are used to model one period. Since the permeability of NdFeB rare earth magnets is approximately unity, the fields from the individual magnet blocks add and finally determine the final field properties. Therefore in the fabrication of the PPM undulators, the characterization of the individual magnets is extremely important .For a given quality of the individual magnets, it is required to adopt magnet sorting and pairing schemes to design good quality undulators. Afterwards a detailed shimming scheme is necessary in order to correct the magnetic field after the magnets have been assembled into the structure. The Halbach field configuration is the standard procedure for PPM undulator technology. The peak field on axis depends on the dimension of the magnet blocks and the gap. For the case of infinitely wide blocks, the field is given by [4],

$$B[T] = 2B_r \frac{\sin(\pi / M)}{(\pi / M)}[1 - e^{\frac{-2\pi h}{\lambda_u}}]e^{\frac{-\pi g}{\lambda_u}} \qquad (1)$$

Where $B[T]$ is the peak magnetic field read on axis , $B_r$ is the remnant magnetic field of the magnetic material, $M$ is the number of blocks per period, $\lambda_u$ is the period, $h$ is the height of the blocks , $g$ is the magnet to magnet gap. The hybrid scheme uses high permeable shims to concentrate the flux provided by the permanent magnets on axis. The hybrid structure uses two horizontal magnets and two poles per period and has the unique advantage as the scheme reduces the sensitivity of the design to magnetic field orientation errors in the individual magnet blocks.

$$B[T] = a\exp[b(g/\lambda_u) + c(g/\lambda_u)^2] \quad (2)$$

$a, b, c$ are different dimensionless coefficients for different pole material and different for NdFeB and $SmCo_5$ magnet material blocks. The equation is valid over a given range. Most of the hybrid undulators use vanadium permendur poles for its high permeability .The abc coefficients are generally found from the data collected from the measurement and fitted with the approximation in Eq.(2). Some typical values are $a = 3.381, b = -4.73, c = 1.198$ [2], with soft iron poles and $a = 3.56, b = -5.25, c = 1.67$ $a = 3.44, b = -5.08, c = 1.54$ with vanadium permendur poles.

In this paper, we describe the design details of a prototype DEVU25.20h, a gap variable hybrid undulator designed, built and measured in house at IDD laboratory, DAVV, Indore, India. The undulator is the NdFeB based hybrid undulator. The minimum gap is 6 *mm* and maximum gap is 50 *mm*. The gap is manually driven. Low cost M35 grade cobalt steel is the pole material of the undulator. The hybrid undulator is designed as asymmetric structure i.e one end starts with a magnet and terminated with a pole at the other end. In section 2, we describe the magnet design with our end design scheme. In section 3, we discuss the Hall probe measurement results and evaluate the performance of the undulator.

## II. DEVU25.20h Magnet Design

The planar DEVU25.20h is a NdFeB based hybrid undulator with two NdFeB magnets and two poles per period. The undulator has twenty five periods and twenty *mm* each period length. The N35 grade NdFeB magnets for the structure is procured from Shanghai strong Magnet Industrial and commercial co Ltd, Shanghai, China. The magnet size is 6.25 *mm* (width), 6.25 *mm* (height) and 50 *mm* in length. The Cobalt steel, M35 is available in the local market and has 0.93% Carbon, 4.20% Chromium, 5.00% Molybdenum, 6.40% Tungsten, 4.80% Cobalt, 1.80% Vanadium, 0.35% Silicon. The magnetic permeability is measured in a 7 Tesla MPMS SQUID vibrating sample magnetometer. The M-H curve and B-H curve of the material is shown in Fig.(1a) and Fig.(1b) respectively. The maximum permeability of the material measures 425 from Fig.(1c). The peak magnetic field (Vertical) is 3000 Gauss ± 150 Gauss measured at 1 mm from the magnet. The H- magnets give a peak field of 2500Gauss ± 90 Gauss. The Cobalt steel, M35 is available in the local market and has 0.93% Carbon, 4.20% Chromium, 5.00% Molybdenum, 6.40% Tungsten, 4.80% Cobalt, 1.80% Vanadium, 0.35% Silicon. The magnetic permeability is measured in a 7 Tesla MPMS SQUID vibrating sample magnetometer. The M-H curve and B-H curve of the material is shown in Fig.(1a) and Fig.(1b) respectively. The maximum permeability of the material measures 425 from Fig.(1c). The pole size is 3.75 *mm* (width), 4.7 *mm* (height) and 50 *mm* in length. The period length is 20 *mm* (6.25 *mm*x2 + 3.75 *mm* x2).

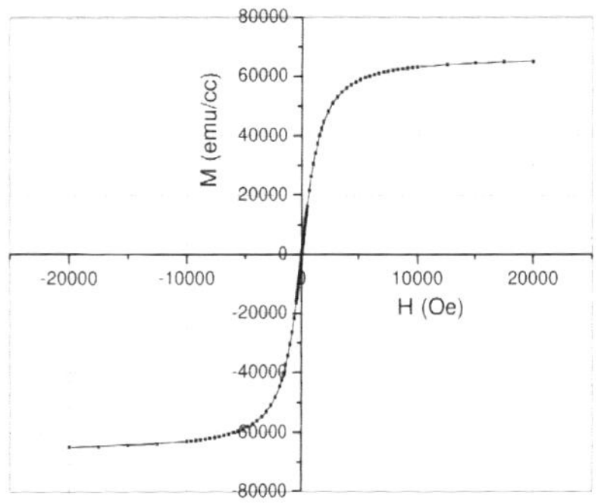


Fig 1(a). magnetization versus magnetic field of the pole material.

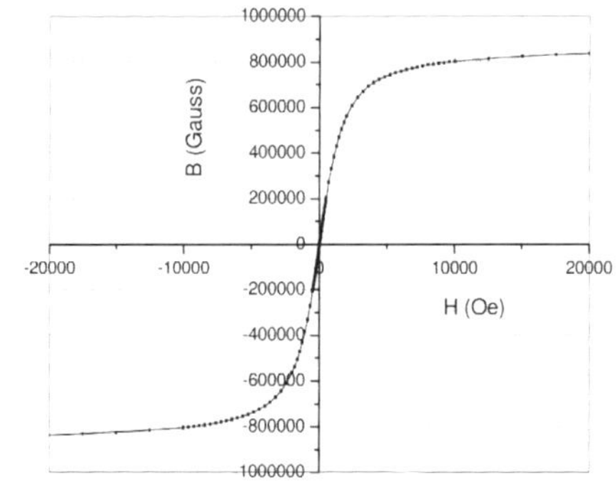


Fig 1(b) B-H curve of the pole material

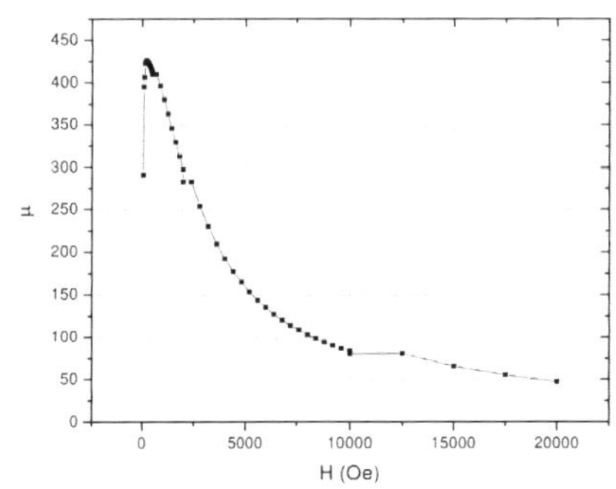


Fig 1c. μ -H curve of the pole material.

The twenty periods are arranged in a 112 *mm* wide, 500 *mm* in length 316 steel jaws, the design considerations are described in Ref[5]. The complete undulator assembly is seen in Fig.( 2). The end design of the hybrid undulator is the critical issue to the magnetic performance of the hybrid undulator. It is required that the field should terminate in the end magnet correctly with zero angle and zero displacement of the electron beam. Different schemes of end designs are classified into five schemes. These are adjusting the end magnet strengths, reduced pole heights, end pole shaping, magnetic shimming and additional end magnets. In this paper, we follow 1:1:1/2 scheme for end termination. According to this scheme the magnets are arranged in -1/2,+1/2,-1,+1......-1,+1,-1/2,+1/2.

Journal of Physics: Conference Series **755** (2016) 012042 doi:10.1088/1742-6596/755/1/012042

Fig 2. complete assembly of the 0.5 meter long Hybrid undulator

## III. FIELD MEASUREMENT, Results and discussion

The two quantities are of interest are the first and second field integrals of the undulator i.e

$$I_1 = \int_0^L B_y dz \tag{3}$$

$$I_2 = \int_0^L dz' \int_0^L B_y dz \tag{4}$$

The above equations when multiplied by $-e/\gamma mc$ gives the angular and trajectory offset respectively. Using $-e/mc = 565\ T^{-1}\text{-}m^{-1}$.In practical units ,

$$-e/\gamma mc = 0.298/E \quad T^{-1}\text{-}m^{-1}, E \text{ is in GeV}.$$
$$-e/\gamma mc = 298/E \quad T^{-1}\text{-}m^{-1}, E \text{ is in MeV}. \tag{5}$$

For a 10 MeV electron beam we have $-e/\gamma mc = 29.8\ \ T^{-1}\text{-}\ m^{-1}$. The magnet design analysis has been carried with RADIA for gaps of $5mm$ to 20 $mm$. The magnetic field and the field integrals are plotted in Fig 3a and Fig3b for different gaps.

Journal of Physics: Conference Series **755** (2016) 012042 doi:10.1088/1742-6596/755/1/012042

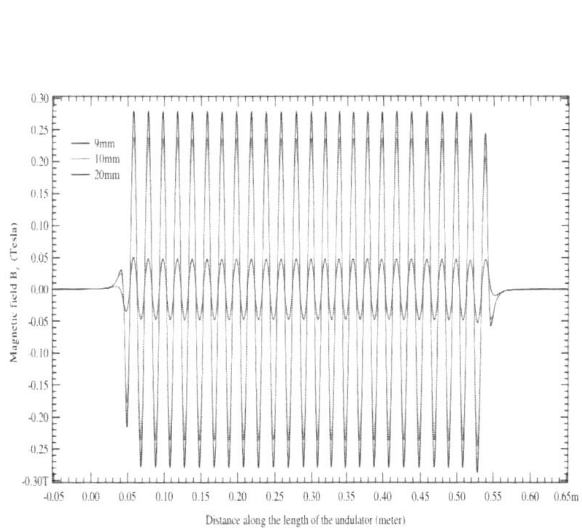


Fig3a. RADIA output for the DEVU25.20h

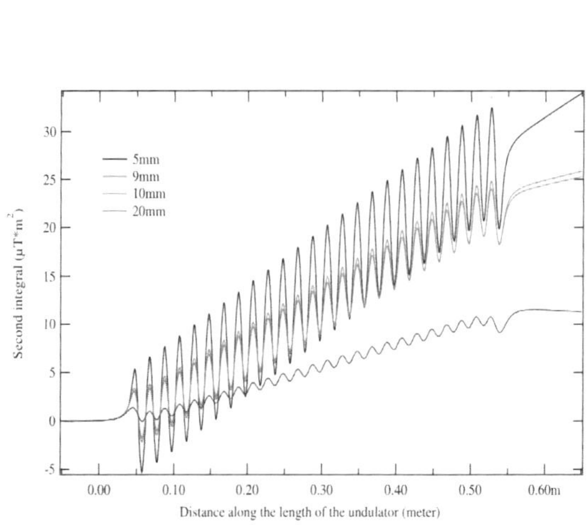


Fig3b.second field integral from RADIA

The performance of the hybrid undulator is evaluated by Hall Probe measurement. The Hall Probe bench is set on a vibration isolation table. The Hall Probe is F.W. Bell make Model No. 8010. The Hall Probe is mounted on a XY –translation stage capable of moving 25 *mm* each. The complete probe fitting sits on a manually driven translation stage of length 1500 *mm* .The Hall probe data is taken at a step length of 500 $\mu m$ .

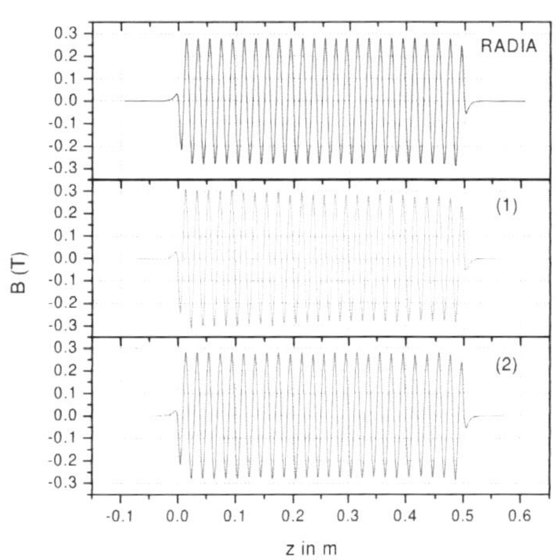


Fig4a. Magnetic field of DEVVU25.20h at 9mm.

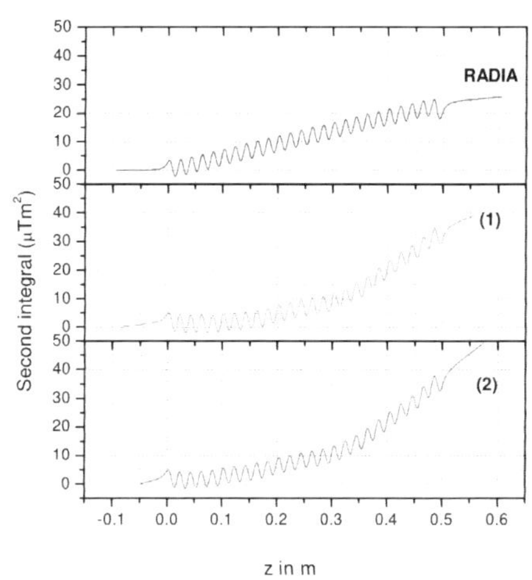


Fig4b. Second field integral at 9mm

The Hall probe data is compared with RADIA in Fig4a for two sets of measurements at 9mm. The first set of data shows that the undulator jaws are tapered towards the end. The structure is then corrected to remove this manufacturing error. The second field integral is plotted in Fig.(4b) at this gap. The RADIA predicts an overall second field integral value of less than 25 $\mu T - m^2$ .From Fig.(4.c) ,the first integral value is 20 $\mu T - m$ .

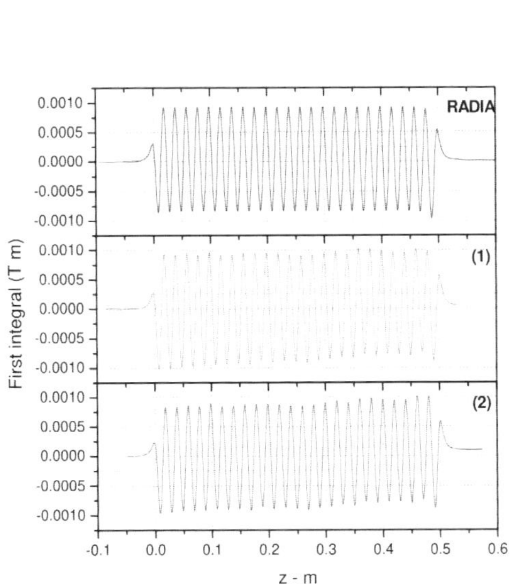


Fig 4c. First field integral at 9mm

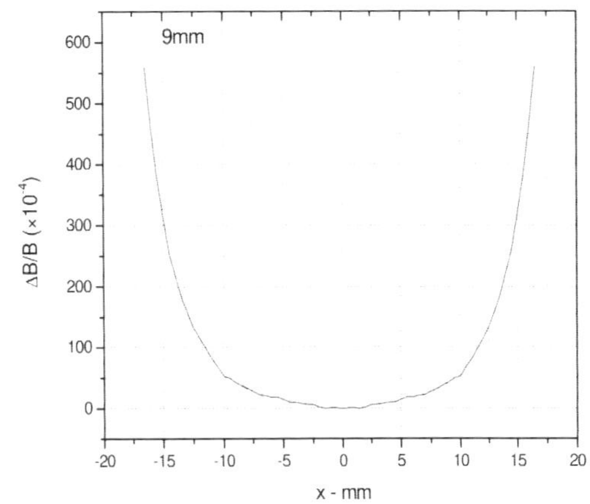


Fig 5 Transverse field Roll off

The measured Hall probe data overshoots this value due to several mechanical fabrication constraints. In both the measurements, the second field integral value reaches up to 35 $\mu T - m^2$ . The second field integral corresponds to a positive value at the exit. This corresponds to a negative value of the electron trajectory at the exit. This implies a negative angle (angle below the on-axis). The FEL process involves overlapping of the electron beam and the radiation field .Any deterorientation of the overlapping reduces the free electron laser gain. Hence the straightness of the electron trajectory is an issue of the undulator. To correct the first integral at the exit, we need to strengthen the end pole by approximately 50 *Gauss-cm*. Looking at the second field integral, a field of -3500 *Gauss* $cm^2$ needs to be corrected at the exit, it is required to apply $\frac{3500Gcm^2}{50cm} = 70G - cm$ . Assuming that we have already applied 50 *G-cm* to correct the first field integral, an additional 20 *G-cm* need to be applied additional to the total length of the undulator. Looking at the trajectory, one can locate the locations where to apply the corrections in order to straighten the trajectory .The trajectory has three linear parts: - 0-10 *cm*, 10-30 *cm*, 30-50 *cm*. It is required to correct the kicks at 10 *cm*, 30 *cm*. The kicks requirement at 10 *cm* can be determined by slope differences of the two straight parts. Consider the middle part i.e. 10-30 *cm* of the trajectory, when slope= $\frac{-900G\ cm^2 - (-200G\ cm^2)}{30 - 10\ cm} = -35G\ cm$ which is required to be compensated at $z = 10cm$. So we require at a kick of 35 *G-cm*. From the last position of the trajectory i.e. 30-50 *cm*, slope= $\frac{-3500G\ cm^2 - (-900G\ cm^2)}{50 - 30\ cm} = -130G\ cm$ . Hence at $z = 30cm$, we need an extra kick of $-\left[(-130G\ cm) - (-35G\ cm)\right] = 95G\ cm$. To be more specific, to add +35 *G-cm* kick at 10 *cm*, one can either strengthen a positive pole by this value or equivalently one can weaken the negative pole by this amount. Similarly at z = 30 *cm*, one needs to look for corrections to a value amounts to 95 *G-cm*.

The performance of the undulator depends on a variety of parameters. Optimization of these parameters requires several iterative processes and procedures. In the following we mention the tolerances of the hybrid undulator.. A range of ± 1 *mm* is conventionally required to accommodate the beam width and to have some margin tolerance for the horizontal alignments of the undulator segments. In the present case from Fig .(5) at undulator gap of 9mm , we read, $\Delta B_y(x)/B_y(x=0) \approx 0.70165\times10^{-4}$ .The period error is an important factor which affects the peak field error and as a consequence contributes to the phase error and beam wander. A large number of undulator periods are required for small line width of the undulator radiation spectrum. Any error in the undulator period lengthens the undulator length. The DEVU25.20h is a 25 period undulator .The errors in each period is presented in Fig.(6).The first few periods are made perfect and towards the end of the undulator the period length introduces errors due to size variations in magnets and poles. It has 0.5 % over all period and rms period error is 0.13 $mm$ .The peak field error is caused by gap errors, period errors and due to inhomogeneous of the material. In the present case due to rms gap error of 0.05156 mm at 9mm Fig.(7) is $\Delta B_{peak}/B_{peak} \le 2.5\%$ .The FEL process is sensitive to electron and radiation field overlapping. The main factor which prohibits this overlapping arises from the random trajectory offsets along the undulator lengths. This is estimated from the field integral calculations. The second field integral of 35 $\mu T-m^2$ gives an orbit displacement of 1043 $\mu m$ at 10 MeV. A first integral value of 50 $\mu T-m$ gives 1490 $\mu$ *radian* as the angular misalignment. Comparing with the RADIA result that we can compensate the mechanical tolerances up to 25 $\mu T-m^2$ will yield axis displacement of 745 $\mu m$ and with a first integral value of 20 $\mu T-m$ , we need adequate compensation to come to 596 $\mu$ *radian*. Setting 100 $\mu m$ as an upper limit for the orbit displacement we have 3 $\mu T-m^2$ at 10MeV electron beam. Similarly a value of 10 $\mu T-m$ will set an angular offset of 0.3 *m radian.* We have designed and developed a hybrid undulator at moderate operating parameters.

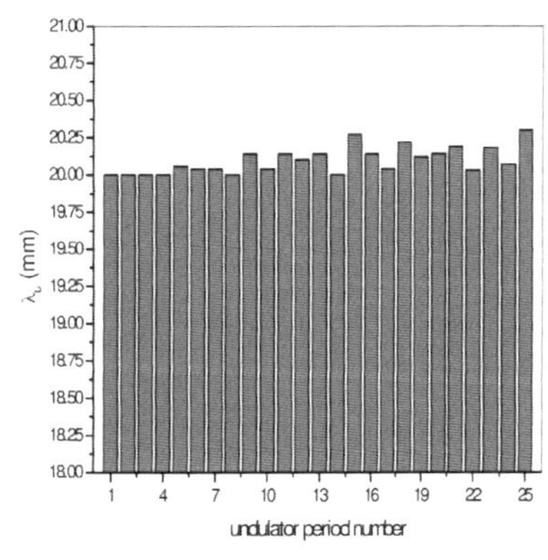


Fig6 period error

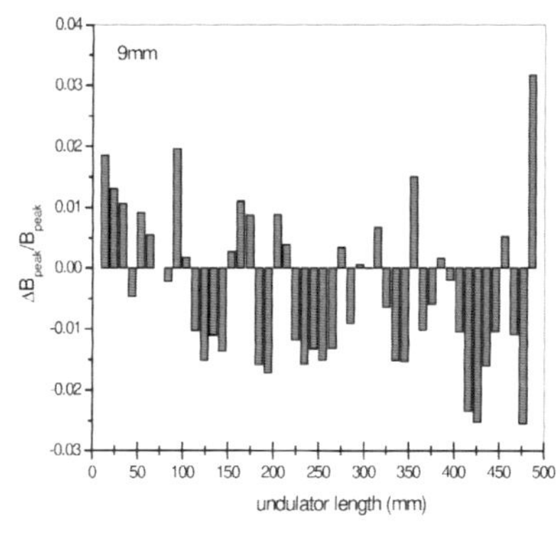


Fig7 peak field error

The measurements are repeated for undulator gap of 10 mm. The magnetic field data, first field integral and the second field integral are presented in Figs.( 8-10) respectively. The errors in peak field and the transverse field roll off are given in Fig.(11) and Fig.(12) respectively.. Table 1 and Table 2 summarize all the undulator parameters and tolerances at both *9mm* and *10 mm* gap measurements. In Fig.(13), we fix the abc coefficients of the hybrid undulator. The data are taken from a gap of 7 mm to 50mm and fitted with Eq.(2).This yields $a = 3.06, b = -5.59, c = 1.123$ . The abc values of the present used poles are comparatively less than the vanadium permendur pole materials. Fig.( 14) studies the force per unit length between the two undulator jaws of the hybrid undulator.

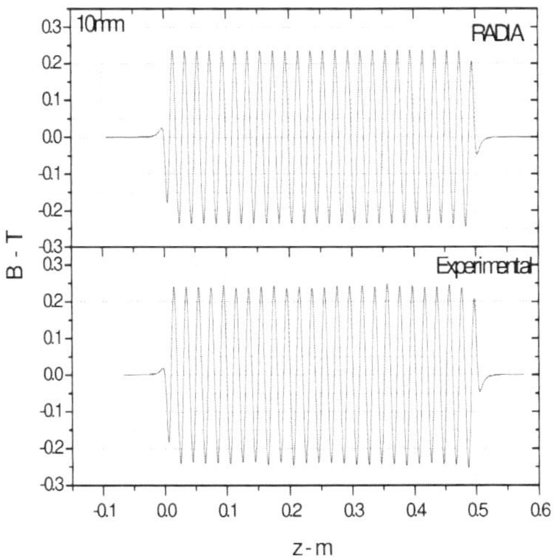


Fig8. Magnetic field measurement

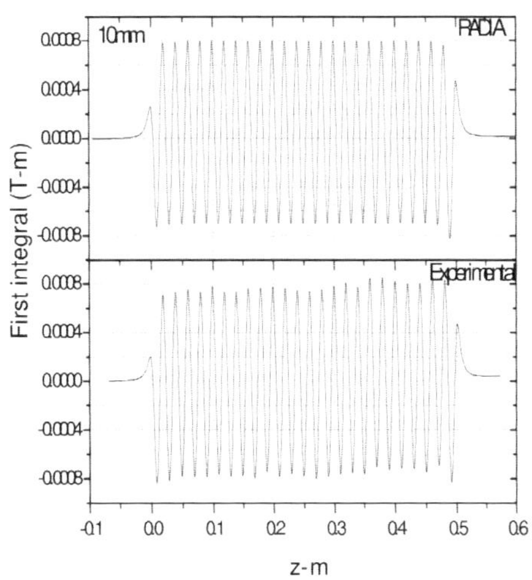


Fig9.First field integral

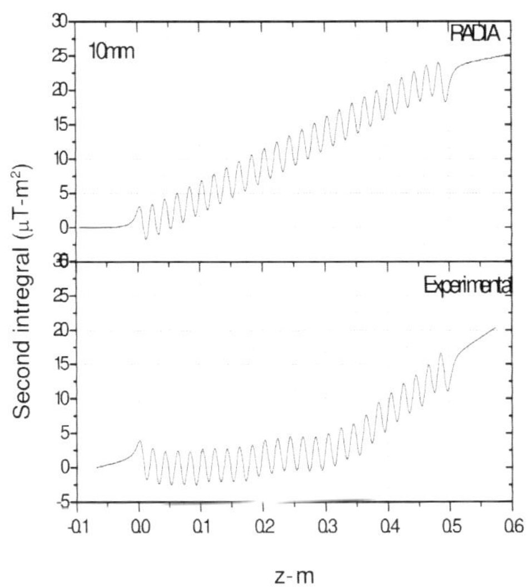


Fig10.second field integral

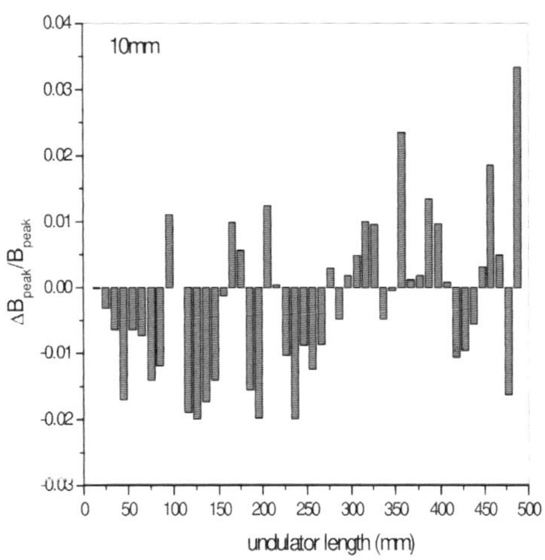


Fig11.peak field error

A conceptual design and development of a hybrid undulator with M35 cobalt steel is undertaken at IDDL, DAVV, Indore, India. The expected performance of the device is estimated by RADIA. The study verified a new pole material for use in hybrid undulator and evaluated its abc coefficients. The analysis reveals that further tuning method is required to improve the field integral data and straightness of the trajectory. The undulator mechanical parameters and performance parameters are tabulated in Table 1 and Table 2 respectively at two undulator gaps i.e 9mm and 10 mm . The gap dependence of the field integrals and phase error are the other two critical parameters that needs to be investigated.

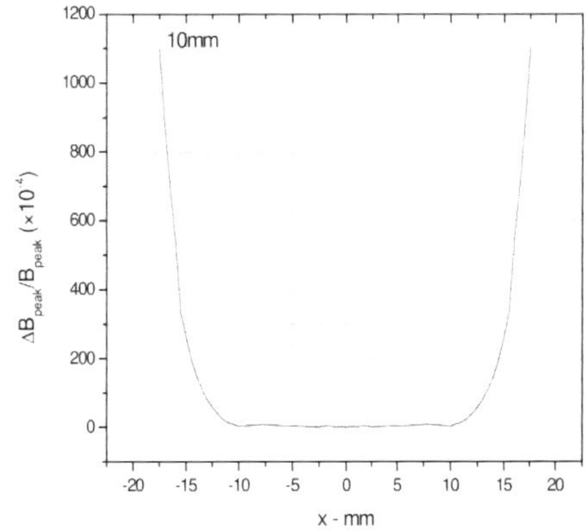


Fig12.transverse field roll off

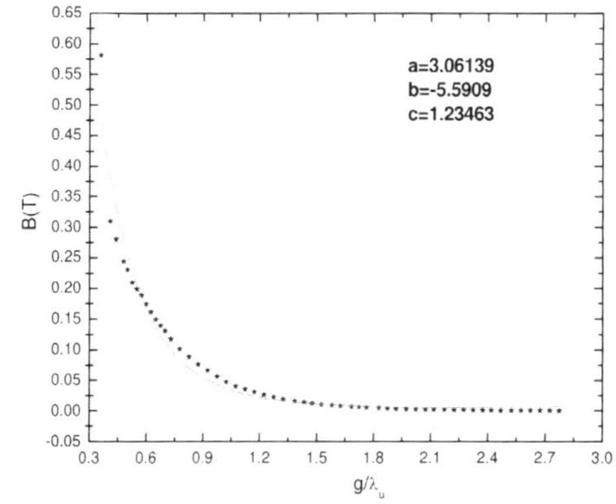


Fig13.magnetic field versus undulator gap

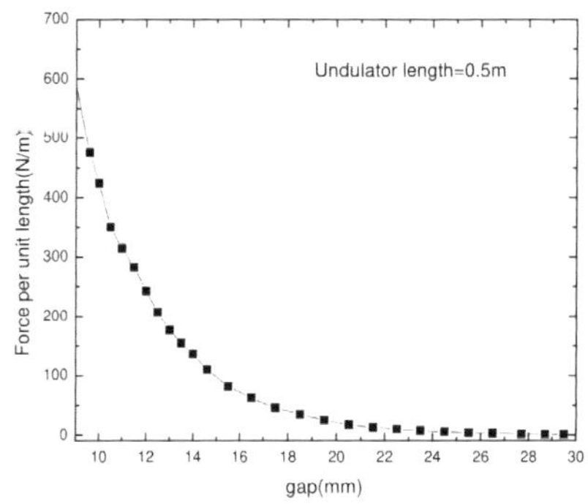


Fig14.Force per unit length between the two undulator jaws

**Table1.**
**DEVU25.20h design parameters**

| | |
|---|---|
| Magnetic structure | asymmetric ,Hybrid |
| Length of the period | *20 mm* |
| rms period error | *0.13* mm |
| Number of periods | *25* |
| Magnet material | NdFeB |
| Number of regular magnets | *98* |
| Pole material | Cobalt steel ,M35 Grade |
| Number of regular poles | *98* |
| gap | variable , *9mm to 20 mm* |
| Regular magnet size | *6.25 mm x6.25 mm x 50 mm* |
| Regular pole size | *3.75 mm x 4.7mmx 50 mm* |
| End scheme | *1:1/2:1/2* |

Table2.
DEVU25.20h Tolerances

| | |
|---|---|
| Peak field | *2777.7* Gauss at *9 mm, 2420.4* Gauss at *10 mm* |
| rms peak field error | *36* Gauss at *9mm,* *29* Gauss at *10mm* |
| Transverse field Roll off at x = ± *1 mm* | *0.70165 x10$^{-4}$* at *9mm* , *0.00396x10$^{-4}$* at *10mm* |
| Transverse field Roll off at x = ± *3 mm* | $6.6657\times10^{-4}$ at *9 mm,* $1.9009\times10^{-4}$ at *10mm* |
| First field integral | *50* $\mu T - m$ at 9 *mm,* *40* $\mu T - m$ at 10 *mm* |
| Second field integral | 35 $\mu T - m^2$ at *9 mm,* 15.6 $\mu T - m^2$ at *10 mm* |
| Transverse Field Roll off at x= ± 5 *mm* | $14.73477\times10^{-4}$ at *9mm* , $4.356\times10^{-4}$ at *10 mm* |
| nominal *K* | *0.51* at *9mm* , *0.45* at *10mm* |
| rms gap error | *0.05156 mm* at *9mm* , *0.025 mm* at *10mm* |
| Magnet size along x | *45 mm* |
| Total length | *500 mm,* length overshoot 2.5 *mm* |

## ACKNOLEDGEMENT

This undulator magnet is designed by a financial support from UGC, Delhi, India. The measurement and analysis is carried out by financial support from SERB, DST, New Delhi, India.The authors are grateful to Dr. Markus Tischer, DESY, Germany for many technical discussions and comments on DEVU25.20h during FEL workshop at DAVV, Indore, India during 4-6, Dec, 2014.

## REFERENCES

[1] Klaus Halback, Nuclear Instruments and Methods in Physics Research A, 187, 1981, pp-109.67.
[2] K. Halbach, J. Phys. (Paris) C1-44, 1983, 211.
[3] R.P.Walker, Nuclear Instruments and Methods in Physics Research A, 237, 1981, p-366 [4] Ma Youwu, Zhang Zhooming, Liu Bo, 1990, "Hybrid Undulator Design and Field error analysis", *Inspire Conf. Proc.* **C900612**, pp-1377-1379.
[4] R.P.Walker, M.W.Poole, D.G.Taylor, Journal de Physique, C1-321, 1984.
[5] G. Mishra, Geetanjali Sharma, Roma Khullar[#], Bramh Prakash, Mona Gehlot,: "Design and fabrication of PPM and hybrid undulator for free electron laser applications", Journal of Physics: Conference Series, IOP publishing, 534 (2014) 012053.

# Effect of deposition time on structural and magnetic properties of pulse laser deposited hard-soft composite films

**M Satalkar[a], S N Kane[a, 1], A Pasko[d], M LoBue[d] and F Mazaleyrat[d]**
[a]School of Physics, D. A. University, Khandwa road Campus, Indore – 452001, India
[c]UGC-DAE CSR, University Campus, Khandwa Road, Indore – 452 001, India
[d]SATIE, ENS de Cachan,CNRS,UniverSud, 61, Av. du Pdt. Wilson, F-94230 Cachan, France

E-mail: kane_sn@yahoo.com (S. N. Kane)

**Abstract.** Hard-soft composite ($BaFe_{12}O_{19}$:$Mg_{0.1}Ni_{0.3}Zn_{0.6}Fe_2O_4$ (2:1) films, were deposited by 'Pulsed Laser Deposition' (PLD) technique on Si (100) substrate using different deposition time – 30, 60, 90 and 120 minutes. Influence of deposition time on structural and magnetic properties were studied via X-ray diffraction (XRD) and vibrating sample magnetometer (VSM). XRD confirms the presence of soft and hard phases in all the prepared thin films. Small amount of secondary phase - $Fe_2O_3$ is also detected in all the thin films except for the deposition time - 90 mins. With deposition time average grain diameter of both hard ($BaFe_{12}O_{19}$) and soft ($Mg_{0.1}Ni_{0.3}Zn_{0.6}Fe_2O_4$) phase increases. Increase in the distance between the magnetic ions ($Ni^{2+}$ and $Fe^{3+}$) at tetrahedral (A) and octahedral [B] site leads to increase in the hopping length at A and B site except for the the deposition time of 60 minutes. Magnetic measurements shows that the coercivity and magnetization of the prepared thin films respectively ranges between 112.07 – 213.03 Oe and $1.4 \times 10^{-7} – 3.15 \times 10^{-7}$ $Am^2$.

## 1. Introduction

Recently, there has been immense attention on the materials having good exchange spring coupling behavior. According to the exchange spring theory proposed by Kneller and Hawig [1], nanocomposite magnets consist of soft and hard phases, having sufficiently exchanged coupling between both the phases which are known as exchange spring magnets. Exchange spring magnets possess large energy product value and are potential candidates for next generation permanent magnet [2]. The technological realization of such materials shows that they should be crystallographically coherent and consequently they are magnetically exchange coupled [1]. The hard magnetic grains provide the high anisotropy and coercive fields while the soft-magnetic grains enhance the magnetic moment. The soft grains are pinned to the hard-magnet grains at the interfaces by the exchange interaction while the center of the soft magnet grains can rotate in a reversed magnetic field. Such magnets are characterized by enhanced remanent magnetization and reversible demagnetization curves since the soft grains will rotate back into alignment with the hard grains when the applied field is removed [3]. Exchange spring behavior is becoming important and it has been predicted that a single hard-phase magnetic behavior is expected for a nano-composite consisting of both soft- and hard-magnetic phases due to high exchange coupling between the two different phases. This, in turn combines high magnetization of the soft phase with high coercivity of the hard phase. Thus, with an aim to enhance the coercivity and magnetization values, the paper focusses on the effect of deposition time on structural and magnetic properties of hard-soft composite films prepared via 'Pulsed Laser Deposition' (PLD) technique.

## 2. Experimental details

For depositing hard-soft composite films on Si (100) soft ($Mg_{0.1}Ni_{0.3}Zn_{0.6}Fe_2O_4$) and hard ($BaFe_{12}O_{19}$ or BaM) ferrite powders were prepared by sol gel auto-combustion method [4]. Composite of hard and soft ferrite was prepared by thoroughly mixing of the individual ferrite components in the ratio 2:1, with an aim to obtain better structural homogeneity and to be able to observe exchange spring behaviour. The composite powder was pressed in the form of a pallet by applying pressure ~ 90 K Newton and, was annealed at 800°C for 24 hrs. Thin films were deposited using the following composite pallet: $BaFe_{12}O_{19}$:$Mg_{0.1}Ni_{0.3}Zn_{0.6}Fe_2O_4$ (2:1) mounted on a rotating axel, on (100) Si substrate at different deposition time:- 30, 60, 90 and 120 minutes. PLD set-up (Lambda Physik model COMPEX-201) uses a KrF excimer laser of wavelength 248 nm and pulse duration of 20 ns. During deposition the laser energy was kept 200 mJ/pulse, with pulse repetition rate of 10 Hz. The substrate temperature and oxygen pressure were maintained respectively at 750 °C and ~ 1mTorr (base pressure ~ $5\times10^{-6}$ Torr). Focussed laser beam was made incident on the target surface at an angle of 45°. The substrate was mounted opposite to the target (on a heater plate) at a distance of 4 cm. After deposition, the sample was cooled slowly to room temperature, in absence of oxygen pressure.
X-ray diffraction (XRD) of the prepared $BaFe_{12}O_{19}$ and $Mg_{0.1}Ni_{0.3}Zn_{0.6}Fe_2O_4$ ferrite shows single phase behaviour of both the hard and soft phases [5]. XRD ($\vartheta$-2$\vartheta$ configuration) measurements of composite hard - soft composite films were done using a Bruker D8 Advance diffractometer, with Cu-$K_\alpha$ radiation ($\lambda$ = 0.1540562 nm). Magnetization measurements on deposited thin films were done using LakeShore Model 7410 vibrating sample magnetometer (VSM) with a maximum applied field of ± 1.7 Tesla.

## 3. Results and discussions

Figure 1 depicts the XRD pattern (Left panel) of the composite $BaFe_{12}O_{19}$:$Mg_{0.1}Ni_{0.3}Zn_{0.6}Fe_2O_4$ (2:1) thin film grown by PLD on silicon (100) substrate for (a) 30 minutes, (b) 60 minutes, (c) 90 minutes

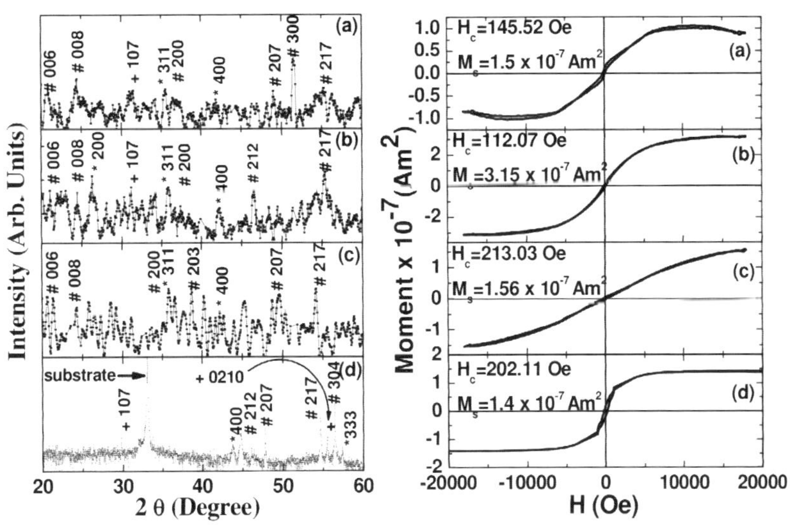


**Figure 1** XRD pattern (Left panel) and hysteresis loop (Right panel) of the composite $BaFe_{12}O_{19}$:$Mg_{0.1}Ni_{0.3}Zn_{0.6}Fe_2O_4$ (2:1) thin film deposited on Si (100) substrate for (a) 30 minutes, (b) 60 minutes, (c) 90 minutes and (d) 120 minutes. *(* Soft phase - $Mg_{0.1}Ni_{0.3}Zn_{0.6}Fe_2O_4$, # Hard phase - $BaFe_{12}O_{19}$, + $Fe_2O_3$)*

and (d) 120 minutes. The planes shown in the XRD pattern of the composite films were matched with the Joint Committee on powder Diffraction Standard (JCPDS) card of $BaFe_{12}O_{19}$, which shows the formation of $BaFe_{12}O_{19}$ (hard) phase. XRD pattern of the grown thin film was also compared with the available data in the literature, suggesting the formation of $Mg_{0.1}Ni_{0.3}Zn_{0.6}Fe_2O_4$ (soft) phase. A small amount of secondary $Fe_2O_3$ phase is also identified in all the deposited thin films except for the deposition time of 90 minutes. No impurity phase is detected in the XRD patterns of the preliminary powders of $BaFe_{12}O_{19}$ and $Mg_{0.1}Ni_{0.3}Zn_{0.6}Fe_2O_4$ ferrite [5], which were used for the preparation of composite films. The presence of secondary phase in the prepared composite films can be ascribed to the oxygen deficiency during the sample preparation. Figure 1 (Right panel) represents the hysteresis loops of the composite $BaFe_{12}O_{19}$:$Mg_{0.1}Ni_{0.3}Zn_{0.6}Fe_2O_4$ (2:1) thin film grown by PLD on silicon (100) substrate for (a) 30 minutes, (b) 60 minutes, (c) 90 minutes and (d) 120 minutes. Coercivity ($\boldsymbol{H_c}$) and magnetization ($\boldsymbol{M_S}$) obtained from the hysteresis loops is represented in figure 1 right panel. Low values of $\boldsymbol{H_c}$ and $\boldsymbol{M_s}$ shows the major contribution of the soft phase in the prepared composite films.

Table 1 shows the structural parameters of the phases formed in the prepared composite films deposited for different time. Perusal of *a* and *c* values reveal the formation of $BaFe_{12}O_{19}$ phase. For soft ferrite value of ***a***, confirms the formation of cubic spinel structure. Average grain diameter (***D***) of the prepared thin films for different deposition time are also given in table 1. Higher grain size for the hard and soft phase is obtained for the thin film deposited for 120 minutes. The hopping length for tetrahedral - A ($\boldsymbol{L_A}$) and octahedral - B ($\boldsymbol{L_B}$) site for the soft spinel ferrite phase is also shown in table1. The hopping length/jump length '*L*' (the distance between the magnetic ions) of electrons influences the physical properties of the soft phase. Electrons those are hopping between B- and A sites are less probable compared to that between B- and B sites, because the distance between the two metal ions placed in B-sites is smaller than if they were placed one in B-sites and the other in A-sites. $\boldsymbol{L_A}$ and $\boldsymbol{L_B}$ for the soft spinel ferrite phase - $Mg_{0.1}Ni_{0.3}Zn_{0.6}Fe_2O_4$ respectively ranges between 0.3589 - 0.3639 nm and 0.2930 -0.2972 nm. $\boldsymbol{L_A}$ and $\boldsymbol{L_B}$ decreases for 60 minutes and then constantly increases up to 120 minutes. The variation in the hopping length at A and B site can be attributed to the increase in the distance between the magnetic ions ($Ni^{2+}$ and $Fe^{3+}$) at A and B site for all the values of the deposition time except for 60 minutes.

Figure 2 (left panel) shows the variation of $\boldsymbol{H_c}$ and $\boldsymbol{M_S}$ with deposition time. Highest value of ***Hc*** and $\boldsymbol{M_S}$ is obtained for the deposition time - 90 (213.03 Oe) and 60 ($3.15 \times 10^{-7}$ $Am^2$) minutes respectively. Higher values of $\boldsymbol{H_c}$ and $\boldsymbol{M_S}$ can be ascribed to the two phase coexistence. It is known that in the PLD process the oxides become more oxygen vacant as the $O_2$ partial pressure decreases. The magnetization values increases as the oxygen pressure increases inside the chamber, due to the filling of more vacancies. The substrate temperature was same for all the deposited hard - soft composite films but the deposition time was different. This can be considered as in-situ annealing of the films at different time. For higher deposition time (120 minutes) decrease in magnetization is due to the

**Table 1:-** XRD Parameters: lattice constant (***a***), Grain Size (***D***), hopping length for A ($\boldsymbol{L_A}$) and B ($\boldsymbol{L_B}$) site for the prepared soft ($Mg_{0.1}Ni_{0.3}Zn_{0.6}Fe_2O_4$) - hard ($BaFe_{12}O_{19}$) composite thin film on Si (100) substrate for different deposition time.

| Deposition time (minutes) | Phase | a (nm) | $L_A$ (nm) | $L_B$ (nm) | D (nm) |
|---|---|---|---|---|---|
| 30 | $BaFe_{12}O_{19}$ | ***a*** =0.5648, ***c*** =2.7347 | - | - | 14.92 |
| | $Mg_{0.1}Ni_{0.3}Zn_{0.6}Fe_2O_4$ | ***a*** =0.8347 | 0.3614 | 0.2951 | 7.38 |
| 60 | $BaFe_{12}O_{19}$ | ***a*** =0.5370, ***c*** =2.4182 | - | - | 16.84 |
| | $Mg_{0.1}Ni_{0.3}Zn_{0.6}Fe_2O_4$ | ***a*** =0.8288 | 0.3589 | 0.2930 | 15.10 |
| 90 | $BaFe_{12}O_{19}$ | ***a*** =0.5680, ***c*** =2.6132 | - | - | 21.56 |
| | $Mg_{0.1}Ni_{0.3}Zn_{0.6}Fe_2O_4$ | ***a*** =0.8309 | 0.3598 | 0.2938 | 32.31 |
| 120 | $BaFe_{12}O_{19}$ | ***a*** =0.5885, ***c*** =2.3462 | - | - | 91.97 |
| | $Mg_{0.1}Ni_{0.3}Zn_{0.6}Fe_2O_4$ | ***a*** =0.8405 | 0.3639 | 0.2972 | 107.05 |

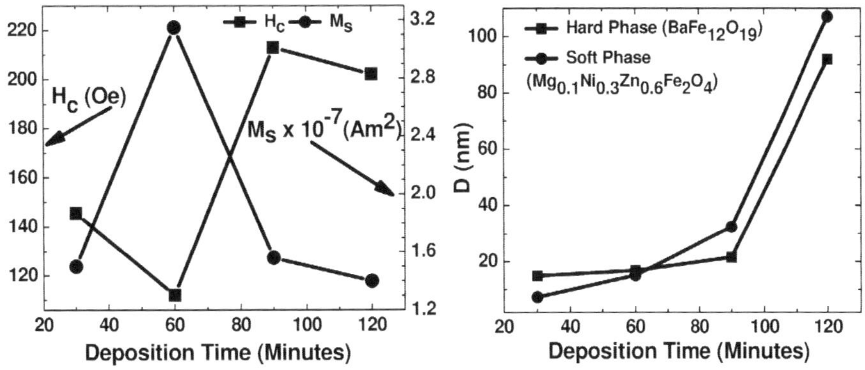


**Figure 2** (Left Panel) Variation of coercivity (***H**$_C$*) and magnetization (***M**$_s$*) with deposition time. (Right Panel) Variation of average grain diameter (***D***) with deposition time.

creation of more oxygen vacancies. Considerable increase in ***M**$_s$* is obtained for the prepared films but expected increase in ***H**$_C$* is not observed, ascribable to lower anisotropy and weak exchange coupling strength between the hard and soft phase. Variation of ***D*** for the hard and soft phase with different deposition time is depicted in figure 2 (right panel). As the deposition time increases ***D*** of both hard and soft phases increases, attributed to crystal growth and morphological evolution. Highest grain size for soft and hard phase is obtained for 120 minutes deposition time. Deposition of the thin film at a particular temperature but for higher time can lead to increase in ***D***.

## 4. Summary

($BaFe_{12}O_{19}$:$Mg_{0.1}Ni_{0.3}Zn_{0.6}Fe_2O_4$ (2:1)) composite films were deposited by PLD on Si (100) substrate at different deposition time – 30, 60, 90 and 120 minutes. Effect of deposition time on structural and magnetic properties were studied. Formation of hard and soft phases in all the prepared thin films for different deposition time was confirmed by XRD. Presence of $Fe_2O_3$ as secondary phase was also detected in all the composite films except for the deposition time - 90 minutes. With deposition time increase in average grain diameter of both the phases $BaFe_{12}O_{19}$ (hard) and $Mg_{0.1}Ni_{0.3}Zn_{0.6}Fe_2O_4$ (soft) were observed, attributed to crystal growth and morphological evolution. Highest value of ***H**$_c$* (213.03 Oe) and ***M**$_S$* ($3.15 \times 10^{-7}$ Am$^2$) is obtained for the deposition time - 90 and 60 minutes respectively, ascribed to the coexistence of two phases. Objective to get better magnetic parameters - ***M**$_s$* is somewhat attained but, substantial increase in ***H**$_c$* is not observed, ascribed to low anisotropy and weak exchange coupling strength between the hard and soft phase.

**Acknowledgments**
This work, M. S. are supported by project No. CSR-I/CSR51/2011-12. Authors thank Dr. M. Gupta, Dr. V. R. Reddy, UGC-DAE CSR, Indore XRD measurements and, Dr. D. M. Phase, Dr. R. J. Chaudhary for their help in pulsed laser deposited samples.

**References**

[1] Kneller E F and Hawig R 1991 *IEEE Trans. Magn.* **27** 3588
[2] Hazra S, Ghosh B, Joshi H R, Patra M K, Jani R K, Vadera S R, Ghosh N N, 2014 *RSC Adv.* **4** 45715
[3] Fullerton E E, Jiang J S, Grimsditch M, Sowers C H, and Bader S D, 1998-II *Physical Review B* **58** 12193
[4] Ghosh A, Satalkar M, Rathod S, Nag S P, Vyas P, Kane S N, Ghodke N, Prasad R, R Dwivedi 2013 *Int. J. of Mod. Phys.: Conf. Series* **22** 28
[5] Satalkar M, Kane S N, Ghosh A, Ghodke N, Phase D M, Chaudhary R J, Pasko A, LoBue M and Mazaleyrat F 2014 *J. of Phys.: Conf. Series* **534** 012043

# Structural, Optical and EXAFS Studies of Nickel Substituted Copper Ferrites Nano-Particle by Sol-Gel Auto Combustion Method

**Ashutosh Mishra, Mehjabeen khan, Kanaka M. Jarabana, Supriya Bisen**
School of Physics, Devi Ahilya University, Indore-452001, India

E-mail:amishra1960@yahoo.co.in, mehjabeen.khan2013@gmail.com

**Abstract.** The Cu-Ni ferrites with general formula $Cu_{1-x}Ni_xFe_2O_4$ (where x=0.0, 0.1, 0.2, 0.3) were prepared by sol gel method. The Extended X-ray absorption fine structure (EXAFS), X-ray diffraction (XRD) and Fourier transform infrared spectroscopy (FTIR) measurements were carried out. EXAFS spectra have been recorded at the k edge of Iron using the dispersive EXAFS (DEXAFS) beam line at 2.5 GeV at Indus -2 synchrotron radiation source RRCAT, Indore, India. The EXAFS data have been analyzed using the computer software Athena. These have been used to determine the bond lengths in these ferrites with the help of three different methods, namely, Levy's, Lytle's and Lytle, Sayers and Stern's (LSS) methods. The phase uncorrected bond lengths have also been obtained from Fourier transformation method and the results obtained have been compared with the results obtained from LSS method, which also gives phase uncorrected bond lengths. XRD shows the structure is the tetragonal, and FTIR was used to determine the nature of the vibrational modes present in the system.

## 1. Introduction

A ferrite is a type of ceramic compound composed of iron oxide ($Fe_2O_3$) combined chemically with one or more additional metallic elements.[1] Ferrites are usually non-conductive ferromagnetic ceramic compounds derived from iron oxides such as hematite ($Fe_2O_3$) or magnetite ($Fe_3O_4$) as well as oxides of other metals. Ferrites technologically essential materials that are used in the fabrication of magnetic, electronic and micro- wave devices. They have gained technological importance by virtue of their high resistivity and negligible eddy current losses [2, 3]. Nickel-copper ferrites play significant role among magnetic materials due to their high electrical resistivity, high saturation magnetization and high magnetic permeability. Ni-Cu ferrites have appealing electrical and magnetic properties, as copper ferrite is one of the most interesting spinel ferrite among all the ferrites [4-6]. It has been reported by several research groups that, the magnetization in Ni-Cu ferrites decreases with increasing Cu content and vice-versa The Cu content was found to have a significant influence on the electromagnetic properties [7-13] .Therefore it would be meaningful to investigate the structural and magnetic properties of Cu –Ni ferrites. Thus, in this paper, XRD, EXAFS studies, FTIR of the prepared Cu-Ni ferrites have been reported.

## 2. Experimental details

To prepare $Cu_{1-x}Ni_xFe_2O_4$ (where x=0.0, 0.1, 0.2 and 0.3), the reactants Cu $(NO_3)_2$, Ni $(NO_3)_2$, Fe $(NO_3)_3$ and citric acid should be combined in a molar proportion of 1:2:2.22. All nitrates and citric acid were dissolved in distilled water and stirred continuously. During the constant stirring ammonia solution was added drop wise to adjust pH7; the solution was heated at 100 $^0$C to transform into gel. The gel burnt in a self propagating combustion manner until all gels were completely burnt to form ashes. The as prepared powder was grind using a mortar and pestle, and then calcined in muffle furnace at 800 for 4 hours and cooled slowly to the room temperature.

The synthesized ferrites were characterized by XRD measurements in order to elucidate their geometry. X-ray diffraction pattern have been recorded by Bruker D8 advance diffractometer at IUC, Indore. The data obtained has been preceding using XRD data analysis program Origin 6.0 Professional. From the experimental measurements, various parameters, e.g., particle size, lattice parameter have been estimated.

Fourier transform infrared spectra of the powders were obtained by Br disc technique by using VERTEX 70 (IUC, INDORE) in the range of 4000-400 cm$^{-1}$.

X-ray absorption spectra of these ferrites have been recorded at BL-8 Dispersive EXAFS beam line at the 2.5-GeV INDUS-2 Synchrotron Source, RRCAT, Indore, India. The experimental data have been analyzed using the available computer software packages Origin 6.0 professional and Athena.

## 3. Result and Discussion

### *3.1. X-ray Diffraction*

XRD patterns are shown in figure 3.1. All the samples were characterized by x-ray diffraction using Cu-kα radiation. The diffraction patterns of samples were recorded between 2 theta (ranging from 10$^o$ to 70$^o$) and intensity. All detectable peaks are indexed as Cu-Ni ferrites with an inverse structure, as shown in the standard data (JCPDS: 340425). The diffraction peak of CuO impurity is observed at diffraction angle 38.7$^o$; it is reported by Asbrink and Waskowska et al [14].

The lattice parameters a and c were calculated from XRD pattern shown in table 1.the values of lattice parameters a and c is close to that reported for $CuFe_2O_4$ (a=5.844 and c=8.630 Angstrom) in the standard data (JCPDS: 340425). The structure of Copper Ferrite is tetragonal. The particle size of the samples is estimated using Debye sheerer formula. The particle size was found to be within in the range 45.6 -58.17 nm.

### *3.2. Fourier Transform Infrared Spectroscopy (FTIR)*

Figure 3.2 depicts FTIR spectra of all the compounds. All compounds exhibit a prominent fine fundamental absorption band at 770 cm$^{-1}$, which represents ferrite phase (attributed to stretching metal-oxygen). Also, the spectrum contains a small peak near 463 cm$^{-1}$ leading to the presence of M-O vibrations. All the samples under investigation contain the multiple peaks (M-O absorption bands) in

**Table 3.1.** Variation of lattice constant and particle size with doping Ni

| Sample | Particle size (in nm) | c/a | Lattice constant α(A$^0$) | c(A$^0$) |
|---|---|---|---|---|
| $CuFe_2O_4$ | 48.6 | 1.53 | 5.7982 | 8.9087 |
| $Cu_{.9}Ni_{.1}Fe_2O_4$ | 55.61 | 1.59 | 5.8451 | 9.3380 |
| $Cu_{.8}Ni_{.2}Fe_2O_4$ | 45.69 | 1.47 | 5.8910 | 8.6357 |
| $Cu_{.7}Ni_{.3}Fe_2O_4$ | 58.17 | 1.42 | 5.902 | 8.4042 |

the frequency region 1500-2625 $cm^{-1}$, which are due to the presence of Ni-O and Cu-O or Fe-O vibrations.

*3.3 EXAFS*

The X-ray absorption spectra have been recorded at BL-8 Dispersive Extended X-ray Absorption Fine Structure (EXAFS) beam line at the 2.5-GeV INDUS-2 Synchrotron Source, RRCAT at Indore, India. The normalized EXAFS spectra have been Fourier transformed. EXAFS maxima and minima positions have been obtained. The bond length has been determined using four different methods, i.e., Levy's, Lytle, Sayers and Stern's (LSS) and Fourier transformation methods. The values of energy E and wave vector k corresponding to these maxima and minima have been shown in Table 3.3 and graph between them shown in figure 3.3(a). The magnitudes of Fourier transform of figure 3.3(b).

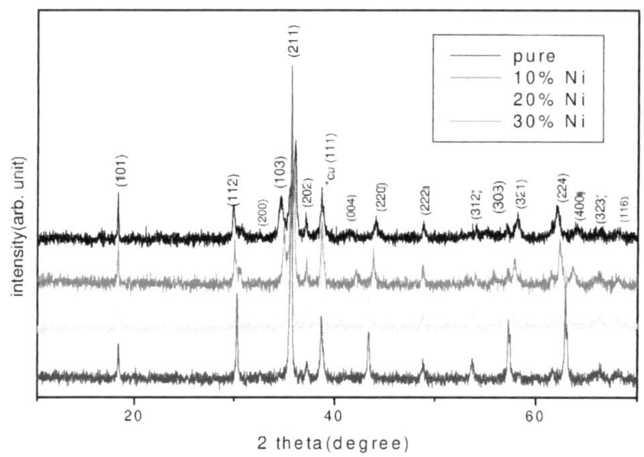


**Figure 3.1**. X-ray diffraction pattern of $Cu_{1-x}Ni_xFe_2O_4$ (where x=0.0, 0.1, 0.2, 0.3) sintered at temperature 800 °C

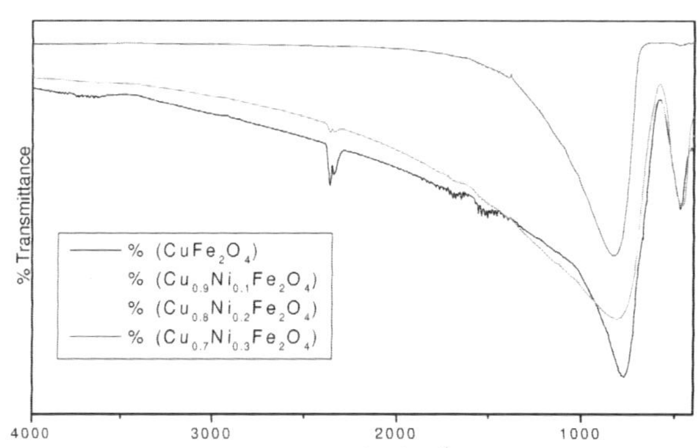


**Figure 3.2.** FTIR spectrum of $Cu_{1-x}Ni_xFe_2O_4$ (where x=0.0, 0.1, 0.2, 0.3)

**Table 3.3(a)**. Energy E and wave vector K for EXAFS maxima and minima at the K-absorption edge of Copper ferrite and their corresponding value of n and Q

| | | | $CuFe_2O_4$ | |
|---|---|---|---|---|
| Structure | n | Q | E (eV) | K($A^{-1}$) |
| A | 0 | 2.04 | 13.72 | 1.90 |
| A | 1 | - | 37.15 | 3.12 |
| B | 2 | 6.04 | 57.83 | 3.99 |
| β | 3 | - | 98.89 | 5.12 |
| C | 4 | 12.00 | 106.8 | 5.30 |
| Γ | 5 | - | 135.0 | 5.96 |

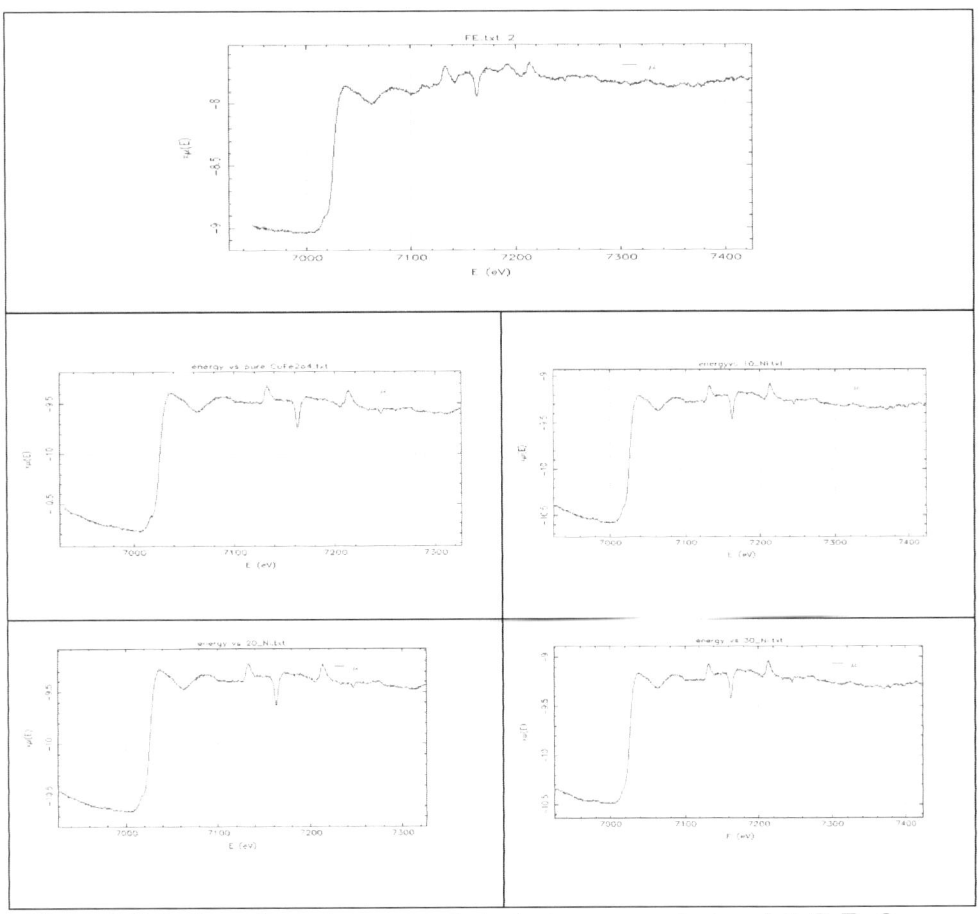


**Figure 3.3(a).** Raw absorption versus photoelectron energy for iron foil, $CuFe_2O_4$, $Cu_{.9}Ni_{.1}$ $Fe_2O_4$, $Cu_{.8}Ni_{.2}$ $Fe_2O_4$ and $Cu_{.7}Ni_{.3}Fe_2O_4$.

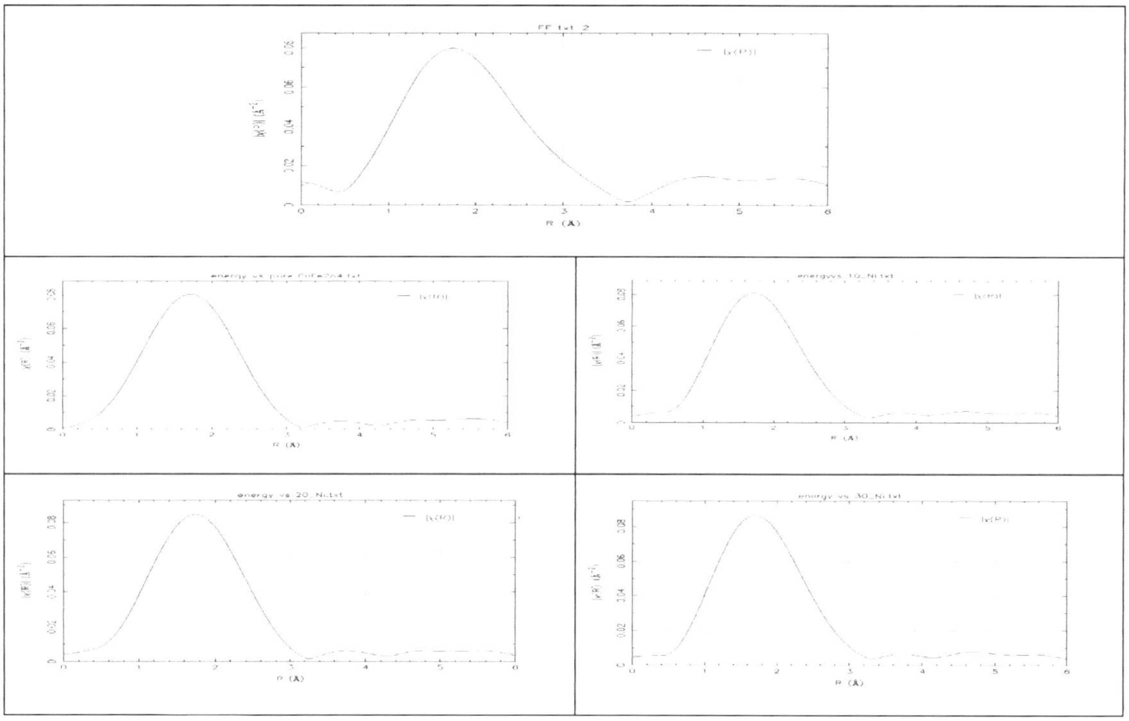

**Figure 3.3(b).** Magnitude of Fourier transform of the $\chi(k)$ versus k curves for iron foil, $CuFe_2O_4$, $Cu_{.9}Ni_{.1}\ Fe_2O_4$, $Cu_{.8}Ni_{.2}\ Fe_2O_4$ and $Cu_{.7}Ni_{.3}Fe_2O_4$

The bond length of Cu-Ni Ferrites all sample were calculated by Fourier transform, has been compared with Levy's, Lytle and L.S.S.methods are shown in Table 3.3(b).

**Table 3.3(b).** Values of first shell bond lengths (in Å) calculated from Levy's, Lytle's, L.S.S. and Fourier transform methods for Cu-Ni ferrites.

| | | | Phase uncorrected | |
|---|---|---|---|---|
| | Levy's method | Lytle's method | L.S.S. method | F.T.method |
| Sample | $R_1$ | $R_S$ | $R_1$-$\alpha_1$ | R |
| Iron foil | 2.35 | 1.97 | 1.95 | 1.70 |
| Pure Cu | 2.53 | 2.01 | 1.88 | 1.70 |
| 10% Ni | 2.52 | 1.96 | 1.83 | 1.70 |
| 20% Ni | 2.51 | 2.01 | 1.89 | 1.73 |
| 30% Ni | 2.46 | 2.01 | 1.86 | 1.69 |

## 4. Conclusion

Copper –Nickel ferrites with general formula $Cu_{1-x}\ Ni_x\ Fe_2O_4$ (where x=0 to x=0.3) were studied using X-ray diffraction. The X ray diffraction pattern (XRD) of the prepared ferrites were obtained using Bruker Advanced D8 Diffractometer at IUC, Indore.XRD diffraction patterns show that all samples have a pure single phase tetragonal spinel structure. From these patterns, the lattice parameters, crystallite size and density of these samples have been calculated. The XRD patterns were compared and indexed using JCPDS. As seen from figure 4.1, XRD peaks are very narrow indicating the higher grain size. The particle size was found to be within in the range 45.6 -58.17 nm.

Infrared spectra of the ferrites were obtained by KBr disc technique by using VERTEX 70 (IUC, Indore) in the range of 4000-400 $cm^{-1}$. Its innovative design results in the highest flexibility and highest performance. All compounds exhibit a prominent fine fundamental absorption band at 770 $cm^{-1}$, which represents ferrite phase (attributed to stretching metal-oxygen). Also, the spectrum contains a small peak near 463 $cm^{-1}$ leading to the presence of M-O vibrations. All the samples under investigation contain the multiple peaks (M-O absorption bands) in the frequency region 1500-2625 $cm^{-1}$, which are due to the presence of Ni-O and Cu-O or Fe-O vibrations.

From the obtained data of EXAFS, the bond length in the complexes has been determined by three different methods viz. Levy's, Lytle's and L.S.S. method. From the Fourier transforms of the EXAFS spectra, the uncorrected bond length (uncorrected for phase shift) has been determined. It has been observed that the value of phase uncorrected bond length, as determined from L.S.S. method and that determined from the Fourier transformation method, are in good agreement with each other.

**Acknowledgment**
The authors are thankful to Dr. Mukul Gupta, UGC-DAE CSR for XRD, Dr. S. N. Jha for his help in recording the EXAFS spectra and Dr. Uday Deshpandey for FTIR measurement, Indore for providing the experimental facilities and for the precious guidance.

**References**

[1] Carter, Barry C; Norton, Grant M 2007 *Ceramic Materials: Science and Engineering. Springer* 212
[2] Venkataraju C, Sathishkumar G and Sivakumar K 2010 *Journal of Magnetism and Magnetic Materials* **322** 230.
[3] Goldman A *"Modern Ferrite Technology,"* Van Nos- trand Reinhold, New York, 1990.
[4] Shirsath S E, Toksha B G and Jadhav K M 2009 *Materials Chemistry and Physics* **117** 163.
[5] Upadhyay C, Mishtra H C D, Verma S and Anand Das R P 2003 *Journal of Magnetism and Magnetic Materials* **260** 188.
[6] Cvejic Z, Rakic S, Kremenovic A, Antic B, Jovale-kic C and Colomban P 2006 *Solid State Sciences* 8 908.
[7] Gabal M A, Angari Y M Al and Al-Juaid S S 2010 *Journal of Alloys and Compounds* **492** 411
[8] Doh G, Kim E B, Lee B H and Oh J H 2004 *Journal of Magnetism and Magnetic Materials* **272** 2238.
[9] Gabal M A, Angari Y M Al and Kadi M W 2011 *Polyhedron* **30** 1185.
[10] Msomi J and Moyo T 2009 *Journal of Magnetism and Magnetic Materials* **321** 1246. Jain S C, Willander M, Narayan J and van Overstraeten R 2000 *J. Appl. Phys.* **87** 965
[11] Azadmanjiri J, Salehani H K, Barati M R and Far-zan F 2007 *Materials Letters* **61** 84.
[12] Roumaih K 2008 *Journal of Alloys and Compounds* **465** 291.
[13] Tan X Y, Li G Y, Zhao Y and Hu C W 2009 *Materials Research Bulletin* **44** 2160.
[14] Asbrink S and Waskowska A 1991 *J. Phys. Condens. Matter.* **3** 8173

# EXAFS spectra using synchrotron radiation of Cu (II) complexes

**Samrath Ninama[1], A Mishra[2]**
[1]Kasturbagram Rural Institute, Kasurbagranm Indore [MP] 452020, India
[2]School of Physics, Vigyan Bhawan, Devi Ahilya University, Indore - 452001, India

E-mail: samrathn@gmail.com

**Abstract**: EXAFS analysis of Cu (II) complex as a ligand of 2-methyl-3-[(bis-aniline(R) phenyl]-3H-1, 5 benzodiazepine. Extended X-ray absorption fine structure (EXAFS) spectra have been recorded at the K-edge of Cu (II) using the energy dispersive EXAFS beam line at 2.5GeV Indus - 2 synchrotron source at RRCAT, Indore, India. A theoretical EXAFS data analysis is also carried out by Fourier analysis of experimental EXAFS data of the copper (II) complexes. This analysis includes details of the Fourier transform of the data and the extraction of metal-ligand bond length. Bond lengths determined from data analysis methods are compared with the bond lengths obtained from several other known techniques, namely, Levy's, Lytle's and Lytle, Sayers and Stern's (LSS) methods. These data have also been calibrated by derivative method and bond lengths have also been obtained from Fourier transformation method and the results have been compared with the each other. The EXAFS data have been analyzed using the computer software *Athena*.

## 1 Introduction

In the present work, the X-ray absorption spectra have been recorded at the K-edge of copper macro-cyclic complexes. The coordination chemistry of macro-cyclic ligand is a fascinating area of research. The extended X-ray absorption fine structure (EXAFS) spectroscopy is an effective technique for selectively investigating the local coordination environment around the metal active site of complexes. The EXAFS K-edge of Cu (II) complexes were recorded at 8980 eV and also determined the bond length between metal-ligand of four complexes namely

(a)Cu1- Cu(II)2-methyl-3[(bis-ortho nitophenyl)]-3H 1, 5 benzodiazepine chloride.
(b)Cu2-Cu(II)2-methyl-3[(bis-meta nitrophenyl)]-3H 1, 5 benzodiazepine chloride.

A method for determining the metal-ligand bond length by an appropriate Fourier analysis of the K-edge data has been presented. This is illustrated by a detailed analysis of the data for a series of metal organic based on pyridine ligand systems, which model active centers of natural copper containing metal proteins [1]. The Fourier transform of the experimental EXAFS data is done using the FEFFIT program [2] and metal-ligand bond length is determined. This is open source program for flexible data reduction/fitting engine that can be used directly in command line mode or called by external programs. *Athena* has nine integrated graphical interfaces that use IFEFFIT for most of the numerical calculations. The noticeable advantage of this technique is that the X-ray absorption spectroscopy may be used as a bulk probe to investigate the surface effects. In particular, the technique permits the analysis of buried interfaces systems that are difficult to be analyzed even with more sophisticated surface investigation equipment. Applications of EXAFS can be found in literature on a variety of fields, namely, chemical reactions at the solid state [3], surface treatments [4] and structural studies of thin films [5].A structural analysis of X-ray

Journal of Physics: Conference Series **755** (2016) 012045 doi:10.1088/1742-6596/755/1/012045

absorption fine structure (XAFS) spectroscopy has been used for a local structural refinement on an atom of interest [6]. In general, the small pre-edge peaks of the K-edge absorption spectra for transition metal compounds have been assigned to the transition from 1s to *n*d orbital, even though it is a dipole forbidden transition. Since the transition is very sensitive to chemical environment, the pre-edge feature in XAFS studies has been used to infer the local structure around the central atom. Some molecular orbital calculation studies have been reported that the pre-edge peak intensity is related to 3d and 4p orbital mixing by perturbation of site symmetry and increases gradually with the departure from a centro-symmetric environment [7]. In this context, the extended X-ray absorption fine structure (EXAFS) spectra have provided many quantitative structural characterizations such as inter-atomic distance, coordination number and Debye-Waller factor [8]. Since the EXAFS refinement does not depend on the long-range order of atomic arrangement and is very sensitive to atomic local sites, it is a powerful technique for local structural analysis of partially substituted inorganic compounds.
EXAFS spectroscopy is a useful method for determining the local structure around a specific atom in disordered systems. This technique provides information about the coordination number, the nature of the scattering atoms surrounding a particular absorbing atom, the inter-atomic distance between the absorbing atom and the backscattering atoms and Debye–Waller factor. The measurements are done with high energy X-rays, which are normally generated by synchrotron radiation sources. The data analysis is facilitated by specially developed program packages suitable for evaluation purposes. EXAFS spectroscopy is employed in several fields for a variety of applications [9-10].

**2.1 Methodology**
Copper Complexes have been prepared by chemical root method. The chemicals used are of pure grade.
**2.2. EXAFS Spectra Recording at Indus-2 Beamline**
The complexes were finely powdered and then absorption screens were prepared by carefully spreading calculating amount of the powdered on 1 cm area of kapton tape. The Copper K-edge EXAFS spectra of these absorption screens have been recorded at BL-8 dispersive EXAFS beamline at 2GeV Indus -2 synchrotron radiation source at Raja Ramanna Center for Advanced Technology (RRCAT), Indore, India [12-14]. Schematic diagram describing basic principal of the beam line is shown in figure 1. The beam line has 460 mm long Si (111) crystal having 2d value equal to 6.2709A° mounted on an elliptical bender, which can bend the crystal to take shape of an ellipse [15].The elliptical optics offers maximum aberration. The radiation transmitted through the sample is detected by a positive-sensitive CCD detector having 2048 × 2048 pixels. The whole absorption spectrum can be recorded simultaneously in even a few microseconds for a rich sample. The beamline has a resolution of 1 eV at the photon energies of 10keV. The plot of absorption *versus* photon energy is obtained by recording intensities $I_0$ and $I_t$, as the CCD outputs, without and with sample, respectively. Using the relation $I_t=I_0\, e^{-\mu x}$, where μ is the absorption coefficient and x is thickness of the absorber. For recording the EXAFS spectrum on the dispersive EXAFS beamline, the crystal bender and the goniometer have been set to cover the energy range of the Copper K-edge absorption spectra.

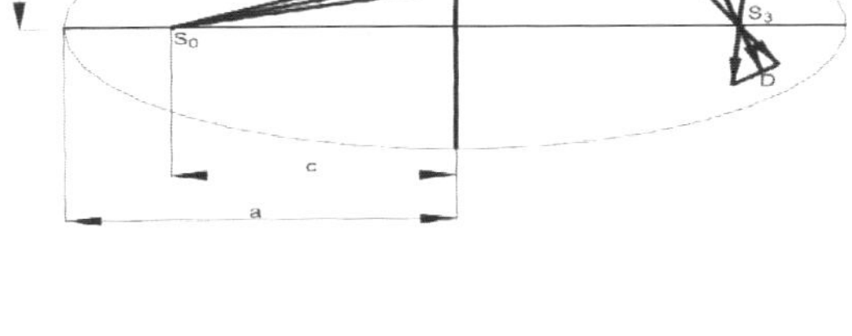

**Figure 1** Schematic showing the working principle of the dispersive EXAFS beamline (BL-8) at Indus - 2 Synchrotron facilities, Indore, India.

**Table 1**. Values of first shell bond length (in Å) for copper complexes

| Complex | Levy's | Lytle | LSS | F.T. |
|---|---|---|---|---|
| Cu1 | 2.48 | 2.17 | 1.55 | 1.71 |
| Cu2 | 2.17 | 1.97 | 1.76 | 1.75 |

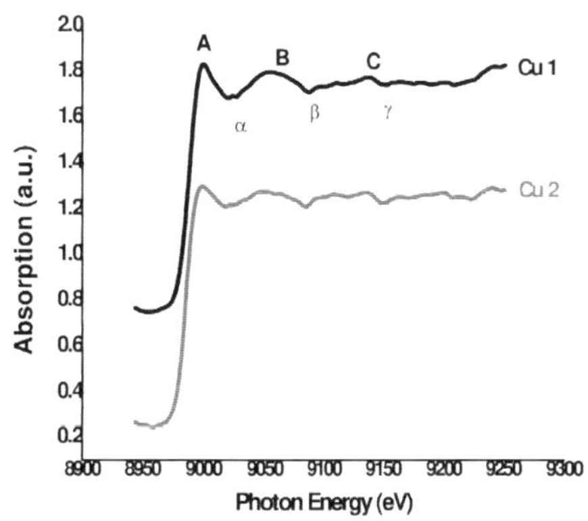


Figure 2 μ (E) vs. E curve of Copper complexes

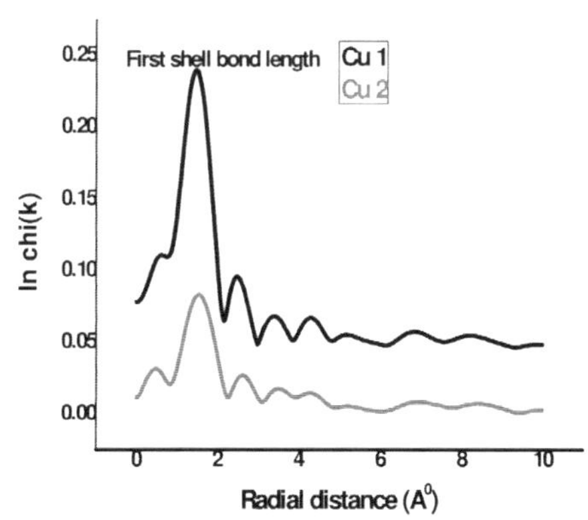


Figure 3 Bond length by FT method

## 3. Results and discussion

### Extended X-ray Absorption fine Structure (EXAFS)

Although, Fourier transform of EXAFS data is a general and powerful technique to determine many physical parameters; however, it is possible to determine nearest-neighbor distances and phase shifts from a simple graphical technique. This is possible because the first shell scattering usually dominates the EXAFS curves; then, if only the major EXAFS peaks are used, the period will be that of the first coordination shell. In this method, the energy positions of the maxima and minima in EXAFS are measured with respect to the inflection point on the main absorption edge, which signifies the Fermi level in metals and the first unoccupied level of suitable symmetry available for the absorption of the ejected electron in the case of compounds.

The X-ray absorption spectra, i.e., normalized μ (E) vs. E curve of Copper is shown in figure 2. The bond lengths have been determined using the following four methods, i.e., Levy's, Lytle, Lytle, Sayers and Stern's (LSS) and Fourier transformation methods and the results obtained are given in table1.

### 3.1 Levy's method

The bond lengths According to Levy's method [17], are given by Bragg relation $R_1 = [151/\Delta E]^{1/2}$ Å, where ΔE is the difference in eV of the energies of the EXAFS maximum B and minimum β and $R_1$ is the radius of the first coordination sphere.

### 3.2 Lytle's method

According to Lytle [18] method the bond length ($R_s$) is evaluate by the radius $R_s$ of equivalent polyhedron through the relation $R_s = [37.60 / M]^{1/2}$.where M is slope of E vs Q plot ,energy values (E) is EXAFS maxima and Q = 2.04, 6.04, 12.0, and 20.0 are constant . The values of $R_s$ calculated with the help of this method are reported in table 1.

### 3.3 Lytle, Sayers and Stern's (LSS) method

In the LSS method [19], for determination of the nearest neighbor distances, gives the value of (1/2+n) $\pi$= $2k(R_1 - \alpha_1)+2\beta_1$ ,where $R_1$ is the bond length. $(R_1 - \alpha_1)$ is the phase uncorrected bond length shown in table 1.

### 3.4 Fourier transformation method

The X-ray absorption spectra were recorded and calculated of copper complexes Fourier transformation is shown in figure 3. The bond lengths of first successive maxima had been determined using Fourier transformation methods and the results obtained are given 1.71 and 1.75 respectively.

The Fourier transforms peaks at the radial distances of the neighboring atoms from the absorbing atom. The distances found in Fourier transform are, however, shorter by 0.2-0.5 Å than the actual distances due to energy dependence of the phase factors in sine function of the theoretical expression for EXAFS, which is known as the EXAFS equation [20]. The peaks in the Fourier transform are shifted towards the origin by an amount $\alpha_j$ and hence the peaks are at distances $R_j$-$\alpha_j$. For the first peak j=1 and hence the position of the first peak determines the distance $R_1$-$\alpha_1$. This distance is given in table 1.

## 4. Conclusion

Calibration of particular setting of the polychromator has to done and it has been shown by derivative method, and after this EXAFS spectrum has been recorded and comparison of the values of bond length $R_s$ is seen to have slightly lesser values than the corresponding bond lengths obtained by Levy's method. This may be described to the fact that Levy's method gives the radius of coordination sphere directly where as Lytle method does not. In later method, the inter atomic spacing is obtained by multiplying $R_s$ with a factor Appropriate to the geometry of the system .the value of the bond length obtained from LSS method are closer to obtained by Levy's method and Fourier transformation method.

## Acknowledgement

The authors are thankful to Dr. S. K. Deb and Dr. Shambhu Nath Jha, RRCAT, Indore for their help in recording the spectra at RRCAT ,Indore, India and also thankful to Dr. Pratibha Sharma, school of chemical science, Devi Ahilya University, Indore - 452001, India for their help in preparation of complexes.

## References

[1] Vlasenko V G, Uraev A I, Zubavichus Ya V, Chernyshov A A, Garnovskii A D, & Mamin R K, 2008 *Bulletin of the Russian Academy of Sciences: Physics* **72(4)** 468

[2] Newville M 2001 *J Synch Rad* **8** 322

[3] Mishra A, Parsai N & Dagonkar N, 2009 *Indian J Pure & Appl Phys* **47** 337

[4] Awate Ruchita, *X-Ray Spectroscopic Studies of Transition Metal Complexes (Copper and cobalt) using Arylazopyiridines as ligands* 2009 Ph D thesis, Devi Ahilya University, Indore.

[5] d'Acapito F, Ghigna P, Alessandri I, Cardelli A and Davoli I 2003 *Nucl Instrum Methods: Phys Res B* **200** 421

[6] Lutzenkirchen H and Frahm R 2005 *Physica B Amsterdam* **357** 213

[7] d'Acapito F , Milita S, Satta A and Colombo L 2007 *J Appl Phys* **102** 043524

[8] Rehr J J, Leon J M, Zabinsky S I, Albers R C 1991 *J Am Chem Soc* **113** 5135

[9] Westre T E, Kennepohl P, DeWitt J G, Hedman B, Hodgson K O, Solomon E I 1997 *J Am Chem Soc* **119** 6297

[10] Ebbinghaus S, Froba M and Reller A 1997 *J Phys Chem B*, **101** 9909

[11] Sayers D E, Stern E A and Lytle F W 1971 *Phys Rev Lett*, **27** 1204

[12] A.Mayura Panchbhai A M, Paliwal L J and Bhave N S 2008 *E-Journal of Chemistry* **5(S1)** 1048

[13] Bhattacharyya D, Poswal A K, Jha S N, Sangeeta, and Sabharwal S C 2009 *Bull. Mater. Sci.* 32

[14] Bhattacharyya D, Poswal A K, Jha S N, Sangeeta, and Sabharwal S C 2009 *Nucl. Instrum.Meth. A* 609

[15] Das N C, Jha S N, Bhattacharyya D, Poswal A K, Sinha A K and Mishra V K 2004 *Sadhana* **29** 5
[16] Lee P L, Beno M A, Jennings G, Ramanathan M, Knapp G S, Huang K, Bai J and Montano P A 1994 *Rev. Sci. Instrum.* 65
[17] Levy R M 1965 *J. Chem. Phys* **43** 1846
[18] Lytle F W 1966 *Advances in X-ray Analysis* **9** 398
[19] Lytle F W, Sayers D E and Stern E A 1975 *Phys. Rev. B* **11** 4825
[20] Stern E A, Sayers D E and Lytle F W 1975 *Phys. Rev. B* **11** 4836

# Characterization and EXAFS Studies of Cobalt Ferrites Synthesized by sol-gel auto Combustion Method

**Ashutosh Mishra[1] , Priya Singh[1], Samrath Ninama[2]**
[1]School of Physics, Devi Ahilya University, Indore-452001, India
[2]Kasturbagram Rural Institute, Kasurbagranm Indore [MP] 452020

E-mail:preeyasingh67@gmail.com

**Abstract.** The Co-Ni ferrites with general formula $Ni_XCo_{1-X}Fe_2O_4$ (where X=0.0, 0.1, 0.15, 0.2, 0.4) were prepared by Sol-gel auto combustion method. X-ray, FTIR and EXAFS measurement were carried out. Extended X ray Absorption Fine Structure (EXAFS) spectra have been recorded at the K-edge of Fe using the Dispersive EXAFS (DEXAFS) beam line-8 Indus-2 synchrotron radiation source RRCAT, Indore, India. The EXAFS data have been analysed using the computer softer Athena. These have been used to determine the Bond length in these ferrites with help of four different methods, namely, Levy's, Lytle's and Lytle, Sayers and stern's (LSS).

## 1. Introduction

A ferrite is a type of ceramic compound composed of iron oxide (Fe2O3) combined with one or more additional metallic elements [1]. Usually ferrites have Spinel structure ($AB_2O_4$ ), where A is divalent and B is a trivalent metal ion. Co-Ni ferrites belong to the group of soft ferrite material characterized by high magnetic permeability's at low losses and have numerous electronic application.[2,3]. Two types of Spinel ferrite, Normal Ferrite $M^{2+}$ are in A site and $Fe^{3+}$ in B site. Inverse Ferrite $M^{2+}$ occupies the B site and the $Fe^{3+}$ ions equally divided between A and B site. Co –Ni ferrites are mixed ferrites and the co-ordination of the Fe+3 ion in the system is very little affected by the changes in compositional parameters [4]. Extended X- ray absorption fine structure, which start approximately from 50 eV to continue up to 1000 eV above the edge. XANES is strongly sensitive to formed oxidation state and coordination chemistry of the absorbing atom while the EXAFS is used to determine the bond distance, Coordination number and species of the neighbours of the absorbing atom [5]. In the present work, the X-ray absorption spectra have been recorded at the K-edge of Iron macro-cyclic complexes. The coordination chemistry of macro-cyclic ligand is a fascinating area of research. The extended X-ray absorption fine structure (EXAFS) spectroscopy is an effective technique for selectively investigating the local coordination environment around the metal active site of complexes.

## 2. Experimental Techniques

The-ray diffraction pattern (XRD) using Bruker D-8 advance UGC DAE CSR , Indore, X-ray powder diffract meter was obtained with Ni- filtered Cu Kα radiation ( $\lambda$= 1.54$A^0$) having voltage 40KV and current 100mA. The X-ray was detected using a counting detector based on silicon strip technology (Bruker Lynx Eye detector 1). Infrared spectra of the complexes were obtained by KBr disc technique by using VERTEX 70 (Bruker), IUC Indore in the range of 4000-400 $cm^{-1}$ .Its innovative design result in the highest flexibility and highest performance. The X-ray absorption spectra have been recorded using Synchrotron radiation source .The X- ray spectroscopy setup is available at Raja Ramanna Centre for Advanced Technology (RRCAT) and is called beam line. This beam line BL-8 has been recently commissioned at the 2.5 GeV Indus-2 synchrotron radiation source.

**3.1 Preparation of $Ni_XCo_{1-X}Fe_2O_4$**

The $CoFe_2O_4$ nano powder was synthesized by the sol- gel auto-combustion method. The proper amount of metal nitrates and citric acid was first dissolved. The molar ratio of Co $(No_3)_2$, $Fe(No_3)_2$ and citric acid ($C_6H_8O_7.H_2O$) is 1:2:22. After preparing a sample it is calcined at temperature $1000^0C$ and $750^0C$for 2 h.

## 4. Result and discussion

### (a) X-ray diffraction

The sample was characterized at room temperature by X-Ray diffraction using CuKα radiation. X-Ray diffraction studies of Ferrite are indicative of their crystalline nature. The diffraction pattern recorded between 2θ ranging from $20^0$ to $90^0$. An analysis of diffraction patterns of all sample confirm the formation of cubic spinel structure. The value of the lattice parameter for $CoFe_2O_4$ ($8.398A^0$) agree with those reported in PCPDF file number 22-1086. The XRD pattern shows in figure 1 show the XRD pattern of pure cobalt ferrite at $1000^0C$ and $750^0C$. The X-ray density ρ calculated using the formula, $\rho=8M/Na^3$ Where M is the molecular weight, N is the Avogadro's number, and a is the lattice constant of the spinel ferrite. The calculated values of lattice parameter, unit cell volume, particle size, lattice spacing and X-ray density are listed in table 1.

**(b) FTIR Studies:** The IR spectra of sample were taken in the range 4000-400cm$^{-1}$ in KBr pellets. IR Spectrums of some of the synthesized sample with metal salts are depicted in Figure 2 . The vibrational spectra of the absorption band of the pure cobalt ferrite nanoparticle, calcined at $1000^0C$ wave observed at $\upsilon_1$ is 761.40cm$^{-1}$ and $\upsilon_2$ is 463.39cm$^{-1}$. The FTIR absorption bands of solid are usually assigned to the vibration of ions in the crystal lattice. The band υ1 (table 2) generally observed in the range 850-780cm$^{-1}$, corresponding to an intrinsic stretching vibration of the metal at the tetrahedral side, $M_{tetra}$-O, whers as the $\upsilon_2$- lowest band, usually observed in the range 450-480cm$^{-1}$ is assigned to octahedral side $M_{octa-O}$ The figure 2 shows a detailed positioning of the peaks in the range 1000-400cm$^{-1}$.The absorption band $\upsilon_1$ and $\upsilon_2$ linearly increases with the incoming content of Nickel ions.

Table 1 the calculated values of lattice parameter, unit cell volume, particle size, lattice spacing and X-ray density

| S.No | Compounds | Lattice constant a(A0) | ρ (Kg/m$^3$) | Cell volume a $^3$(A$^{03}$) | Lattice spacing d |
|---|---|---|---|---|---|
| 1 | $CoFe_2O_4$ | 8.398 | 5261.58 | 592.280 | 0.2532 |
| 2 | $Ni_{0.1}Co_{0.9}Fe_2O_4$ | 8.334 | 5383.20 | 578.842 | 0.2513 |
| 3 | $Ni_{0.15}Co_{0.85}Fe_2O_4$ | 8.340 | 5371.34 | 580.090 | 0.2514 |
| 4 | $Ni_{0.2}Co_{0.8}Fe_2O_4$ | 8.345 | 5427.11 | 581.304 | 0.2516 |

Journal of Physics: Conference Series **755** (2016) 012046 doi:10.1088/1742-6596/755/1/012046

Table 2 FTIR absorption bands values of υ1 and υ2.

| Compound | υ1 ($cm^{-1}$) | υ2 ($cm^{-1}$) |
|---|---|---|
| $CoFe_2O_4$ | 767.40 | 462.01 |
| $Ni_{0.1}Co_{0.9}Fe_2O_4$ | 780.27 | 457.33 |
| $Ni_{0.15}Co_{0.89}Fe_2O_4$ | 811.28 | 468.44 |
| $Ni_{0.2}Co_{0.8}Fe_2O_4$ | 760.32 | 476.05 |
| $Ni_{0.4}Co_{0.6}Fe_2O_4$ | 814.74 | 472.54 |

Table 3 - Value of bond lengths calculated from Levy's, Lytle's, LSS, Fourier transformation method

| Sample | R - Levy's $A^0$ | R -Lytle's $A^0$ | R-LSS $A^0$ | R -FT $A^0$ |
|---|---|---|---|---|
| Fe | 2.80 | 2.01 | 1.85 | 1.95 |
| $CoFe_2O_4$ | 2.01 | 2.03 | 1.87 | 1.71 |
| $Ni_{0.1}Co_{0.9}Fe_2O_4$ | 1.85 | 2.05 | 1.88 | 1.82 |
| $Ni_{0.15}Co_{0.85}Fe_2O_4$ | 1.92 | 2.06 | 1.88 | 1.77 |
| $Ni_{0.2}Co_{0.8}Fe_2O_4$ | 2.01 | 2.06 | 1.93 | 1.79 |
| $Ni_{0.4}Co_{0.6}Fe_2O_4$ | 2.00 | 2.07 | 1.91 | 1.76 |

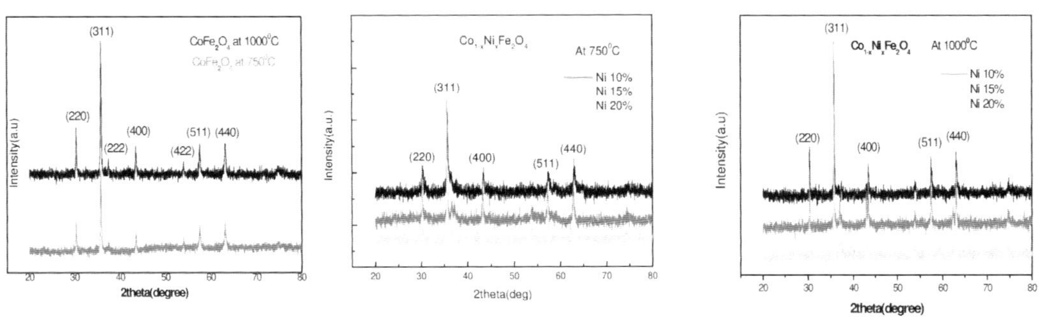


Figure 1 The XRD pattern shows in figure 1a show the XRD pattern of pure cobalt ferrite at $1000^0C$ and $750^0C$

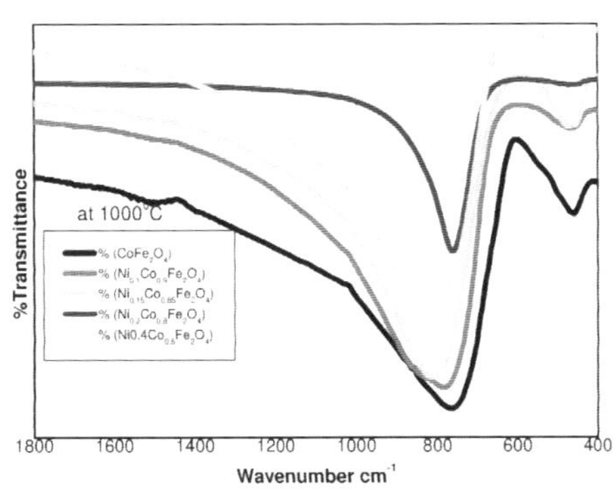


Figure 2 shows a detailed positioning of the peaks in the range 1000-400$cm^{-1}$of FTIR spectra.

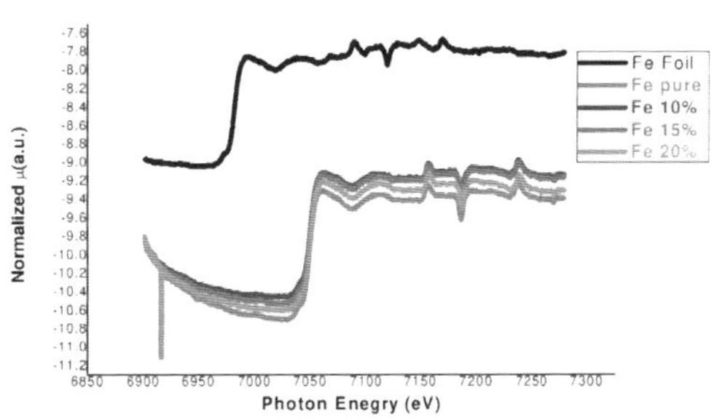


Figure 3 μ (E) vs. E curve of all samples

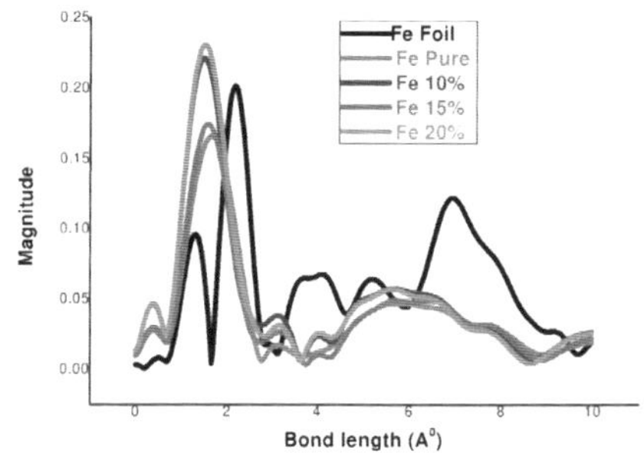


Figure 4 Bond length by FT method for all samples.

### (c) EXAFS Study

**Levy's Method:** In this method, the bond are calculated by using the relation $R_1 = \left(\frac{151}{\Delta E}\right)^{1/2}$ Å. The values of bond length are given in table 3. Normalization and absorption curve are shown in figure 3[5].

**Lytle's method:** The energy values (E) of the EXAFS maxima, and the Q values given by Lytle for p symmetry, i.e., Q = 2.04, 6.0, 12.0, and 20.0. The (E, Q) plots have been found to be linear. The slope M of the E versus Q plots has been used to evaluate the radius Rs of equivalent polyhedron, by using the relation, $Rs = [37.60 / M]^{1/2}$ the value of bond length are given in table 3[6-7].

**LSS method:** The values of the wave vector k (Å-1) for EXAFS maxima (n = 0, 2, 4....) and minima (n =1, 3, 5....), for all the iron complexes, are presented in table 3 In the Lytle, Sayers and Stern's (L.S.S.) method for determination of the nearest neighbour distances[8-9].

**By Fourier Transform of EXAFS spectra:** The EXAFS spectra are obtained using the synchrotron DEXAFS Beamline set-up. The position of the first peak in the Fourier transform gives the value of $R_1$-$\alpha_1$ and the values are given in table 3 for all material. Magnitudes of Fourier transform are shown in figure 4 [10].

## 5. Conclusion

XRD pattern shows that all the sample crystalline in the cubic spinel structure. It is clear that a spinel structure of cobalt ferrite labelled in the JCPDS file no#22-1086 was detected in this sample whose space group is Fd3m(227).The lattice parameter , density, cell volume and lattice spacing are given in table. The band υ1, shift towards the higher frequency site, while the band υ2 remains constant with increasing Ni concentration. FTIR spectra explain tetrahedral and octahedral cluster and also conform the presence of M-O stretching band in ferrite. It has been observed that the value of the phase corrected bond length, i.e., R1 as determined from Levy's method and Rs as determined from the Lytle's method are in good agreement with each other, i.e., both method give nearly the same value of the phase corrected bond length.

**Acknowledgement**

Authors are thankful to Dr.M. Gupta, UGC DAE CSR, Indore for XRD measurements and Dr. Uday Deshpande for FT-IR measurement. Also thankful to Dr. S.N Jha for providing the beam line and other necessary facilities at Indus-2, RRCAT, Indore, India.

**References**

[1] Carter, C. Barry; Norton, Grant M 2007 *Ceramic Materials: Science Engineering Springer* **212**–15
[2] Malik-F A, Zahid M and Hassan M A 1995 *Proceeding 4$^{th}$ international symposium on Advanced materials* 618
[3] Rana-M U, Islam-M U, Ahmad-I and Abbas I 1997 *proceeding 5th International symposium on Advanced materials* 283
[4] Scherrer P, Go¨ttinger Nachrichten. 1990 **2** 98
[5] Levy R M 1965 *J. Chem. Phys.* **43** 1846
[6] Lytle F W 1966 *Advances in X-ray Analysis* **9** 398
[7] Lytle F W, Sayers D E and Stern E A 1975(a) *Phys. Rev. B* **11** 4825
[8] Sayers D E, Stern E A, and Lytle F W 1971 *Phys. Rev. Lett.* **27** 1204
[9] Stern E A 1974 *Phys. Rev. B* **10** 3027
[10] Lee P A and Beni G 1977 *Phys. Rev. B* **15** 2862

Journal of Physics: Conference Series **755** (2016) 012047 doi:10.1088/1742-6596/755/1/012047

# On structural studies and cation distribution of La added Zn-Ni-Mg-Cu spinel nano ferrite

**M Satalkar[a], S N Kane[a,1]**
[a]Magnetic Materials Laboratory, School of Physics, D. A. University, Khandwa road, Indore-452001, India

E-mail: kane_sn@yahoo.com

**Abstract.** $Zn_{0.75-x}Ni_xMg_{0.15}Cu_{0.1}La_{0.02}Fe_{1.98}O_4$ (x = 0.0, 0.15, 0.30, 0.60, 0.75) was synthesized by sol-gel auto-combustion technique. As-burnt samples were thermally annealed at 500°C for 3 hrs. The as-burnt and annealed powder were characterized by X-ray diffraction (XRD) technique. The lattice constant ($a_{exp.}$), unit cell volume ($V$) and hopping length at A and B ($L_A$ and $L_B$) deceases with increase in $Ni^{2+}$ content (x). Scherrer's grain diameter ($D$) for as-burnt and annealed samples respectively ranges between 13.26 - 22.20 nm and 23.61 - 30.93 nm. After annealing the specific surface area of the particles decreases which can be attributed to increase in the particle size. Annealing leads to variation in cation distribution at tetrahedral (A) and octahedral [B] site, which affects theoretical lattice parameter ($a_{th.}$), ionic radii of A-site ($r_A$) and B-site ($r_B$), oxygen positional parameter ($u$), tetrahedral and octahedral bond length ($R_A$, $R_B$), shared tetrahedral and octahedral edge ($d_{AE}$, $d_{BE}$) and Neel's magnetic moment ($n_B^N$). The bond angles $\theta_1$, $\theta_2$, $\theta_3$, $\theta_4$, $\theta_5$ and the distance between the metal ions at tetrahedral/octahedral site and anion ($p$, $q$, $r$) gives information about the A-A, A-B, B-B magnetic interaction. For lower values of Ni content ($0.0 \leq x \leq 0.3$), effective bond length ($R_{eff.}$) increases with annealing and for higher Ni content ($0.30 < x \leq .75$) $R_{eff}$ decreases.

## 1. Introduction

Nanocrystalline spinel ferrites are a class of magnetic materials which have been the most attractive area of the research since the last 5 decades due to its remarkable applications in magnetic switches, magnetic resonance imaging, drug delivery, magnetic ferro-fluids, magnetic bulk cores, high-density data storage, micro-wave absorbers, the high frequency and power devices especially for electromagnetic interference suppression (EMIS) [1, 2]. It is believed that the production of ferrites will increase year by year as their applications become more and more diverse. Spinel ferrites have many versatile structural and magnetic properties owing to its feasibility to make a huge number of solid solutions of different metal cations and a large compositional variability. As the crystal structure of spinel ferrite is face centered cubic (fcc) with anions ($O^{2-}$) linked with two sub-lattices namely tetrahedral (A) and octahedral [B] sites. Distribution of cations over A and B sites has a profound effect on the structural and magnetic properties of spinel ferrites. According to the distribution of cations, there are normal, inverse and mixed/disorder spinels structures, which depends on the type of ions occupying A and B sites. Occupation of A site entirely with a divalent transition metal ion produces normal spinel, while occupation of B site with a divalent transition metal ion produces inverse spinel structure. If divalent transition metal ions are present at both A and B sites, then the structure is mixed or disordered. Thus, $AB_2O_4$ type of compounds with spinel structure show interesting structural, electrical and magnetic properties, which vary with the nature of the ions, their charge and site distribution among tetrahedral and octahedral sites

It is well known that the desired structural and magnetic properties of soft ferrites can be tailored by the addition of suitable divalent, trivalent or tetravalent cations in to the spinel lattice. Incorporation of small amount of rare earth metal ions into the spinel ferrites may lead to structural distortion due to their large ionic radius and to induce strain and significantly modify the micro structural, electrical and magnetic properties. The rare-earth ions commonly reside at octahedral sites and have limited solubility in the spinel lattice due to their large ionic radii. But, the precise value of their solubility in the spinel lattice is not well known [3, 4]. Many researchers have studied the influence of various rare earth doping atoms on the properties of Li–Ni, Ni–Zn, Mn–Zn, Mg–Cu, Cu–Zn ferrites, etc. [5-9] and observed that different rare-earth ions behave differently in spinel ferrite and yield different structural properties. The structural properties and cation distribution of the spinel ferrites are governed by the chemical composition and the method of preparation. Altering the chemical composition is achieved by doping spinel ferrites with different transition metals and rare earth (R) ions. Both the type of the dopant and its amount are critical in changing the cation distribution and consequently the properties of the doped spinel ferrite [10, 11]. In spinel ferrites the proper choice of rare earth cation in spinel ferrite has a large influence upon the magnetic anisotropy of the system making the spinel ferrite an excellent option in place of hexaferrites or garnets. Thus, from the application point of view, the properties of ferrite nanoparticles can be tuned by the inclusion of suitable rare earth ions. Therefore, the main aim of the present work is to study the structural properties and cation distribution of as-prepared and thermally treated $La^{3+}$ added Zn-Ni-Mg-Cu ferrite prepared by sol-gel auto-combustion method.

## 2. Experiment details

The compositions with the general formula $Zn_{0.75-x}Ni_xMg_{0.15}Cu_{0.1}Fe_2O_4$ spinel ferrite powders with x = 0.00, 0.15, 0.30, 0.45, 0.60, 0.75 with 2% addition of $La^{3+}$ were prepared by the sol–gel auto-combustion method. For the synthesis, stoichiometric amount of citrate–nitrate/acetate pre-cursors [Zinc Nitrate - $Zn(NO_3)_2.6H_2O$, Nickel Acetate – $Ni(CH_3COO)_2.4H_2O$, Magnesium Acetate – $(CH_3COO)_2.Mg.4H_2O$, Copper Nitrate - $Cu(NO_3)_2.3H_2O$, Lanthanum nitrate - $La(NO_3)_3.6H_2O$ and Ferric Nitrate - $Fe(NO_3)_3.9H_2O$] were mixed with citric acid as fuel. Metal salt to fuel ratio was taken as 1:1. All the pre-cursors were dissolved in de-ionized water. Ammonia was then slowly added to the solution to adjust the pH at 7. Then the solution was heated at 120 $^{o}$C in air till the loose powder (fluffy) was formed called as 'dry gel or as-burnt powder' which was then annealed at 500$^{o}$C for 3 hrs.

The characterization of as-burnt and annealed powders were performed by powder X-ray diffraction (θ-2θ configuration) on a Bruker D8 advance diffractometer (equipped with a fast counting detector based on Silicon strip technology (Bruker LynxEye detector) using Cu-$K_\alpha$ radiation (wavelength 'λ' =0.1540562 nm). Experimental lattice parameter ($a_{exp.}$), unit cell volume ($V$), x-ray density ($\rho_{XRD}$), Scherrer's grain diameter ($D$), surface area ($S$) of the particles and hopping length at A ($L_A$) and B ($L_B$) site of the studied samples were calculated as described in [12-15].

Cation distribution of all the studied samples was estimated using x-ray diffraction intensities, which directly depend on the position of atoms in the spinel unit cell while the position of the x-ray diffraction peaks depend on the shape and size of the unit cell. In this work, the Bertaut method [16] is used to determine the cation distribution. This method selects a few pairs of reflections according to the expression: $\frac{I_{hkl}^{obs}}{I_{h'k'l'}^{obs}} = \frac{I_{hkl}^{cal}}{I_{h'k'l'}^{cal}}$. Where, $I_{hkl}^{obs}$ and $I_{hkl}^{cal}$ are the observed and calculated intensities for the reflection (hkl), respectively. The reflections (220), (400), (440) and (422) were used to calculate x-ray intensity as these reflections are sensitive to cation distribution among tetrahedral and octahedral sites of the spinel lattice. The calculated and observed intensity ratios were compared for several combinations of cations distribution at (A) and [B] sites as describe in [17]. The best cation distribution among the tetrahedral and octahedral sites for which theoretical and experimental ratios agree clearly, is taken to be the correct one. With the help of obtained cation distribution ionic radii of A-site ($r_A$) and B-site ($r_B$), theoretical lattice parameter ($a_{th.}$), oxygen positional parameter ($u$), tetrahedral/octahedral bond length ($R_A$, $R_B$), shared tetrahedral/octahedral edge ($d_{AE}$, $d_{BE}$), interionic distances between cation and anion ($M_e$-O) ($p$, $q$, $r$), the bond angles ($\theta_1$, $\theta_2$, $\theta_3$, $\theta_4$, $\theta_5$) and effective bond length ($R_{eff.}$) were calculated as described in [14, 18, 19]. The radius of the ions involved in the above calculation has been taken from that reported by Shannon [20].

## 3. Results and discussions

The x-ray diffraction (XRD) patterns of as-burnt and annealed $Zn_{0.75-x}Ni_xMg_{0.15}Cu_{0.1}La_{0.02}Fe_{1.98}O_4$ spinel ferrite powders with x = 0.00, 0.15, 0.30, 0.45, 0.60, 0.75 are presented in Fig. 1a and 1b respectively. The diffractograms of all the studied samples reveal the formation of cubic spinel structure. No diffraction lines attributed to the presence of impurities or metal oxides were observed. It is observed from the XRD patterns of annealed samples (fig. 1b) that the peak intensity of each reflection increases, which indicate the complete crystallization of the samples.

Table 1 depicts the experimental lattice parameter ($a_{exp.}$), unit cell volume ($V$), hopping length for A-site ($L_A$) and B-site ($L_B$), x-ray density ($\rho_{XRD}$), Scherrer's grain diameter ($D$) and specific surface area ($S$) with Ni content (x) of as-burnt and ann. 500°C/3hrs $Zn_{0.75-x}Ni_xMg_{0.15}Cu_{0.1}La_{0.02}Fe_{1.98}O_4$. The observed decrease in the $a_{exp.}$, $V$, $L_A$ and $L_B$ with increasing Ni content of La added spinel ferrite for as-burnt and annealed system is ascribed to the replacement of an ion with higher ionic radius ($Zn^{2+}$: 0.060 nm) by an ion with lower ionic radius ($Ni^{2+}$: 0.055 nm). The variation in the lattice constant is in accordance with Vegard's law [21], which suggests that in solid solution spinels within the miscibility range, a linear change in lattice constant with the concentration of the components is observed. It can be observed from table 1 that with annealing $a_{exp.}$, $V$, $L_A$ and $L_B$ decreases (except for x = 0.00) because of the change in reaction kinetics during the annealing process of $Zn_{0.75-x}Ni_xMg_{0.15}Cu_{0.1}La_{0.02}Fe_{1.98}O_4$ nanoparticles. The decrease in the values of above mentioned parameters can also be ascribed to nanosized particles and increase in their ligand field. The increase in $a_{exp.}$, $V$, $L_A$ and $L_B$ after annealing for the un doped sample can be explained using surface disorder in the nanoparticles. Nanoparticle's surface plays an important role in determining its physical properties due to large surface to volume ratio. The surface disorder in ferrite nanoparticles is due to the chemical bond frustration and oxygen

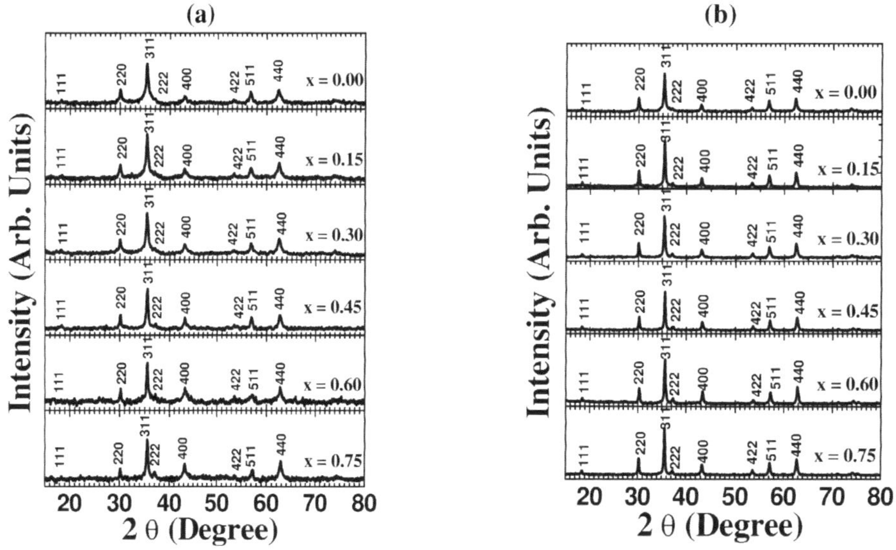


**Figure 1** XRD pattern of (a) as-burnt, (b) annealed $Zn_{0.75-x}Ni_xMg_{0.15}Cu_{0.1}La_{0.02}Fe_{1.98}O_4$.

**Table 1.** Experimental lattice parameter (***$a_{exp.}$***), unit cell volume (***V***), hopping length for A-site (***$L_A$***) and B-site (***$L_B$***), x-ray density (***$\rho_{XRD}$***), Scherrer's grain diameter (***D***) and specific surface area (***S***) with Ni content (**x**) of as-burnt and annealed 500°C /3hrs $Zn_{0.75-x}Ni_xMg_{0.15}Cu_{0.1}La_{0.02}Fe_{1.98}O_4$ spinel ferrite.

| x | $a_{exp.}$ (nm) | V (nm$^3$) | $L_A$ (nm) | $L_B$ (nm) | $\rho_{xrd}$ (Kg/m$^3$) | D (nm) | S (m$^2$/g) |
|---|---|---|---|---|---|---|---|
| **As-burnt** | | | | | | | |
| 0.00 | 0.8410 | 0.5948 | 0.3642 | 0.2973 | 5278.67 | 13.26 | 85.72 |
| 0.15 | 0.8403 | 0.5933 | 0.3639 | 0.2971 | 5269.50 | 16.61 | 68.55 |
| 0.30 | 0.8398 | 0.5923 | 0.3636 | 0.2969 | 5256.51 | 16.14 | 70.72 |
| 0.45 | 0.8381 | 0.5887 | 0.3629 | 0.2963 | 5266.02 | 21.36 | 53.34 |
| 0.60 | 0.8380 | 0.5885 | 0.3629 | 0.2963 | 5245.36 | 22.20 | 51.53 |
| 0.75 | 0.8378 | 0.5881 | 0.3628 | 0.2962 | 5226.55 | 21.66 | 53.00 |
| **(Ann. 500°C /3hrs)** | | | | | | | |
| 0.00 | 0.8426 | 0.5982 | 0.3649 | 0.2979 | 5248.65 | 23.61 | 48.42 |
| 0.15 | 0.8398 | 0.5923 | 0.3636 | 0.2969 | 5278.92 | 27.87 | 40.78 |
| 0.30 | 0.8400 | 0.5927 | 0.3637 | 0.2970 | 5252.76 | 26.00 | 43.93 |
| 0.45 | 0.8364 | 0.5851 | 0.3622 | 0.2957 | 5298.20 | 30.93 | 36.61 |
| 0.60 | 0.8354 | 0.5830 | 0.3617 | 0.2954 | 5294.48 | 27.96 | 40.53 |
| 0.75 | 0.8345 | 0.5811 | 0.3613 | 0.2950 | 5288.80 | 29.07 | 39.03 |

vacancies at different sub-lattices on the surface. With decreasing nanoparticle size, such surface disorder becomes more pronounced. The particle size for the un doped sample is the smallest among all the annealed particles. For annealed nanoparticles, the terminated surface unit cells get completed after getting atmospheric oxygen during the annealing process and reduce surface stresses/strains and hence, $a_{exp.}$, $V$, $L_A$ and $L_B$ increases. X-ray density ($\rho_{xrd}$) depends upon the molecular weight and on the lattice constant of the composition. $\rho_{xrd}$ for as-burnt samples decreases with increase in Ni content (x). The decrease in x-ray density can be ascribed to the replacement of an ion (Zn) with higher mass by an ion (Ni) with lower mass. But non liner variation in the x-ray density is observed for annealed samples. $\rho_{xrd}$ initially increases and then decreases for x = 0.30 and with further increase in Ni content $\rho_{xrd}$ again increases for x = 0.45 and finally decrease for ($0.45 < x \leq 0.75$). Increase in the x-ray density can be attributed to the fact that the decrease in the volume of the unit cell overtakes the decrease in mass of the unit cell. And decrease in the x-ray density can be ascribable to the fact that the decrease in the mass of the unit cell overtakes the decrease in volume of the unit cell. With annealing $\rho_{xrd}$ increases, (except for x = 0.0, 0.3) because during the sintering process, the thermal energy generates a force that drives the grain boundaries to grow over pores, thereby decreasing the pore volume and densifying the material. The specific surface area of the particles is the summation of the areas of the exposed surfaces of the particles per unit mass. Ferrites, offer new opportunities for enhancing the performances of solid catalysts [22] due to the much higher surface to bulk ratio compared to coarse micro-grained materials. High specific surface areas of particles are extremely important for heterogeneous catalytic processes. High surface area of ferrites is priority for catalytic purposes. There is an inverse relationship of particle size and density with surface area. In general, high surface areas imply small particle sizes. The smaller is the particle size, the larger is the surface area. Specific surface area of the as-burnt particles decreases with increasing in Ni content but the specific surface area of the annealed particles does not show any monotonic change with the doping content. After annealing the specific surface area of the particles decreases which can be attributed to increase in the particle size. Thus, as-burnt samples are more suitable for catalysts because of its high surface area. Non-monotonic behaviour of average crystallite size ($D$) with increasing Ni concentration is observed for both as-burnt and annealed systems. Such behaviour may be due to inhomogeneous stirring, diff gel/dry gel formation time etc. The crystal growth in the solution depends on various parameters

**Table 2.** Cation distribution (for A and B site), theoretical lattice parameter ($a_{th.}$) for as-burnt and annealed $Zn_{0.75-x}Ni_xMg_{0.15}Cu_{0.1}La_{0.02}Fe_{1.98}O_4$ system as a function of Ni content (x).

| | | **As-burnt** | |
|---|---|---|---|
| | **A - site** | **B - site** | **$a_{th.}$ (nm)** |
| 0.00 | $(Zn^{2+}_{0.70}Mg^{2+}_{0.075}Cu^{2+}_{0.05}Fe^{3+}_{0.175})$ | $[Zn^{2+}_{0.05}Mg^{2+}_{0.075}Cu^{2+}_{0.05}La^{3+}_{0.02}Fe^{3+}_{1.805}]$ | 0.8425 |
| 0.15 | $(Zn^{2+}_{0.60}Mg^{2+}_{0.10}Cu^{2+}_{0.05}Fe^{3+}_{0.25})$ | $[Ni^{2+}_{0.15}Mg^{2+}_{0.05}Cu^{2+}_{0.05}La^{3+}_{0.02}Fe^{3+}_{1.73}]$ | 0.8404 |
| 0.30 | $(Zn^{2+}_{0.45}Mg^{2+}_{0.10}Cu^{2+}_{0.05}Fe^{3+}_{0.40})$ | $[Ni^{2+}_{0.30}Mg^{2+}_{0.05}Cu^{2+}_{0.10}La^{3+}_{0.02}Fe^{3+}_{1.58}]$ | 0.8388 |
| 0.45 | $(Zn^{2+}_{0.30}Mg^{2+}_{0.10}Cu^{2+}_{0.05}Fe^{3+}_{0.55})$ | $[Ni^{2+}_{0.45}Mg^{2+}_{0.05}Cu^{2+}_{0.10}La^{3+}_{0.02}Fe^{3+}_{1.43}]$ | 0.8372 |
| 0.60 | $(Zn^{2+}_{0.10}Ni^{2+}_{0.30}Mg^{2+}_{0.10}Cu^{2+}_{0.05}Fe^{3+}_{0.45})$ | $[Zn^{2+}_{0.05}Ni^{2+}_{0.30}Mg^{2+}_{0.05}Cu^{2+}_{0.05}La^{3+}_{0.02}Fe^{3+}_{1.53}]$ | 0.8363 |
| 0.75 | $(Mg^{2+}_{0.10}Fe^{3+}_{0.90})$ | $[Ni^{2+}_{0.75}Mg^{2+}_{0.05}Cu^{2+}_{0.10}La^{3+}_{0.02}Fe^{3+}_{1.08}]$ | 0.8363 |
| | | **(Ann. 500°C /3hrs)** | |
| 0.00 | $(Zn^{2+}_{0.70}Mg^{2+}_{0.075}Cu^{2+}_{0.05}Fe^{3+}_{0.175})$ | $[Zn^{2+}_{0.05}Mg^{2+}_{0.075}Cu^{2+}_{0.05}La^{3+}_{0.02}Fe^{3+}_{1.805}]$ | 0.8425 |
| 0.15 | $(Zn^{2+}_{0.60}Ni^{2+}_{0.15}Cu^{2+}_{0.05}Fe^{3+}_{0.20})$ | $[Mg^{2+}_{0.15}Cu^{2+}_{0.05}La^{3+}_{0.02}Fe^{3+}_{1.78}]$ | 0.8407 |
| 0.30 | $(Zn^{2+}_{0.45}Ni^{2+}_{0.30}Mg^{2+}_{0.10}Cu^{2+}_{0.05}Fe^{3+}_{0.10})$ | $[Mg^{2+}_{0.05}Cu^{2+}_{0.10}La^{3+}_{0.02}Fe^{3+}_{1.88}]$ | 0.8398 |
| 0.45 | $(Mg^{2+}_{0.10}Cu^{2+}_{0.05}Fe^{3+}_{0.85})$ | $[Zn^{2+}_{0.30}Ni^{2+}_{0.45}Mg^{2+}_{0.05}Cu^{2+}_{0.10}La^{3+}_{0.02}Fe^{3+}_{1.13}]$ | 0.8359 |
| 0.60 | $(Zn^{2+}_{0.10}Mg^{2+}_{0.10}Cu^{2+}_{0.05}Fe^{3+}_{0.75})$ | $[Zn^{2+}_{0.05}Ni^{2+}_{0.60}Mg^{2+}_{0.05}Cu^{2+}_{0.05}La^{3+}_{0.02}Fe^{3+}_{1.23}]$ | 0.8353 |
| 0.75 | $(Mg^{2+}_{0.15}Fe^{3+}_{0.85})$ | $[Ni^{2+}_{0.75}Cu^{2+}_{0.10}La^{3+}_{0.02}Fe^{3+}_{1.13}]$ | 0.8339 |

including extrinsic parameters (molecular concentration of the material, pH and temperature) and intrinsic parameters (heat of formation, site preferences, electronic configuration). Both the parameters can affect the growth process of the crystal. Increase in the average grain diameter for the synthesized samples with annealing temperature can be attributed to the crystal growth during annealing process. During thermal annealing, solid vapour surface of the crystals are replaced by solid - solid interface via diffusion to reduce the overall surface energy which leads to the expansion of volume. During annealing the activation energy essential for the nucleation is lowered which enhances the crystallite growth. At elevated temperature small grains begin to grow and the free energy of the system decreases as the energy of the grain boundaries is reduced.

Table 2 illustrates the cation distribution for tetrahedral (A) and octahedral [B] site. With increasing Ni content for as-burnt samples up to x = 0.45, $Ni^{2+}$ ions occupy B site forcing $Zn^{2+}$ ions to migrate completely from B to A site. For x = 0.60, 50% of Ni content occupies both A and B site and with further increase in Ni content i. e. for x = 0.75, $Ni^{2+}$ ions migrate completely to B site. For x =0.75, $Cu^{2+}$ ions occupy only B site. When $Ni^{2+}$ ion is doped for $Zn^{2+}$ ion, 66.7% $Mg^{2+}$ ions occupy A site and 33.3% $Mg^{2+}$ ions occupy B site. With annealing there is redistribution of cations from one site to another. With initial increase in Ni concentration (x = 0.15) for annealed samples, $Ni^{2+}$ ions occupy A site forcing $Mg^{2+}$ ions to migrate completely from A to B site. For x = 0.30, $Ni^{2+}$ ions occupy A site and 66.7 % $Mg^{2+}$ ions migrate to A site. 66.7 % $Mg^{2+}$ ions occupy A site up to x = 0.60. But for x 0.75, $Mg^{2+}$ ions occupy only A site. $Ni^{2+}$ ions are distributed only on B site for $0.30 < x \leq 0.75$. No changes in the distribution of $Cu^{2+}$ ions is observed for A and B site. But for x =0.75 $Cu^{2+}$ ions occupy only B site. With increase in Ni content, $Zn^{2+}$ ions occupy only A site up to x = 0.30 and for x = 0.45, $Zn^{2+}$ ions migrate completely from A to B site. For x = 0.60, 66.7% $Zn^{2+}$ ions occupy A site and the remaining $Zn^{2+}$ ions occupy B site. The rare earth ion ($La^{3+}$) occupies only octahedral site for both as-burnt and annealed system. Perusal of table 1 shows that the cation distribution obtained from the x-ray diffraction is in close conformity with the real distribution, as there is realistic agreement between theoretical ($a_{th.}$) and experimental ($a_{exp.}$) lattice parameter (table 1).

A correlation between ionic radii of A-site ($r_A$), oxygen positional parameter or anion parameter ($u$), shared tetrahedral edge ($d_{AE}$) and tetrahedral bond length ($R_A$) has been established in table 3. The

**Table 3.** Ionic radii of tetrahedral/octahedral site (***$r_A$, $r_B$***), oxygen positional parameter (***u***), shared tetrahedral/octahedral edge (***$d_{AE}$, $d_{BE}$***) and tetrahedral/octahedral bond length (***$R_A$, $R_B$***) of as-burnt and annealed $Zn_{0.75-x}Ni_xMg_{0.15}Cu_{0.1}La_{0.02}Fe_{1.98}O_4$ system.

| **$r_A$ (nm)** | $u^{\bar{4}3m}$ | **$d_{AE}$ (nm)** | **$R_A$ (nm)** | **$r_B$ (nm)** | **$d_{BE}$ (nm)** | **$R_B$ (nm)** |
|---|---|---|---|---|---|---|
| | | | **As-Burnt** | | | |
| 0.0577 | 0.3843 | 0.3195 | 0.1956 | 0.0650 | 0.2752 | 0.2027 |
| 0.0568 | 0.3838 | 0.3180 | 0.1947 | 0.0647 | 0.2762 | 0.2029 |
| 0.0552 | 0.3828 | 0.3154 | 0.1932 | 0.0650 | 0.2784 | 0.2036 |
| 0.0535 | 0.3819 | 0.3127 | 0.1915 | 0.0654 | 0.2800 | 0.2039 |
| 0.0531 | 0.3817 | 0.3122 | 0.1912 | 0.0653 | 0.2804 | 0.2040 |
| 0.0498 | 0.3794 | 0.3066 | 0.1878 | 0.0672 | 0.2858 | 0.2058 |
| | | | **(Ann. 500°C /3hrs)** | | | |
| 0.0577 | 0.3841 | 0.3196 | 0.1957 | 0.0650 | 0.2762 | 0.2033 |
| 0.0569 | 0.3840 | 0.3183 | 0.1949 | 0.0647 | 0.2755 | 0.2027 |
| 0.0570 | 0.3840 | 0.3184 | 0.1950 | 0.0644 | 0.2756 | 0.2027 |
| 0.0502 | 0.3799 | 0.3073 | 0.1882 | 0.0668 | 0.2841 | 0.2051 |
| 0.0513 | 0.3808 | 0.3091 | 0.1893 | 0.0660 | 0.2817 | 0.2041 |
| 0.0502 | 0.3802 | 0.3073 | 0.1882 | 0.0661 | 0.2828 | 0.2044 |

variation of $u$, $d_{AE}$ and $R_A$ has been studied as a function of ionic radii of A-site. It is observed from table 3 that with increase in $r_A$, for both as-burnt and annealed system $u$ parameter, shared tetrahedral edge and tetrahedral bond length increases. As ionic radii of tetrahedral ions increases the tetrahedral site will expand in such a way that anions associated with tetrahedral site will move away from the tetrahedrally coordinated A-site cations, which results in the increase in u parameter, distance between unshared anions ($d_{AE}$) and the tetrahedral bond length ($R_A$). Similarly the correlation between ionic radii of B-site ($r_B$), oxygen positional parameter ($u$), shared octahedral edge ($d_{BE}$) and octahedral bond length ($R_B$) is also premeditated in table 3 for as-burnt and annealed La added Zn-Ni-Mg-Cu ferrite. The variation of oxygen positional parameter, shared octahedral edge and octahedral bond length has been studied as a function of ionic radii of B-site. As $r_B$ increases $d_{BE}$ and $R_B$ increases but u parameter decreases. When $r_B$ increases, the anions will move away from the octahedral site thus, increasing the distance between octahedral cation-anion ($R_B$) and the distance between the "shared anions ($d_{BE}$). Increase in $r_B$ implies that the anions associated with B site will move towards tetrahedral site, resulting a decrease in the u parameter, as u is the value which represents the movement of tetrahedral oxygen. The tetrahedral bond length is smaller than octahedral bond length. The smaller values of $R_A$ can be ascribed to a larger overlapping of orbitals of $Fe^{3+}$ and $O^{2-}$ at the A-site. There exists an inverse relationship between the covalent character of spinel and bond lengths. If the bond length increases/decreases with content of Ni, it can be concluded that there is a decrease/increase of iono-covalent character of the spinel with Ni content [23].

Magnetic information can also be extracted through structural data. Information about the magnetic interaction can also be obtained with the help of bond angles ($\theta_1$, $\theta_2$, $\theta_3$, $\theta_4$, $\theta_5$) and the distance between the cation and anion ($p$, $q$, $r$). Figure 2 (a, b) illustrates respectively the variation of the bond angles $\theta_1$, $\theta_2$ and the distance between the cation and anion ($q$) with Ni content (x). The bond angles $\theta_1$, $\theta_2$ are the angles between A-O-B and $q$ is the distance between the metal ions at tetrahedral site and anion. For all the studied samples, the bond angles $\theta_1$, $\theta_2$ increases and the distance between the metal ions at tetrahedral site and anion ($q$) decreases with Ni content, this suggests strengthening of A-B interaction. Figure 2 (c, d) respectively describes the variation of the bond angle between A-O-A ($\theta_5$)

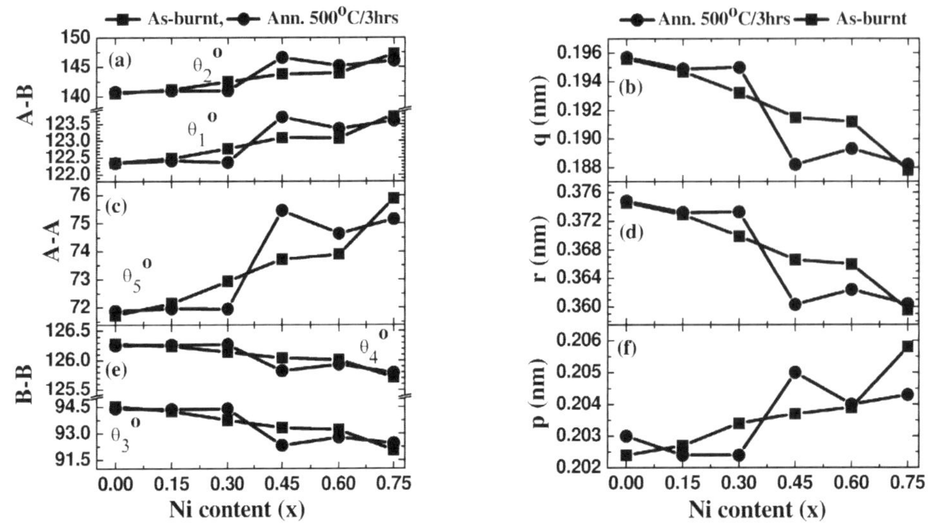

**Figure 2.** Variation of the bond angles **(a)** ($\theta_1^{A\text{-}O\text{-}B}$, $\theta_2^{A\text{-}O\text{-}B}$), **(c)** $\theta_5^{A\text{-}O\text{-}A}$, **(e)** ($\theta_3^{B\text{-}O\text{-}B}$, $\theta_4^{B\text{-}O\text{-}B}$) and the distance between cation and anion **(b)** q, **(d)** r, **(f)** p with Ni content (x) for as-burnt and annealed $Zn_{0.75-x}Ni_xMg_{0.15}Cu_{0.1}La_{0.02}Fe_{1.98}O_4$ system. Line connecting points are guide to the eye.

and the distance between the cation and anion (*r*) with Ni content (x). For as-burnt and annealed samples, $\theta_5$, increases and *r* decreases with Ni content. Increase in the bond angles $\theta_5$ and decrease in r suggests strengthening of A-A interaction. The variation of the bond angle between B-O-B ($\theta_3$, $\theta_4$) and the distance between the octahedral cation and anion (*p*) with Ni content is represented in fig. 2 (e, f) respectively. For as-burnt system $\theta_3$, $\theta_4$ decreases and the distance p increases with Ni content. But such linear trend is not observed in annealed samples. In annealed samples, the Ni content for which the bond angles between B-O-B decreases for the same value of x, *p* increases and for the value Ni content (x) for which $\theta_3$, $\theta_4$ increases for the same value of x, *p* decreases. Decrease in the bond angles $\theta_3$, $\theta_4$ and increase in p implies weakening of B-B interaction because the magnetic interaction between B-B is directly proportional to the bond angle $\theta_3$, $\theta_4$ and inversely proportional to *p*. And increase in the bond angles $\theta_3$, $\theta_4$ and decrease in p indicate strengthening of B-B interaction. The overall strength of the magnetic interactions (A–B and A–A) depends upon the bond length and bond angles between the cations and cation - anion. The strength is directly proportional to bond angle but inversely proportional to bond length. The same is proved in the figure 2. With annealing $\theta_1$, $\theta_2$, $\theta_5$ decreases and $\theta_3$, $\theta_4$ increases except for x = 0.45, 0.60. This indicates weakening of A-B and A-A interaction and strengthening of the B-B interaction. And for x = 0.45 and 0.60, $\theta_1$, $\theta_2$, $\theta_5$ increases and $\theta_3$, $\theta_4$ decreases with annealing indicating strengthening of A-B and A-A interaction and weakening of the B-B interaction. During annealing cations migrate from A site to B site or vice a verse and hence during migration of cations, interionic distances between cations and between cation and anion changes which bring variation in the bond angle between them.

The spinel ferrites with cubic crystal structure have two different crystallographic sub-lattices for magnetic ions, tetrahedral (A) and octahedral (B) sub-lattices. The magnetic order of the spinel ferrites mainly results from super-exchange interaction between the magnetic ions in the A and B sub-lattices mediated by oxygen ions. According to the Ne´el's two-sub-lattice model of ferrimagnetism, the magnetic moments of ions on the A and B sub-lattices are aligned antiparallel to each other and their spins have a collinear structure. Therefore, total magnetization is $M_{oct}$ - $M_{tet}$ ($M_{oct}$ and $M_{tet}$ are the sum of the magnetic moments on B sub-lattice and A sub-lattice, respectively) [24]. Néel magnetic moment

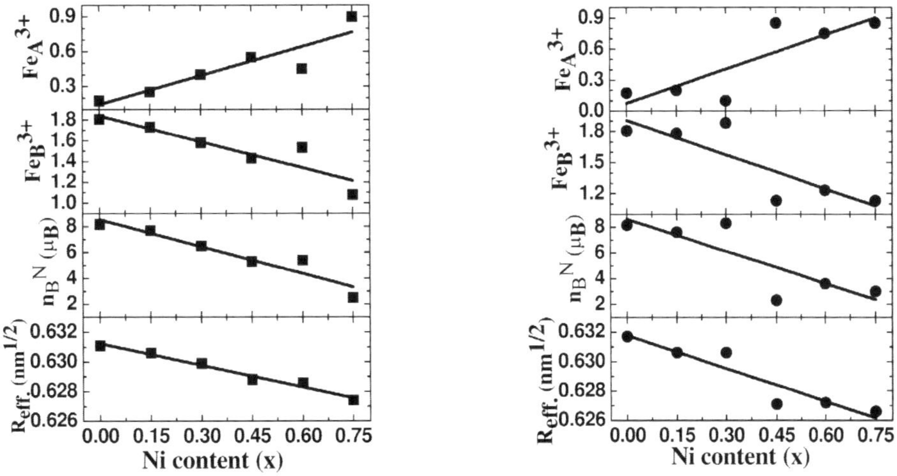


**Figure 3.** Variation of $Fe^{3+}$ ions on A site ($Fe_A^{3+}$) and B site ($Fe_B^{3+}$), Néel magnetic moment ($n_B^N$) and effective bond length ($R_{eff.}$) with Ni content (x) for as-burnt (left panel) and annealed (right panel) $Zn_{0.75-x}Ni_xMg_{0.15}Cu_{0.1}La_{0.02}Fe_{1.98}O_4$ system. Line connecting points is linear fit to the experimental data.

($n_B^N$) per formula unit in $\mu_B$ is calculated using the cation distribution (given in table 2) and and the ionic magnetic moments of $Mg^{2+}$, $Ni^{2+}$, $Zn^{2+}$, $Cu^{2+}$, $La^{3+}$ and $Fe^{3+}$ - $0\mu_B$, $2\mu_B$, $0\mu_B$, $1\mu_B$, $0\mu_B$ and $5\mu_B$ respectively. The variation of $n_B^N$ with respect to Ni content is represented in fig. 3 for as-burnt (left panel) and annealed (right panel) $La^{3+}$ added Zn-Ni-Mg-Cu spinel ferrite system. $n_B^N$ decreases linearly with increase in Ni content because of increase in the concentration of $Fe^{3+}$ ions on A site and decrease of $Fe^{3+}$ ions on B site. The values of $Fe^{3+}$ ions on A site ($Fe_A^{3+}$), B site ($Fe_B^{3+}$) and effective bond length ($R_{eff.}$) are shown in fig. 3 for as-burnt (left panel) and annealed (right panel) ferrite system. It can be clearly observed from fig.3 that $Fe_A^{3+}$ increases and $Fe_B^{3+}$ decreases linearly with Ni doping. With annealing there is redistribution of cations at tetrahedral (A) and octahedral (B) site, which changes the concentration of $Fe^{3+}$ ions on A and B site and hence, $n_B^N$ changes with annealing. The variation of the effective bond length ($R_{eff.}$) which is the distance between two cations, also affects the magnetic properties of a material. For the studied as-burnt and annealed samples $R_{eff.}$ decreases linearly with increase in Ni content. For lower values of Ni content ($0.0 \leq x \leq 0.3$) effective bond length increases with annealing and for higher Ni content ($0.30 < x \leq .75$) effective bond length decreases. Decrease in the effective bond length suggests decrease of the exchange coupling and increase in the effective bond length suggests increase of the exchange coupling.

## 4. Summary

Nanocrystalline Ni-substituted Zn-Mg-Cu-La ferrites were synthesized using sol-gel auto-combustion method. To study the thermal effect on the structural properties and cation distribution the as-burnt samples were annealed at 500°C for 3hrs. X-ray diffraction patterns of as-burnt and annealed samples suggest the formation of single cubic spinel phase. Lattice constant ($a_{exp.}$), unit cell volume ($V$) and hopping length at A and B ($L_A$ and $L_B$) of the studied samples decreases with increase in Ni content. Scherrer's grain diameter ($D$) for as-burnt and annealed samples respectively ranges between 13.26 - 22.20 nm and 23.61 - 30.93 nm. Relation between oxygen positional parameter ($u$), tetrahedral bond length ($R_A$), shared tetrahedral ($d_{AE}$) is interpreted by ionic radii of A-site and the relation between oxygen positional parameter ($u$), octahedral bond length ($R_B$) and shared octahedral edge ($d_{BE}$) have been deduced via ionic radii of B-site ($r_B$). Neel's magnetic moment ($n_B^N$) decreases linearly with increase in Ni content because of increase in the concentration of $Fe^{3+}$ ions on A site and decrease of

$Fe^{3+}$ ions on B site. With annealing $\theta_1$, $\theta_2$, $\theta_5$ decreases and $\theta_3$, $\theta_4$ increases except for x = 0.45, 0.60. This suggests weakening of A-B and A-A interaction and strengthening of the B-B interaction.

**Acknowledgments**
Authors thank Dr. M. Gupta, UGC-DAE CSR, Indore India for performing XRD measurements. This work is supported by UGC-DAE CSR, Indore [CSR-IC/CRS-74/2014-15/2104].

***References***
[1] Pallai V and Shah D O 1996 *J. Magn. Magn. Mater.* **163** 243
[2] Skomski R 2003 *J. Phys. – Condens. Matter.* **15** R1
[3] Roy P K and Bera J 2007 *Mater. Res. Bull.* **42** 77
[4] Jiang J, Li L and Xu F 2007 *J. Rare Earths* **25** 79
[5] Al-Hilli M F, Li S and Kassim K S 2009 *Mater. Sci. Eng. B* **158** 1
[6] Costa A C F M, Diniz A P A, de Melo A G B, Kiminami R H G A, Cornejo D R, Costad A and Gama A L 2008 *J. Magn. Magn. Mater.* **320** 742
[7] Ravinder D and Vijaya Kumar K 2001 *Bull. Mater. Sc.* **24** 505
[8] Rezlescu N, Rezlescu E, Tudorache F and Popa P D 2004 *J. Optoelectr. Adv. Mater.* **6** 695
[9] Jiang J, Li L, Xu F and Xie Y 2007 *Mater. Sci. Eng. B* **137** 166
[10] Yehia M, Ismail S M and Hashhash A 2014 *J. Supercond. Nov. Magn.* **27** 771
[11] Hashhash A, Yehia M, Ismail S M and Ata-Allah S S 2014 *J. Supercond. Nov. Magn.* **27** 2305
[12] Cullity B D 1978 Elements of X-ray diffraction, second ed., Addison-Wesley Publishing Co.Philippines Ch. 10 p. 338
[13] Satalkar M, Kane S N, Ghosh A, Ghodke N, Barrera G, Celegato F, Coisson M, Tiberto P and Vinai F 2014 *J. Alloys Comp.* **615** S313
[14] Smit J and Wijn H P J 1959 Ferrites, Philips Technical Library, Eindhoven, The Netherlands, Ch. 8, p.144
[15] Qi X, Zhou J, Yue Z, Gui Z and Li L 2003 *Mater. Sci. Eng. B* **99** 278
[16] Weil L, Bertaut E F and Bochirol L 1950 *J. Phys. Radium* **11** 208
[17] Tanna R and Joshi H H 2013 *World Academy of Sc. And Eng. Tech.* **75** 78
[18] Lakhani V K, Pathak T K, Vasoya N H and Modi K B 2011 *Solid State Sc.* **13** 539
[19] Sickafus K E, Wills J M and Grimes N W 1999 *J. Am. Ceram. Soc.* **82** 3279
[20] Shannon R D 1976 *Acta Cryst.,* **A32** 751
[21] Vegard L 1969 *Zeitscchrift fur Physik*, **5** 17
[22] Hirabayashi D, Yosikawa T, Kawamoto Y, Mockizuki K and Sukuki K 2006 *Adv. in Sc. Tech.* 45 2169
[23] Levine B F 1973 *Phys. Rev.* **B7** 2591
[24] Néel L 1950 *C R Acad. Sci. Paris* **230** 375

# Structural, Optical and Dielectric properties of Nickel substituted Barium Titanate ceramics.

**Supriya Bisen, Ashutosh Mishra, Kanaka M. Jarabana**
School of Physics, Devi Ahilya University, Khandwa Road, Indore (M.P.) - 452001, India

E-mail: sbisen.sop@gmail.com

**Abstract.** In present study, this paper describes the fabrication of $Ba_{1-x}Ni_xTiO_3$ (X=0, 0.05, 0.10) nanomaterial's synthesized by chemical route sol-gel method. Their structural, optical and dielectric properties have been investigated as a function of sintering temperature. Dielectric measurements of the sintered sample were carried out for the composition over the frequency range of 10Hz - $10^7$Hz. X-ray diffraction technique reveals that the formation of single phase crystalline structure. The presence of O-H stretching, $NO_3^-$, C-H bending modes and C=O stretching were verified by the Fourier transform infra-red spectroscopy for Ni substituted $BaTiO_3$.

## 1. Introduction

Since the discovery of $BaTiO_3$ in early 1940s, this material has been used for a wide range of scientific and industrial applications such as capacitors, ultrasonic transducers, pyroelectric infrared sensors, and positive temperature coefficient (PTC) resistors, etc [1-3]. For the application of doped $BaTiO_3$ ceramics as a capacitor material apart from a good density, a high dielectric constant and low loss factor have to be achieved. Barium titanate is a classical example of a ferroelectric crystalline material. Firstly, it is chemically and mechanically very stable, secondly it exhibits ferroelectric properties at and above room temperature and finally it can be easily prepared and used in the form of ceramic polycrystalline samples [4]. However, the effect of the specific impurity on the electrical conductivity depends on the substitution site as well as on the nature of the defect/electronic charge compensating defect. In the present work we have studied that the structural, dielectric and FTIR properties of $BaTiO_3$ and doped $BaTiO_3$ with one transition element $Ni^{2+}$ (0.05, 0.10 respectively) [5].

## 2. Experimental Details

Polycrystalline sample with composition of $Ba_{1-x}Ni_xTiO_3$ (x=0.05, 0.10) have been prepared by chemical route sol-gel method. The precursor Barium nitrate, Titanium nitrate, Nickel nitrate were taken in stoichiometric ratio. We mixed all materials in beaker and dissolved in de-ionized water. The mixture is slowly heated on magnetic stirrer and citric acid is added. The pH value is adjusted 7 by adding ammonia solution. If pH value is below 7, no gel is formed. The formed gel was dried on furnance and then gets a powder form which calcined at $900^0$C for 2 hrs.

## 3. Result and Discussion

### *3.1. Structural Analysis*

The room temperature X-ray diffraction pattern (Fig 1) and XRD data (Table 1) of $Ba_{1-x}Ni_xTiO_3$ (x=0.05, 0.10) confirm the single phase tetragonal Barium Titanate structure. All peaks are identified as tetragonal $BaTiO_3$ in space group P4mm. The lattice parameters $BaTiO_3$ were refined to a=b=4.01A$^0$, c=4.04A$^0$ and $Ba_{1-x}Ni_xTiO_3$ where (x=0.05, 0.10) were refined to a=b=4.01A$^0$, c=3.99A$^0$ and a=b=4.01A$^0$, c=4.00A$^0$ and reported value of a=b=3.994A$^0$, c=4.033A$^0$ corresponding to tetragonal-$BaTiO_3$ are compared and Nickel doped BTO are seen to reduce the constant cell volume (Table 1).

### *3.2. Fourier transform Infra-red Spectroscopy Analysis*

FTIR analysis was carried out at 900$^0$C for detection the presence of the functional groups. By using this analysis the reaction mechanisms during the sol gel process can be detected. The FT-IR spectrum of the undoped and doped powder presented in Fig 2 shows the presence of the band located at 542 cm$^{-1}$ specific to the vibration mode of the Ti-O bond in the $TiO_6$ octahedra. The characteristic absorption band at 3486 cm$^{-1}$ is assigned to the O-H stretching vibration in water and confirms that the existence of the water in the $BaTiO_3$ and doped $BaTiO_3$ ceramics. The main bands specific to the ionic $NO_3^-$ (866 cm$^{-1}$ and 1425 cm$^{-1}$) introduced in the system by the catalyst and by the Barium and Nickel source, $Ba(NO_3)_2$, $Ni(NO_3)_2 \cdot 6H_2O$ in the case of the doped precursors, were also identified. The absorption bands at 1013, 1042, 1062 cm$^{-1}$ can be considered as C-H bending modes and the very small band at 1739, 1749 cm$^{-1}$ assigned to the C=O stretching mode of an ester bond. These two bonds can be related to the formation of the citrate complex during synthesis.

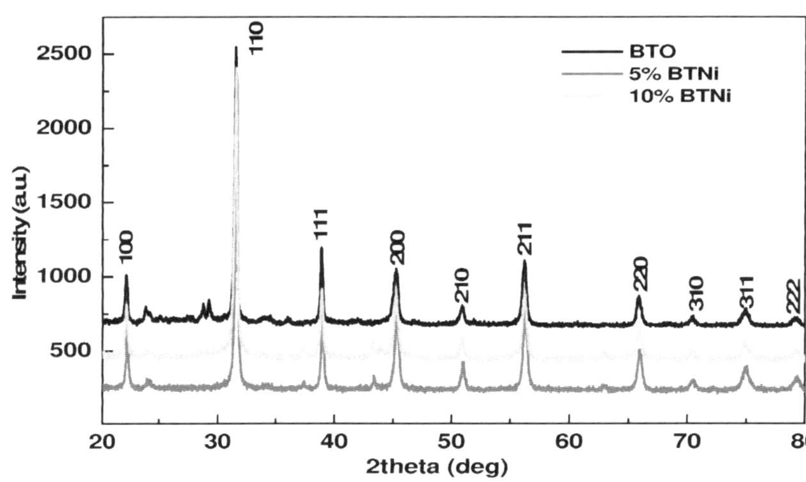


**Figure 1.** XRD pattern of pure and Ni doped BTO (0.05, 0.10) at 900$^0$C.

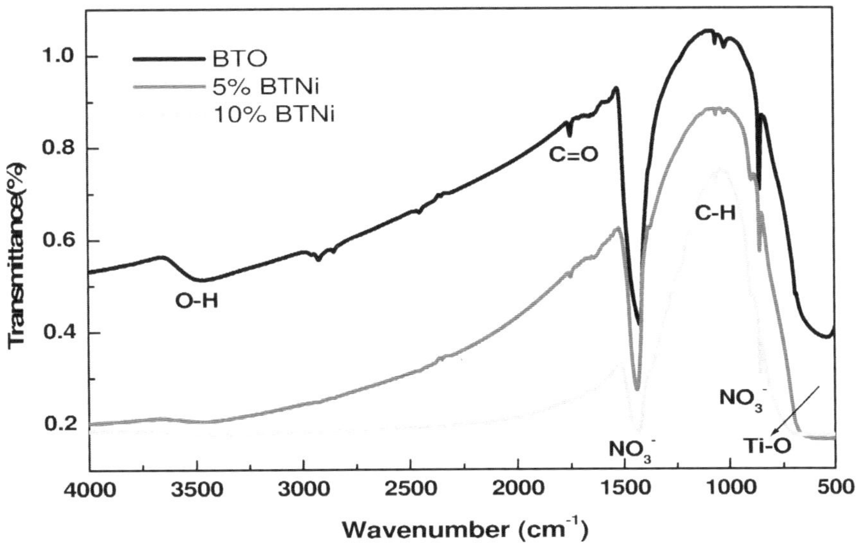


**Figure 2.** FTIR spectrum of pure and Ni doped BTO (0.05, 0.10) at $900^0$C.

**Table 1.** Lattice parameter, unit cell volume, density and crystal size of Pure and Iron doped $BaTiO_3$.

| | Lattice Parameter (a, b ,c) | *c/a* *($A^0$)* | Volume $(A^0)^3$ | *density* *$\rho(g/cm^3)$* | Crystal Size (nm) |
|---|---|---|---|---|---|
| **$BaTiO_3$** | a=b=4.01, c=4.04 | 1.007 | 64.96 | 5.969 | 26.48 |
| **BTNi5** | a=b=4.01, c=3.99 | 0.995 | 64.15 | 5.965 | 27.66 |
| **BTNi10** | a=b=4.01, c=4.00 | 0.997 | 64.32 | 5.783 | 24.97 |

*3.3. Dielectric Studies*

Dielectric studies have been performed in a wide temperature range. The electric permittivity $\varepsilon'$ and dielectric loss tanδ have been measured at different frequencies for Ni (x=0.05, 0.10) doped $BaTiO_3$. The variation of Dielectric Constant with temperature for sample with different compositions fig shows that the Nickel doped $BaTiO_3$ have a prominent influence on dielectric properties.

The temperature dependence ($40^0$C t0 $180^0$C) of the dielectric parameters has been measured for Ni doped BTO (sintered at $900^0$C) as a function of frequency. No evidence of any structural transition in this temperature range were found which is in contrast with doped $BaTiO_3$. The result indicates that the dielectric constant decreased as a frequency increased from 1Hz to $10^5$Hz. Generally, the dielectric constant at low frequency is higher than those obtained at the higher frequency. The decrease in dielectric constant is due to the dipoles which cannot follow the alternation of the applied ac electric field. Loss tangent (tanδ) variation with temperature at different frequencies shows similar characteristics.

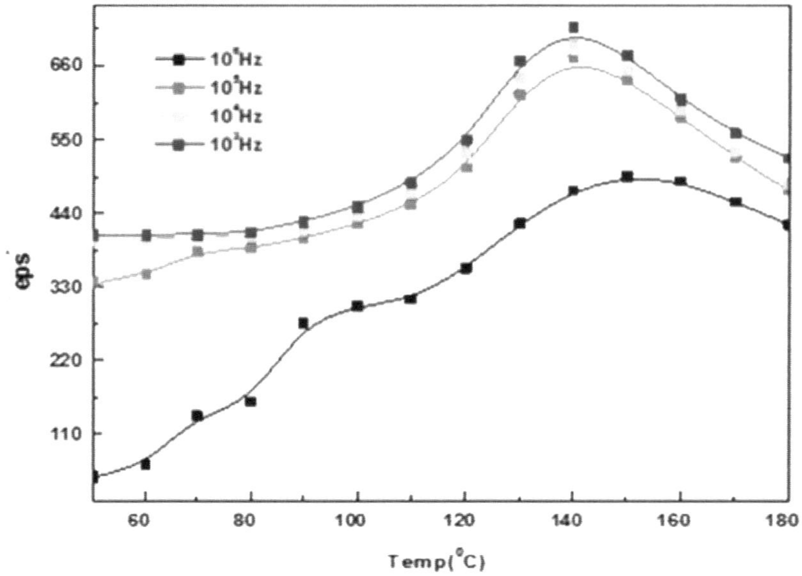


**Figure 5.** Temperature dependent Dielectric Constant of $BaTiO_3$ at $900^0C$.

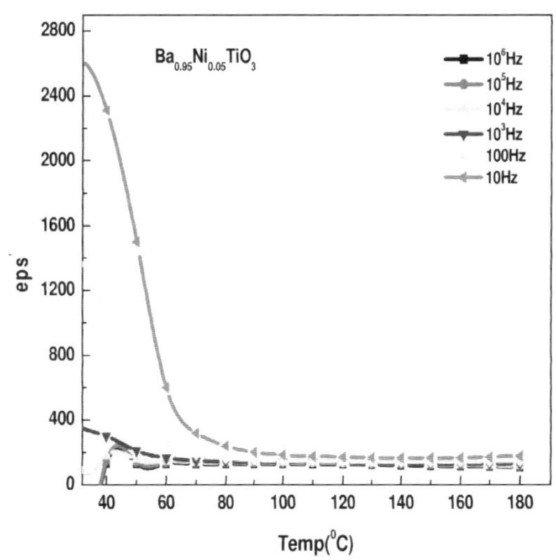


**Figure 4(a).** Temperature dependent Dielectric Constant of $Ba_{0.95}Ni_{0.05}TiO_3$ at $900^0C$.

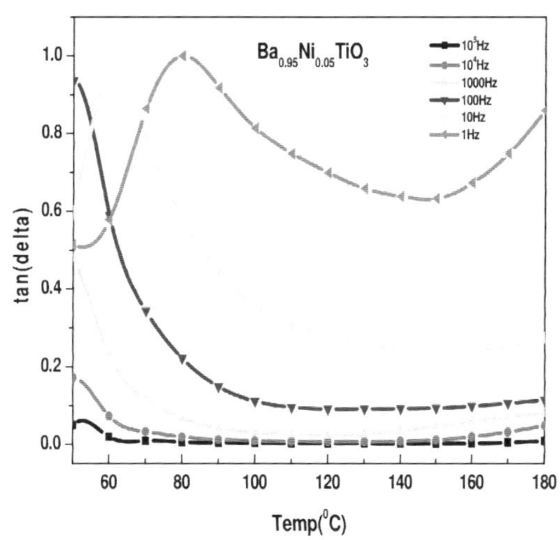


**Figure 4(b).** Temperature dependent Dielectric tanδ of $Ba_{0.95}Ni_{0.05}TiO_3$ at $900^0C$.

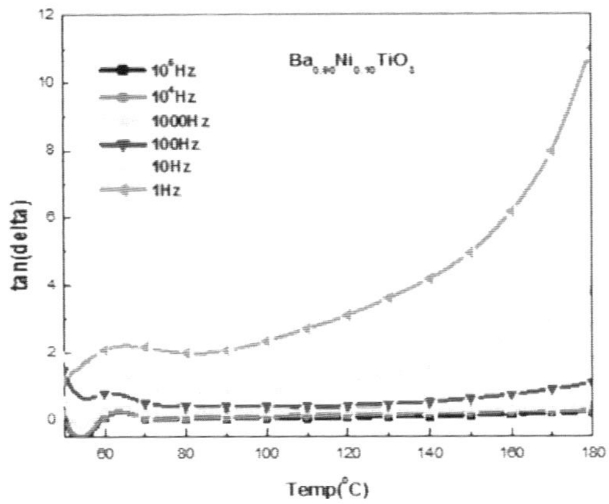


**Figure 5(a).** Temperature dependent Dielectric Constant of $Ba_{0.90}Ni_{0.10}TiO_3$ at $900^0C$.

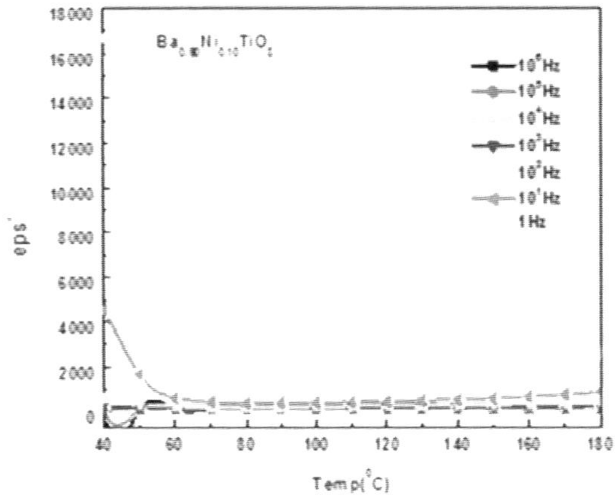


**Figure 5(b).** Temperature dependent Dielectric tanδ of $Ba_{0.90}Ni_{0.10}TiO_3$ at $900^0C$.

## 4. Conclusion

Barium Titanate and Ni (x=0.05, 0.1) doped BTO powders were prepared by Sol gel method. The synthesized powders were characterized by X-ray diffraction using a Bruker D8 Advance X-ray diffractometer. XRD pattern were obtained Cu-Kα radiation and the wavelength of X-ray was $1.54A^0$.The XRD patterns showed that this material crystallizes in a perovskite-type single phase tetragonal structure with space group p4mm. The XRD data has concluded that all major peaks of BTO and Ni-BTO are matching with reported XRD data, and sample is in single phase. The FTIR spectrum was measured at room temperature. The FTIR spectra of Pure $BaTiO_3$ and doped with Nickel, by the FT-IR data we get the intensity of nitrate ($NO_3^-$) and hydroxyl (O-H) peak reduced with doping of Nickel in pure BTO.

In dielectric studies no evidance of any structural transition in this temperature range were found which is in contrast with doped $BaTiO_3$. The result indicates that the dielectric constant decreased as a frequency increased from 1Hz to $10^5$Hz. The decrease in dielectric constant is due to the dipoles which cannot follow the alternation of the applied ac electric field.

**Acknowledgment**

The authors are thankful to Central Director Dr. V. Ganeshan, IUC-UGC-DAE CSR Indore for providing the research fellowship. I am also thankful to Dr. R. Prasad whose under the guidance of preparing sample, Dr. Mukul Gupta for XRD measurement, Dr. Uday Deshpande for FTIR measurement and Dr. A. M. Awasthi for Dielectric measurement, UGC-DAE CSR Indore for providing the experimental facilities and for the precious guidance.

**References**

[1] Haertling G H 1999 J. Amer. *Ceram*. Soc. **82** 797
[2] Jones R E, Mainar P D, Olowolafe J O, Campbell J O and Mogab C 1992 J *appl. Phys. Lett.* **60** 1022
[3] Frey M H and Payne D A 1963 *ibid.* **63** 2753
[4] Uchino K, Sadanaga E and Hirose T 1989 J. Am. *Ceram Soc.* **72** 1555
[5] G. Bush 1987 Ferroelectrics **4** 267

# Synthesis and characterization of $Co_{0.8}Fe_{2.2}O_4$ nano ferrite

**S Raghuvanshi[a], S N Kane[a,*], N P Lalla[b] and V R Reddy[b]**

[a]Magnetic Materials Laboratory, School of Physics, Devi Ahilya University, Khandwa Road Campus, Indore-452001, India
[c]UGC-DAE CSR, University Campus, Khandwa Road, Indore-452001, India

[*]Corresponding author: kane_sn@yahoo.com

**Abstract.** $Co_{0.8}Fe_{2.2}O_4$ nano ferrite has been synthesized by sol gel auto-combustion method. X-ray diffraction (XRD) and Mössbauer spectroscopy were used to characterize the synthesized as burnt and thermally annealed (600 °C for 3 hours ) samples. Both XRD and Mössbauer measurements confirm the formation of spinel phase with Scherrer's grain diameter ($D_s$) ≈ 37 nm. Presence of α-Fe and, $Fe_2O_3$ was also observed. Thermal annealing induced decrease of $\rho_{xrd}$ (from 5267.9 to 5262.7 Kg/m$^3$ ) for the annealed samples is ascribable to the fact that, increase of unit cell volume overtakes the increase in mass of the unit cell. Comparison of dry gel and thermally annealed sample shows that in dry gel, B site is more populated with Fe ions, whereas thermal annealing leads to migration of Co from A to B site with simultaneous migration of Fe from B to A site, thus dropping the Néel magnetic moment. Bond angle values clearly reveal that thermal annealing leads to strengthening of B-O-B interaction with simultaneous wreaking of A-O-B, and, A-O-A interaction.

## 1. Introduction

Nanosized spinel ferrites have been a topic of intense research owing to broad possibilities of their use in various technological applications e. g.- high density data storage [1], biomedical applications [2] etc. and, fundamental understanding of these materials. Spinel ferrites exhibit face centred cubic (fcc) structure belongs to $Fd_3m$ space group. The lattice consists of 32 divalent oxygen ions forming a closed pack face centred cubic arrangement with 64 tetrahedral interstitial sites (A sites) and 32 octahedral interstitial sites (B sites). Out of these, only 8 tetrahedral (A sites) and 16 octahedral (B sites) sites are occupied by the divalent and, trivalent cations and, rest remain empty. Thus, large fraction of empty interstitial sites, makes the crystal structure rather unfilled, encouraging migration of cations. It is well-known fact that the properties of ferrites are strongly influenced by their composition, microstructure and, are also sensitive to the preparation methodology used in their synthesis [3].

Phase transition is an important subject of study in magnetic materials and, temperature is one of the important parameters known to influence properties of magnetic materials [4]. In magnetic materials, the applied magnetic field also has been shown to be a variable parameter [4,5], highlighting the need of structural studies in presence of magnetic field, i. e. doing x-ray diffraction (XRD) measurements in the presence of magnetic field, (*usually available only at synchrotron sources* [6]), to be able to study magneto-structural phase transition. For ease in the experiments, at the same time keeping the quality of the information on magneto-structural phase transition in magnetic materials, there is a need for a laboratory source based XRD setup, capable of performing measurements at high magnetic fields as well as at low temperatures. This need is fulfilled by the recent development of low temperature and high magnetic field (LTHM) powder x-ray diffraction (XRD) [7] and, now is being used for LTHM XRD measurements to study magneto-structural phase transition.

Co ferrite has emerged as quite promising material for the development of rare-earth free hard magnets, with high coercivity (~16.7 kOe) , reasonable magnetization (~ 89.1 emu/g) and, rather high magneto-crystalline anisotropy (~ $11.6\times10^6$ erg/cc), which can be controlled by grain diameter [8]. It also shows remarkable chemical stability and, mechanical hardness [9]. $CoFe_2O_4$ has face cantered cubic (*fcc*) structure belonging to $Fd_3m$ space group and has two inter-penetrating sub-lattices A (*tetrahedral*) and B (*octahedral*) and, displays inverse spinal structure shown as : $(Fe^{3+})^A$ $[Co^{2+} Fe^{3+}]^B$ $O_4^{2-}$[10]. Co-ferrite is also magnetostrictive, shows magnetostrictive constant $\lambda_s \approx 400$ ppm) [11], so application of high magnetic field would lead to changes in the sample dimension (i. e. generating *deformation* in the sample), which is expected to be reflected in changes in the line intensities of the XRD pattern [12,13], indicating changes in cation distribution,. Recently [11] via magnetic measurements, anisotropy driven transition (*caused by rotation of the magnetization vector jumping over an energy barrier*) is reported in single crystal Co-ferrite - $Co_{0.8}Fe_{2.2}O_4$, although structural studies were not reported. XRD measurements under applied magnetic field (*where changes in the intensity of XRD peaks can be observed*) can yield information on magneto-structural phase transition, as was also reported in earlier studies in $TbVO_4$ system [12].

Consequently, as a first step to study magnetic field induced phase transition in poly-crystalline Co nano-ferrite, in the present work report the synthesis of poly-crystalline $Co_{0.8}Fe_{2.2}O_4$, spinel ferrite by sol-gel auto combustion method and, its preliminary characterization using XRD and Mössbauer measurements. In the second step of our studies, we intend to perform low temperaturer XRD measurements under applied magnetic field, which will be a subject of separate paper.

## 2. Experimental Details

### *2.1 Material synthesis and, characterization*

$Co_{0.8}Fe_{2.2}O_4$ specimen was synthesized by sol–gel auto-combustion technique by utilizing nitrate-Citrate precursors: [$Co(NO3)_2$ $6H_2O$ – Cobalt Nitrate and $Fe(NO_3)_3.9H_2O$ - Ferric Nitrate]. All the precursors were taken in stoichiometric ratio, were dissolved in de-ionized water and citric acid was used as fuel, taking metal salt to fuel mass ratio as 1:1. Solution pH was maintained to 7 by adding Ammonia solution ($NH_4OH$). Afterwards the solution was heated at 110 °C to obtain fluffy powder (dry gel). Obtained dry gel powder was annealed at 600 °C for 3 hours. Both dry gel and annealed samples were used for x-ray diffraction and, Mössbauer studies. Schematic diagram of the synthesis process is shown in figure 1.

Room temperature x-ray diffraction (XRD) measurements ($\theta$ - $2\theta$ configuration) were done by 18 kW rotating anode source (Rigaku) utilizing CuK$\alpha$ ($\lambda$ = 0.1540562 nm) radiation. X-ray tube is focused, made parallel by using a parabolic mirror (Xenox). Scattered x-rays were detected by NaI detector. Room temperature Mössbauer spectra were recorded in transmission geometry using a $^{57}$Co:Rh source and, the spectra were computer fitted to obtain hyperfine parameters.

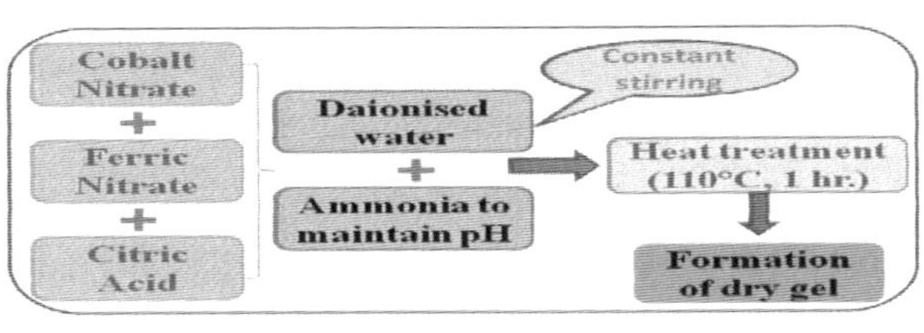


**Figure 1:** Schematic diagram of the synthesis of $Co_{0.8}Fe_{2.2}O_4$ by sol-gel auto combustion method.

*2.2 Data analysis*

Lattice parameter ($a_{exp}$) corresponding to [311] reflection was obtained by: $a_{exp} = d\ (h^2+k^2+l^2)^{1/2}$, where d - inter-planer spacing and (h, k, l) – Miller indices. $a_{exp}$ was used to obtain the x-ray density ($\rho_{XRD}$) of the prepared samples using the equation $\rho_{XRD} = 8M_w/N(a_{exp})^3$ , where $M_w$ – Molecular weight, N – Avagrado's number. Grain diameter was calculated by the line width of [311] reflection, using Scherrer's formula: $D_S = 0.9\lambda / \beta \cos\theta$, Where $\lambda$ - wavelength of the x-ray used, $\beta$ - line width, $\theta$ peak position (in 2θ scale). Cation distribution of the studied samples was estimated using XRD peak intensities. The calculated and, observed intensity ratios were compared for several combinations of cations distribution at (A) and [B] sites [14]. The best cation distribution amongst the tetrahedral and octahedral sites for which theoretical and experimental ratios agree clearly, is taken to be the correct one. Ionic radii of tetrahedral ($r_A$), octahedral sites ($r_B$), theoretical lattice parameter ($\boldsymbol{a}_{\mathbf{th.}}$), oxygen positional parameter ($u$), inter-ionic distances between cations ($M_e$-$M_e$) (*b, c, d, e, f*) and cation anion ($M_e$-O) (*p, q, r, s*), bond angles ($\theta_1$, $\theta_2$ , $\theta_3$ , $\theta_4$, $\theta_5$) were calculated as described elsewhere [15,16]. Néel magnetic moment ($n_N$) was calculated by using expression : $n_N = M_B - M_A$, where $M_B$ and $M_A$ respectively are magnetic moments at B and A site. Both $M_A$ and $M_B$ were obtained from cationic distribution. Full profile Rietveld analysis of XRD pattern is done by MAUD (*Materials Analysis Using Diffraction*) software [17]. The program continues the refinement till convergence is obtained with the values of the quality factor GOF (goodness of fit) is close to 1, confirming the goodness of refinement.

## 2. Results and Discussion

Fig. 2 depicts Rietveld refined XRD plot of (a) $Co_{0.8}Fe_{2.2}O_4$ (Dry gel), (b) $Co_{0.8}Fe_{2.2}O_4$ (ann. at 600°C/3hours.). XRD confirms the formation of nano-crystalline cubic spinel structure. XRD analysis provides - lattice parameter, grain diameter etc. , theoretical lattice parameter ($a_{th}$), oxygen positional parameter (u), interionic distances between cation and anion (Me-O) (p, q, r, s), interionic distance between cations (Me-Me)(b, c, d, e, f), bond angle between cation and anion ($\theta_1$, $\theta_2$, $\theta_3$, $\theta_4$, $\theta_5$). Néel magnetic moment ($n_N$) were calculated from cation distribution.

Table 1 shows the calculated values of experimental and, theoretical lattice parameters, ($a_{exp}$, $a_{th}$), average Scherrer's grain diameter (***D***), x-ray density ($\boldsymbol{\rho_{xrd}}$), and specific surface area (S) for dry gel and annealed samples. Lattice constant does not change appreciable for the studied samples. The average Scherrer's grain diameter ($\boldsymbol{D_s}$) increases from 37.3 to 37.6 nm. Observed minor change in ***D*** can be ascribed to thermal annealing induced grain growth. Thermal annealing induced decrease of $\boldsymbol{\rho_{xrd}}$ (from 5267.9 to 5262.7 $Kg/m^3$ ) for the annealed samples is ascribable to the fact that, increase of unit cell volume overtakes the increase in mass of the unit cell.

Table 2 gives cation distribution and, Néel magnetic moment of the studied samples. Perusal of table 2 shows that thermal annealing leads to migration of Co from A to B site with simultaneous migration of Fe from B to A site. In case of dry gel sample, more population of Fe ions on B site leads to higher Néel magnetic moment.

Table 3 illustrates the variation of ionic radii of A-site ($\mathbf{r_A}$) and B-site ($\mathbf{r_B}$), oxygen positional parameter ($u$), inter-ionic distances between cations ($M_e$-$M_e$) (**b, c, d, e, f**) and cations anions (Me-O)

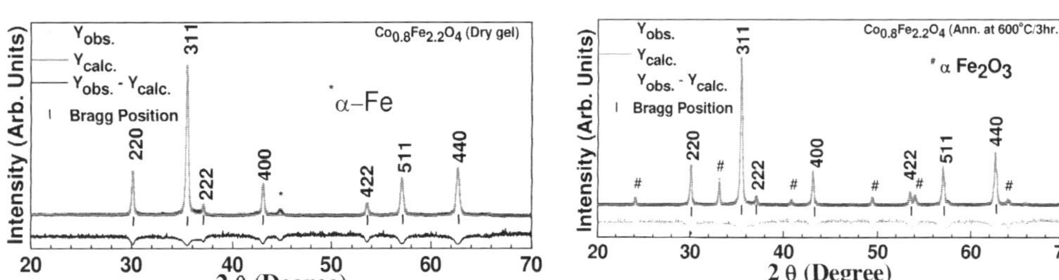


**Figure 2**: Rietveld refined plot of (a) $Co_{0.8}Fe_{2.2}O_4$ - Dry gel , (b) $Co_{0.8}Fe_{2.2}O_4$ - Ann. at 600 °C / 3hours.

**Table 1**.Variation of experimental, theoretical lattice parameters, ($a_{exp}$, $a_{th}$), Scherrer's grain diameter ($D_s$), x-ray density ($\rho_{xrd}$) and specific surface area ($S$).

| Parameters | $Co_{0.8}Fe_{2.2}O_4$ Dry gel | $Co_{0.8}Fe_{2.2}O_4$ Ann.600°C/3h |
|---|---|---|
| $a_{exp}$ (nm) | 0.8387 | 0.8390 |
| $a_{th}$ (nm) | 0.8386 | 0.8387 |
| $D_s$ (nm) | 37.3 | 37.6 |
| $\rho_{XRD}$ (Kg/m$^3$) | 5267.9 | 5262.7 |
| S (m$^2$/g) | 30.6 | 30.4 |

**Table 2**. Cation distribution, and Neel magnetic moment ($n^N$).

| Sample | Distribution | $n^N(\mu_B)$ |
|---|---|---|
| $Co_{0.8}Fe_{2.2}O_4$ | $(Co_{0.06}^{2+} Fe_{0.94}^{3+})^A$ | 3.7 |
| *(Dry gel)* | $[Co_{0.74}^{2+}Fe_{1.26}^{3+}]^B$ | |
| $Co_{0.8}Fe_{2.2}O_4$ | $(Co_{0.0}^{2+} Fe_{1.0}^{3+})^A$ | 3.4 |
| *(Ann. 600°C/3h)* | $[Co_{0.8}^{2+}Fe_{1.20}^{3+}]^B$ | |

**p, q, r, s**) bond angles (**$\theta_1$, $\theta_2$, $\theta_3$, $\theta_4$, $\theta_5$**) for effective magnetic interactions (**A-O-A, A-O-B, B-O-B**). It is worth noting that strength of the magnetic interactions (A-O-B, B-O-B and, A-O-A) strongly depend on bond length and bond angles between the cations and, cation-anion and is directly proportional to the bond angle, but inversely proportional to the bond length [13]. Bond angle values (shown in table 3) clearly show that thermal annealing leads to strengthening of B-O-B interaction with simultaneous wreaking of A-O-B, and, A-O-A interaction, which is reflected in bond length and Néel magnetic moment. Oxygen positional parameter (***u***) is a measure of distortion in the structure and, its reduction after thermal annealing suggests reduction of structural disorder. Variation of ionic radii ($\boldsymbol{r_A}$, $\boldsymbol{r_B}$) in the studied samples can be explained as follows : due to presence of more Co on B site (shown in table 2), in thermally annealed sample shows higher $\boldsymbol{r_B}$ in comparison to dry gel sample, whereas $\boldsymbol{r_A}$ shows opposite behaviour as that of $\boldsymbol{r_B}$.

Mössbauer measurements (spectra not shown here) also depict the formation of spinel phase and, the magnetic nature of the samples, as reflected in Mössbauer parameters shown in table 4. Perusal of table 4 shows that in dry gel apart from ferrite presence of α-Fe is also seen (also observed in XRD measurements) and, in annealed sample, formation of $Fe_2O_3$ is observed. Obtained isomer shift values for ferrite component indicates that iron present in ferrite component is $Fe^{3+}$.

**Table 3**. Variation of ionic radii of A site ($r_A$) and B site ($r_B$), oxygen positional parameter (u), inter-ionic distance between cations (*Me-Me*) (*b, c, d, e, f*), between cation and anion (*Me-O*) (*p, q, r, s*) and bond angles ($\Theta_1$, $\Theta_2$, $\Theta_3$, $\Theta_4$, $\Theta_5$).

| Parameters | $Co_{0.8}Fe_{2.2}O_4$ *(Dry gel)* | $Co_{0.8}Fe_{2.2}O_4$ *(Ann.600°C/3h)* |
|---|---|---|
| $r_A$ (nm) | 0.0495 | 0.0490 |
| $r_B$ (nm) | 0.0682 | 0.0685 |
| $u$ | 0.3791 | 0.3787 |
| $p$ (nm) | 0.2062 | 0.2067 |
| $q$ (nm) | 0.1875 | 0.1870 |
| $r$ (nm) | 0.3591 | 0.3581 |
| $s$ (nm) | 0.3652 | 0.3651 |
| $b$ (nm) | 0.2965 | 0.2966 |
| $c$ (nm) | 0.3477 | 0.3478 |
| $d$ (nm) | 0.3632 | 0.3633 |
| $e$ (nm) | 0.5448 | 0.5450 |
| $f$ (nm) | 0.5136 | 0.5138 |
| *(A-O-B)* $\theta_1$ (°) | 123.94 | 124.08 |
| *(A-O-B)* $\theta_2$ (°) | 147.76 | 148.41 |
| *(B-O-B)* $\theta_3$ (°) | 91.92 | 91.72 |
| *(B-O-B)* $\theta_4$ (°) | 125.71 | 125.66 |
| *(A-O-A)* $\theta_5$ (°) | 76.16 | 76.53 |

**Table 4**. Isomer shift (I.S.), Average hyperfine field ($B_{hf}$) and, components present in the studied samples.

| **Sample** | **I. S. (mm/s)** | **$B_{hf}$ (T)** | **Component** |
|---|---|---|---|
| $Co_{0.8}Fe_{2.2}O_4$ (Dry gel) | 0.485 ± 0.036 | 33.33 ±0.10 | α-Fe |
| | 0.335 ± 0.0004 | 51.13 ±0.08 | Ferrite* |
| | 0.306 ± 0.002 | 49.65 ±0.06 | Ferrite# |
| $Co_{0.8}Fe_{2.2}O_4$ (Ann 600 °C/3h) | 0.432 ± 0.031 | 51.98 ±0.15 | $Fe_2O_3$ |
| | 0.266 ± 0.022 | 51.84 ±0.09 | Ferrite* |
| | 0.289 ± 0.007 | 49.07 ±0.09 | Ferrite# |

*Octahedral (*B*) site, #Tetrahedral (*A*) site

Based on the hyperfine field values, magnetic sextets are assigned to Fe ions located on tetrahedral (A) and octahedral(B) sites.

To summarize, XRD, Mössbauer study dry gel and thermally annealed (600 °C for 3 hours ) $Co_{0.8}Fe_{2.2}O_4$ nano ferrite (grain diameter ~ 37 nm) prepared by sol gel auto-combustion method is reported. Studies reveal the formation spinel phase and, also the presence of α-Fe and, $Fe_2O_3$ phase. Thermal annealing leads to : i) minor changes in x-ray density, ii) migration of Co from A to B site with simultaneous migration of Fe from B to A site, resulting in reduction of Néel magnetic moment and, iii) illustrate strengthening of B-O-B interaction with simultaneous wreaking of A-O-B, and, A-O-A interaction.

## Acknowledgement

This work is supported by CRS project : CSR-IC/CRS-149/2015-16/06 dated 26 March 2016. SR is financially supported by Project: CSR-IC/CRS-74/2014-15/2104.

## References

[1] Frey N A, Peng S, Cheng K and Sun S 2009 *Chem. Soc. Rev.* 2532
[2] Colombo M, Carregal-Romero S, Casula M F, Gutierez L, Morales M P, Bohm I B, Heverhagen J T, Prosperi D, Parak W J 2012 *Chem. Soc. Rev.* **41** 4306
[3] Verma A, Goel T C, Mendiratta R G, Alam M I 1999 *Mater. Sci. Eng. B* **60** 156
[4] Pecharsky V K, Holm A P, Gschneidner, Jr. K A and Rink R *Phys. Rev. Lett.* 91 (2003) 197204
[5] Chaddah P, Kumar K and Banerjee 2008 A *Phys. Rev. B* **77** 100402(R)
[6] Islam Z, Nojiri H, Ruff J P C, Das R, Captina D and Lang J C *Synch.Rad. News* 25 (2012)
[7] Shahee A, Sharma S, Singh K, Lalla N P and Chaddah P *AIP Conf. Proc.* 1665 (2015) 060004-1
[8] López-Ortega A, Lottini E, de Julián F and Sangregorio C 2015 *Chem. Mater.* **27** 4048
[9] Jadhav S S, Shirsath S E, Patange S M and Jadhav K M 2010 *J.Appl. Phys.* **108** 093920
[10] Pettit G A, Forester D W 1971 *Physical Review B* **4** 3912
[11] Kriegisch M, Ren W, Sato-Turtelli R, Muller H, Grossinger R, Zhang Z 2012 *J. Appl. Phys* **111** 07E308
[12] Frings P, Vanacken J, Detlefs C, Duc F, Lorenzo J E, Nardone M, Billette J, Zitouni A, Bras W and Rikken G L J A 2006 *Review of Scientific Instruments* **77** 063903
[13] J. Smit, H.P.J. Wijn, Ferites John Wiley & Sons, New York, 1959
[14] Weil L, Bertaut E F, Bochirol L 1950 *J. Phys. Radium* **11** 208
[15] Varalaxmi N, Reddy N R, Ramana M V, Rajagopal E, Murthy V R and Sivakumar K V 2008 *J. Mater. Sci: Mater. Electron.* **19** 399
[16] Lakhani V K, Pathak T K, Vasoya N H and Modi K B 2011 *Solid State Sci.* **13** 539
[17] Lutterotti L, Matthies S and Wenk H R *MAUD (Material Analysis Using Diffraction): a user friendly Java program for Rietveld Texture Analysis and more*, Proceeding of the 1999 *Twelfth International Conference on Textures of Materials* (ICOTOM-12), Vol. **1**, 1599.

# On the study of Structural properties and Cation distribution of $Zn_{0.75-x}Ni_xMg_{0.15}Cu_{0.1}Fe_2O_4$ nano ferrite: Effect of Ni addition

**M Satalkar[a], S N Kane[a,1]**
[a]Magnetic Materials Laboratory, School of Physics, D. A. University, Khandwa road, Indore-452001, India

E-mail: kane_sn@yahoo.com (S. N. Kane)

**Abstract.** Effect of Ni addition on structural properties and, cation distribution of $Zn_{0.75-x}Ni_xMg_{0.15}Cu_{0.1}Fe_2O_4$ (x = 0.0, 0.15, 0.30, 0.60, 0.75) ferrites, prepared using sol-gel auto-combustion method was studied using X-ray diffraction (XRD) technique. XRD analysis reveals the formation of spinel phase even in dry gel form. The lattice constant ($a_{exp.}$) decreases with $Ni^{2+}$ substitution and follows Vegard's law. Scherrer's grain diameter ($D$) lies within the range of 17.33 – 26.47 nm. Due to difference in the ionic radii of $Zn^{2+}$ and $Ni^{2+}$ significant changes are observed in the structural parameters - unit cell volume ($V$), hopping length at A ($L_A$) and B ($L_B$) site, and x-ray density ($\rho_{XRD}$). Linear increase in Neel magnetic moment ($n_B^N$) with oxygen positional parameter is observed in the studied samples, attributed to simultaneous weakening of A–B interaction and strengthening of the B–B interaction.

## 1. Introduction

Spinel ferrite, $MFe_2O_4$ (M = $Mn^{2+}$, $Co^{2+}$, $Zn^{2+}$, $Mg^{2+}$, etc.), is one of the most important magnetic oxides, where oxygen has fcc close packing and $M^{2+}$ and $Fe^{3+}$ ions occupy either tetrahedral (A) or octahedral (B) interstitial sites. The spinel ferrite is a very promising candidate for understanding and controlling the magnetic properties of nanoparticles at the atomic level. They are the most widely used structural material having low cost, high performance for high-frequency applications. Several methods, [1, 2] have been used to synthesize spinel ferrite nanoparticles. Among all these synthesis methods, sol-gel auto-combustion has been proved to be a simple and economic way to prepare nanoscale powders. In this technique, a thermally induced anionic redox reaction takes place. The energy from the exothermic reaction between oxidant and reductant can be high enough to form a spinel ferrite phase within a very short period of time i. e. even in dry gel/as-burnt form (i. e. the samples are without any thermal/sintering treatment). The literature also reports about the spinel phase formation in dry gel form in Cr doped Ni-Cu-Zn [3], Ni-Cu-Zn [4], Zn substituted in Mg-Ni ferrite [5], Ni doped Zn-Mg-Cu ferrite [6].

Substituted spinel ferrites with different divalent cations, are important candidate for both the fundamental studies and application point of view. Spinel ferrites are used in various applications, such as drug delivery [7] bio-separation, magnetic refrigeration systems [8], gas sensing [9], catalytic application [10], absorbent materials [11], ultra high density recording, magnetic resonance imaging, data processing devices, microwave and radio frequency devices, quantitative immunoassay [12], multilayer chip inductor and electromagnetic interference (EMI) suppression [13]. These applications of ferrites depend on its structural, magnetic and electrical properties. The origin of such properties of spinel ferrites is attributed to the cations distribution among tetrahedrally coordinated A sites (8 per unit cell) and octahedrally coordinated B sites (16 per unit cell). In general, magnetic characteristics of spinel ferrites are largely governed by three types of anti-ferromagnetic superexchange interaction, namely, $J_{AA}$ (A–O–A), $J_{AB}$ (A–O–B), and $J_{BB}$ (B–O–B), between cations on the A and B sites, each mediated by oxygen ions. Among them, $J_{AB} >> J_{BB}$, whereas $J_{AA}$ is small due to the large separation between two A-site ions. Owing to the difference in the size of cations, any alteration in their distribution changes the lattice constant and the oxygen parameter, thereby altering spin interactions that determine $J_{AB}$.

According to authors best knowledge no literature is available which reports the cation distribution of as-burnt Zn-Ni-Mg-Cu ferrite. Bachhava, et al [14] studied similar composition: $Ni_xMg_{0.5-x}Cu_{0.1}Zn_{0.4}Fe_2O_4$ (x = 0.1, 0.2, 0.3, 0.4 and 0.5) annealed at 1050°C for 24 hours and, observe decrease in the cation distribution parameters: bond lengths ($R_A$, $R_B$) and site radii ($r_A$, $r_B$) with nickel content. Although the above mentioned paper gives the parameters obtained by cation distribution, but does not report the obtained cation distribution for the studied system: $Ni_xMg_{0.5-x}Cu_{0.1}Zn_{0.4}Fe_2O_4$ ferrite [14]. Cation distribution can give an enhanced understanding of the magnetic interaction in spinel ferrites. Therefore, a clear understanding of the cation distribution in spinel ferrites is essential for studying and controlling their structural, magnetic properties [15] and to establish a better relationship between them so that the properties of spinel ferrite can be easily tuned just by varying the distribution of cations at A and B site.

Therefore, the present work focuses on the structural properties and, cation distribution of as-burnt $Zn_{0.75-x}Ni_xMg_{0.15}Cu_{0.1}Fe_2O_4$ (x = 0.00, 0.15, 0.30, 0.60, 0.75) powder synthesized via sol-gel auto-combustion method.

## 2. Experiment details

$Zn_{0.75-x}Ni_xMg_{0.15}Cu_{0.1}Fe_2O$ spinel ferrite powders with x = 0.00, 0.15, 0.30, 0.60, 0.75 were prepared by sol-gel auto-combustion technique. For the synthesis, citrate–nitrate/acetate pre-cursors [Zinc Nitrate – $Zn(NO_3)_2.6H_2O$, Nickel Acetate – $Ni(CH_3COO)_2.4H_2O$, Magnesium Acetate – $(CH_3COO)_2.Mg.4H_2O$, Copper Nitrate – $Cu(NO_3)_2.3H_2O$ and Ferric Nitrate – $Fe(NO_3)_3.9H_2O$] were mixed in stoichiometric ratio with citric acid as fuel. Metal salt to fuel ratio was taken as 1:1. All the pre-cursors were dissolved in de-ionized water to form aqueous solution. The pH value of solution was adjusted to 7 using ammonia ($NH_4OH$). Subsequently the aqueous solution was heated and evaporated under intensive stirring to transform into xerogel. The auto-combustion reaction was initiated by heating xerogel in air till the loose powder was formed called as 'dry gel or as-burnt powder' which was used for structural measurements.

## 3. Calculations

Experimental Lattice parameter ($a_{exp.}$) was determined using the relation - $a = d\sqrt{h^2+k^2+l^2}$, where d – inter-planer spacing and, (h, k, l) - miller indices. X-ray density ($\rho_{XRD}$) was obtained by [16]: $\rho_{XRD} = 8M_w / Na^3_{exp.}$, where $M_w$ – Molecular weight, N – Avagrado's Number. Scherrer's grain diameter ($D$) of the spinel powders was calculated from the x-ray broadening of the diffraction peak [311] using the Scherrer formula [17]: $D = 0.9\lambda / \beta\cos\theta$, where λ - wavelength of the x-ray used, β - line width, θ - peak position (in 2θ scale). The jump/hopping length '$L$' (the distance between the magnetic ions) for tetrahedral (A) and octahedral [B] site was calculated by using the expressions [18]: $L_A = (a_{exp.}\sqrt{3})/4$ and, $L_B = (a_{exp.}\sqrt{2})/4$, where $L_A$ and $L_B$ are respectively the hoping length for site A and site B. The specific surface area of the particles ($S$), which is the summation of the areas of the exposed surfaces of the particles per unit mass and is calculated by using the formula [19]: $S = 6/(D \times \rho_{XRD})$, where D = Scherre's grain diameter (nm) and $\rho_{XRD}$ = x-ray density (g/m$^3$).

Cation distribution in spinel ferrite can be obtained from the analysis of XRD pattern. In this work, the Bertaut method [20] is used to determine the cation distribution. This method selects a few pairs of reflections according to the expression: $I_{hkl}^{obs}/I_{h'k'l'}^{obs} = I_{hkl}^{cal}/I_{h'k'l'}^{cal}$, where $I_{hkl}{}^{obs}$ and $I_{hkl}{}^{cal}$ are the observed and calculated intensities for reflection (hkl), respectively. Best information regarding cation distribution can be obtained on comparing experimental and calculated intensity ratios for reflections which does not depend of the oxygen positional parameter ($u$). In the present study, (220) (422) (400) (440) planes are considered for cation distribution as these planes are sensitive to distribution of cation among tetrahedral (A) and octahedral [B] sites of the spinel lattice. The calculated and observed intensity ratios were compared for several combinations of cations distribution at (A) and [B] sites as described in [21]. The best cation distribution among the tetrahedral and octahedral sites for which, theoretical and experimental intensities ratios agree clearly, is taken to be the correct one. Observed intensities of the planes: 220, 440 and 311 were used for the calculation of the orientation degree. The orientation degree of the hk0 ($\boldsymbol{O_{hk0}}$) and 220 ($\boldsymbol{O_{220}}$) plane was calculated by using the formula: $I_{220}+I_{440}/I_{311}$ and $I_{220}/I_{311}$ respectively [22].

With the help of obtained cation distribution, ionic radii of A-site ($\boldsymbol{r_A}$) and B-site ($\boldsymbol{r_B}$) can be calculated using the relations [23]:

$r_A = (C_{AMg}{}^{2+} \times r_{Mg}{}^{2+}) + (C_{ANi}{}^{2+} \times r_{Ni}{}^{2+}) + (C_{AZn}{}^{2+} \times r_{Zn}{}^{2+}) + (C_{ACu}{}^{2+} \times r_{Cu}{}^{2+}) + (C_{AFe}{}^{3+} \times r_{Fe}{}^{3+})$

$r_B = [(C_{B\,Mg}{}^{2+} \times r_{Mg}{}^{2+}) + (C_{BNi}{}^{2+} \times r_{Ni}{}^{2+}) + (C_{BZn}{}^{2+} \times r_{Zn}{}^{2+}) + (C_{BCu}{}^{2+} \times r_{BCu}{}^{2+}) + (C_{BFe}{}^{3+} \times r_{Fe}{}^{3+})]/2$

Where, $C_A$ and $C_B$ are respectively the ionic concentration at A and B sites, $r_{Mg}{}^{2+}$, $r_{Ni}{}^{2+}$, $r_{Zn}{}^{2+}$, $r_{Cu}{}^{2+}$, $r_{Fe}{}^{3+}$ are the ionic radius of $Mg^{2+}$, $Ni^{2+}$, $Zn^{2+}$, $Cu^{2+}$ and $Fe^{3+}$ respectively, which are taken from the work of Shannon [24]. Using the ionic radii of A-site ($r_A$) and B-site ($r_B$), theoretical lattice parameter ($a_{th}$) is calculated using the formula [23]: $a_{th} = (8/3\sqrt{3})\left[(r_A + R_0) + \sqrt{3}(r_B + R_0)\right]$, where $R_o$ is the ionic radius of oxygen ion - 0.138nm.

The spinel structure consists of cubic close packed arrangement (fcc) of $O^{2-}$ ions. Each unit cell contains eight formula units of $AB_2O_4$ with space group Fd3m. The location of the atoms in the spinel depends on the choice of origin in the Fd3m space group. The conventional choices for the origin in the spinel unit cell are either an A-site cation with $\bar{4}3m$ or an octahedral vacancy with $\bar{3}m$ [25]. Oxygen positional parameter or anion parameter ($u$), which is the distance between the oxygen ion and the face of the cube edge along the cube diagonal of the spinel lattice is calculated by [16] : $u^{\bar{4}3m} = ((r_a + R_o)/(\sqrt{3} * a_{exp})) + (1/4)$ (for the unit-cell origin at either $\bar{4}3m$ on an A-site cation ) and $u^{\bar{3}m} = u^{\bar{4}3m} - (1/8)$ (for the unit-cell origin at $\bar{3}m$ on an octahedral vacancy). Where, $r_A$ = ionic radii of A-site, $R_o$ = 0.138 nm (ionic radius of oxygen ion) and $a_{exp.}$ = experimental lattice parameter. For a perfect fcc structure, ideal oxygen positional parameter ($u_{ideal}$) for origin at $\bar{4}3m$ and $\bar{3}m$ respectively is $u_{ideal}{}^{\bar{4}3m} = 0.375$ and $u_{ideal}{}^{\bar{3}m} = 0.250$. In spinel ferrite the distances between the unshared anions (shared tetrahedral edge - $d_{AE}$) and shared anions (shared octahedral edge - $d_{BE}$) can also be calculated by using respective formula [25]: $d_{AE} = a_{exp.}\sqrt{2}(2u^{\bar{4}3m} - 0.5)$ and $d_{BE} = a_{exp.}\sqrt{2}(1 - 2u^{\bar{4}3m})$. The distance between the tetrahedral cation and anion, known as tetrahedral bond length ($R_A$) and the distance between the octahedral cation and anion, known as octahedral bond length ($R_B$) was calculated by using the expressions [25]: $R_A = a_{exp.}\sqrt{3}(u^{\bar{4}3m} - (1/4))$ and, $R_B = a_{exp.}\sqrt{3(u^{\bar{4}3m})^2 - (11/4)u^{\bar{4}3m} + (43/64)}$. The interionic distances between cations ($M_e$-$M_e$) ($b$, $c$, $d$, $e$, $f$) was calculated using the formula: $b = (a_{exp.}/4)\sqrt{2}$, $c = (a_{exp.}/8)\sqrt{11}$, $d = (a_{exp.}/4)\sqrt{3}$, $e = (3a_{exp.}/8)\sqrt{3}$, $f = (a_{exp.}/4)\sqrt{6}$. The bond angles ($\theta_1$, $\theta_2$, $\theta_3$, $\theta_4$, $\theta_5$) were calculated using the subsequent expressions [25]. $\theta_1 = \cos^{-1}\left[(p^2 + q^2 - c^2)/2pq\right]$, $\theta_2 = \cos^{-1}\left[(p^2 + r^2 - e^2)/2pr\right]$, $\theta_3 = \cos^{-1}\left[(2p^2 - b^2)/2p^2\right]$, $\theta_4 = \cos^{-1}\left[(p^2 + s^2 - f^2)/2ps\right]$, $\theta_5 = \cos^{-1}\left[(r^2 + q^2 - d^2)/2rq\right]$. Where, (b, c, d, e, f) and (p, q, r, s) are the interionic distances between cations ($M_e$-$M_e$) and between cation and anion ($M_e$-O) respectively.

## 4. Results and discussions

Figure 1 (a) represents the X-ray diffraction (XRD) pattern of all the samples of $Zn_{0.75-x}Ni_xMg_{0.15}Cu_{0.1}Fe_2O$ (x= 0.00, 0.15, 0.30, 0.60, 0.75) spinel ferrite system. Fig. 1 (b, c) respectively depicts the Reitveld refinement for the composition x = 0.3 and x = 0.6, done via MAUD (Material Analysis Using Diffraction) software [26]. The presence of the strong diffraction peaks in the as-burnt powder corresponding to the planes (111), (220), (311), (222), (400), (422), (511/333), (440), (533) confirms the presence of cubic spinel ferrite phase similar to JCPDS card No. 08-0234. Hence, no further sintering is required in order to obtain the desire spinel ferrite phase. The as-burnt samples are in single phase form except for x = 0.75, slight presence of $\alpha$-$Fe_2O_3$ (JCPDS card No. 87-1166) is observed as a secondary phase. Formation of a secondary phase may be ascribed to the fact that the combustion process is taking place in an oxygen-rich environment. $\alpha$-$Fe_2O_3$ phase was also observed earlier in [5, 6, 27].

The linear variation of lattice parameter with $Ni^{2+}$ substitution for various compositions is shown in figure 2. It is observed that the lattice parameter of the $Ni^{2+}$ substituted Zn-Mg-Cu ferrite decreases linearly with the increasing the doping content, attributed to the replacement of larger $Zn^{2+}$ (0.060 nm) ion by smaller $Ni^{2+}$ (0.055 nm) ion. The partial replacement of $Zn^{2+}$ by $Ni^{2+}$ ion on the tetrahedral site results in the shrinkage of the unit cell. Decrease in the lattice parameter and shift of 311 peak towards higher angle with increase in nickel concentration confirms that the doped $Ni^{2+}$ ions have incorporated into the spinel structure. The linear variation of lattice constant with the doping content obeys Vegard's law [28]. Dependence of cation distribution on the experimental lattice parameter is shown in fig. 2 inset. Lattice parameter decreases linearly with increase in $Ni^{2+}$ ion occupation at tetrahedral A site. Thus, $Ni^{2+}$ ions occupying A site are mainly responsible for the shrinkage of the crystal lattice.

Intensities of the diffracted x-ray peaks directly depends on the atomic position in the spinel unit cell while the XRD peaks position depends on the shape and size of the unit cell. A closest agreement between the observed and calculated integrated intensity ratios - $I_{400/422}$, $I_{220/400}$ (Table 1) suggest that the estimated cation distribution is in good agreement with the real distribution. The orientation degree of the hk0 ($O_{hk0}$) and 220 ($O_{220}$) plane of the as-burnt $Zn_{0.75-x}Ni_xMg_{0.15}Cu_{0.1}Fe_2O$ ferrite system is also presented in table 1. The larger values of the $O_{hk0}$ and $O_{220}$ predicts the abundance of crystallites oriented in the [hk0] and [220] direction respectively. $O_{hk0}$ value first decreases for x = 0.15 and then increases for all values of x. $O_{220}$ value decreases for all values of x,

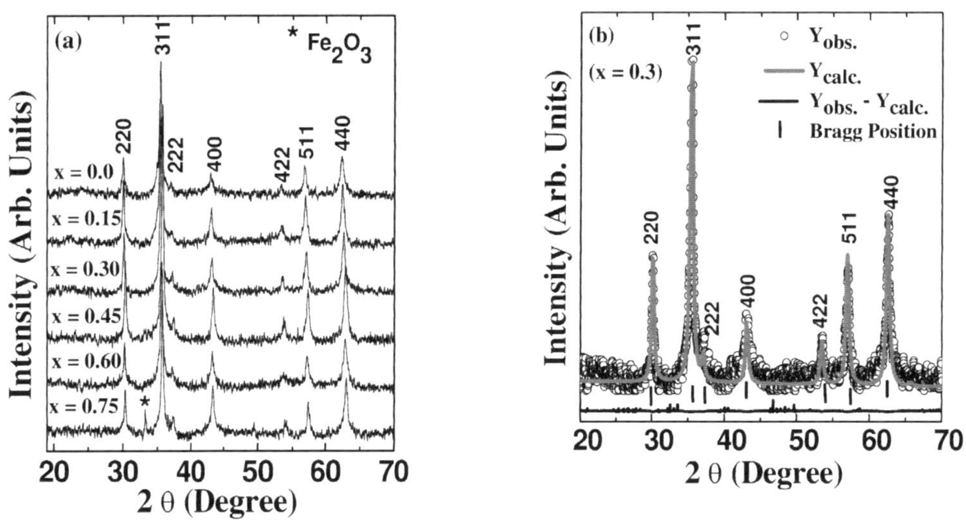


**Figure 1.** **(a)** XRD pattern of as-burnt $Zn_{0.75-x}Ni_xMg_{0.15}Cu_{0.1}Fe_2O_4$ system. **(b)** Reitveld refinement for the composition x = 0.3.

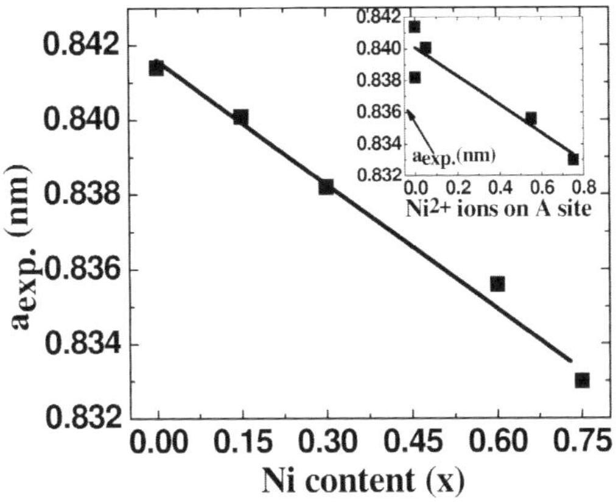


**Figure 2.** Linear decrease of experimental lattice parameter ($a_{exp.}$) with Ni doping and occupation of $Ni^{2+}$ ions at tetrahedral A site (inset).

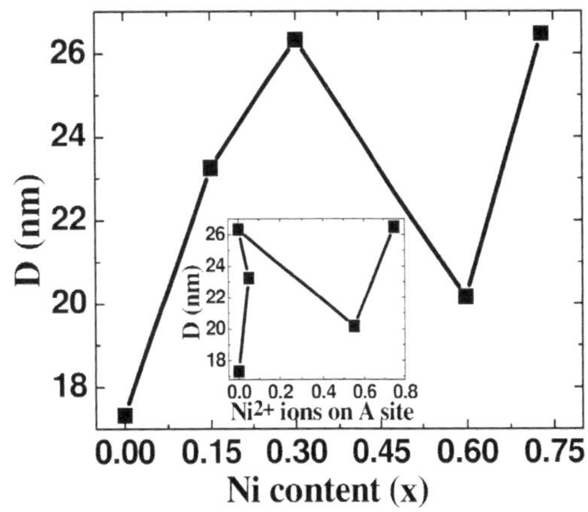


**Figure 3.** Variation of Scherrer's grain diameter ($D$) with Ni content. Inset: Dependence of $D$ on $Ni^{2+}$ ion occupation at tetrahedral A site.

except for x = 0.60. The variation in the values of $O_{hk0}$ and $O_{220}$ may be ascribed to the distribution of cations at A and B site.

Figure 3 describes the variation of Scherrer's grain diameter - $D$ (17.33 – 26.47 nm) with Ni (x) content. Though all the samples were prepared under identical condition, the particle size was not the same for all the $Ni^{2+}$ doped samples. D increases up to x = 0.30, decreases for x = 0.60 and again increases for x = 0.75. Such non-monotonic behavior of average crystallite size with increasing Ni concentration can be ascribed to the different dry gel formation time of the spinel ferrite, which favored the formation of new nuclei and preventing further growth of particles. It is of value to note that the highest value of D is observed for x = 0.75, which may be attributed to the presence of $Fe_2O_3$ phase [29]. Fig. 3 inset shows the particle size dependence of $Ni^{2+}$ ion occupation at tetrahedral A site. Perusal of fig. 3 shows the similar variation of particle size with Ni content and concentration of $Ni^{2+}$ ions at A site. Thus, D is mostly affected by the $Ni^{2+}$ ions occupying A site.

Table 2 displays the XRD parameters: the cell volume ($V$), hopping length at tetrahedral - A ($L_A$) and octahedral - B site ($L_B$), x-ray density ($\rho_{xrd}$) and specific surface area of the particles ($S$) of

**Table 1.** Cation distribution (for A and B site), observed, calculated intensity ratios for $I_{400/422}$, $I_{220/400}$ and orientation degree of the hk0 ($O_{hk0}$) and 220 ($O_{220}$) plane of $Zn_{0.75-x}Ni_xMg_{0.15}Cu_{0.1}Fe_2O_4$ as a function of Ni content (x).

| x | Cation distribution | $I_{400/422}$ Cal. | $I_{400/422}$ Obs. | $I_{220/400}$ Cal. | $I_{220/400}$ Obs. | $O_{hk0}$ | $O_{220}$ |
|---|---|---|---|---|---|---|---|
| 0.00 | $(Zn^{2+}_{0.75}Mg_{0.12}Cu_{0.05}Fe^{3+}_{0.08})_A$ $[Mg^{2+}_{0.03}Cu_{0.05}Fe^{3+}_{1.92}]_B$ | 1.72 | 1.76 | 1.65 | 1.65 | 0.80 | 0.39 |
| 0.15 | $(Zn^{2+}_{0.60}Ni^{2+}_{0.05}Cu_{0.05}Fe^{3+}_{0.30})_A$ $[Ni^{2+}_{0.10}Mg^{2+}_{0.15}Cu_{0.05}Fe^{3+}_{1.70}]_B$ | 1.43 | 1.41 | 1.97 | 1.65 | 0.70 | 0.36 |
| 0.30 | $(Zn^{2+}_{0.45}Fe^{3+}_{0.55})_A$ $[Ni^{2+}_{0.30}Mg^{2+}_{0.15}Cu_{0.10}Fe^{3+}_{1.45}]_B$ | 1.67 | 1.67 | 1.73 | 1.53 | 0.71 | 0.35 |
| 0.60 | $(Zn^{2+}_{0.03}Ni^{2+}_{0.55}Cu_{0.05}Fe^{3+}_{0.37})_A$ $[Zn^{2+}_{0.12}Ni^{2+}_{0.05}Mg^{2+}_{0.15}Cu_{0.05}Fe^{3+}_{1.63}]_B$ | 1.73 | 1.69 | 1.65 | 1.63 | 0.78 | 0.38 |
| 0.75 | $(Ni^{2+}_{0.75}Fe^{3+}_{0.25})_A$ $[Mg^{2+}_{0.15}Cu^{2+}_{0.10}Fe^{3+}_{1.75}]_B$ | 1.68 | 1.79 | 1.71 | 1.88 | 0.79 | 0.36 |

**Table 2.** Variation of cell volume (*V*), hopping length for A-site ($L_A$) and B-site ($L_B$), x-ray density ($\rho_{xrd}$) and specific surface area (*S*) with Ni content (x) $Zn_{0.75-x}Ni_xMg_{0.15}Cu_{0.1}Fe_2O_4$.

| x | V (nm$^3$) | $L_A$ (nm) | $L_B$ (nm) | $\rho_{xrd}$ (Kg/m$^3$) | S (m$^2$/g) |
|---|---|---|---|---|---|
| 0.00 | 0.5957 | 0.3643 | 0.2975 | 5234.32 | 66.14 |
| 0.15 | 0.5929 | 0.3638 | 0.2970 | 5236.27 | 49.28 |
| 0.30 | 0.5889 | 0.3630 | 0.2963 | 5249.42 | 43.43 |
| 0.60 | 0.5834 | 0.3618 | 0.2954 | 5253.09 | 56.66 |
| 0.75 | 0.5780 | 0.3607 | 0.2945 | 5279.47 | 42.93 |

$Zn_{0.75-x}Ni_xMg_{0.15}Cu_{0.1}Fe_2O_4$ (x = 0.00, 0.15, 0.30, 0.60, 0.75) spinel ferrite. Observed decrease in $L_A$, $L_B$ and V can be explained via replacement of an ion with higher ionic radius ($Zn^{2+}$) by an ion with lower ionic radius ($Ni^{2+}$). Decrease in the unit cell volume contribute towards the x-ray density ($\rho_{xrd}$) of Zn-Ni-Mg-Cu spinel ferrite. $\rho_{xrd}$ increases with Ni (x) content which can be attributed to the fact that the decrease in the volume of the unit cell overtakes the decrease in mass of the unit cell. Specific surface area of the as-burnt samples decreases for all values of x except for x = 0.6. High specific surface areas of particles are exceptionally essential for catalytic processes. There is an inverse relationship of particle size and density with surface area. In general, high surface areas suggest small particle sizes. The smaller is the particle size, the larger is the surface area. The catalytic property of the $Zn_{0.75-x}Ni_xMg_{0.15}Cu_{0.1}Fe_2O_4$ ferrite particles decreases with increase in Ni content.

Table 3 shows the interionic distances between cations ($M_e$-$M_e$) (*b, c, d, e, f*) for Zn-Ni-Mg-Cu as-burnt ferrite system. The observed decrease in interionic distances between cation can be can be understood by the fact that an element with higher ionic radius ($Zn^{2+}$ = 0.060 nm) is being replaced by an ion with lower ionic radius ($Ni^{2+}$ = 0.055 nm).

Figure 4 (a) illustrates the variation of the ionic radii of tetrahedral site ($r_A$) with Ni content. Decrease in $r_A$ up to x = 0.60 is due to the increase in the concentration of smaller $Fe^{3+}$ (0.049 nm) ions and decrease in the larger $Zn^{2+}$ (0.060 nm) ions on tetrahedral (A) site. Increase in $r_A$ for x = 0.75 can be attributed to the increase in the concentration of $Ni^{2+}$ (0.055 nm) ions on A site, as $Ni^{2+}$ ions migrates from octahedral (B) site to A site. Also the concentration of smaller $Fe^{3+}$ ions on A site decreases by a large amount. It is observed from fig. 4 (a, b, c, d) respectively that the ionic radii of tetrahedral site ($r_A$), oxygen positional parameter ($u$ ($\bar{4}3m$)) and distance between unshared anions ($d_{AE}$) and the tetrahedral bond length ($R_A$) varies in the same trend with Ni content (x). Variation of $u$, $d_{AE}$ and $R_A$ can be explained on the basis of $r_A$. Decrease in $r_A$ (up to x = 0.60), results in the contraction of the tetrahedral site in such a way that anions will move closer to the tetrahedrally coordinated A-site cations, which results in the decrease in oxygen positional parameter ($u$), distance between unshared anions ($d_{AE}$) and the tetrahedral bond length ($R_A$). And when $r_A$ increases with Ni content, $u$, $d_{AE}$ and $R_A$ also increases (for x = 0.75) because the tetrahedral site will expand as the anions associated with tetrahedral site move away from A-site cations. It is of value to note that the $u$ parameter value is greater than the ideal value for all the compositions of $Zn_{0.75-x}Ni_xMg_{0.15}Cu_{0.1}Fe_2O_4$. Deviation of u from its ideal value reveals distortion in the spinel structure.

Figure 5 (a) represents the variation of the ionic radii of octahedral site ($r_B$) with Ni content. With increase in Ni content $r_B$ increases up to x = 0.30 due to continues decrease in the concentration of

**Table 3.** Interionic distances between cations ($M_e$-$M_e$) (*b, c, d, e, f*) of as-burnt $Zn_{0.75-x}Ni_xMg_{0.15}Cu_{0.1}Fe_2O_4$ ferrite system.

| x | b (nm) | c (nm) | d (nm) | e (nm) | f (nm) |
|---|---|---|---|---|---|
| 0.00 | 0.2975 | 0.3488 | 0.3643 | 0.5465 | 0.5153 |
| 0.15 | 0.2970 | 0.3483 | 0.3638 | 0.5457 | 0.5145 |
| 0.30 | 0.2963 | 0.3475 | 0.3630 | 0.5444 | 0.5133 |
| 0.60 | 0.2954 | 0.3464 | 0.3618 | 0.5427 | 0.5117 |
| 0.75 | 0.2945 | 0.3453 | 0.3607 | 0.5410 | 0.5101 |

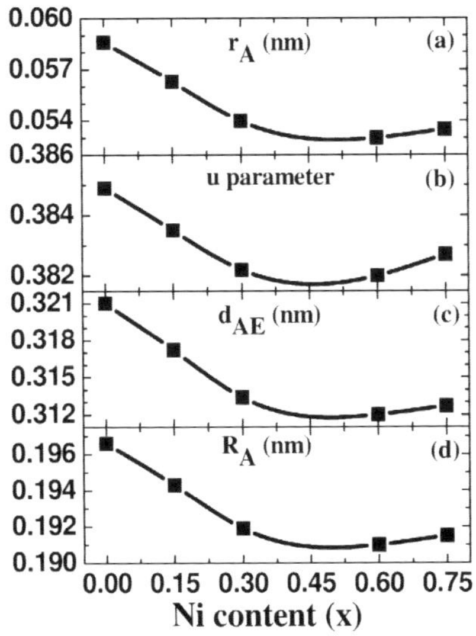


**Figure 4.** Ni content dependence of (**a**) ionic radii of A site ($r_A$), (**b**) oxygen parameter ($u$), (**c**) shared tetrahedral edge ($d_{AE}$) and (**d**) tetrahedral bond length ($R_A$) for as-burnt $Zn_{0.75-x}Ni_xMg_{0.15}Cu_{0.1}Fe_2O_4$.

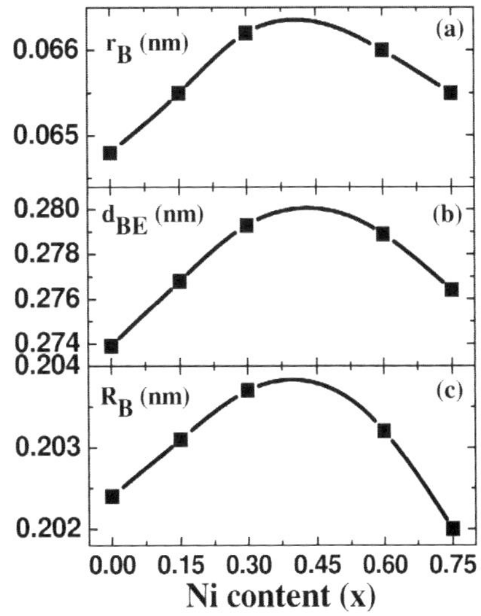


**Figure 5.** Ni content dependence of (**a**) ionic radii of B site ($r_B$), (**b**) shared octahedral edge ($d_{BE}$) and (**c**) octahedral bond length ($R_B$) for as-burnt $Zn_{0.75-x}Ni_xMg_{0.15}Cu_{0.1}Fe_2O_4$.

smaller $Fe^{3+}$ (0.0645 nm) ions and increase in the larger $Ni^{2+}$ (0.069 nm) ions on B site. Rapid decrease in $r_B$ for x ≥ 0.60 can be ascribed to the increase in the concentration of smaller $Fe^{3+}$ ions and decrease of larger $Ni^{2+}$ ions on B site. The same variation is observed in shared octahedral edge ($d_{BE}$) and octahedral bond length ($R_B$) with Ni content as depicted in fig. 5 (b) and (c) respectively. $d_{BE}$ and $R_B$ increases up to x = 0.30 and decreases for x ≥ 0.60. The variation in $d_{BE}$ and $R_B$ can be explain on the basis of ionic radii of B site. When $r_B$ increases, the anions associated with octahedral site move away from it thus, increasing the distance between octahedral cation-anion ($R_B$) and the distance between the "shared anions ($d_{BE}$). Similarly, when $r_B$ decreases, the anions associated with octahedral site will move closer to it thus, decreasing $R_B$ and $d_{BE}$.

It is of value to note that, in case of spinel ferrites XRD pattern not only gives information about structural data but, it also provides the magnetic information. One can predict the magnetic information in the form of Neel magnetic moment per unit Bohr Magneton ($n_B^N$ ($\mu_B$)) from the cation distribution (calculated as: $M_B$ - $M_A$, where $M_B$ and $M_A$ are respectively magnetic moment of B and A site) which is estimated from the x-ray diffraction intensities. Such type of magnetic information extracted from the diffraction pattern is also specified in our previous work [30].

Cation distribution parameters: oxygen positional parameter ($u$), relative bond angles ($\theta_1^{A\text{-}O\text{-}B}$, $\theta_2^{A\text{-}O\text{-}B}$, $\theta_3^{B\text{-}O\text{-}B}$, $\theta_4^{B\text{-}O\text{-}B}$, $\theta_5^{A\text{-}O\text{-}A}$) and Neel's magnetic moment affects the magnetic interaction between tetrahedral (A) and octahedral [B] site of spinel ferrite. The magnetic properties of the spinel ferrite materials originate from the spin magnetic moment of the unpaired transition-metal 3d electrons, coupled by the super-exchange interaction via oxygen ions separating the magnetic ions. The magnetic properties of ferrites are also greatly influenced by the exchange interactions. In ferrites, the exchange interaction takes place by the participation of oxygen 'anions' and is called super-exchange interaction. Since in ferrites magnetic ions occupy (A) and (B) sites, so there are three possible super-exchange interactions namely: $J_{AA}$ (A–O–A), $J_{AB}$ (A–O–B), and $J_{BB}$ (B–O–B). According to Neel's two sub lattice model of ferrimagnetism, $J_{AB}$ >> $J_{BB}$, whereas $J_{AA}$ is small (*hence, not consider in the calculation*) due to the large separation between two A-site ions. Neel's magnetic moment (Net

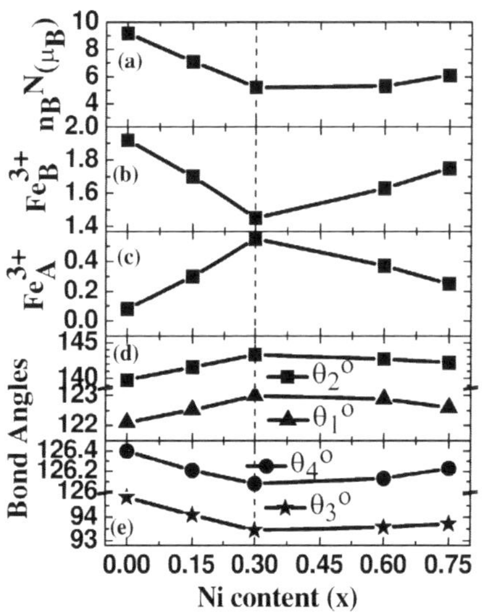


**Figure 6.** Variation of (**a**) Neel magnetic moment ($n_B^N$ ($\mu_B$)), (**b**) $Fe^{3+}$ content at tetrahedral site ($Fe^{3+}{}_A$), (**c**) $Fe^{3+}$ content at octahedral site ($Fe^{3+}{}_B$), (**d**) bond angles between A-O-B ($\theta_1$, $\theta_2$) and (**e**) bond angles between B-O-B ($\theta_3$, $\theta_4$) with Ni content (x). Dotted vertical line ( at x = 0.3 ) depicts change in the trend in the above mentioned parameters.

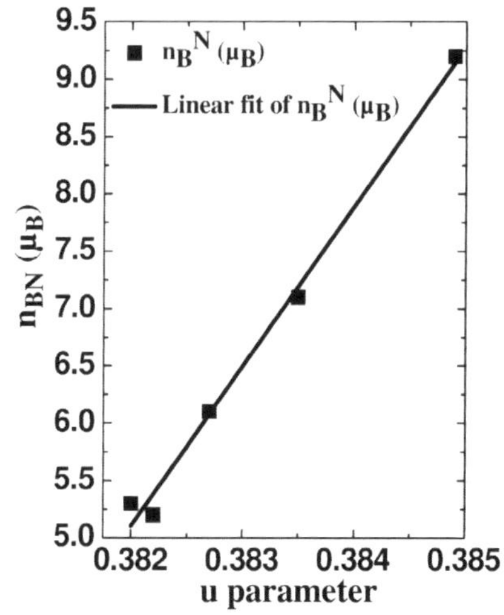


**Figure 7.** Variation of Neel magnetic moment ($n_B^N$ ($\mu_B$)) with oxygen positional parameter ($u$).

magnetic moment) of lattice is therefore, difference between the magnetic moment of B ($M_B$) and A ($M_A$) sub lattice [31]: $n_B^N = M_B - M_A$. The variation of Neel's magnetic moment with Ni (x) content is represented in fig. 6 (a). The value of Neel's magnetic moment were calculated considering the magnetic moments for $Mg^{2+}$, $Ni^{2+}$, $Zn^{2+}$, $Cu^{2+}$ and $Fe^{3+}$ - $0\mu_B$, $2\mu_B$, $0\mu_B$, $1\mu_B$ and $5\mu_B$ respectively. Since, the magnetic moment of $Fe^{3+}$ ions is much greater than the magnetic moment of the rest of the cations ($Mg^{2+}$, $Ni^{2+}$, $Zn^{2+}$, $Cu^{2+}$) therefore, the concentration of $Fe^{3+}$ ions occupying A and B site are mainly responsible for Neel's magnetic moment. $Fe^{3+}$ concentration at B ($Fe^{3+}{}_B$) and A ($Fe^{3+}{}_A$) is also depicted in fig. 6 (b,c). $n_B^N$ decreases up to x = 0.3 because of the decrease in the magnetic moment at B site ($M_B$) and increase in the magnetic moment at A site ($M_A$) and hence, the net magnetic moment ($M_B$-$M_A$) decreases. The magnetic moment at B site decreases due to decrease in the $Fe^{3+}$ concentration at B site. Increase in $Fe^{3+}$ content on A site, increases the magnetic moment at A site. Similarly, $n_B^N$ increases for x $\geq$ 0.60 because of increase in the magnetic moment at B site (as $Fe^{3+}$ at B site increases) and decrease in the magnetic moment at A site (as $Fe^{3+}$ at A site decreases) and hence, the magnetic moment decreases. Figure 6 (d) also illustrates the variation of the bond angles $\theta_1$, $\theta_2$ with Ni content (x). The bond angles $\theta_1$, $\theta_2$ are the angles between A-O-B. The bond angles $\theta_1$, $\theta_2$ initially increases up to x = 0.3 and thereafter, decreases up to x = 0.75. Increase in the bond angles $\theta_1$, $\theta_2$ suggests strengthening of A-B interaction (for 0 < x $\leq$ 0.3) and decrease in the bond angles $\theta_1$, $\theta_2$ indicate weakening of A-B interaction (for 0.3 < x $\leq$ 0.75). Similarly, the variation of the bond angle between B-O-B ($\theta_3$, $\theta_4$) is depicted in fig. 6 (e). Perusal of fig. 6 (e) shows that the bond angles $\theta_3$, $\theta_4$ initially decreases up to x = 0.3, and increases up to x = 0.75. Increase in the bond angles $\theta_3$, $\theta_4$ indicate strengthening of B-B interaction (for 0 < x $\leq$ 0.3) and decrease in $\theta_3$, $\theta_4$ suggests weakening of B-B interaction (for 0.3 < x $\leq$ 0.75). The variation of the bond angles: $\theta_1$, $\theta_2$, $\theta_3$, $\theta_4$ with Ni content are consistent with that of Neel magnetic moment. Due to decrease in $Fe^{3+}$ ions at B site up to x = 0.3, B-B-interaction decreases ($\theta_3$, $\theta_4$ decreases) and A- B interaction increases ($\theta_1$, $\theta_2$ increases) which in turn decreases the Neel magnetic moment and for 0.3 < x $\leq$ 0.75, B-B-interaction increases ($\theta_3$, $\theta_4$ increases) and A- B interaction decreases ($\theta_1$, $\theta_2$ decreases) which in turn increases the Neel magnetic moment. Thus increase/decrease in B-B interaction is responsible for increase/decrease in Neel magnetic moment.

Neel's magnetic moment also depends on the oxygen positional parameter (*u*). Whenever distortion is created in the crystal, cation redistribution occurs at A and B site, which alters the magnetic moment of A and B site. Fig. 7 shows a linear increase in oxygen positional parameter (*u*) with Neel's magnetic moment. As u parameter increases distortion is created in the spinel structure which forces more $Fe^{3+}$ ions to remain on B site, contributing in the enhancement of magnetic moment.

## 5. Summary

$Zn_{0.75-x}Ni_xMg_{0.15}Cu_{0.1}Fe_2O_4$ with x = 0.0, 0.15, 0.30, 0.60, 0.75 ferrites were synthesized by sol-gel auto-combustion technique without any heat treatment. X-ray diffraction analysis confirms the cubic spinel phase formation with a secondary phase of $Fe_2O_3$ for x = 0.75. Broadness of the diffraction peaks (311) indicates smaller particle formation. With increase in Ni content catalytic property of the $Zn_{0.75-x}Ni_xMg_{0.15}Cu_{0.1}Fe_2O_4$ ferrite particles decreases because of the decrease in specific surface area of the particles. The lattice parameter decreases linearly with Ni content and follows Vegard's law in the entire studied range of composition. Other structural parameters - unit cell volume (*V*), hopping length at A ($L_A$) and B site ($L_B$), x-ray density ($\rho_{XRD}$) decreases with increase in Ni content. Close occurrence between calculated and observed intensities ratio shows that the estimated cation distribution is in good agreement with the actual one. The interionic distances between cations ($M_e$-$M_e$) (*b, c, d, e, f*) decreases with Ni substitution. The migration of cations between A and B site bring perceptible changes in theoretical lattice parameter ($a_{th.}$), ionic radii of A-site ($r_A$) and B-site ($r_B$), oxygen positional parameter (*u*), shared tetrahedral and octahedral edge ($d_{AE}$, $d_{BE}$). Magnetic information in the form of Neel magnetic moment ($n_B^N$) is also extracted with the help of structural data. The B-B interaction is responsible for the variation of Neel magnetic moment with Ni content (x).

## Acknowledgments

This work is supported by MPCST project No.1783/CST/R & D/Phy and, Engg Sc and by UGC-DAE CSR, Indore [CSR-IC/CRS-74/2014-15/2104]. Authors thank Dr. M. Gupta, UGC-DAE CSR, Indore for XRD measurements.

## References

[1] Zhang J, Shi J and Gong M 2009 *J. Solid State Chem.* **182** 2135
[2] Gao X, Du Y, Liu X, Xu P and Han X 20111 *Mater. Res. Bull.* **46** 643
[3] Gabal M A and Al Angari Y M 2010 *J. Magn. Magn. Mater.* **322** 3159
[4] Yue Z, Guo W, Zhou J, Gui Z and Li L 2004 *J. Magn. Magn. Mater.* **270** 216
[5] Ghosh A, Satalkar M, Rathod S, Nag S P, Vyas P, Kane S N, Ghodke N, Prasad R and Dwivedi R, *International J. of Modern Physics: Conf. Series* **22** (2013) 28
[6] Satalkar M, Kane S N, A. Ghosh, N. Ghodke, G. Barrera, F. Celegato, M. Coisson, P. Tiberto, F. Vinai 2014 *J. Alloys Comp.* **615** S313
[7] Sun S, Zeng H, Robinson D B, Raoux S, Rice P M, Wang S X and Li G 2004 *J. Am. Chem. Soc.* **126** 273
[8] Chen Q and Zhang Z J 1998 *Appl. Phys. Lett.* **73** 3156
[9] Niu X, Du W and Du W 2004 *Sensor Actuators B* **99** 405
[10] Costa A C F M, Lula R T, Kiminami R H G A, Gama L F V, de Jesus A A and Andrade H M C 2006 *J. Mater. Sci.* **41** 4871
[11] Kobayashi M, Shirai H and Nunokawa M 2002 *Ind. Eng. Chem. Res.* **41** 2903
[12] Dey S, Dey S K, Ghosh B, Reddy V R and Kumar S 2013 *Mater. Chem. Phys.* **138** 833
[13] Zhang H, Ma Z, J.Zhou, Z. Yue, L. Li and Z. Gui 2000 *J. Magn. Magn. Mater.* **213** 304

[14] Bachhava S G, Patil R S, Ahirrao P B, Patil A M and Patil D R 2011 *Mater. Chem. Phys.* **129** 1104
[15] Chinnasamy C N, Yang A, Yoon S D, Hsu K, Shultz M D, Carpenter E E, Mukerjee S, Vittoria C and Harris V G 2007 *J. Appl. Phys.* **101** 09M509
[16] Smit J and Wijn H P J 1959 *Ferites* (Eindhoven, The Netherlands: Philips Technical Library)
[17] Cullity B D 1978 *Elements of X-ray diffraction* (Philippines: Addison-Wesley Publishing Co.)
[18] Amer M A and Hiti M L 2001 *J. Magn. Magn. Mater.* **234** 118
[19] Qi X, Zhou J, Yue Z, Gui Z and Li L 2003 *Mater. Sc. and Eng. B* **99** 278
[20] Weil L, Bertaut E F and Bochirol L 1950 *J. Phys. Radium* **11** 208
[21] Tanna A R and Joshi H H 2013 *World Academy of Sc. And Eng. Tech.* **75** 334
[22] Schwartz R W, Clem P G, Voigt J A, Byhoff E R, Stry M V, Headley T J and Missert N A 1999 *J. Am. Ceram. Soc.* **82** 2359
[23] Lakhani V K, Pathak T K, Vasoya N H and Modi K B 2011 *Solid State Sc.* **13** 539
[24] Shannon R D 1976 *Acta Cryst* **A32** 751
[25] Sickafus K E, Wills J M and Grimes N W 1999 *J. Am. Ceram. Soc.* **82** 3279
[26] Lutterotti L and Scardi P 1990 *J Appl Cryst* **23** 246
[27] López J G, Rodriguez-Senín E, Pastor J Y, Paris M A, Martín A, Levenfeld B and Várez A 2011 *Powder Technology* **210** 29
[28] Vegard L 1969 *Zeitscchrift fur Physik* **5** 17
[29] Sun K, Zhongwen L, Zhong Y, Xiona J and Jiaomin H 2011 *J. Magn. Magn. Mater.* **323** 927
[30] Satalkar M, Kane S N, Kulriya P K and Avasthi D K 2016 *Nucl. Instrum. Methods B* **379** 235
[31] Gadkari A B, Shinde T J and Vasambekar P N 2010 *J. Magn. Magn. Mater.* **322** 3823

# Studies of EXAFSSpectra using Copper (II) Schiff Base complexes and Determination of Bond lengths Using Synchrotron Radiation

**A Mishra[1], V Vibhute[1], S Ninama[1], N Parsai[1], S N Jha[2], P Sharma[3]**
[1]School of Physics Devi Ahilya University, Khandwa Road, Indore [MP]
[2]Applied Spectroscopy Division, Bhabha Atomic Research Centre, Mumbai, India
[3]School of Chemical Sciences, Devi Ahilya University, Khandwa Road, Indore [MP]

Email: vishakhavibhute28@gmail.com, mishraa1960@yahoo.in

**Abstract:**X-ray absorption fine structure (XAFS) at the K-edge of copper has been studied in some copper (II) complexes with substituted anilines like (2Cl, 4Br, $2NO_2$, $4NO_2$ and pure aniline) with o-PDA (orthophenylenediamine) as ligand. The X-ray absorption measurements have been performed at the recently developed BL-8 dispersive EXAFS beam line at 2.5 GeV Indus-2 Synchrotron Source at RRCAT, Indore, India. The data obtained has been processed using EXAFS data analysis program Athena.The graphical method gives the useful information about bond length and also the environment of the absorbing atom. The theoretical bond lengths of the complexes were calculated by using interactive fitting of EXAFS using fast Fourier inverse transformation (IFEFFIT) method. This method is also called as Fourier transform method. The Lytle, Sayers and Stern method and Levy's method have been used for determination of bond lengths experimentally of the studied complexes. The results of both methods have been compared with theoretical IFEFFIT method.

## 1. Introduction

Ligand, a metal surrounded by a cluster of ions or molecule, is used for preparation of complex compounds named as Schiff base[1]. Metal complexes of Schiff bases play an essential role in agriculture, pharmaceutical and industrial chemistry and biology. The wide range of uses of o-PDA and its Schiff bases urged us to study its Schiff base. Large number of Schiff bases and their complexes are of significant interest and attention is drawn to them because of their biological activity including anti-tumor, antibacterial, fungicidal and anti-carcinogenic properties. Since the Schiff bases play an important role in the field of science and technology and due to their biological and industrial applications.

X-ray absorption fine structure (XAFS) refers to modulation in X-ray absorption coefficient just above the X-ray absorption edge. XAFS is often divided into two regions, XANES (which lies within the first 30 eV from the edge position) and EXAFS (which lies beyond 30 eV from the absorption edge).EXAFS is applicable to both condensed matter and gases. Single crystal is not required. Information available from EXAFS is average distance between the absorbing atom and its neighboring atoms, mean square variation in the distance and its coordination number and metal- ligand bond length.

In the present paper the bond lengths of the following copper complexes have been calculated experimentally by using Levy's, Lytle's and LSS (Lytle, Sayers and Stern) methods.
1.Copper(II) Pureaniline ($C_{32}H_{28}N_8Cl_2$), namely(Cu1) 2.Copper(II) 2Cl Aniline($C_{32}H_{26}N_8Cl_4$)namely (Cu2) 3.Copper(II)2$NO_2$aniline($C_{32}H_{26}N_{10}O_4Cl_2$)namely(Cu3) 4.Copper(II)4$NO_2$aniline($C_{32}H_{26}N_{10}O_4Cl_2$) namely (Cu4)

## 2. Experimental Techniques

In the present investigations, the X-Ray absorption spectra have been recorded using synchrotron radiation. The X-Ray spectroscopic setup is available at RRCAT and is called beam line. This beam line has been recently commissioned at the 2.5 GeV Indus-2 synchrotron radiation sources [1-3]. The beamline used in the present investigation has been designed in the dispersive mode and is called dispersive EXAFS (DEXAFS) BL-8 beamline. The computer software *Athena* version 0.8.056 has been used for analysis of the digital spectral data. All the complexes were prepared according to the standard methods and their purity was checked. For preparing the pellets, the complexes are mixed uniformly with a binder material like cellulose. Now the pellets of well mixed samples are made using the press (pelletizer).

**Sample Preparation:**

**Preparation of ligand:** Takeproperly weighedsubstituted aniline, 16 cc HCland 16 ccwater in a 500 c.c. beaker. Fit a thermometer in the solution and place the beaker in a through of crushed ice. Cool until the temperature of solution has fallen below $5^0$c.Dissolve 4 gm of sodium nitrite in 20cc of water and cool it also in ice bath. Add sodium nitrite solution in installments of 2-3 cc. each time to the well cooled solution of aniline and stir the mixture well with keeping the temperature low.The mixture obtained is added to a third beaker containing Acetylacetone, sodium acetate and alcohol by constantly stirring in crushed ice. Reflux this solution for 6 hours and allow it to dry. Again this dried solution is refluxed for 6-7 hours with o-PDA (orthophenylelediamine). Finally, allow it to come in the form of precipitates.

**Preparation of metal complex:** The metal salt ($CuCl_2.2H_20$) and the above prepared ligand were dissolved in Ethanol and refluxed for 8 - 12 hours. The contents we are then allowed to cool at room temperature for one day. The solution was thendecanted in watch glass andallowed to evaporate. The precipitates were then washed and dried inair [4].

## 3. Results and Discussion

The bond length in the copper complexes has been calculated with the help of Levy's, Lytle's and LSS (Lytle, Sayers and Stern) methods. The three methods are described as below:

### (A)Levy's method (Levy, 1965)

Levy has outlined a simple method of determining bond length (Levy, 1965)[5]. According to this method the bond length $R_1$ is given by-

$$R_1 = \left(\frac{151}{\Delta E}\right)^{1/2} \text{ Å}$$

Where $\Delta E$ is energy difference between the first EXAFS maxima B and first EXAFS minima β.

### (B) Lytle's method

Lytle employed a "particle in a box" theory to calculate the bond length [6]. In this method the bond length ($R_s$) is calculated by the radius $R_s$ of equivalent polyhedron through the relation

$$R_s(\text{Å}) = \{37.60/M\}^{\frac{1}{2}}$$

Where M is slope of E vs Q plot, energy values E is maxima and Q=2.04, 6.04, 12 and 20 are constants. The values of $R_s$ calculated with the help of this method are reported in table 1.

**(C)L.S.S. method**

In the the LSS method, for determination of the nearest neighbor distances, formula is given by $(1/2+n)\,\pi = 2k(R_1-\alpha_1) + 2\beta_1$, where $R_1$ is the bond length. $(R_1-\alpha_1)$ is the phase uncorrected bond length [7].

**(D)Fourier transforms (FT) method**

The magnitude of Fourier transform spectra are shown in Fig 1.The Fourier transform peaks at the radial distances of the neighboring atoms from the absorbing atoms. The distances found in Fourier Transform are, However, shorter by 0.2- 0.5Å than the actual distances due to energy dependence of the phase factors in sine functionof the theoretical expression for EXAFS, known as EXAFS equation[8]. The peaks shifted in the Fourier Transform ar shifted towards the origin by an amount $\alpha_j$ and hence the peaks are at distances $R_j-\alpha_j$. For the first peak j=1 and hence the position of the first peak determines the distance $R_1-\alpha_1$. The distance is given in table 1.

**Table1:** Values of first shell bond lengths (in Å) calculated from Levy's, Lytle's, L.S.S. and Fourier transform methods for copper (II) complexes.

| S. No. | Complex | Phase corrected | | Phase uncorrected | |
|---|---|---|---|---|---|
| | | Levy's method $R_1$ | Lytle's method $R_s$ | L.S.S. method $R_1-\alpha_1$ | F.T. method R |
| 1. | Cu1 | 1.89 | 1.55 | 1.40 | 1.43 |
| 2. | Cu2 | 1.52 | 1.56 | 1.34 | 1.30 |
| 3. | Cu3 | 1.61 | 1.82 | 1.30 | 1.29 |
| 4. | Cu4 | 1.61 | 1.82 | 1.34 | 1.32 |

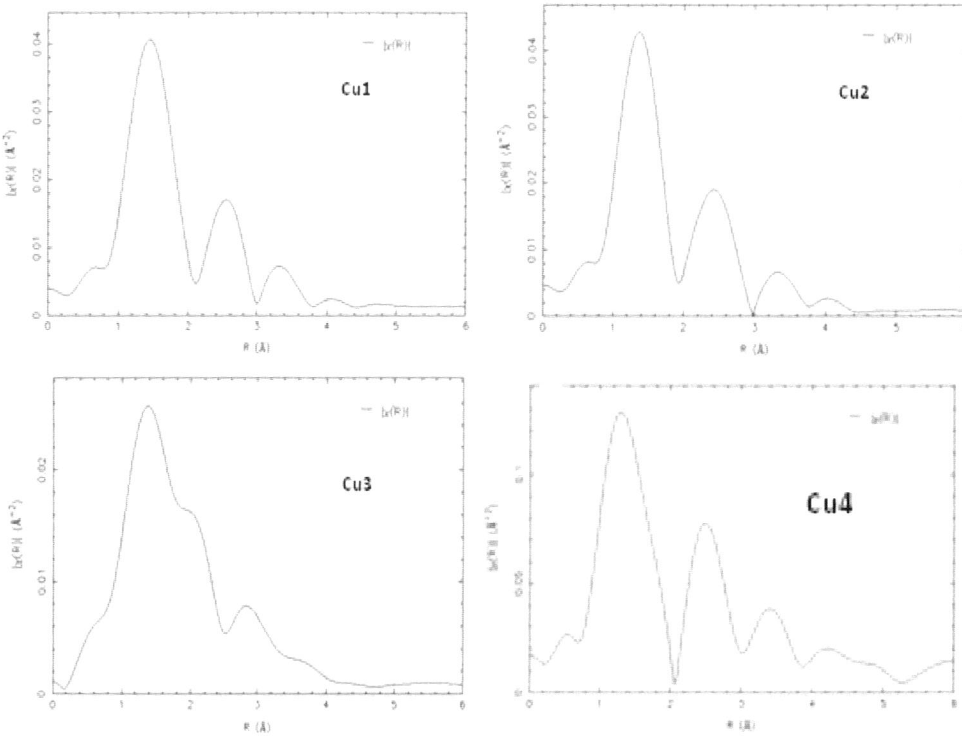


**Figure 1** Magnitude of Fourier transform bond length

## 5. Conclusions

X-ray absorption spectra at the K-edge of copper in copper (II) complexes have been recorded at BL-8 dispersive EXAFS beam line at 2.5 GeV Indus-2 Synchrotron Source at RRCAT, Indore, India.From the positions of EXAFS maxima and minima, the bond lengths in the complexes have been determined by four methods viz. Levy's, Lytle's and LSS methods and the Fourier transforms method. The bond lengths determined by LSS and Fourier transform method are uncorrected for phase shift. It has been observed that the values of $R_1$-$\alpha_1$ as determined from the LSS method and that determined from the Fourier transformation method are in good agreement with each other.

## Reference

[1] Dhar D N &Taploo C L 1982 *J SciInd Res* **41** 501
[2] Bhattachrya D, Poswal A K, Jha S N, Sangeeta and Sabharwal S C 2009 *Bull mater Sci* **52 (1)** 103
[3] Das N C, Jha S N, Bhattacharyya D, Poswal A K, Sinha A K and Mishra V K 2004 *Sadhana* **29** 545
[4] Raman N, Raja Y P and Kulundaisamy A 2009 Proc. *Indian Acad. Sci* **11 (3)** 183
[5] Levy R M 1965 *J. Chem. Phys.* **43** 1846
[6] Lytle F W 1966 *Advances in X-ray Analysis* **9** 398
[7] Lytle F W, Sayers D E and Stern E A 1975 *Phys. Rev. B.* **11** 4825
[8] Stern E A, Sayers D E and Lytle F W 1975 *Phys. Rev. B.* **11** 4836

Journal of Physics: Conference Series 755 (2016) 012052 doi:10.1088/1742-6596/755/1/012052

# Temporal evolution of electron beam generated Argon plasma in pasotron device

**Neha Khandelwal**[1], **U N Pal**[1], **Ram Prakash**[2] **and** Y **Choyal**[1]
[1]School of Physics, Devi Ahilya University, Indore (M.P.), India
[2]Microwave Tube Division, CSIR-Central Electronics Engineering Research Institute, Pilani (Raj.), India

E-mail: neha202k@gmail.com

**Abstract.** The plasma- assisted slow wave oscillator (PASOTRON) is a high power microwave source in which the electron beam in the interaction region is confined by the background plasma. The plasma is generated by impact ionization of background gas with the electron beam. A model has been developed for temporal evolution of Argon plasma in pasotron device. In this model, we consider electron beam of energy E interacting with Argon gas. The resulting ionization creates quasi neutral argon plasma composed of argon Ar atoms, singly ionized ions $Ar^{+1}$and electrons having energy from 0 to E. Electron impact excitation, ionization, radiative decay, radiative recombination and three body recombination processes are considered in this model. Population of ground and excited states of argon atom, ground state of argon ion as well as the population of electron energy groups is calculated by solving time dependent rate equations. Temporal evolution of electron beam generated plasma is given.

## 1. Introduction

The plasma assisted slow wave oscillator (PASOTRON) [1,2],is a unique high power microwave source that operates in the absence of external magnetic fields that makes these device light in weight and therefore is attractive for airborne applications. The transport of an intense electron beam in the pasotron is based on the focusing effect of beam generated plasma which neutralizes the beam space charge. Initially the interaction is region is filled with a neutral argon gas, so there is no plasma and beam diverges due to its own self fields [3]. The plasma neutralizes the space charge field of the electron beam and consequently an external magnetic field to guide the beam is not needed.

The paper is organized as follows. In section 2, kinetics and rate equations of electron beam generated plasma for pasotron is described. Section 3 contains the simulation results and in the final section we summarize our results.

## 2. Kinetics of electron beam generated argon plasma

In our model, we consider impact ionization of argon gas with an electron beam of energy E. The resulting ionization creates quasi neutral argon plasma composed of argon Ar atoms, argon singly ionized ions Ar+ and electrons having energy from 0 to E. We ignore higher ionization states of Ar. When an energetic electron interacts with gas, it loses its energy through excitation, ionization and secondary electron production. The important processes are spontaneous radiative transition and collisional transitions induced by electron impact. The processes included in the model are:

(i) Electron impact excitation: Ar atom can be excited from one energy level to a higher energy level due to inelastic collision with incident electrons.

$$e + Ar \rightarrow Ar^* + e$$

(ii) Electron impact ionization: Electrons with sufficient energy can remove an electron from an atom and produce one extra electron and ion.

$$e + Ar \rightarrow Ar^+ + e + e$$

(iii) Spontaneous emission: Once an atom is excited, it will de-excite to the ground state without any external influence. This process is called spontaneous emission.

$$Ar^* \rightarrow Ar + h\vartheta$$

(iv) Radiative Recombination: $Ar^+$ can recombine with secondary electrons creating Ar atom and a photon.

$$e + Ar^+ \rightarrow Ar + h\vartheta$$

(v) Three body recombination: $Ar^+$ can recombine with secondary electrons in the presence of Ar atom.

$$e + Ar^+ + Ar \rightarrow Ar + Ar$$

During these processes primary electrons loose energy while the secondary electrons gain energy. The secondary electrons will then generate tertiary electrons and the avalanche process will follow. To obtain the energy distribution, the energy range 0-E is discretized into N subgroups, i.e., energy bins, denoted by index g, of different energies [4]. The flow of electrons in energy bins for secondary electrons due to collision with atoms, ions and electron is calculated in a top to down order. An appropriate rate equation system is needed to describe the energy distribution of argon plasma. These equations involve the process of excitation, spontaneous emission, ionization, radiative recombination and three body recombination processes of populating and depopulating the levels. Beam induced field effect is not considered in this model.

*2.1 Rate Equations*

The rate equation for the population density N(n) of the n discrete state can be expressed in the general form [4]:

$$\frac{dN(n)}{dt} = G - L, \tag{1}$$

where G is the gain term and L is the loss term.

The rate equation begins with the ground state of argon atom:

$$\frac{dN1}{dt} = \sum_{k>1} A_{k,1} N_k(t) - \sum_{g_i=1}^{N_g} \sum_{g_j=g_i+1}^{N_g} \alpha(g_i, g_j) N_e(g_i, t) - \sum_{k\le 6} \sum_{g_i}^{N_g-1} R_{1k}(g_i)\, N_1(t) + \left[ \sum_{g_i=1}^{N_g} \beta_r(g_i)\, N_e(g_i, t) + \sum_{g_i=1}^{N_g} \beta_3(g_i)\, N_e(g_i, t) N_1(t) \right] N_7(t) \tag{2}$$

In Eq. (2), the first term is the spontaneous radiative decay rate from excited energy state of argon atom $k$ to ground state of argon atom, by which ground state are populated. The second term is product of the ionization coefficient and the electron number density at time t. Third term is the product of the excitation rate for generating excitation from ground state to energy state $k$ in argon atom due to electron impact excitations by the beam electrons. Last two terms are due to recombination processes.

The rate equations for excited states for neutral argon atom are given as follow:

$$\frac{dN_l}{dt} = \sum_{l=2}^{6} \sum_{g_i}^{N_g-1} R_{1l}(g_i)\, N_1(t) - \sum_{l=2}^{6} A_{l1} N_l(t) \tag{3}$$

Excited states are populated by electrons of higher energy created from ionizing collisions that scatter down into the excited states and population density reduces by spontaneous emission and product of ionization coefficient and number density of that excited level. The rate equation for the ground level of singly ionized argon is

$$\frac{dN_7}{dt} = \sum_{g_i=1}^{N_g-1} \sum_{g_i=g_i+1}^{N_g-2} \alpha(g_i, g_j) N_e(g_i, t) - \left[ \sum_{g_i=1}^{N_g-1} \beta_r(g_i) + \sum_{g_i=1}^{N_g} \beta_3(g_i)\, N_1(t) \right] N_e(g_i, t) N_7(t).$$

(4)

The rate equation for secondary electrons of energy bins $g_j$ is given by following expression:

$$\frac{dN_e(g_j,t)}{dt} = \sum_{g_i=1}^{N_g-1} \alpha(g_i,g_j)N_e(g_i,t) - [\beta_r(g_j) + \beta_3(g_j)N_1(t)]N_e(g_j,t)N_7(t) - \sum_{g_p=g_j}^{N_g+7} \chi(g_j,g_p)N_e(g_j,t) + \sum_{g_p=g_{j-1}}^{1} \chi(g_j,g_p)N_e(g_{j-1},t) \quad (5)$$

The gain term in secondary electron rate equation is due to electrons of higher energy created from ionizing collisions that scatter down into the energy bin. The loss term is due to electrons which suffer scattering collisions which cause them to lose sufficient energy to drop down to a lower energy bin as well as electrons in the bin which experience recombination with argon ion.

*2.2 Transition Rates*

In the study of relativistic electron beam produced plasma, several methods of energy apportionment have been developed to calculate the production yields from degradation of the incident electron energy. We have used Peterson- Green integral equation, in which a continuous energy loss[5] is assumed. The energy spectrum of the secondaries is rather broad, it extends from 0 up to $(E_p\text{-}I_j)/2$, where is $E_p$ incident kinetic energy and $I_j$ the ionization potential of the $j^{th}$ shell.

Excitation rates are calculated by product of loss function $L(g_i)$, velocity $v$, fractional energy loss $\mu_k$ and secondary electron number density $Ne$ of that energy group is

$$R_{1k}(g_i) = \frac{L(g_i)v(g_i)\mu_k(g_i)Ne(g_i)}{E_{1k}}.$$

Ionization rates for electrons in group $g_i$ causing an ionization of Ar in ground state resulting in an emitted electron in energy group $g_j$ can be expressed as

$$\alpha(g_i,g_j) = \frac{L(g_i)\ v(g_i)\mu_{gi}(g_i)N_1(t)}{\Delta E}.$$

where $\Delta E$ is the difference of energy group $g_i$ and group $g_{i+1}$.

Spontaneous radiative decay rates are taken from standard data table [6]. Recombination rate coefficient is calculated using formula [7]

$$\beta_r = 3.8X10^{-9}T_e^{-4.5}N_e + 1.55X10^{-10}T_e^{-0.63}.$$

Three body recombination rate coefficients for electrons in group g recombining with argon ion to create argon atom is calculated using formula [8]

$$\beta_3 = 3.68X10^{-32}((\frac{135300}{Te}) + 2)\exp(47800/Te)$$

## 3. Numerical Results

We have Ng + 7 set of coupled differential equations given by Eqs. (2) – (5) that are solved for number densities of different energy levels. The results are given for the following set of parameters: Ng=8000, electron beam energy E= 40 keV, beam duration 100ns and beam current 50mA in presence argon gas at pressure 10 mtorr. The beam radius is 0.12mm.

The electron number density can be calculated by sum of number density of each energy groups. The result of are given in Fig.1. The numerical value $3.204 \times 10^{14} cm^{-3}$ of the electron density is nearly equal to the gas density $3.2126 \times 10^{14} cm^{-3}$ indicating that almost whole gas converted into plasma. Energy distribution of the secondary electrons for various times is given in Fig. 2.

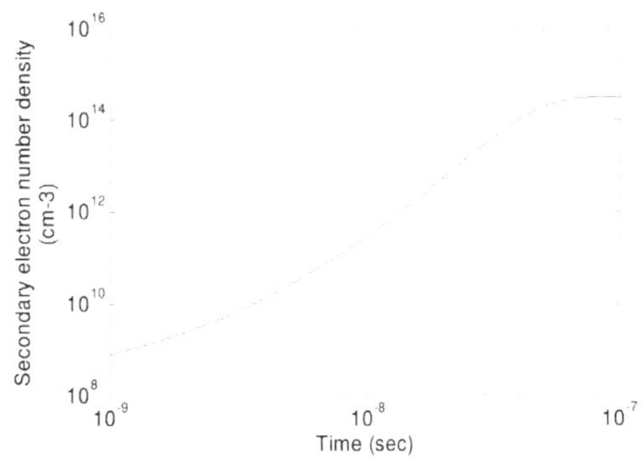


*Fig.1 Variation of plasma density with time*

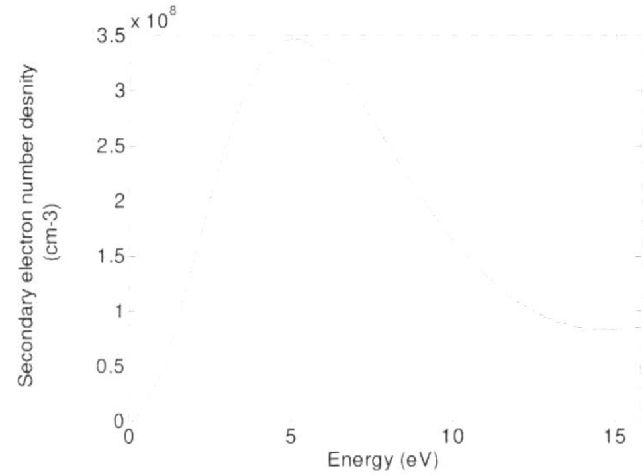


*Fig 2. Distribution of plasma density in different energy bins at different time*

The flow of electrons due to collision with atoms, ions and electron is calculated in a top-down order. An electron may collide with an electron, an atom or an ion and thus impart some of its energy to other particle in either elastic or inelastic manner. This causes a loss of an electron from the bin corresponding to initial energy and a production of an electron at some lower energy bin. Energy bin corresponding to energy 5.0 eV has maximum secondary electron number density. Similar results for different set of parameters and variation of plasma density at different beam radius has been presented elsewhere [9].

## 4. Conclusions and Discussions

Kinetics of relativistic electron beam interacting with neutral argon gas is developed. The results provide a general qualitative picture of the physical processes in a temporal evolution of electron beam generated argon plasma and electron energy distribution, which are consistent with general trend.

**Acknowledgment**
This work is supported by CSIR – Delhi, Grant No. 80(0081)/13/EMR-II.

## References

[1] Schumacher R W, Harvey R, Hyman J and Santoru J 1990 *U.S. Patent 4912367*
[2] Nusinovich G S, Carmel Y, Shkvarunets A G, Rodgers J C, Thomas M A, Granstein V L, Bilokh Y P, Geobel D M and Verboncoeur J P 2005 *IEEE Trans. On Electron devices* **52** 845
[3] Bilokh Y P and Nusinovich G S 2001 *IEEE Trans. of plasma science* **29**
[4] McGarrah D B and Brake M L 1990 *Laser and Particle Beams* **8** 507
[5] Peterson L R and Green A E S 1968 *J. Phys. B. (Proc. Phys. Soc.) Ser. 2* **1**
[6] NSRDA 1969 *National Stand. Ref. Data Ser., Nat. Bur. Stand. (U.S.)* **22** 306
[7] Skrzypkowski M P, Johnsen R, Rosati R E and Golde M F 2004 *Sci. Direct, Chemical Physics* **296** 23-27
[8] Owano T G and Kruger C H 1993 *AIAA Journal* **31**
[9] Khandelwal N, Maheshwari K P, Choyal Y, Pal U N and Ram Prakash 2015 *Proceedings of National Conference on Emerging Trends in Vacuum Electronic Devices and Applications* 149

# Effect of oxygen partial pressure on the structural and optical properties of ion beam sputtered $TiO_2$ thin films

**Firdous A Tantray[1], Romita Chouhan[1], Swati Rajput[2], Arpana Agrawal[1], Joseph T Andrews[2], Pranay K Sen[2], Mukul Gupta[3] and Pratima Sen[1]**

[1]Nonlinear optics lab, Laser Bhawan, School of Physics, Devi Ahilya University, Indore-452001, India
[2]Department of Applied Physics and Optoelectronics, Shri G.S. Institute of Science and Technology, Indore-452003, India
[3]UGC-DAE Consortium for Scientific Research, Khandwa road, Indore-452001, India

**Abstract.** We report the effect of oxygen partial pressure on the structural, electronic and nonlinear optical properties of ion beam sputtered $TiO_2$ thin films deposited on glass substrate at 40% of oxygen (S1) and 20% of oxygen (S2) partial pressure. XRD data shows the crystalline nature of S1 film while the film S2 was amorphous in nature. The energy band gap of the thin films calculated from their UV-Vis spectra was found to be 3.63 eV (S1) and 3.56 eV (S2). The decrease in the band gap with decrease in oxygen partial pressure may be attributed to the amorphous nature of the film. The nonlinear refractive indices for both the films were obtained from the closed aperture Z-scan experiment performed using a cw He-Ne laser source operating at 632.8 nm and were found to be $17.6\times10^{-9}$ $m^2/W$ and $-5.64\times10^{-9}$ $m^2/W$ for S1and S2 films, respectively. The reversal in the sign of the nonlinear refractive index may also be ascribed to the crystallinity of the grown films.

## 1. Introduction

$TiO_2$ thin films are well known transparent conducting oxides and the knowledge of their nonlinear refractive index makes them potential candidate material for optoelectronic device applications. As a semiconductor material with long-term stability, non-toxic environmental acceptability and broadly low cost availability, $TiO_2$ has also been taken into account for photovoltaic applications. However, due to optical gaps slightly above 3 eV, and high refractive index, natural $TiO_2$ is only photoactive in the UV region of the electromagnetic spectrum and an efficient active solar cell material [1-3]. The structural and optical properties of these films are highly sensitive to the environment in which they are grown and are dependent on many parameters such as thickness, film structure, oxygen partial pressure and substrate temperature.

In view of the above, the present manuscript focuses on the study of the effect of oxygen partial pressure on the structural and optical properties of $TiO_2$ thin films.

## 2. Experimental details

$TiO_2$ thin films designated as S1 and S2 were deposited on corning glass substrate by ion beam sputtering technique. Before deposition, substrates were ultrasonically cleaned in acetone. The base pressure was $6\times10^{-7}$ Torr and the films were deposited at a working pressure of $1.4\times10^{-5}$ Torr. The

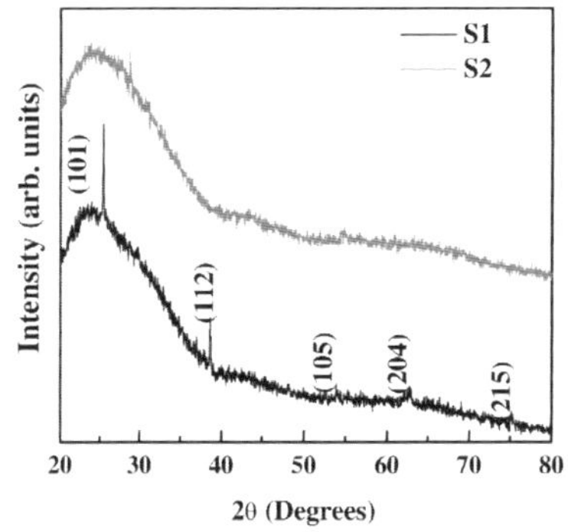


**Figure 1.** X-ray diffraction patterns of S1and S2 $TiO_2$ thin films.

deposition time for both the films was 90 minutes. Pure Oxygen and Argon were used as reactive and sputtering gas, respectively. Sample S1 is deposited at 3 sccm Ar gas and 2 sccm Oxygen gas while the film S2 is deposited at 4 sccm Ar gas and1 sccm Oxygen gas. The thickness of both the films was measured by stylus profilometer and was found to be 200 nm. The grown films were characterized by X-ray diffraction (XRD) technique for the structural analysis. The energy band gaps were determined from ultraviolet visible spectroscopy (UV-Vis). The nonlinear optical properties were investigated by performing the Z-scan technique using a 20 mW cw He-Ne laser operating at wavelength of 632.8 nm. The closed aperture Z-scan experiment was done to determine the nonlinear refractive index of the deposited films. The laser beam waist was precisely determined by performing knife edge experiment and was found to be $w_o \approx 26.77$ μm, and the corresponding Rayleigh range ($z_R = \pi w_0^2/\lambda$) is 3.55 mm. Thickness of the films is much smaller than $z_R$ and ensures that the film experiences uniform intensity throughout its thickness. The film area was large (1×1 $cm^2$) as compared to the spot size and thereby allowing us to neglect the probable diffraction errors in the experiment. Initially, the Z-scan experiment was performed on bare quartz substrate and negligible contribution for nonlinear refraction was found at the input intensity used.

## 3. Results and discussions

Figure 1 shows the XRD patterns of the grown films. No sharp peak was observed in S2 film (amorphous nature) as compared to S1 (crystalline nature) film which suggests that decrease in the oxygen partial pressure reduces the crystallinity of the $TiO_2$ thin films [4]. Since S1 film is crystalline in nature, we have determined the particle size using Scherrer formula and found it to be 51.8 nm.

In figure 2, we have represented the $(\alpha h\nu)^2$ vs photon energy ($h\nu$) for both the films. The energy band gap was determined by Tauc plot and was found to be 3.63 eV and 3.56 eV for S1 and S2 films, respectively. The decrease in the band gap of S2 film compared to S1 film by decreasing oxygen partial pressure may be attributed to the amorphous nature of the S2 film and thus it is assumed as indirect band gap. This result is consistent with the earlier reported work of Hassan et.al. [5].

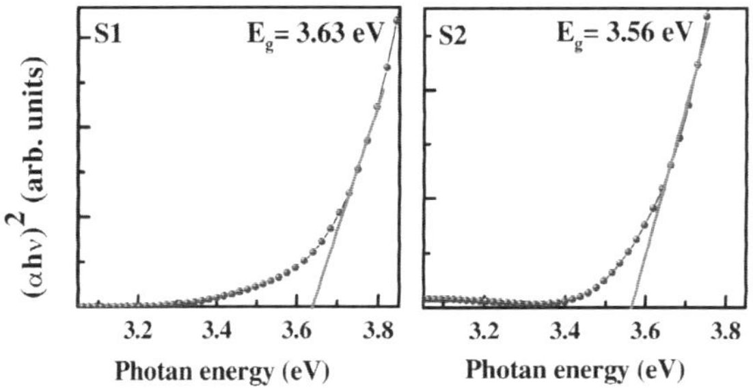


**Figure 2.** $(\alpha h\nu)^2$ versus photon energy ($h\nu$) plot of S1 and S2 $TiO_2$ thin films

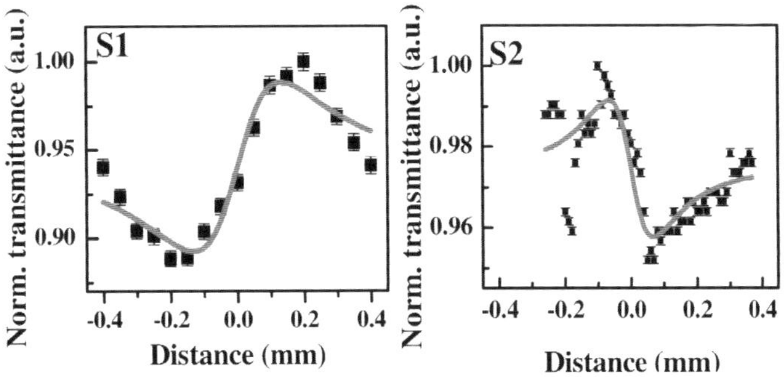


**Figure 3.** Closed aperture z-scan measurements (at 632 nm) of the grown films. Symbols represent the experimental curve and the solid lines shows theoretical fit using eq. (1)

Figure 3 exhibits the normalized transmittance as a function of distance near the focal region of the lens. Symbols in the figure indicate the experimental data while the solid lines show the theoretical fit to the experimental data. The appearance of transmittance peak followed by valley or vice versa in closed aperture (CA) experimental curve decides the nature of the sample lens and signatures the sign of nonlinear refractive index (NLRI). A peak (valley) followed by a valley (peak) in the CA curve is indicative of negative (positive) lens behaviour of the sample and consequently negative (positive) NLRI. The experimental curves obtained by us exhibit a prefocal transmittance minima (maxima) (valley/peak) followed by postfocal transmittance maxima (minima) (peak/valley) for S1 (S2) films and confirm the positive (negative) NLRI of the films.
Under CA z-scan condition, the normalized transmittance as a function of distance is given by [6, 7]

$$T[x, \Delta\phi_o]=1 + \frac{4\,\Delta\phi_o x}{[x^2+1][x^2+9]} \tag{1}$$

Here, x is the dimensionless sample position, $\Delta\phi_o$ is the induced phase shift ($\Delta\phi_o=n_2I_okL_{eff}$) treated as the fitting parameter, k is the propagation vector, $L_{eff}$ is the effective length of the samples, $I_o$ is the intensity at the focal plane and $n_2$ is the nonlinear refractive index.

Using the value of $\Delta\phi_o$, we have determined the values of nonlinear refractive index using equation (1) for both the films and was found to be 17.6 x $10^{-9}$ $m^2/W$ and -5.64 x $10^{-9}$ $m^2/W$ for S1 sample the S2 sample, respectively.

The origin of NLRI is mainly due to electronic and thermal effects. The electronic effect occurs due to band-to-band transition and can be measured only by ultrashort pulse excitation. Laser irradiation also gives rise to thermal lensing effect leading to thermal nonlinearity. In the present experimental situation, we have employed the cw He-Ne laser and have attributed the cause of nonlinear refractive index in the thermal lensing effect [8, 9]. However, we have also observed change in the sign of NLRI from positive (S1) to negative (S2) with the decrease in the oxygen partial pressure. Such a reversal of sign of NLRI may be attributed to crystallinity of the grown films.

## 4. Conclusion

In conclusion, we have grown the $TiO_2$ thin films at different oxygen partial pressure using ion beam sputtering technique. The film grown at higher oxygen partial pressure shows the crystalline nature whereas the crystallinity destroys as oxygen partial pressure is decreased during the deposition process. Energy band gap of the films decreases with decrease in oxygen partial pressure. The nonlinear refractive index was determined using closed aperture Z-scan technique. The positive and negative nonlinear refractive indices were observed for S1 and S2 films, respectively. The origin of nonlinear refractive index is ascribed to the thermal lensing effect and the change in the sign is attributed to the crystalline nature of the films. The observed nonlinearity suggest the utility of the grown $TiO_2$ films for optoelectronic device applications.

**Acknowledgement**
Authors are thankful to Dr. T. Shripathi and Dr. U.P. Deshpande from UGC-DAE CSR, Indore for the UV-Vis measurements. The financial support provided by MPCST, Bhopal and UGC-DAE CSR, Indore are acknowledged herewith

**References**
[1] Tang H, Prasad K, Sanjinbs R, Schmid P E and Levy F 1994 *J. Appl. Phys.* **75** 2042
[2] Dakka A, Lafait J, Abd-Lefdil M and Sella C 1995 M.J. *Condens. Matter Phys.* **2** 153
[3] Saleh Arwaa F, Balawa Batool D and Hateef Areej A 2010 Al- Mustansiriya *J. Sci* **21** 132
[4] Bao-Xing Zhao, Ji-Cheng Zhou and Lin-yon Rong 2010 *Trans. Nonferrous Met. Soc. China* **20** 1429
[5] Hasan Nahida B, Haider Adawiya J and Al-Amar Mohammed A 2012 *European J. Scientific Research* **69** 520
[6] Sheik-Bahae M, Said A A and Stryland E W Van 1989 *Opt. Lett.* **14** 955
[7] Singh P and Aghamakar P 2014 *Appl. Phys. Lett.* **104** 111112
[8] Dar T A, Agrawal A, Sen P K, Choudhary R J and Sen P 2016 *Thin Solid Films*, http://dx.doi.org/10.1016/j.tsf.2016.01.059
[9] Agrawal A, Dar T A, Solanki R, Phase D M and Sen P 2015 *Phys. Status Solidi B* **252** 1848

# Development of Stretched wire measurement bench at IDDL, DAVV Indore

**Mona Gehlot, G Mishra**

Insertion Device Development Lab, School of Physics, DAVV, Indore.

Corresponding author: # mona_gehlot@yahoo.com

**Abstract:** A stretched wire magnetic measurement bench is under development at IDDL, DAVV, Indore. In this method a multistrend wire consisting of N turns is stretched inside the undulator to measure the field integrals of the undulators. The wire moved with constant velocity of translation measures the first integral of the undulator field. The cross motion of the wire at the undulator ends measures the second field integral. The measurement accuracy depends on the wire conditions and material properties. In this paper we follow an analytical approach to find the voltage fluctuations due to wire vibrations during the field measurement. It is shown that the voltage fluctuations depend on undulator gap, magnitude of the impulse on the wire. The mass density and the length of the wire also cause sizeable voltage fluctuations. The analytical derived expression is analysed to optimize system parameters for minimum errors during the measurement.

## 1. Introduction

The IDD Laboratory at DAVV, Indore is in the stage of fabrication of different types of undulator like 1m long PPM undulator, 0.5m hybride undulator. For the perfect fabrication of undulator a precise measurement procedure is mandatory to decide its suitability for FEL experiments. The IDD Laboratory has developed a Hall Probe Bench[1,2] and a Pulsed Wire Bench[1,2] for the undulator measurement. The Hall probe method for magnetic field measurement is a point to point measuring system. In this method the probe is moved along the undulator axis to measure the field at each point of the undulator axis. By the integration of the field we get the first and second field integral of the undulator. The first and second field integral are the quantities which gives the angle and electron trajectory along the length of the undulator. In quality undulators ,both these quantities hold to a minimum value. In the Pulsed wire method a wire is stretched along the length of the undulator axis. An appropriate pulse is provided to the wire and because of field of the undulator mechanical vibrations are produced in the wire and these vibrations directly gives the first and second field integral.

Now a days in the laboratory development of stretched wire bench[3-5] is in progress. In this method the field integrals are measured by integrating the voltage induced during the motion of the stretched wire inside the undulator. By moving the wire with the constant speed, induced voltage is proportional to first integral of the undulator field and the cross motion of the wire ends induces the voltage proportional to second integral of the field. In this paper we are describing the details of experimental setup and analytical approach to find the voltage fluctuations due to wire vibrations during the field measurement.

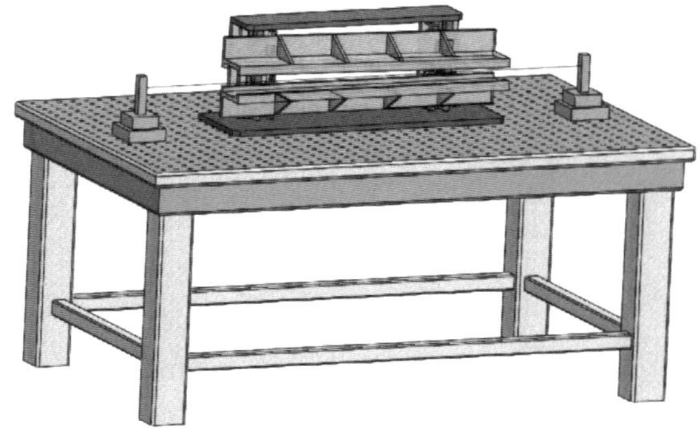

Fig 1.Stretched wire bench.

**2. Details of experimental setup:**

The details of components being used in the stretched wire bench is describing in this section. The measuring setup is under development stage which consists of two XYZ translation stages. The XY stages will be used for positioning and translating the stretch wire. The stages being used are Holmarc make model no. LMS-100-100. These stages are driven by motion controller. The travel range is upto 50mm. The Z stages are manually driven Holmarc make model no. TS-120-MU-10-01 and will be used for stretching the wire to reduce sag. The induced voltage will be measured by 2182A Keithley nanovoltmeter. It has the accuracy of typically just 15nVp-p noise at 1s response time, 45-50nV p-p noise at 60ms.

The XY stages and the voltmeter will be interfaced to a PC with LABVIEW software by National Instruments. Titanium, Cu-Be, Tungsten wire of diameter 125μm will be used for the measurements. The entire stretched wire setup with the undulator will be supported on a Honeycomb table top vibration isolation structure. The pneumatic vibration isolation system is a steel structure framework for supporting the honeycomb breadboard nonmagnetic table top, isolating the tabletop from vertical and horizontal disturbances. The physical dimension of the table top is 2000 mm × 1200 mm × 200 mm .

**3. Theory for wire vibration induced field integral:**

If the wire is moved with a constant velocity of translation in the transverse direction, we find an induced voltage as,

$$V_1 = \frac{d}{dt}\iint B_y dz dx = v I_y \tag{1}$$

Where $v = dx/dt$ is the velocity of translation of the wire in $x$ –direction. The second field integral is measured with cross motion of the ends i.e. by moving the ends of the wire in opposite directions and getting the induced voltage as,

$$V_2 = \frac{d}{dt}\iint B_y dz dx = \frac{2v}{L} II_y \tag{2}$$

Vibrations in the wire cause and introduce fluctuations in integral field measurements. In order to determine the effects of these vibrations , we consider a wire of length $L$ with fixed ends at $z=0$ and $z=L$ under a tension $T$. Using the impulse solution [4] we can write the Eq(1) as

$$V_1(t) = v\int_0^L B_u dz + \int_0^L B_u \sum_n \frac{2I}{\mu L}\sin(\alpha_n z)\cos(\omega_n t) dz \tag{3}$$

In Eq.(3) is obtained in the approximation that the undulator field does not change too much in the transverse direction . The undulator field is represented as , $B_u = \hat{y}B_0 \sin(k_u z)$, $k_u = 2\pi/\lambda_u$ , $\lambda_u$ is the undulator period. Eq.(3) can be expressed as,

$$V_1(t) = vI_y + \frac{IB_0}{\mu L}\sum_n I_n \sin(\alpha_n \xi)\cos(\omega_n t) \qquad (4)$$

$$I_n = \int_0^L 2\sin(k_u z)\sin(\alpha_n z)dz \qquad (5)$$

Eq.(4) is evaluated by averaging *V(t)* for a sufficient time in order to have the translation motion in the transverse direction in order of few *mm*, thus remaining in the approximation that the undulator field is independent of $x$ and solving the integral in Eq.(5) $\Delta\bar{t} = \Delta t C_s / L$ , using $\cos n\pi = (-1)^n$ , $\xi = L/2, \sin(\alpha_n \xi) = 1$ for $n =$ odd , $\xi = L/2, \sin(\alpha_n \xi) = 0$ for $n =$ even, we rewrite Eq.(4) as ,

$$\int_0^{\Delta t} V_1(t)dt = \Delta x I_y + \frac{2IB_0 L}{\mu C_s}\sum_{n=odd}\frac{\sin(k_u L)}{(n^2\pi^2 - k_u^2 L^2)}\sin(n\pi\Delta\bar{t}) \qquad (6)$$

The effects of the wire vibrations on second field integral can be estimated following a similar procedure. The voltage induced can be read as, Substituting and solving the Eq.(5) using the same quantities as used for Eq(6) ,we get

$$\int_0^{\Delta t} V_2(t)dt = \frac{2\Delta x}{L} II_y + 2\sum_{n=odd}\frac{2IB_0 L}{\mu C_s}\frac{\sin(k_u L)}{(n^2\pi^2 - k_u^2 L^2)}\sin(n\pi\Delta\bar{t}) \qquad (7)$$

## 4. Results and discussions:

Fig. 1 shows the design of stretched wire bench. In this paper ,we have followed an analytic approach to estimate the voltage fluctuations due to wire vibrations in a stretched wire measuring bench. Denoting

$$b_n = \frac{2IB_0 L}{\mu C_s}\frac{\sin(k_u L)}{(n^2\pi^2 - k_u^2 L^2)}\sin(n\pi\Delta\bar{t})$$

In Fig .2, we plot error for several values of impulse in the range from $2.0\times10^{-7}$ to $2.0\times10^{-1}$ *N-s* For $I = 2\times10^{-7}$ *N-sec* , $\Delta t = 5$ *sec*, $\mu = 4.14\times10^{-5}$ *Kg/m*, T=2.5 *N* , $C_s$=245.5 *m/sec*. For example consider an undulator with $\lambda_u = 50$ *mm* ,N=20. The plot shows one important feature of the effects of the wire vibrations. The wire vibrations contains several modes of vibrations. Whenever the wire resonant frequency corresponding to a particular mode equals the undulator frequency ( $\omega_n = \omega_u$ ) ,the peak value of the wire oscillations increases and brings a large fluctuations in voltage recording causing a large error. From Fig.(2), It is seen that it corresponds to the wire length i.e $l = L_u + n\lambda_u / 2$ where $n$ is an odd number i.e 1,3,5,..... $L_u = N\lambda_u$ is the undulator length. For our system it occurs at 1025 *mm*,1075 *mm*. Fig 3 ,4and 5, we calculate the effects of the wire vibrations on the induced voltage for different values of ,impulse, peak magnetic field and tension respectively. Where we define $\bar{V}_1(t) = \int_0^{\Delta t} V_1(t)dt$. For the parameters of study there is a voltage fluctuation of 0.3 to 5nVoltssec.

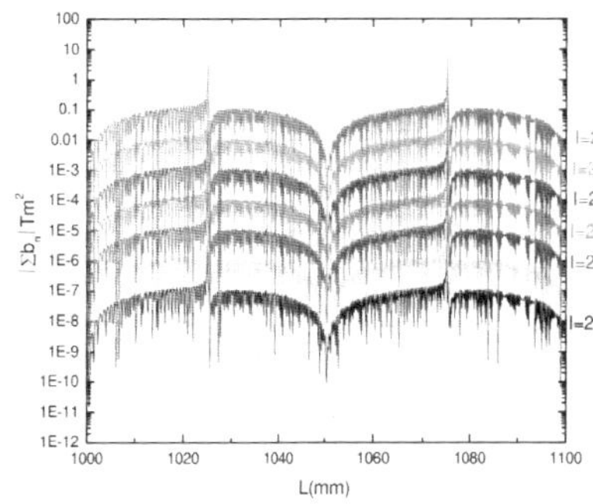


Fig 2 Integrated voltage errors versus wire length with several impulse magnitudes

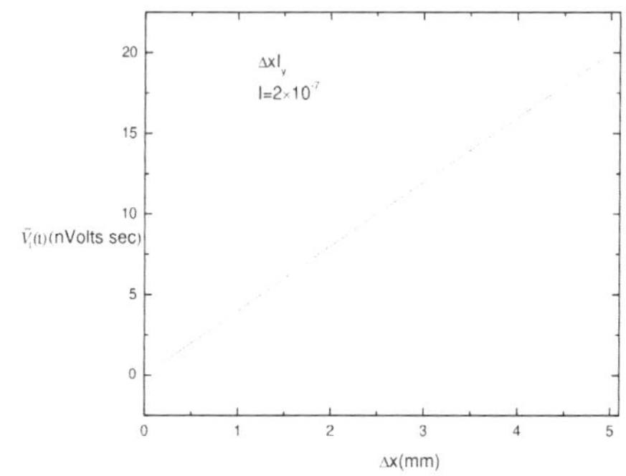


Fig. 3 Integrated voltage versus transverse wire displacement.

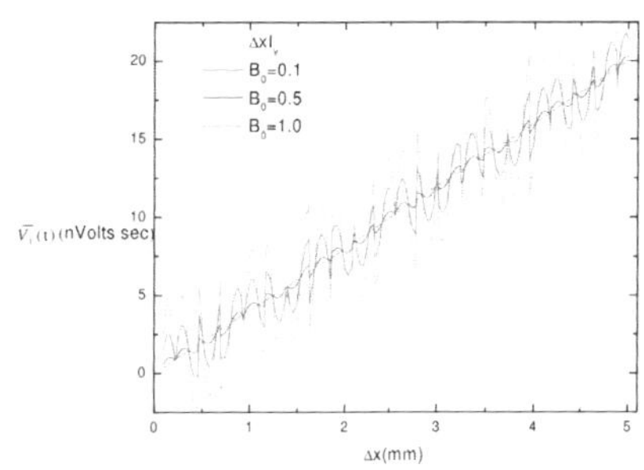


Fig. 4 Integrated voltage versus transverse wire displacement.

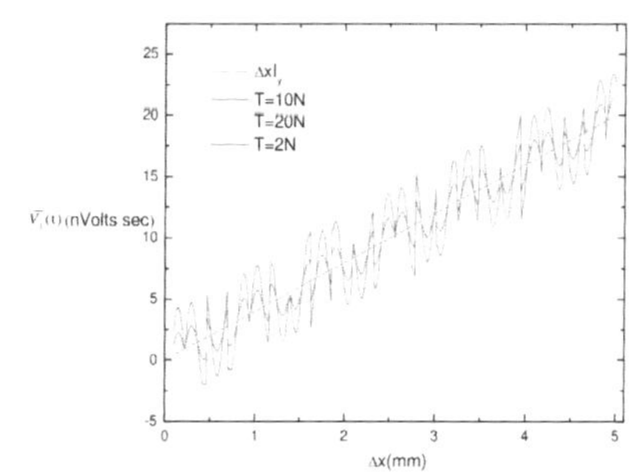


Fig. 5 Integrated voltage versus transverse wire displacement.

**Acknoledgement**
This work is financially supported by SERB,DST, Delhi, Govt. of India through a Grant No.EMR /2014/000120. Dr. Mona Gehlot is grateful to UGC, Delhi for woman post doc fellowship to join IDDL, DAVV, Indore, India.

**References**

[1] Sharma G, Gehlot M, Mishra G 2014 *World Scientific Journals* **3** 1450001
[2] Sharma G, Mishra G and Gehlot M 2016 *Measurement* **82** 334
[3] Gehlot M, Mishra G (Communicated 2016 ) *Nuclear Instruments and Meth.s in Phy. Re.s A*
[4] Ciocci F, Dattoli G and Sabia E 2006 *SPARC-FEL*-**06/001**
[5] Zangrando D and Walker R P 1996 *Nuclear Instruments and Meth.s in Phy. Re.s A* **376** 275

# Low-Power and High-Speed Technique for logic Gates in 20nm Double-Gate FinFET Technology

**A Priydarshi[1], M K Chattopadhyay[2]**

[1]School of Nanotechnology, RGPV Bhopal
[2]School of Electronics, Devi Ahilya University Indore

[1]achintyapriydarshi@gmail.com
[2]mkorwal@yahoo.com

**Abstract :** The FinFET is the leading example of multigate MOSFETS to substitute conventional single gate MOSFETs for ultimate scaling [1]. The FinFET structure is a combination of a thin channel region and a double gate to suppress the short channel effects (SCEs) and Vthvariation [2]. By using FinFET,figure of merits viz. ION, IOFF, output resistance, propagation delay, noise margin and leakage power, can be improved for ultra low power and high performance applications[3].

In this paper, a new high speed low power dynamic circuit design technique has been proposed using 20nm FinFETs. By applying the appropriate clock and sleep signal to the back gates of the FinFETs, the proposed circuit can efficiently control the dynamic power. During the pre-charging period, Vth of PMOS is controlled low so that a fast precharging can occur;

*Keywords- Power dissipation, Delay, PDP, CMOS families- AVL, Domino and Body Bias technique.*

## 1. Introduction

The FinFET (Fig.1.1) is the leading example of multigate MOSFETS to substitute conventional single gate MOSFETs for ultimate scaling [1]. The FinFET technology has been proposed by ITRS as a possible sacling path for low power and high performance CMOS technologies [2]. The FinFET structure is a combination of a thin channel region (which eliminates subsurface leakage path) with a double gate structure (which increases the capacitive coupling between the gate and the channel) to suppress the short channel effects (SCEs) and $V_{th}$ variation [3].

The channel of FinFET is a slab (fin) of undoped silicon perpendicular to substrate .At least two sides of the fin are wrapped around by oxide simultaneously so that the active regions are broken up into several fins and a gate overlaps the channel region of the fin on either side. Consequently increased electrostatic control of gate over channel causes very high $I_{ON}$ / $I_{OFF}$ ratios. In device designing the design parameters such as retrograde channel doping profile, length, width and height of the gated channel are important for improving the performance of the device. The leakage and delay of the device can be improved by taking less steep retrograde channel doping profile [4].

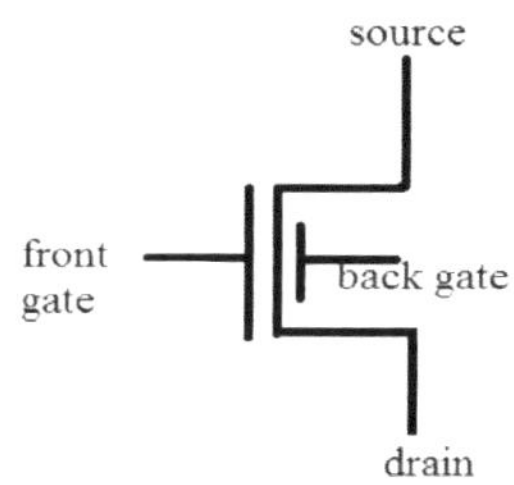


**Fig.1.1 : Symbol Double-Gate FinFET**

Gate control of the channel potential can be enhance by gating of the channel sidewall surfaces to suppress SCEs and, hence variability. Thus the FinFET can achieve superior electrostatic integrity (hence, scalability), as compare with other conventional MOSFETs [5], without requiring the formation of sub-gate length or high aspect ratio feature.

For low power digital circuit applications, the figure of merit are propagation delay, noise margin and power delay product (PDP). But the low ON current and very high gate drain capacitance [4-5] are the factors responsible for a poor digital circuit performance. Therefore, in this regard, FinFET can be use to overcome the low ON current, and reducing gate drain capacitance for improving switching activity [6-8]. Therefore, by using FinFET ,figure of merits such as $I_{ON,}$ $I_{OFF}$ , output resistance, propagation delay, noise margin and leakage power, can be improve for ultra low power and high performance applications[9].

## 2. Implementation of Inverter

### 2.1 Implementation of Domino inverter

A standard FinFET based domino logic circuit shown in Fig. 2.1 consists of an n-type dynamic logic block. During pre-charge, the output of the dynamic gate is charged to VDD and the output of the inverter is set to 0. During evaluation, the inverter makes conditional transition from 0 to 1. If the output of the domino gate is fed to other domino gates, then it must be ensured that all inputs are set to 0 at the end of the pre-charge phase and the transitions during evaluation are only 0 to 1[10]. Hence the dynamic node discharges only when the previous stage evaluates to 1.

### 2.2 Implementation of AVL inverter

To verify the functionality and benefit of AVL technique in domino logic, domino inverter as shown in Fig.2.3 is considered for simulation and is implemented in 20nm technology using CADENCE SPECTRE simulator with a supply voltage of VDD = 1V[12-15]. The advantage of using AVL circuit is that the load circuits can operate quickly when they are in active mode due to the increase in drain

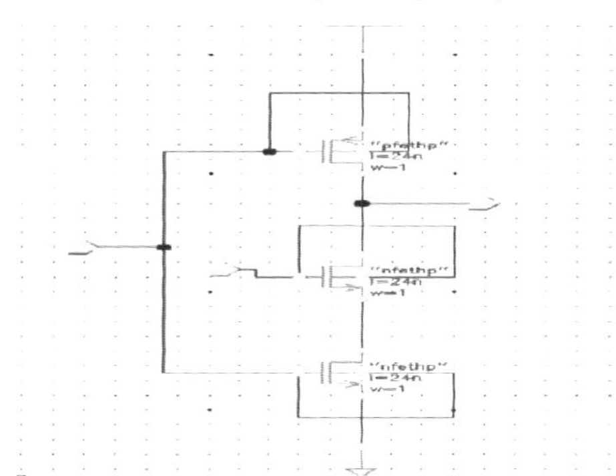

Fig. 2.1: Schematic of Domino inverter

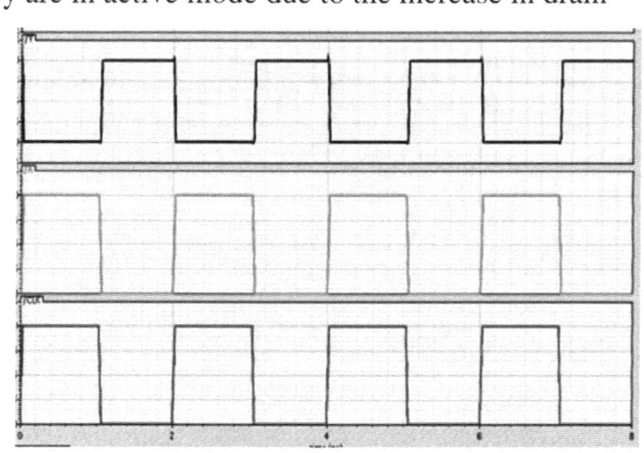

Fig. 2.2: Output of Domino inverter

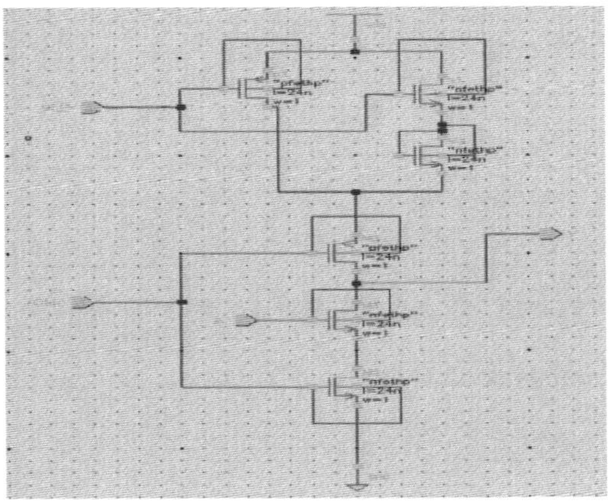

**Fig. 2.3: Schematic of AVL inverter**

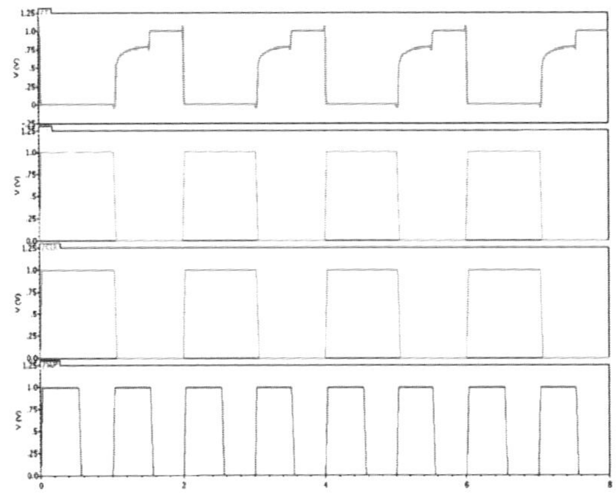

**Fig. 2.4: Output of AVL inverter**

-source current as the AVL circuit supplies the maximum drain-source voltage VDS to the on-MOSFETS through on-switches.

On the other hand, during standby mode, it supplies a slightly lower voltage through the weakly-on switches. Hence the sub-threshold leakage current of the off-FinFET decrease and the standby power gets reduced. It also produces high noise immunity. During normal mode of operation, the control signal sleep is low and the transistor PMOS is ON and NMOS is cutoff. Hence a voltage VDD appears across the circuit. That is VDC = VDD. But in standby mode, sleep is high turning off transistor PMOS and NMOS is turned ON. So VDD is applied to the load circuit through three weakly-on NMOS transistors. Hence the output voltage of the AVL circuit gets reduced, which reduces the leakage current of the circuit [16].

### 2.3 Implementation of proposed inverter

In the proposed method (fig. 2.5) to reduce the delay, and power dissipation as well as to improve the noise margin we have used Body Biasing technique with AVL circuit technique. In this method we have taken Domino inverter with AVL circuit and body biasing. Here the back gate of FinFET is connected to the clock. The simulations are performed using Tanner CADENCE Tools at 20 nm technology. The advantage of using AVL circuit is that load circuit will operate quickly when they are in active mode due to increase in drain source current but in standby mode due to less drain source voltage, threshold voltage will increase which reduces the leakage current, through the circuit[14].

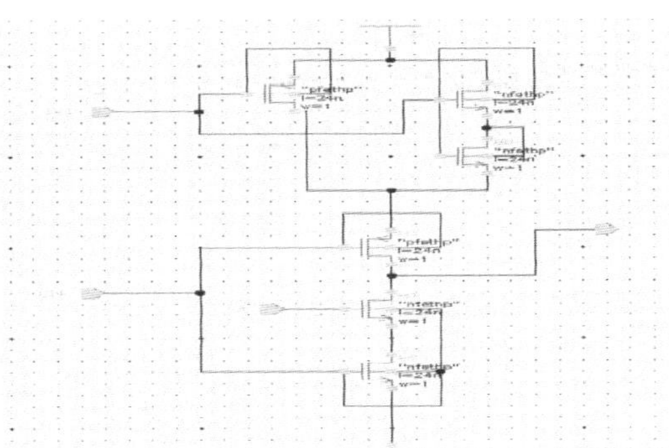

**Fig. 2.5: Schematic of proposed inverter**

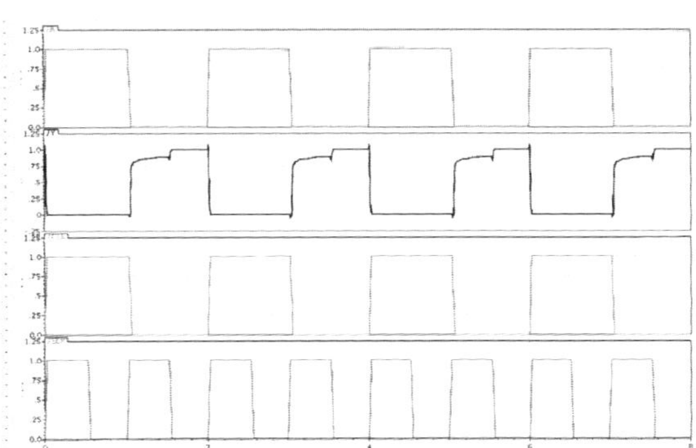

**Fig. 2.6: Output of proposed inverter**

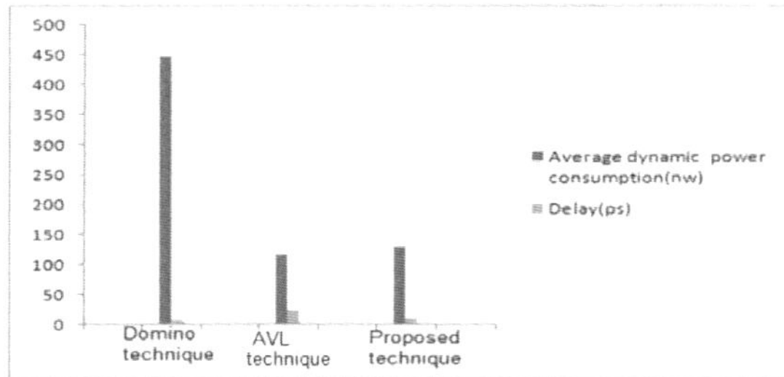


**Fig. 3.7: Graphical comparison of parameters**

**Table 3.1: Performance Parameter Results (at VDD=1v, Temperature =27$^0$ c)**

| Logic Techniques | Average dynamic power consump-tion(nw) | Delay( ps) | Power-delay product ($10^{-18}$J) |
|---|---|---|---|
| Domino inverter | 448 | 9.52 | 4.26 |
| AVL inverter | 117 | 24.26 | 2.83 |
| Proposed inverter | 129.9 | 11.1 | 1.44 |

## 3. Implementation of Nand Gate

### 3.1 Implementation of Domino NAND

A standard domino NAND logic circuit consists of an n-type dynamic logic block. Fig. 3.1 shows schematic of domino NAND logic gate. During pre-charge, the output of the dynamic gate is charged to VDD and the output of the inverter is set to 0. During evaluation, the inverter makes conditional transition from 0 to 1. If the output of the domino gate is fed to other domino gates, then it must be ensured that all inputs are set to 0 at the end of the pre-charge phase and the transitions during evaluation are only 0 to 1[15]. Hence the dynamic node discharges only when the previous stage evaluates to 1.

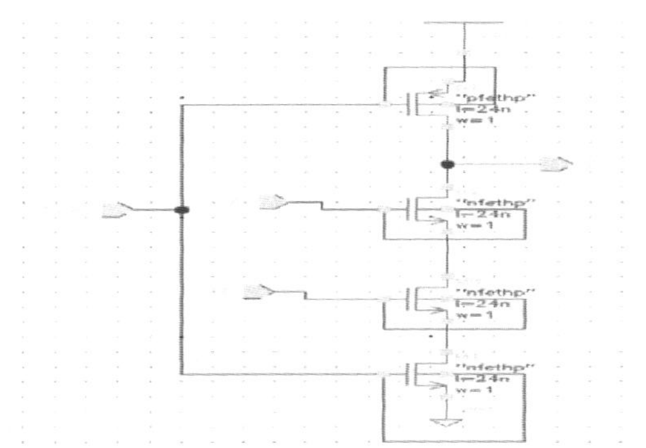

**Fig. 3.1: Schematic of Domino NAND**

**Fig. 3.2: Output of Domino NAND**

### 3.2 Implementation of AVL NAND

The advantage of using AVL circuit is that the load circuits can operate quickly when they are in active mode due to the increase in drain-source current as the AVL circuit supplies the maximum drain-source voltage VDS to the on-MOSFETS through on-switches. NAND circuits are the sub-circuits that are mostly used in various circuit especially-Arithmetic circuits (Full adder and multipliers), Compressors, Comparators, Parity checkers, Code converters, Error-detecting or Error-correcting codes and Phase detector. The performance of complex logic circuit is enhanced by the individual performances of the NAND-NOR circuits [20]. Several designs are available to realize the NAND function using different logic styles [11]. The proper selection of NAND circuit can add to the performance of large number of circuits. It optimizes the design for reduced delay, lesser PDP and lesser degradation on output voltage level. The aim to design NAND gate is to obtain low power consumption and delay in the critical path and full output voltage swing with less number of transistors to implement it [10].

### 3.3 Implementation of proposed NAND gate

Figures 3.5 show the proposed method for designing dynamic NAND circuits using FinFET, in which Vth depends on the back gate voltage. For a NMOS FinFET, a high VGb makes Vth low and a low VGb makes Vth high [11]. For a PMOS FinFET, the reverse scenarios are applicable. Usually the inputs of the circuits are connected to the front and back gates. For a NMOS FinFET, if the input is high, then Vth will be reduced by a high Vgb (gate voltage of the back gate) to make a fast signal transition [13]. If the input is low, then Vth will be high due to a low Vgb. However by connecting the front and back gates together more leakage current will be generated in the presence of many high input states (this occurs because these inputs make Vth low).

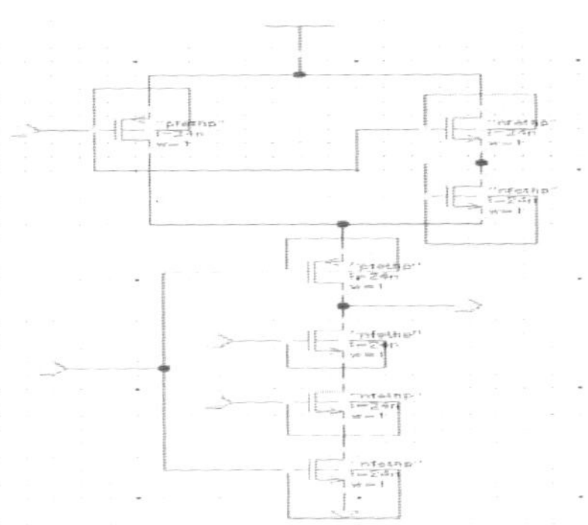

**Fig. 3.3: Schematic of AVL NAND**

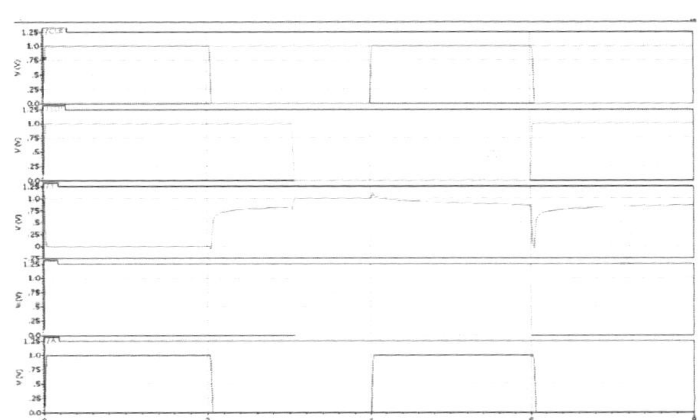

**Fig. 3.4: Output of AVL NAND**

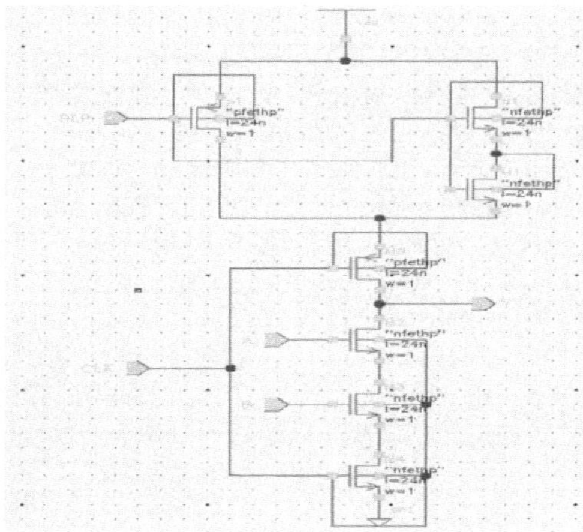

**Fig. 3.5: Schematic of proposed NAND**

**Fig. 3.6: Output of proposed NAND**

## 4. Implementation of Nor Gate

### 4.1 Implementation of Domino NOR

A standard domino circuit consists of an n-type dynamic logic block as shown in Fig. 4.1. During pre-charge, the output of the dynamic gate is charged to VDD and the output of the inverter is set to 0. During evaluation, the inverter makes conditional transition from 0 to 1. If the output of the domino gate is fed to other domino gates, then it must be ensured that all inputs are set to 0 at the end of the pre-charge phase and the transitions during evaluation are only 0 to 1. Hence the dynamic node discharges only when the previous stage evaluates to 1 and a high fan-out is achieved.

**Table 3.1: Performance Parameter Results (at VDD=1v, Temperature =27$^{0}$ c)**

| Logic Techniques | Average dynamic power consumption(nw) | Delay(ps) | Power-delay product($10^{-18}$J) |
|---|---|---|---|
| Domino NAND | 98.99 | 2.98 | 294.91 |
| AVL NAND | 21.87 | 28.12 | 614.98 |
| Proposed NAND | 26.1 | 3.04 | 79.34 |

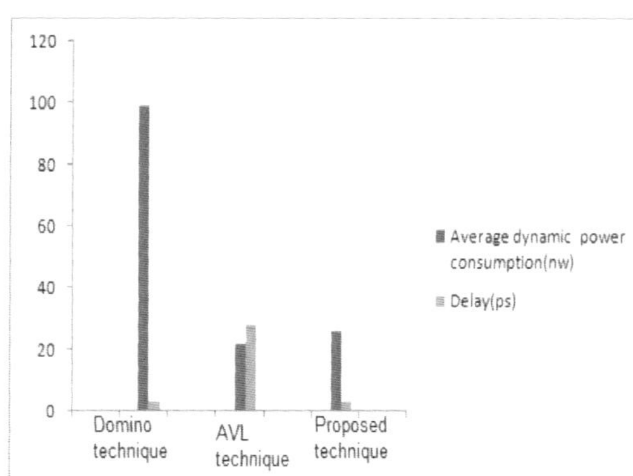


**Fig. 3.7: Graphical comparison of parameters**

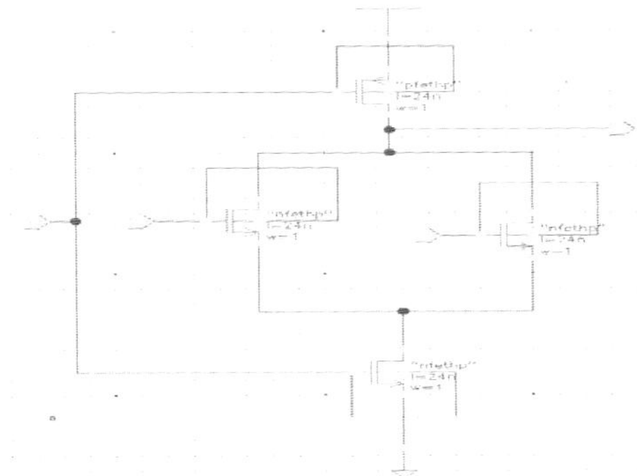

**Fig. 4.1: Schematic of Domino NOR**

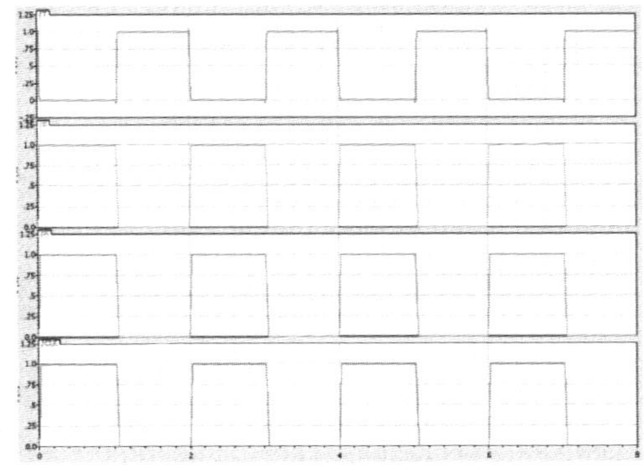

**Fig. 4.2: Output of Domino NOR**

## 4.2 Implementation of AVL NOR

To verify the functionality and benefit of AVL technique in domino logic, domino NOR as shown in Fig. 5.3 is considered for simulation and is implemented in 20nm technology using SPECTRE simulator with a supply voltage of VDD = 1V. The advantage of using AVL circuit is that the load circuits can operate quickly when they are in active mode due to the increase in drain-source current as the AVL circuit supplies the maximum drain-source voltage VDS to the on-MOSFETS through on-switches. On the other hand, during standby mode, it supplies a slightly lower voltage through the weakly-on switches. Hence the sub-threshold leakage current of the off-MOS transistors decrease and the standby power gets reduced.

## 4.3 Implementation of proposed NOR

In this method we have taken domino NOR gate with AVL circuit and body biasing techniques, where the back gate of PMOS FinFET is connected to the clock and the back gate of NMOS connected to its source terminal as shown in Fig. 4.5. In the proposed method to reduce the delay, and power dissipation we have used SG technique with AVL circuit technique. In this method we have taken domino NOR gate with AVL circuit and body biasing. Here the back gate of PMOS and NMOS is connected to the clock. The simulations are performed using SPECTRE simulator at 20 nm technology. The advantage of using AVL circuit is that load circuit will operate quickly when they are

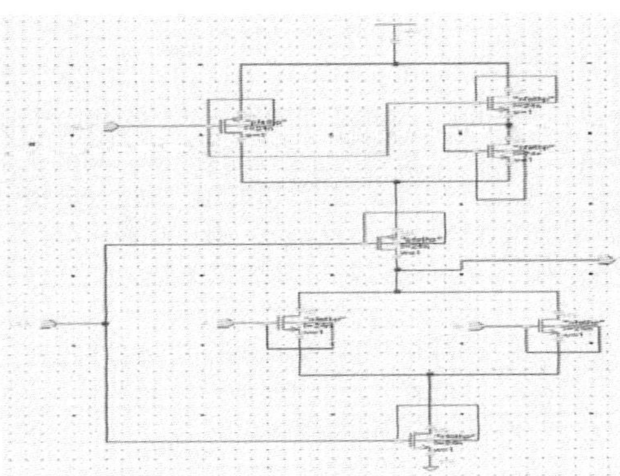

**Fig. 4.3: Schematic of AVL NOR**

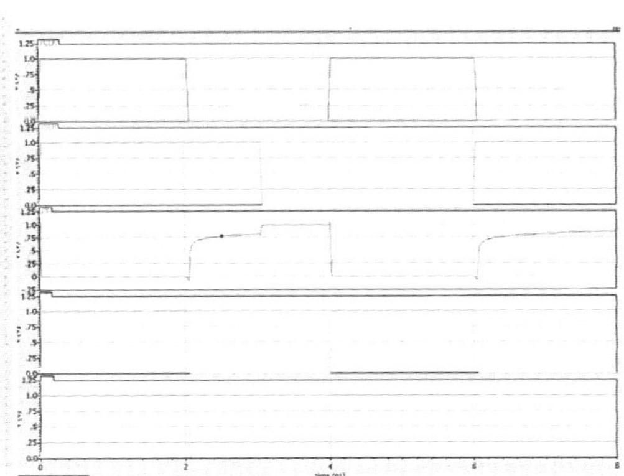

**Fig. 4.4: Output of AVL NOR**

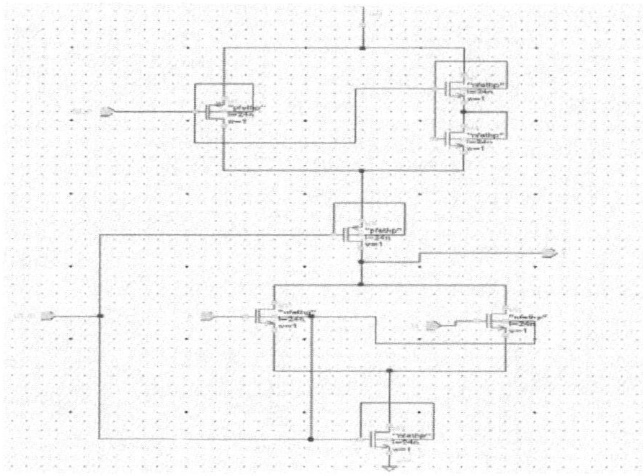

**Fig. 4.5: Schematic of proposed NOR**

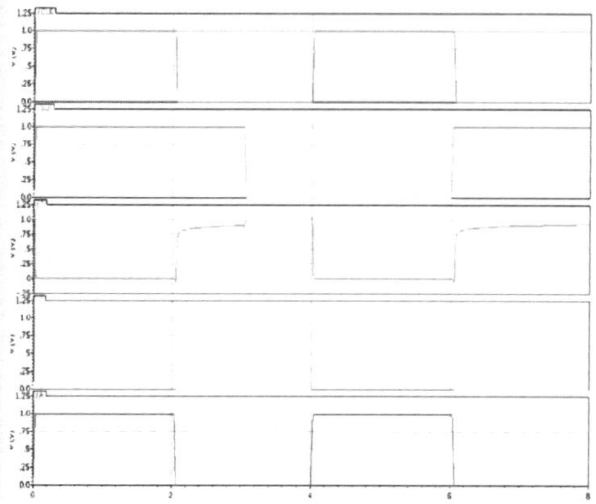

**Fig. 4.6: Output of proposed NOR**

**Table 4.1: Performance Parameter Results (at VDD=1v, Temperature =27$^0$ c)**

| Logic Techniques | Average dynamic power consumption(nw) | Delay(ps) | Power-delay product($10^{-18}$J) |
|---|---|---|---|
| Domino NOR | 235.00 | 3.49 | 821.32 |
| AVL NOR | 40.92 | 22.11 | 904.74 |
| Proposed NOR | 77.07 | 12.81 | 987.26 |

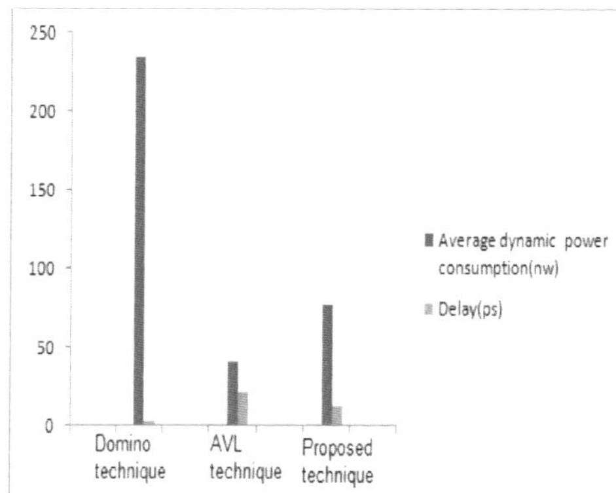


**Fig. 4.7: Graphical comparison of parameters**

in active mode due to increase in drain source current but in standby mode due to less drain source voltage, threshold voltage will increase

which reduces the leakage current, through the circuit.

## 5. Conclusion

We have discussed various logic techniques for low power and high speed FinFET based inverter, NAND gate and NOR gate. A new technique is investigated using Cadence SPECTRE simulator at 20nm node. Our simulation results indicate that on an average, our proposed technique has 68% low dynamic power consumption than the Domino technique and 70% low delay than and AVL technique. We conclude that for low power and high speed application, our proposed technique is better than the Domino and AVL techniques at 20nm FinFET technology.

## Acknowledgement
We would like to thank Dr. S.K. Vishvakarma for his support throughout this research work at IIT indore.

**Reference**

[1] Colinge J P, FinFETs and other multi-gate transistors, Springer, 2008
[2] Chiang M H, Kim K, Chaung C T and Tretz C 2006 *IEEE Trans.Electron Devices* **53** 2370
[3] Tawfik S A and Kursun V 2008 *IEEE Trans.Electron Devices* **55**
[4] Moselund K E ,Ghoneim H, Bjork M T, Schmid H, Karg S, Lortscher E, Riess W and Riel H 2009 *in Proc. Eur. Solid State DeviceRes. Conf.* 448
[5] Kang C Y, Choi R, Hussain M M, Wang J, Suh Y J, Floresca H C, Kim M J, Kim J, Lee B H and Jammy R 2007 *Appl. Phys. Lett.* **91** 033511
[6] Goncalves N F, Deman H J 1983 *IEEE Journal of Solid State Circuits*, **18** 261
[7] Madhuban K 2000 *5th IEEE International Symposium on Defect and Fault-Tolerance in VLSI Systems (DFT 2000)* 25-27
[8] http://www.computerhistory.org
[9] Krambeck R H, Lee C M, Law H S 1982 *IEEE Journal of Solid State Circuits* **SC-17** 614
[10] Fulde M, Engelst adter J P, Knoblinger G, and Schmitt-Landsiedel D 2007 *Adv. Radio Sci* **5** 287
[11] Kedzierski J, Ieong M, Kanarsky T, Zhang Y, and Philip Wong H S 2004 *IEEE transactions on electron devices* **51** 2115
[12] Wang M C 2010 *IAENG Transactions on Engineering Technologies* **4**
[13] Tawfik S A and Kursun V 2008 *IEEE Transactions On Electron Devices* **55**
[14] Moradi F, Gupta S K, Panagopoulos G, Wisland D T, Mahmoodi H, and Roy K 2011 *IEEE transactions on electron devices* **58**
[15] Datta A, Goel A, Cakici R T, Mahmoodi H, Lekshmanan D, and Roy K 2007 *IEEE Transactions On Computer-Aided Design Of Integrated Circuits And Systems* **26**, 1957
[16] Sharma D and Vishvakarma S K 2012 *Microelectronics Journal, Elsevier* **43** 358

# Cosmic Ray Daily Variation on Anomalous Days

**Arvind Dubey[a], Santosh Kumar[b], and S K Dubey[c]**

[a]Department of Physics,Seva Sadan College, Burhanpur - 450331 (M.P.), India
[b]H.No. 992/3036, Jabali Dant Chikitsalaya,Survey of India Chouraha, behind SBI qrts, Vijay Nagar, Jabalpur-482002 (M.P.), India
[c]Govt. model science college,(Autonomous) Jabalpur - 482001 (M.P.), India

Email Id – arvinddubey3768@gmail.com

**Abstract-** The Characteristics of the daily variation of cosmic ray intensity on different types of anomalous days have been studied by using data of Neutron Monitors. It is observed that Cosmic Ray intensity remains statistically low during the period of 1996-1998 on both Beijing and Moscow Neutron Monitor stations. These data are subjected to Harmonic Analysis and Fourier Techniques for entire period of consideration that is 1996-2006. The amplitude and phase of diurnal and semidiurnal anisotropies on these stations particularly on Anomalous days have been discussed. The occurrence of these events are dominated during solar activity minimum years. The number of days of occurrence is more anti-correlated with solar cycle.

**Keywords:** anomalous cosmic rays, solar poloidal magnetic field, daily variation.

## 1. Introduction

Cosmic ray daily variation (diurnal, semi-diurnal and tri-diurnal) arise from spatial anisotropies in interplanetary space. Ground-based detectors record these once every day as their asymptotic cone of acceptance sweeps through the direction containing the spatial anisotropy, co-rotational direction. The asymptotic cone of acceptance of a detector is the solid angle that contains all the asymptotic direction of approach of particles of various energies, which make a significant contribution to the counting rate of the detector. In addition to the diurnal component, the daily variation is composed of at least two more contribution of lesser amplitudes, i.e., semi-diurnal and tri-diurnal components.

The solar diurnal variation of the cosmic ray intensity is interpreted initially on the basis of an outward radial convection and an inward diffusion along the IMF. The balance between the convection and diffusion generates an energy independent anisotropic flow of cosmic ray particles from the 18-hour co-rotational direction.

The continuous outward low of solar wind and frozen-in magnetic field produces time variations in cosmic ray intensity of different periodicities, viz. 22 years, 11 years, 27 days and 24 hours. The systematic study of the time variation of relativistic cosmic rays started some 60 years ago by using the ground based detectors. The ground based observations, its anisotropies and their relationship with other geomagnetic and interplanetary parameters, provide the based to understand the time variation characteristics of cosmic ray intensity.

[1]. In their study of diurnal anisotropy on day to day concluded that on an average basis the diurnal anisotropy of cosmic radiation is completely understood as a superposition of simple convection and field aligned diffusion. Cosmic ray intensity observed on the ground is subject to the solar semi-diurnal variation of extraterrestrial origin. The variation is due to the second order anisotropy produced by the diffusion-convection of cosmic rays in interplanetary space. [2, 3] Studies of the solar semi-diurnal variation have been made by [4, 5] to obtain information about solar modulation in various conditions of the heliosphere. [6, 7] have investigated the existence of the tri-diurnal variation i.e., the third harmonic of the daily variation in the recorded cosmic ray intensity. The results of power spectrum and harmonic analysis for different worldwide cosmic ray station have showed that the observed tri-diurnal variations are of extraterrestrial origin and arises from an ecliptic plane anisotropy in free space.

Solar diurnal variation of cosmic ray intensity shows a large day-to-day variability. This variability is a reflection of the continually changing conditions in interplanetary space [8]. The average diurnal anisotropy of cosmic radiation being explained in terms of azimuthal corotation [9]. The systematic and significant deviations of amplitude as well as phase for diurnal/semi-diurnal anisotropies from the average values are known to occur in association with strong geomagnetic activity [10]. The distinguishing features of these systematic deviations are the unusually low or high amplitude and usually, through not always, a shift in the phase towards earlier hours [11].

The average characteristics of cosmic ray diurnal anisotropy are adequately explained by co-rotational concept. [12, 13]. This concept supports means diurnal amplitude in space of 0.4% along the 1800 Hr direction using the worldwide neutron monitor data. Though ,the day-to-day deviation both in amplitudes and phase and the abnormally large amplitudes or abnormally low amplitudes of consecutive days cannot be explained in co-rotational terms. Many scientists [14] and [16] used a new concept for the interpretation of the diurnal variation. [17] first suggested the extension of this new concept from the solar cosmic events to the observed diurnal variation and the theoretical formulation is provided by [18]. On the basis of this mechanism, the diurnal variation can be explained in terms of radial convection together with diffusion, which is mainly along the magnetic field line. The co-rotational concept is a special case of the convective-diffusive model with which we can explain the characteristics of the diurnal variation even on a day-to-day basis. The phase shift of the diurnal anisotropy to earlier hours is well understood in terms of the convective-diffusive mechanism [15, 16] have noted that the non-field-aligned diffusion on the days of nominal diurnal amplitude which are influenced by magnetic sector passages.

## 2. Data Sources and analysis

The anisotropic events are identified using the hourly plots of cosmic ray intensity recorded at ground based Deep River neutron monitoring station (data from http://spidr.ngdc.noaa.gov/NeutronMonitor).

In earlier studies on cosmic rays variation ,it has been reported that the tri-diurnal amplitude generally decreases by a factor of 4 to 5 from the semi-diurnal amplitude. The signal-to-noise ratio is generally found to be poor even on annual average basis. [19, 20] have reported the detailed characteristics of tri-diurnal anisotropy of cosmic ray intensity for the period 1973-1975 and 1976-1980, respectively. They have reported a positive correlation between semi-diurnal and tri-diurnal amplitudes. Their results suggest that solar polar coronal holes may influence both the solar tri-diurnal and semi-diurnal variation of galactic cosmic ray intensity. [21] have reported a significant relationship of first two harmonics of cosmic ray daily variation with solar activity. They found a significant and positive correlation of diurnal amplitude and phase with sunspot numbers is also reported for semi-diurnal phase. Recently, [22] have reported the long-term trend of the first three harmonics of the daily variation of cosmic ray intensity for the period 1991-2004, using the Haleakala neutron monitor data. They have reported the 11-year cycle trend in diurnal and semi-diurnal component but did not find any long-term variational trend in tri-diurnal component. [23] reported the long-term behaviour of the diurnal wave of cosmic ray anisotropy in relation with interplanetary magnetic field. Long-term characteristics of cosmic ray diurnal variation are also reported in recent publications. In the present analysis, the amplitude and phase have been derived first on daily basis and then on average basis, for the Beijing and Moscow neutron monitors covering different cutoff rigidities.

The 11-year (1996-2006) harmonic analysis alongwith the geographic co-ordinates of the two stations are given in Table-1.

The first station Beijing and second Moscow station are at some different latitude. These two stations consist of high and low cutoff rigidities, which respond to different energy range of cosmic ray particles.

## 3. Result

In recent years a new component in low energy cosmic rays (1-30 MeV/amu) is discovered in the vicinity of Earth and at distances upto 18 B.U. [24, 25]. The component has anomalous He, N, O, Ne abundances, that are quite different from solar and galactic cosmic rays. The IMP 7 and 8, Pioneer 10 and 11 and Voyager 1 and 2 Spacecrafts have provided new and interesting information on the competition, time variation ,redial and perpendicular gradient of this component. Pioneer 10 was at 20 A.U. in mid 1979 and travelling at the rate of 3 A.U. per year is likely to reach close to the hemisphere boundary at 50 A.U. by about this new component.

The origin of these particles is not definitely known at present. Several attempts have been made to explain its origin [26] .The modal of [27] have some attractive features which could partly account for the composition of the anomalous component. However, recent result show very large time variation of this component and the intensity reached very low value in early 1979 as discussed in details the long-term modulation of the anomalous component and they pointed

out that hemispheric model of [27] cannot account for the long-term variation observed recently and these variation can be easily understood from the stellar model of [25] they interpret the modulation as occurring due to the polarity reversal of the polar magnetic field of the Sun, one such reversal having occurred in the period of 1969-71.This interpretation is based on the hypothesis that when polar magnetic field of the Sun is nearly parallel to galactic magnetic field, they both could easily connect with each other are hence the low energy galactic rays could penetrate more easily into the hemispheric along the magnetic lines of force, as compared with those in the antiparallel state of the magnetic field. This leads to a 22 year variation in cosmic ray intensity because polarity reversal occurs around every solar maximum. These authors point out the sudden appearance of the anomalous component in 1972 and its absence in 1975 could be easily understood by the hypothesis of [25] as a polarity reversal. Further, these authors point out the observation presented by the Chicago group that the anomalous component have abnormally large density gradient perpendicular to the solar equatorial plane can also be understood by the stellar origin model of [25] .

The modal by Fisk, Koslovsky and Ramaty (hereinafter FK and R) Predicts that there should be ACR contributions to species that are mainly or partially neutral in the interstellar medium. The observed abundance of He, C, O, Ne and Ar generally consistent with this picture and provide a means of measuring the composition of the natural interstellar medium For example, that low abundance of carbon in ACRs implies that only~1%of the carbon in the interstellar medium is in a neutral state. The model of FK and R also predicts that ACRs should be singly charged, in contrast to galactic cosmic ray, which are essentially fully stripped and there is now abundant evidence that the bulk of ARCs with~10MeV/nuc are singly charged [28].

As pioneer 10 and 11 and later voyager 1 and 2, begin to explore that outer solar system they

Table 1: 11- year cosmic ray daily variation

| **Stations** | **Geographic Latitude (Deg.)** | **Geographic Longitude (Deg.)** | **Cutoff rigidity, (GV)** |
|---|---|---|---|
| Beijing | 40 | 116 | 9.56 |
| Moscow | 55 | 37 | 2.42 |

found that the intensity of ACRs increased with distance from the Sun, and the distribution of ACRs in the hemisphere has now been measured out of 60 AU, and to latitudes as high as $80^{0}$.ulysses has recently measured the abundances of the "pick-up'' ions that are the seed population for ACR acceleration. It is now believed that the bulk of ACR acceleration takes place that the solar wind termination shock estimate to be at a distance of~80 to 100 AU from the Sun. Because the access of low energy cosmic rays to inner solar system is strongly affected by interplanetary condition ("Solar modulation''), ACRs are detectable at 1AU only near solar minimum.

Figure.1 Shows plot of annual mean diurnal amplitude and phase of Beijing and Moscow neutron for the period 1996-2006.Most of the value are statistical as shown in Figure. The amplitude of cosmic ray daily variation are invariant in different time scale. The values of amplitude are found much larger during 2002-2006.**RDVV, Jabalpur.**

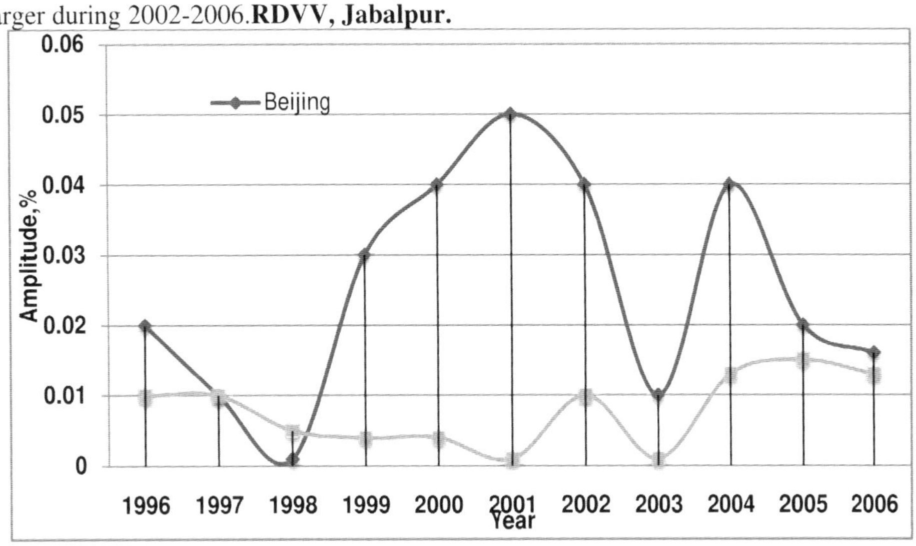


Figure. 1 — Amplitudes (%) for the annual average diurnal variation for Moscow and Beijing neutron monitors.

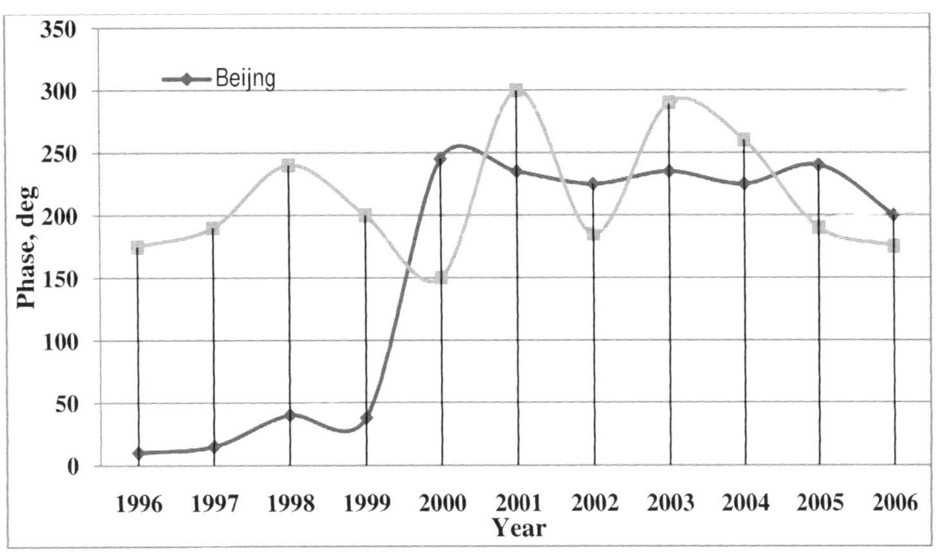


Figure 2 – Phases (degree) for the annual average diurnal variation for Moscow and Beijing neutron monitors

Change in phases is larger from year to year. Linear plots for Moscow and Beijing station for phases(in degree)are shown in Figure. 2. **RDVV, Jabalpur.**

The station Beijing and Moscow are situated in different latitudes. These station consist of low and high cutoff rigidities which respond to different energy ranges of cosmic ray particles.

## 4. Conclusion

On the basis of the present investigation the following conclusions have emerged.

- The amplitude significantly deviates from the annual average value of diurnal anisotropy. The time of maximum of the diurnal anisotropy has shifted toward earlier hours for the low amplitude anisotropic wave events.
- The long-term behaviour of the amplitude of the diurnal anisotropy can be explained in term of the occurrence of low amplitude events.
- The occurrence of low amplitude anisotropy wave train events is dominant during solar activity minimum years.
- The amplitude of the diurnal anisotropy is correlated with the solar cycle but the direction of the anisotropy is not correlated with the solar cycle and shows a systemic shift to earlier hours.
- The long-term behaviour of the time of maximum of the diurnal anisotropy vectors could be explained in term of co-rotational (1800 Hr) component and 1200 Hr component.
- The anisotropy does not show any long-term variation.

**References :**

1. Ananth A G, Agrawal S P and Rao U R 1974 *Pramana*, **3** 74.
2. Quenby J J and B. Lietti 1968 *Planet. Space Sci.* **16** 11209.
3. Munakata K and Nagashima K 1986 *Planet. Space Sci.* **34** 99.
4. Ahluwalia H S and Fikani M M 1996a *J. Geophys. Res.* **101** 11075.
5. Ahluwalia H S and Fikani M M 1996b *J. Geophys. Res.* **101** 11087.
6. Mori S, Yasue S and Ichinose M 1971 *12th Int. Cosmic ray Conf. Hobart* **2** 673 .
7. Nagashima K, Fujimoto K, Fujii Z, Ueno H and Kondo I 1971 *Reprint DPNU Nagoya* **10.**
8. Fluckiger E O 1991 *22nd Int. Cosmic Ray Conf. Dublin* **5** 273.
9. Venkatesan D and Badruddin 1990 and references theirin *Space. Sci. Rev.* **52** 273.
10. Kumar S, Agarwar R, Mishra R and Dubey S K 2002 *Int. J. Mod. Phys.* **11** 1243.
11. Hashim A and Thambyahpillai H 1969 *Planet. Space Sci.* **17** 1879.
12. Parker E N 1964 *Planet Space Sci.* **12** 735.
13. Axford W I 1965 *Planet Space Sci.* **13** 115.
14. Rao U R, Ananth A G and Agrawal S P 1972 *Planet Space Sci.* **20** 1799.
15. Kane R P 1974 *J. Geophys. Res.***79** 1321.
16. Owens A J and Kash M M 1976 *J. Geophys. Res* **81** 3471.
17. McCracken K G, Rao U R and Fowler B G 1965 *IQSY Instruction Manual No* **10.**
18. Forman M A and Glesson L J 1975 *Astrophys. Space Sci.Rev* **32** 77.
19. Agrawal S P 1981 *J. Geophys. Res.***10** 10115.
20. Shrivastava P K 1985 *Ph D Thesis A P S University Rewa.*
21. Tiwari C M, Tiwari D P, Agrawal S P and Shrivastava P K 2004 *Indian J. Radio Space Phys.* **33** 167.
22. Pandey G K, Tiwari C M and Shrivastava P K 2005 *29th Int. Cosmic ray Conf. Pune* **2** 167.
23. Kudela K, Firoz K A, Langer R D and Kollor V 2008 *21th European Cosmic Ray Conf.* 374.
24. Biswas S, Durgaprasad N, Nevatia J, Venkatavaradan V S, Goswami J N, Jayanthi U B, Lal D and Mattoo S K 1975 Refrences therein *Astrophy. Sapce Sci.* **33** 337.
25. Biswas S, Durgaprasad N and Trivedi S 1981 Refrences therein *17th Int. Cosmic Ray Conf. Paris* **2** 314.
26. Biswas S, Durgaprasad N and Trivedi S 1981 Refrences therein *Proc.Earth Planet Sci.* **90** 337.
27. Biswas S, Durgaprasad N 1980 Refrences therein *Space Sci Rev.* **25** 285.
28. Klecker B, McNab M C, Blake J B, Hovestsdt D, Kastl H,Hamilton, Looper M D, Mason G M, Mazur J E and Scholer M 1995 Refrences therein *J. Astrophys.* **442** L69.

# Analysis of dispersion characteristics of long period fiber grating

**Vishal Jain[1], Santosh Pawar[2], S Kumbhaj[3] and P K Sen[4]**

[1]Department of Applied Physics, Malwa Institute of Technology, Indore, 452016, India
[2]Department of Electronics & Communication, Dr. A P J Abdul Kalam University, Indore, 452016, India
[3,4]Department of Applied Physics, S G S Institute of Technology & Science, Indore, 452003, India

E-mail: vishaljainphysics@gmail.com

**Abstract.** Present work deals with theoretical analysis of dispersion characteristics of long period fiber grating using straight forward coupled mode theory. Simple analytical solutions are obtained for co propagating core and cladding modes under linear regime. These solutions are used to derive expressions for transmission coefficient ($t_{LPG}$), phase ($\phi_L$), delay ($\tau_\rho$) and group velocity dispersion ($D_\rho$) for proposed grating structure. Attention is paid to study the delay response of the grating, by varrying physical parameters like incident wavelength and coupling strength of grating. Negative values of group delay for certain value of coupling strength shows that long period fiber can be used as dispersion compansator device in optical fiber communication link.

## 1. Introduction

Dispersion is an important optical characteristic in the optical fiber, which cause serious limit on the transmission capacity and bandwidth of optical fiber. The major concern of optical fiber communication link has been shifted from compensation of loss to compensation of dispersion. Several techniques have been proposed for dispersion compensation in optical fiber communication [1-4].Nowadays fiber Bragg gratings (FBG) are used for chromatic dispersion compensation by recompressing the dispersed optical signal [5-9].

long period fiber gratings (LPFGs) are of considerable research interest for various light wave communication and sensors application such as gain flattening filters for erbium doped fiber amplifiers, dispersion compensator, add/drop multiplexer, optical fiber polarizer, and strain & temperature sensors [10-14]; because of their simple structure, low insertion loss and negligible delay ripple. The device is very suitable for optical communication link since it works on transmission mode and avoids the use of circulator and chirping the grating. The most significant properties of the LPFG is its tunability for the desired transmission characteristics , by changing the grating parameters like refractive index, modulation depth, grating length and period etc. Dispersion characteristics of LPFG devices play a significant role for high-bit-rate optical communication applications and using as an inline optical device for dispersion compensation. M. das and K. Thyagarajan have proposed the use of uniform LPG fabricated on relatively high refractive index difference fibers as efficient dispersion compensators [11].

In the present work, we have presented a simple theoretical model to analyze delay and dispersion characteristics of long period grating. The analytical expression for amplitude and phase of complex transmission coefficient, delay and dispersion parameter of the grating has been obtained under the linear regime with the help of coupled mode theory. Finally, specific attention is paid to study delay experienced by different frequency components and dispersion compensation capabilities of LPFGs.

## 2.Theorotical Analysis

Long-period gratings can be considered as a special class of fibre Bragg grating in which the period of the index modulation is such that it satisfies a phase matching condition between the fundamental core mode and a forward propagating cladding mode of an optical fibre. To study the spectral characteristics of grating in linear regime for a continuous wave, following coupled mode equations have been used [15, 16]

$$\frac{\partial A_{core}}{\partial z} = i\delta A_{core} + i\kappa A_{cladd} \quad (1) \qquad \frac{\partial A_{cladd}}{\partial z} = -i\delta A_{cladd} + i\kappa A_{core} \quad (2)$$

Here δ and Κ are the detuning parameter and coupling coefficient of grating respectively, define as

$$\delta = \frac{1}{2}\left(\beta_{core} - \beta_{clad} - \frac{2\pi}{\wedge}\right) \quad (3) \qquad \kappa = \frac{\pi n_g}{\lambda_c} \quad (4)$$

A general solution of the these linear equation takes the form

$$A_{core}(z) = C_1 e^{iqz} + C_2 e^{-iqz} \quad (5) \qquad A_{clad}(z) = D_1 e^{iqz} + D_2 e^{-iqz} \quad (6)$$

Where q is the dispersion parameter of the grating and defined as

$$q = \left(\delta^2 + \kappa^2\right)^{\frac{1}{2}} \quad (7)$$

In terms of an effective transmission coefficient $t(q)$ equations (8) and (9) can be written as

$$A_{core}(z) = D_1 e^{iqz} t(q) + C_2 e^{-iqz} \quad (8) \qquad A_{clad}(z) = D_1 e^{iqz} - C_2 e^{-iqz} t(q) \quad (9)$$

$$\text{Where } t(q) = \left(\frac{\kappa}{q+\delta}\right) = \left(\frac{q-\delta}{\kappa}\right) \quad (10)$$

The 'q' dependency of $t(q)$ and the dispersion relation (Eq. 10) indicates that both the magnitude and phase of the transmitted wave depends upon the frequency 'ω' of propagating beam in the LPFG.
Using above Equations and on applying the boundary conditions

$$A_{core}(z=0) = 1, A_{core}(z=L) = 1 \quad (11)$$

The complex transmission amplitude and phase response of LPFG are calculated as

$$t_{LPG} = \cos(qL) + i\frac{\delta \sin(qL)}{q} \quad (12) \qquad \phi_L(\omega) = \tan^{-1}\left(\frac{\delta}{q}\tan(qL)\right) \quad (13)$$

The rate of change in phase with respect to frequency (ω) gives the delay experience by the frequency component of transmitted wave and calculated as

$$\tau_\rho = \frac{d\phi_L}{d\omega} = -\frac{\lambda^2}{2\pi c}\frac{d\phi_L}{d\lambda} = -\frac{\lambda^2}{2\pi c}\left[\frac{\left(\frac{\delta}{q}\right)\sec(qL)^2\left(\frac{dq}{d\lambda}\right) + \tan(qL)\frac{d}{d\lambda}\left(\frac{\delta}{q}\right)}{1+\left(\frac{\delta}{q}\right)^2\tan(qL)^2}\right] \quad (14)$$

It is observed from these equations that if we choose proper physical parameters of long period grating such as grating length and magnitude of induced index change, it is possible to vary the delay experienced by the frequency components.

Now the rate of change of group delay with frequency determines the group velocity dispersion experienced by the frequency component. Thus the group velocity dispersion can be calculated as

$$D_\rho = \frac{d\tau_\rho}{d\lambda} = -\frac{\lambda}{\pi c}\frac{d\phi_L}{d\lambda} - \frac{\lambda^2}{2\pi c}\frac{d^2\phi_L}{d\lambda^2} \tag{15}$$

This equation shows the dependency of group velocity dispersion on the wavelength of incident wave inside the grating .The Wavelength component having minimum detuning would be the central wavelength $\lambda_c$ of the transmission spectrum. on either side of $\lambda_c$, detuning increases in magnitude with positive or negative sign and With increasing magnitude of detuning the frequency components suffer lesser coupling to cladding region as a result maximum power flow confined to the core region. Since the group velocity in the core region is lesser than in cladding, these sideband wavelengths would suffer larger delay in propagation compared to the central wavelength, with increasing delay on both sides. This results in a positive dispersion corresponding to one band edge and negative dispersion corresponding to the other band edge. Magnitude of dispersion would depend upon the slope of the delay variation with frequency. Hence steeper the delay variation, larger would be the dispersion [11].

## 3. Result and Discussion

The dispersion relation of long period fiber grating exhibit an important property of LPFG known as dispersion property as seen in Figure (1), where detuning parameter δ is plotted as a function of dispersion parameter q for both uniform medium (dashed line) and a periodic medium (solid line). A figure 1 shows that, in case of uniform medium the slope is constant thus the dispersion is negligible. This dispersion is modified on introducing a grating. If the dispersion parameter lies in the range of $0 < q/\kappa < 1$ the core and cladding modes are coupled in the grating structure and are attenuated.

To study the dispersion capability of LPFG we have considered a LPFG of effective core index 1.458, effective cladding index 1.450, grating index $n_g \approx 1\times10^{-4}$ and centre wavelengths $\lambda_D \approx 1550$ nm. Figure 2 shows the delay response of this LPFG as a function of coupling strength $\kappa L$ for three different values of incident wavelength (a) λ = 1550 nm (b) λ = 1550.25 nm and (c) λ = 1550.50 nm. Figure 2(a) plotted at incident wavelength 1550 nm, we obtain three negative values of delay as (1) -240.6 ps at $\kappa L = 1.60$ (2) -248.55 ps at $\kappa L = 4.76$ and (3) -257 ps at $\kappa L = 7.90$ and delay ripple is suddenly jumps from positive to negative value for these values of $\kappa L$ and remains almost linear for other values. Figures 2(b) and 2(c) are plotted for incident wavelengths 1550.25 nm and 1550.50 nm, respectively. From these figures it is observed that the variation in delay is in positive manner having sinusoidal or stepwise nature.

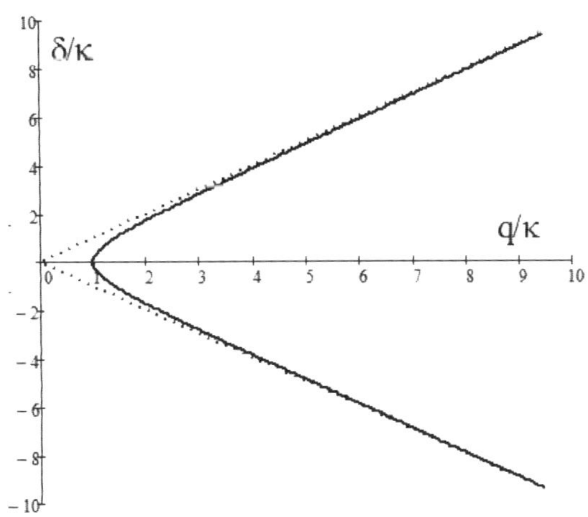


**Figure 1.** Dispersion curves showing variation of dispersion parameter q with detuning parameter δ

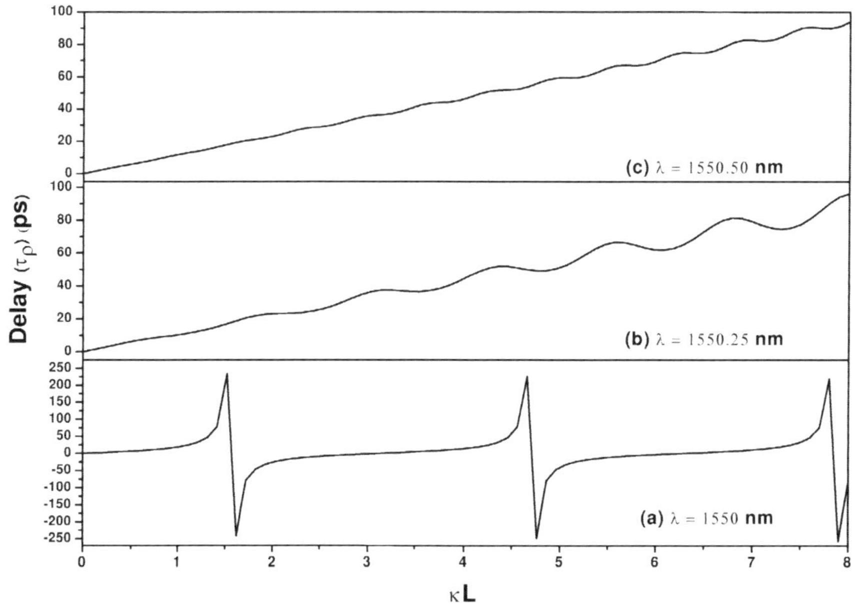


**Figure 2**. Delay Response of the long period gratings as a function of grating coupling strength for different values of incident wavelength (a) λ = 1550 nm (b) λ = 1550.25 nm and (c) λ = 1550.50 nm.

Above results shown that the delay becomes negative for certain values of κL = 1.60, 4.76 and 7.90 with design wavelength $\lambda_D$=1550 nm.With these physical parameter long period grating can be utilized for the application of dispersion compensation in optical communication system.

## 4. Conclusion

By solving coupled mode equation we have derived expressions for transmission coefficient ($t_{LPG}$), phase ($\phi_L$), delay ($\tau_\rho$) and group velocity dispersion ($D_\rho$) for proposed grating structure and the effect of grating strength on delay spectra of LPFG has been studied in detail. We have shown that, by precisely optimizing the grating length, average refractive index deviation and coupling strength of the grating, LPFG can be used as a dispersion compensating device.

## References

[1] Vengaskar A M, Lemaire P J and Judkins J B 1996 *J. Lightw. Technol.* **14** 58
[2] Oullette F 1987 *Opt.Lett.***12** 847
[3] Weiming M and S Chengcheng 1999 *Tsinghua Science and Technol*. **4** 1624
[4] Toulouse J 2005 *Journal of Lightwave Technol.* **23** 3625
[5] Devra G K S 2011 *International Journal of Engineering and Information Techno.* **3** 1
[6] Sun C, Cao J, and Li C 2002 *Tsinghua Science and Technol.* **7** 305
[7] Yang A, Wang Z, Zhang Z, Chen Z and Wu D 2003 *Journal of Electronics* **20** 461
[8] Islam J and Islam R 2012 *Procedia Engineering* **30** 821
[9] Litchinitser N M, Eggleton B J and Agrawal G P1998 J. *Lightw. Technol.* **16** 1523
[10] Singh R, singh S and Sharma E K 2004 *Opt. Commun.* **240** 123
[11] Das M and Thyagarajan K 2001 *Opt. Commun.* **190** 159
[12] Huang J F, He Y J and Lo Y L 2006 *Opt. Eng.* **45** 12
[13] Ortega B, Dong L, W F Liu and Reekie L 1997 *Technol. Lett.* **9** 1370
[14] Bhatia V and Vengsarkar A M 1996 *Opt. Lett.* **21** 692
[15] Erdogan T 1997 *J. Light wave Technol*. **15** 1277
[16] Kashyap R 1999 *Fiber Bragg Gratings* (Academic press, San Diego: CA)

# LSPR based fiber optic sensor for fluoride impurity sensing in potable water

**Abhay Tambe[1], S Kumbhaj[2], N P Lalla[3] and P Sen[4]**

[1]Physis Departement, Chamelidevi group of Institute, Indore- 452020 India
[2]Shri G.S.Institute of Technology and Science, Indore-452003 India
[3]UGC-DAE-Consortium for Scientific Research, Indore-452001 India
[4]School of Physics, Devi Ahilya University, Indore-452001 India

E-mail: abhay18.2008@gmail.com

**Abstract.** We have designed localised surface plasmon resonance (LSPR) based fiber optic sensor. Silver nanoparticles are deposited on a few centimetre length of bare core at the middle part of plastic clad silica fiber by means of a simple and low cost laser induced nanoparticle deposition technique. The nanoparticle deposition was confirmed by TEM analysis. The nanoparticle coated fiber is used to design the sensor and the response of sensor was studied to sense fluoride impurity in water.

## 1. Introduction

Potable water is one of the most valuable natural resource that our planet has. Water is essential to sustain life and a safe, adequate and accessible supply must be available to all. Improving access to safe drinking-water can result in tangible benefits to health. Fluoride has beneficial effects on teeth at low concentrations (below 1 ppm) in drinking-water, but excessive exposure to fluoride in drinking-water give rise to a number of adverse effects. These range from mild dental fluorosis to crippling skeletal fluorosis as the level and period of exposure increases. Crippling skeletal fluorosis is a significant cause of morbidity in a number of regions of the world. Fluoride levels between 0.3 and 6.9 mg $l^{-1}$ was found in four villages in the Jind district of Haryana, India [1]. Choubisa et al. examined the prevalence of skeletal fluorosis in Rajasthan in adults exposed to fluoride levels of 1.4 & 6 mg $l^{-1}$ [2] The recommended level of fluoride in drinking water is 1-1.5mg/l as directed by World Health Organization (WHO) [3]. There are various chemical as well as optical techniques used for determination of fluoride impurity in water [4] such as ion selective electrode, colorimetric method [5], atomic absorption spectroscopic method [6], ion chromatography [7] etc. These techniques in general require expensive and standard laboratory equipment and sample pre-treatment apart from the trained staff. Consequently for quick analysis of fluoride in potable water there is a need to develop a simple sensitive and convenient method.

In recent years, localised surface plasmon resonance (LSPR) based fiber optic sensors have fascinated many researchers for various applications in physical chemical and biochemical field [8-9] due to their simplicity in design, low cost, high sensitivity, accuracy, on line monitoring, remote sensing etc. LSPR based fiber optic sensor probe was fabricated by using simple and low cost laser induced nanoparticle deposition technique [10]

## 2. Theory

Localised surface plasmons (LSPs) are quanta of charge density oscillation in metal nanoparticles. Surface plasmons are excited when the frequency of incident light becomes equal to that of the oscillation frequency of conduction electrons of metal nanoparticles. This results in the absorbance of incident light and is called localised surface plasmon resonance (LSPR). The LSP resonance wavelength is highly dependent on the shape, size, material and surrounding medium of the nanoparticles [11] which can be utilised for sensing applications. In our case the unclad portion of a multimode PCS fiber is coated with silver nanoparticles. The nanoparticles are surrounded by sensing medium. Light is launched from one end of the fiber. If $P_0$ is the power transmitted by the fiber in the absence of a sensing medium, then the power transmitted along the fiber in the presence of a sensing medium is given by [12].

$$P = P_0 e^{-\gamma L} \tag{1}$$

Here $\gamma$ is the evanescent absorption coefficient of the sensing medium, L is the length of sensing medium. The evanescent field interact with the nanoparticles resulting in wavelength dependent absorption. The evanescent absorption coefficient of a ray making angle $\theta$ with the normal to the interface is given by [13].

$$\gamma = \frac{NE\lambda n_2 \cos\theta \cot\theta}{2\pi\rho L n_1^2 \cos\alpha \cos^2\theta_c (\sin^2\theta - \sin^2\theta_c)^{1/2}} \tag{2}$$

Here E is the extinction coefficient of a single metal nanoparticle, $\rho$ is the radius of core of the fiber, N is the total number of nanoparticles that can be attached to the bare core of the fiber, $n_2$ is the refractive index of the metal nanoparticle, $n_1$ is the refractive index of core of the fiber, $\theta_c = Sin^{-1}(n_2/n_1)$ is the critical angle for sensing region and $\alpha$ is the skewness parameter. The extinction coefficient of metal nanoparticle is derived using Mie theory and is given by [11].

$$E = \frac{24\pi R \varepsilon_m^{3/2}}{\lambda} \frac{\varepsilon_2(\lambda,R)}{(\varepsilon_1+2\varepsilon_m)^2+\varepsilon_2(\lambda,R)} \tag{3}$$

$\varepsilon_1$,$\varepsilon_2$ being the real and imaginary part of metal dielectric constant, respectively. $\varepsilon_m$ is the dielectric constant of surrounding medium and R is the radius of nanoparticle. A sharp absorption peak will be observed in the absorption spectrum when $\varepsilon_1 = -2\varepsilon_m$ , called as the resonance condition. For different values of the dielectric constant of sensing medium ($\varepsilon_m$), the resonance condition will be satisfied for different values of the wavelength and this is the principle of the working of a sensor.

## 3. Experimental

To fabricate the sensor, cladding of known length (2-3cm) is removed from the middle portion of multimode plastic clad silica (PCS) fiber having diameter 125μm. The bare core of the fiber is dipped in solution of silver nitrate and trisodium citrate for the deposition of silver nanoparticles on the core.

A water cell of length 5cm and diameter 1.5cm with inlet and outlet provisions is designed, holes are provided at the two ends through which fiber is inserted in the water cell such that the bare core should be inside the water cell and remain straight.(as shown in photograph)

**Figure 1.** Experimental set-up for nanoparticle deposition

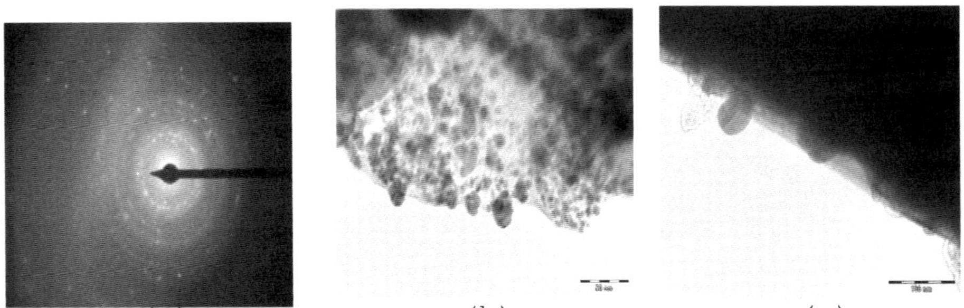

(a) (b) (c)

**Figure 2.** TEM Image of deposited silver nanoparticles.

Light beam from He-Ne laser wavelength 633nm was focussed on bare core immersed in the solution of silver nitrate and sodium citrate. Photochemical reaction causes the formation of silver nanoparticles which deposit on the core of unclad fiber. To exploit LSPR phenomena light from tungsten halogen lamp (THL) of 50watt power was coupled at one end of the fiber through microscope objective (MO) having same numerical aperture. The output intensity is detected by using a photomultiplier tube (PMT).The preliminary test of nanoparticle deposition is confirmed by the reduction in the output intensity. Continuous measurement of intensity vs wavelength at the output end was taken via spectrum analyser and the dip at the surface plasmon resonance in the output spectrum was observed. The deposition of nanoparticles was finally confirmed via transmission electron microscopy (TEM) as shown in figure 2. Designed sensor was tested for the detection of fluoride impurity in water. Standard water samples having fluoride concentrations 30ppm, 50ppm, and 100ppm were prepared by dissolving sodium

Fluoride (NaF) in distil water and the output current is measured for different wavelength. The output spectrum shows the red shift in resonance wavelength with increase in fluoride concentration.

## 4. Result

Figure 2(a) shows the diffraction image of deposited nanoparticles which confirms the deposition of silver particles. We note that silver nanoparticles grow and deposit on the fiber core after laser illumination. During the process of laser induction, silver particles are reduced in the solution under illumination. The illumination time was 6 minute. The deposited nanoparticles are of irregular shape having diameter up to 50nm.

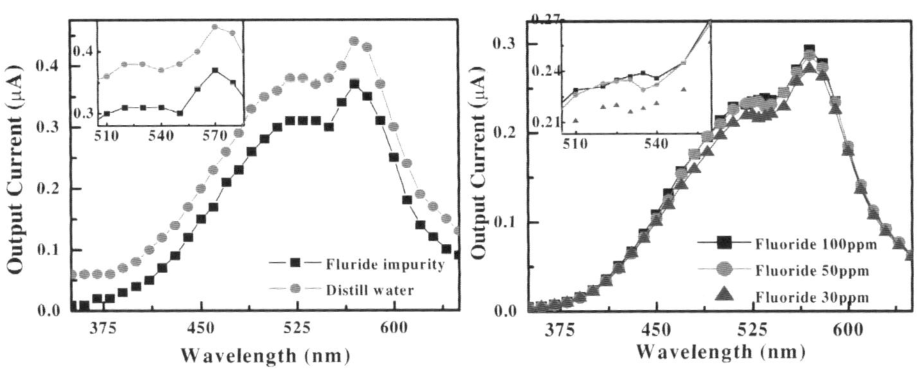


**Figure 3.** Output intensity for distil water and fluoride solution

The output spectrum with distil water shows dip at 541nm wavelength. We observed a shift in the dip position in presence of known concentration of fluoride solution.The output intensity was plotted for different known concentrations (30ppm, 50ppm, 100ppm) of fluoride in distil water. We observed that the surface Plasmon resonance dip shifted to longer wavelength with increasing fluoride impurity.

## 5. Conclusion

We have proposed design of fiber optic sensor based on localised surface Plasmon resonance (LSPR) to detect fluoride impurity in water. Silver nanoparticles were deposited by simple and low cost laser induced nanoparticle deposition technique. The experimental results exhibit the usefulness of the technique for detection of impurities in water.

## References

[1] Garg V K, Kavita, and Malik A, 2004 *Journal of Hazardous Materials* **106B** 85
[2] Choubisa S L, Choubisa D K, Joshi S C and Choubisa L 1997 *Fluoride* **30** 223
[3] Carton R J, 2006 *Fluoride* **39** 163
[4] Apha, Awwa and Wpcfa 1985 *standard methods for the examination of water*
[5] Chang-Quing Z, Jin-Long C, 2005 *Analytica chimica Acta* **539** 311
[6] Faraj-zadeh M A and Kalhor E G 2001 *Microchim Acta* **137**
[7] Weiss J and Weiss T 2005 *John Wiley and Sons* **3** 931
[8] Alexandre R, Paula M P, Ana Carolina M S Dias and Isabel C S 2013 *Optical Express* **21** 27023
[9] Gabriel Ortega-Mendoza J, Padilla-Vivanco A and Toxqui-Quitl C 2014 Sensors **14** 18701
[10] Ming-Shan L and Chang-Xi Y 2010 *Chin. Phys. Lett.* **27** 44202
[11] Link S and El-Sayed M A 1999 *J. Phys. Chem. B* 103 4212
[12] Srivastava S K, Verma R K and Gupta B D 2009 *Appl Optics* **48** 3796
[13] Gupta B D, Sharma A and Singh C D 1993 *Int. J. Optoelectron.* **8** 409S